# Organic Chemistry: Structure and Function

# Organic Chemistry: Structure and Function

Editor: Saul Rowen

NY RESEARCH PRESS

New York

Published by NY Research Press
118-35 Queens Blvd., Suite 400,
Forest Hills, NY 11375, USA
www.nyresearchpress.com

Organic Chemistry: Structure and Function
Edited by Saul Rowen

**Cataloging-in-Publication Data**

Organic chemistry : structure and function / edited by Saul Rowen.
    p. cm.
Includes bibliographical references and index.
ISBN 978-1-63238-638-0
1. Chemistry, Organic. 2. Chemistry. I. Rowen, Saul.
QD251.3 .O74 2019
547--dc23

# Contents

# Preface

This book presents researches and studies performed by experts across the globe in the field of organic chemistry. The scientific study of structures, functions and properties of organic compounds falls under the domain of organic chemistry. Organic chemistry has applications for other purposes such as development of anti-biotics, detecting food adulteration, disease diagnosis, etc. This book is compiled to provide a thorough understanding of the field by explaining the latest concepts and theories related to this area of study. Most of the topics introduced in this book cover new techniques and the applications of this field. It consists of contributions made by international experts and will enable the readers to develop deeper insights into the subject. Coherent flow of topics, student-friendly language and extensive use of examples make this book an invaluable source of knowledge.

This book is a comprehensive compilation of works of different researchers from varied parts of the world. It includes valuable experiences of the researchers with the sole objective of providing the readers (learners) with a proper knowledge of the concerned field. This book will be beneficial in evoking inspiration and enhancing the knowledge of the interested readers.

In the end, I would like to extend my heartiest thanks to the authors who worked with great determination on their chapters. I also appreciate the publisher's support in the course of the book. I would also like to deeply acknowledge my family who stood by me as a source of inspiration during the project.

Editor

# 2-Methyl-1,4-Benzodioxan Substituted Ag(I)-N-Heterocyclic Carbene Complexes and Ru(II)-N-Heterocyclic Carbene Complexes: Synthesis, Crystal Structure and Transfer Hydrogenation of Aromatic Ketones

**Aydın Aktaş[1]\*, Yetkin Gök[1], Mehmet Akkurt[2] and Namık Özdemir[3]**

[1]Department of Chemistry, Faculty of Arts and Sciences, İnönü University, Malatya, Turkey
[2]Department of Physics, Faculty of Sciences, Erciyes University, Kayseri, Turkey
[3]Department of Physics, Faculty of Arts and Sciences, Ondokuz Mayıs University, Samsun, Turkey

## Abstract

In this study a series of unsymmetrically silver(I)-*N*-heterocyclic carbene (NHC) and ruthenium(II)-NHC complexes were synthesised. The Ag(I)-NHC complexes (1a-f) were synthesized in dichloromethane at room temperature from the imidazolium salts and $Ag_2O$. The Ru(II)-NHC complexes (2a-f) were prepared from Ag(I)-NHC complexes by using transmetallation method. All compounds were characterized by spectroscopic techniques (NMR and FT-IR) and elemental analyses. The catalytic activities of the Ru(II)-NHC complexes that were investigated in the transfer hydrogenation reactions showed excellent activity. Also, the 2a complex was characterized by single crystal X-ray crystallography. The Ru atom in the 2a complex, $[RuCl_2(\eta^6\text{-}C_{10}H_{14})(C_{19}H_{20}N_2O_2)]$ exhibit a pseudo-octahedral geometry, with the arene occupying three adjacent sites of the octahedron, two Cl atoms and one carbene group. The six-membered ring of the *p*-cymene is essentially planar [rms deviation=0.008 Å].

**Keywords:** *N*-Heterocyclic carbene; Silver and ruthenium complexes; Homogeneous catalysis; Transfer hydrogenation

## Introduction

In the last decades, using NHC's as ligands has increased in homogeneous catalysis [1-3]. Since the free NHC once isolated in 1991 by Arduengo et al. [4-6], these ligands in coordination chemistry and organometallic catalysis have received considerable interest [7]. The structure of NHC ligands is very diverse. However, the most commonly used catalytic complexes: are imidazol-2-ylidenes, imidazolin-2-ylidenes, thiazo-2-ylidenes and triazo-5-ylidenes. These complexes containe functional ligands (actor ligand, e.g., hydride, alkylidene) and a variety of other ligands are spectator (e.g., ancillary). The protective steric substituents in heterocyclic structure have showed high temperature stability, strong σ-donor and weak π-acceptor character and in very durable metal-NHC bond [8-11]. However, specific features, especially electronic, steric properties, reactivity and catalytic activities of these ligands can be set. Because of the these properties, NHC ligands have been shown as an alternative to phosphines in organometallic catalysis [12]. In recent years, the applications of organic transformations including hydrogenation [13], hydrosilylation [14], copolymerization [15], hydroformylation [16], olefin metathesis [17], C-N [18] and C-C [19] coupling reactions and asymmetric transformations [20] have involved a wide range of metal-NHC complexes as catalysts. The Ag(I)-NHC complexes have attracted continuous attention [21]. Many other metal-NHC complexes have been obtained by using appropriate transfer reagents Ag(I)-NHC complexes. For example, Pd, Ru, Ni, Rh and Ir NHC complexes successfully have been transferred by using this method [22-26]. In the commonly used method in the synthesis of Ag(I)-NHC complexes, a silver base has been used as agent of deprotonation. In this context silver bases ($Ag_2O$, AgOAc, and $Ag_2CO_3$) and the solvents (1,2-dichloroethane, acetone, acetonitrile, DMF, DMSO and methanol) have been used [27-30]. Ruthenium complexes, recently playing an influential role in a variety of studies on catalytic processes have made a great progress. After these developments, a lot of ruthenium complexes were synthesized, designed successfully and used in chemical reactions. Ruthenium complexes with organic or inorganic ligands are positioned around the metal core, with a suitable choice of the electronic and steric environment (e.g., cloride, phosphane, NHC and schiff bases) sometimes arene, alkylidene, vinyli-

dene, allenylidene or cumulenydene substitutes have taken place [31-36]. Among these substituents, NHCs are stable to air and moisture, as great as to make durable bond with the metal which is remarkable property [37,38]. The structural motifs of Ru(II)-NHC complexes have found wide application in catalytic processes [39,40]. The positive features of other known of Ru(II)-NHC complexes, ruthenium presents in the coordination geometry and it has a wide range of oxidation [41]. Catalytic transfer hydrogenation in homogeneous phase has become a common tool in synthetic chemistry. Officially, transfer hydrogenation is the reaction of an unsaturated reagent with molecule $H_2$. In most publications, 2-propanol by addition of a suitable base is used as hydrogen donor. Occasionally, system formic acid/triethylamine is used. 2-propanole is oxidized to acetone in the process of transfer hydrogenation [42]. Usually in industrial processes, transfer hydrogenation is used instead of hydrogen because of its safety and low price. Our studies in this paper involve the use of a new series of 2-methyl-1,4-benzodioxan substituted NHC ligands, we here report the synthesis and structures of a number of Ag(I)-NHC and Ru(II)-NHC complexes. The catalytic activities of the Ru(II)-NHC complexes have been investigated as catalysts for the transfer hydrogenation of a range of ketons in the presence of 2-propanole and KOH as base. The moleculer and crystal structure of dichloro-[1-(2-methyl-1,4-benzodioxan)-3-benzylimidazolidin-2-ylidene](*p*-cymene)ruthenium(II) complex 2a has been determined by single crystal X ray diffraction technique.

---

**\*Corresponding author:** Aydın Aktaş, Department of Chemistry, Faculty of Arts and Sciences, İnönü University, 44280, Malatya, Turkey
E-mail: aydinaktash@hotmail.com

## Experimental Section

All syntheses of Ag(I)-NHC complexes 1a-f and Ru(II)-NHC complexes 2a-f were carried out under an inert atmosphere in flame-dried glassware using standard Schlenk techniques. Solvents were purified by distillation over the drying agents indicated and were transferred under Ar: Et$_2$O (Na/K alloy), CH$_2$Cl$_2$ (P$_4$O$_{10}$), hexane (Na). All other reagents were obtained commercially from Aldrich and used without further purification. Melting points were identified in glass capillaries under air with an Electrothermal-9200 melting point apparatus. FT-IR spectra were recorded as KBr pellets in the range 400-4000 cm$^{-1}$ on an AT, UNICAM 1000 spectrometer. Proton ($^1$H) and Carbon ($^{13}$C) NMR spectra were recorded using a Varian AS 400 Merkur spectrometer operating at 400 MHz ($^1$H) or 100 MHz ($^{13}$C) in CDCl$_3$ and DMSO-d$_6$ with tetramethylsilane as an internal reference. Products were investigated with an Agilent 6890 N GC system by GC-FID with a HP-5 column of 30 m length, 0.32 mm diameter and 0.25 μm film thickness. Column chromatography was performed using silica gel 60 (70-230 mesh). Elemental analyses were performed by İnönü University Scientific and Technology Center. X-ray diffraction data for 2a were collected X-AREA [43]. Cell refinement: X-AREA [43]. Data reduction: X-RED32 [43]. Program(s) used to solve structure: SHELXS97 [44]. Program(s) used to refine structure: SHELXL97 [44]. Molecular graphics: ORTEP-3 for Windows [45]. Software used to prepare material for publication: WinGX [45]. Details of the crystallographic data and structure refinement for 2a is listed Table 1.

## General procedure for the preparation of the silver-NHC complexes, 1a-f

To a solution of 2-methyl-1,4-benzodioxan substituted imidazolidinium salt (1.0 mmol) in dichloromethane (30 mL), Ag$_2$O (0.5 mmol) and activated 4 molecular sieves was added. The reaction mixture was stirred for 24 hours at room temperature in dark condition. The reaction mixture was filtered through celite and the solvent were evaporated under vacuum to afford the product as a white solid. The crude product was recrystallized from dichloromethane/diethyl ether (1:3) at room temperature.

## Synthesis of bromo[1-(2-methyl-1,4-benzodioxan)-3-benzyl-imidazol-2-ylidene]silver(I), 1a

To a solution of 1-(2-methyl-1,4-benzodioxan)-3-benzylimidazol bromide (0.47 g., 1.2 mmol) in dichloromethane (30 mL), Ag$_2$O (0.139 g., 0.6 mmol) and activated 4 molecular sieves was added. The reaction mixture was stirred for 24 hours at room temperature in dark condition. The reaction mixture was filtered through celite and the solvent were evaporated under vacuum to afford the product as a white solid. The crude product was recrystallized from dichloromethane/diethyl ether (1:3) at room temperature. Yield: % 86 (0.51 g)

*Analytical data for bromo[1-(2-methyl-1,4-benzodioxan)-3-benzylimidazol-2-ylidene]silver(I), 1a* $^1$H NMR (300 MHz, DMSO), δ 3.71 and 3.80 (m, 4H, -NCH$_2$CH$_2$N-); 4.00-4.04 (m, 2H, -NCH$_2$CHOCH$_2$-); 4.31-4.38 (m, 1H, -CH$_2$CHOCH$_2$-); 4.48-4.50 (m, 2H, -OCHCH$_2$O-); 4.69 (s, 2H, -NCH$_2$C$_6$H$_5$); 6.84-6.90 (m, 4H, -OC$_6$H$_4$O- Ar-*H*); 7.22-7.40 (m, 5H, -CH$_2$C$_6$H$_5$ Ar-*H*). $^{13}$C NMR (300 MHz, DMSO), δ 47.1 and 47.6 (-NCH$_2$CH$_2$N-); 49.1 (-NCH$_2$CHOCH$_2$-); 54.3 (-NCH$_2$C$_6$H$_5$); 66.5 (-CH$_2$CHOCH$_2$-); 71.5 (-OCHCH$_2$O-); 117.5, 121.5, 121.8, 122.1, 127.0, 127.6, 128.1, 129.1, 136.6, 137.9, 142.7 and 143.4. (Ar-*C*); 205.3 (*C*-Ag). m.p.: 153-155°C; ν$_{(CN)}$ 1666.8 cm$^{-1}$. Anal. Calcd. for C$_{19}$H$_{20}$AgBrN$_2$O$_2$: C: 45.95; H: 4.03; N: 5.64. Found: C: 45.98; H: 4.02; N: 5.63.

## Synthesis of bromo[1-(2-methyl-1,4-benzodioxan)-3-(2-methylbenzyl)imidazol-2-ylidene]silver(I), 1b

The synthesis of 1b was carried out in the same way as that described for 1a, but 1-(2-methyl-1,4-benzodioxan)-3-(2-methylbenzyl) imidazol bromide (0.48 g., 1.2 mmol) was used instead of 1-(2-methyl-1,4-benzodioxan)-3-benzylimidazol bromide. Yield: % 85 (0.52 g)
*Analytical data for bromo[1-(2-methyl-1,4-benzodioxan)-3-(2-methylbenzyl)imidazol-2-ylidene]silver(I), 1b* $^1$H NMR (300 MHz, DMSO), δ 2.31 (s, 3H, -CH$_2$C$_6$H$_4$CH$_3$); 3.79-3.82 (m, 4H, -NCH$_2$CH$_2$N-); 4.01-4.05 (m, 2H, -NCH$_2$CHOCH$_2$-); 4.30-4.34 (m, 1H, -CH$_2$CHOCH$_2$-); 4.52-4.55 (m, 2H, -OCHCH$_2$O-); 4.71 (s, 2H, -NCH$_2$C$_6$H$_4$); 6.83-6.88 (m, 4H, -OC$_6$H$_4$O- Ar-*H*); 7.22-7.35 (m, 4H, -CH$_2$C$_6$H$_4$(CH$_3$) Ar-*H*). $^{13}$C NMR (300 MHz, DMSO), δ 19.7 (-CH$_2$C$_6$H$_4$CH$_3$); 47.8 and 48.7 (-NCH$_2$CH$_2$N-); 49.6 (-NCH$_2$CHOCH$_2$-); 52.4 (-NCH$_2$C$_6$H$_4$); 65.4 (CH$_2$CHOCH$_2$-); 71.2 (-OCHCH$_2$O-); 117.5, 121.9, 126.6, 126.8, 128.3, 129.8, 131.0, 132.0, 134.4, 137.4, 142.5 and 143.3. (Ar-*C*); 205.7 (*C*-Ag). m.p.: 174-177°C; ν$_{(CN)}$ 1645.9 cm$^{-1}$. Anal. Calcd for C$_{20}$H$_{22}$AgBrN$_2$O$_2$: C: 47.04; H: 4.31; N: 5.49. Found: C: 47.06; H: 4.33; N: 5.47.

## Synthesis of bromo[1-(2-methyl-1,4-benzodioxan)-3-(3-methylbenzyl)imidazol-2-ylidene]silver(I), 1c

The synthesis of 1c was carried out in the same way as that described for 1a, but 1-(2-methyl-1,4-benzodioxan)-3-(3-methylbenzyl) imidazol bromide (0.48 g., 1.2 mmol) was used instead of 1-(2-methyl-1,4-benzodioxan)-3-benzylimidazol bromide. Yield: % 87 (0.53 g)
*Analytical data for bromo[1-(2-methyl-1,4-benzodioxan)-3-(3-methylbenzyl)imidazol-2-ylidene]silver(I), 1c* $^1$H NMR (300 MHz, DMSO), δ 2.29 (s, 3H, -CH$_2$C$_6$H$_4$CH$_3$); 3.76-3.82 (m, 4H, -NCH$_2$CH$_2$N-); 3.98-4.03 (m, 2H, -NCH$_2$CHOCH$_2$-); 4.31-4.33 (m, 1H, -CH$_2$CHOCH$_2$-); 4.53-4.55 (m, 2H, -OCHCH$_2$O-); 4.65 (s, 2H, -NCH$_2$C$_6$H$_4$); 6.82-6.88 (m, 4H, -OC$_6$H$_4$O- Ar-*H*); 7.10-7.30 (m, 4H, -CH$_2$C$_6$H$_4$(CH$_3$) Ar-*H*). $^{13}$C NMR (300 MHz, DMSO), δ 21.1 (-CH$_2$C$_6$H$_4$CH$_3$); 49.2 (-NCH$_2$CHOCH$_2$-); 50.7 and 51.8 (-NCH$_2$CH$_2$N-); 52.8 (-NCH$_2$C$_6$H$_4$(CH$_3$);

| Cystal data | |
|---|---|
| C$_{29}$H$_{34}$Cl$_2$N$_2$O$_2$Ru | Z=4 |
| $M_r$=614.55 | $D_x$=1.429 Mg m$^{-3}$ |
| Monoclinic, P2$_1$/c | Mo Kα radiation |
| a=13.2709 (9) Å | Cell parameters from 16009 reflections |
| b=18.5843 (9) Å | θ=1.7-28.0° |
| c=12.7436 (8) Å | μ=0.76 mm$^{-1}$ |
| β=114.639 (5)° | T=296 (2) K |
| V=2856.8 (3) Å$^3$ | Prism, brown |
| **Data collection** | |
| STOE IPDS 2 diffractomer | 2277 reflections with I > 2 σ (I) |
| ω-scans | $R_{int}$=0.091 |
| Absorption correction: integration | θ$_{max}$=25.0° |
| $T_{min}$=0.782, $T_{max}$= 0.873 | h=-15 → 15 |
| 18190 measured reflections | k=-21 → 21 |
| 5028 independent reflections | l=-15 → 13 |
| **Refinement** | |
| Refinement on $F^2$ | H atoms constrained to parent site |
| R[$F^2$ > 2σ($F^2$)]=0.054 | Calculated weights w=1/[σ$^2$($F_o^2$) + (0.0502P)$^2$] where P=($F_o^2$ + 2$F_c^2$)/3 |
| wR($F^2$)=0.123 | (Δ/σ)$_{max}$=0.001 |
| S=0.82 | Δρ$_{max}$=0.79 e Å$^{-1}$ |
| 5028 reflections | Δρ$_{min}$=-0.42 e Å$^{-1}$ |
| 301 parameters | Extinction correction: none |

**Table 1:** Crystallographic data and structure refinement for 2a.

66.1 (CH$_2$CHOCH$_2$-); 71.5 (-OCHCH$_2$O-); 117.4, 117.5, 121.5, 122.1, 128.4, 129.1, 133.7, 137.4, 142.7, 137.4, 143.0 and 143.5. (Ar-$C$); 205.2 ($C$-Ag). m.p.: 182-185°C; ν$_{(CN)}$: 1661.5 cm$^{-1}$. Anal. Calcd for C$_{20}$H$_{22}$AgBrN$_2$O$_2$: C: 47.04; H: 4.31; N: 5.49. Found: C: 47.02; H: 4.33; N: 5.46.

## Synthesis of bromo[1-(2-methyl-1,4-benzodioxan)-3-(4-methylbenzyl)imidazol-2-ylidene]silver(I), 1d

The synthesis of 1d was carried out in the same way as that described for 1a, but 1-(2-methyl-1,4-benzodioxan)-3-(3-methylbenzyl) imidazol bromide (0.48 g., 1.2 mmol) was used instead of 1-(2-methyl-1,4-benzodioxan)-3-benzylimidazol bromide. Yield: % 91 (0.56 g) *Analytical data for bromo[1-(2-methyl-1,4-benzodioxan)-3-(4-methylbenzyl)imidazol-2-ylidene]silver(I), 1d* $^1$H NMR (300 MHz, DMSO), δ 2.26 (s, 3H, -CH$_2$C$_6$H$_4$CH$_3$); 3.76-3.82 (m, 4H, -NCH$_2$CH$_2$N-); 4.12-4.15 (m, 2H, -NCH$_2$CHOCH$_2$-); 4.30-4.34 (m, 1H, -CH$_2$CHOCH$_2$-); 4.42-4.45 (m, 2H, -OCHCH$_2$O-); 4.66 (s, 2H, -NCH$_2$C$_6$H$_4$); 6.81-6.89 (m, 4H, -OC$_6$H$_4$O- Ar-$H$); 7.08-7.27 (m, 4H, -CH$_2$C$_6$H$_4$(CH$_3$) Ar-$H$). $^{13}$C NMR (300 MHz, DMSO), δ 21.1 (-CH$_2$C$_6$H$_4$CH$_3$); 47.5 (-NCH$_2$CHOCH$_2$-); 49.3 and 50.8 (-NCH$_2$CH$_2$N-); 52.8 (-NCH$_2$C$_6$H$_4$); 66.5 (CH$_2$CHOCH$_2$-); 73.0(-OCHCH$_2$O-); 117.4, 121.5, 121.9, 128.4, 129.0, 129.9, 133.5, 134.9, 136.0, 137.4, 138.6 and 143.6. (Ar-$C$); 205.7 ($C$-Ag). m.p.: 193-195°C; ν$_{(CN)}$: 1662.6 cm$^{-1}$. Anal. Calcd for C$_{20}$H$_{22}$AgBrN$_2$O$_2$: C: 47.04; H: 4.31; N: 5.49. Found: C: 47.07; H: 4.33; N: 5.47.

## Synthesis of bromo[1-(2-methyl-1,4-benzodioxan)-3-(2,4,6-trimethylbenzyl)imidazol-2-ylidene]silver(I), 1e

The synthesis of 1e was carried out in the same way as that described for 1a, but 1-(2-methyl-1,4-benzodioxan)-3-(2,4,6-trimethylbenzyl)imidazol bromide (0.52 g., 1.2 mmol) was used instead of 1-(2-methyl-1,4-benzodioxan)-3-benzylimidazol bromide. Yield: % 90 (0.58 g) *Analytical data for bromo[1-(2-methyl-1,4-benzodioxan)-3-(2,4,6-trimethylbenzyl)imidazol-2-ylidene]silver(I), 1e* $^1$H NMR (300 MHz, DMSO), δ 2.14 and 2.26 (s, 9H, -CH$_2$C$_6$H$_2$(CH$_3$)$_3$); 3.76-3.83 (m, 4H, -NCH$_2$CH$_2$N-); 3.93-3.96 (m, 2H, -NCH$_2$CHOCH$_2$-); 4.33-4.36 (m, 1H, -CH$_2$CHOCH$_2$-); 4.48-4.51 (m, 2H, -OCHCH$_2$O-); 4.78 (s, 2H, -NCH$_2$C$_6$H$_2$); 6.81-6.87(m, 4H, -OC$_6$H$_4$O- Ar-$H$); 7.63-7.66 (m, 2H, -CH$_2$C$_6$H$_2$(CH$_3$)$_3$ Ar-$H$). $^{13}$C NMR (300 MHz, DMSO), δ 15.6 and 19.6 (-CH$_2$C$_6$H$_2$(CH$_3$)$_3$); 47.4 (-NCH$_2$CHOCH$_2$-); 48.3 (-NCH$_2$CH$_2$N-); 49.3 (-NCH$_2$C$_6$H$_2$); 66.5 (CH$_2$CHOCH$_2$-); 72.9 (-OCHCH$_2$O-); 117.4, 121.8, 122.1, 126.7, 128.9, 129.5, 133.6, 136.9, 138.3, 143.5 and 145.1. (Ar-$C$); 204.9 ($C$-Ag). m.p.: 182-185°C; ν$_{(CN)}$: 1665.9 cm$^{-1}$. Anal. Calcd for C$_{22}$H$_{26}$AgBrN$_2$O$_2$: C: 49.05; H: 4.82; N: 5.20. Found: C: 49.08; H: 4.85; N: 5.19.

## Synthesis of bromo[1-(2-methyl-1,4-benzodioxan)-3-(2,3,5,6-tetramethylbenzyl)imidazol-2-ylidene]silver(I), 1f

The synthesis of 1f was carried out in the same way as that described for 1a, but 1-(2-methyl-1,4-benzodioxan)-3-(2,3,5,6-tetramethylbenzyl)imidazol bromide (0.53 g., 1.2 mmol) was used instead of 1-(2-methyl-1,4-benzodioxan)-3-benzylimidazol bromide. Yield: % 86 (0.57 g) *Analytical data for bromo[1-(2-methyl-1,4-benzodioxan)-3-(2,3,5,6-tetramethylbenzyl)imidazol-2-ylidene]silver(I), 1f* $^1$H NMR (300 MHz, DMSO), δ 2.18 (s, 12H, -CH$_2$C$_6$H(CH$_3$)$_4$); 3.64-3.74 (m, 4H, -NCH$_2$CH$_2$N-); 3.94-4.02 (m, 2H, -NCH$_2$CHOCH$_2$-); 4.28-4.38 (m, 1H, -CH$_2$CHOCH$_2$-); 4.46-4.51 (m, 2H, -OCHCH$_2$O-); 4.63 (s, 2H, -NCH$_2$C$_6$H); 6.79-6.88(m, 4H, -OC$_6$H$_4$O- Ar-$H$); 6.96 (s, 1H, -CH$_2$C$_6$H(CH$_3$)$_4$ Ar-$H$). $^{13}$C NMR (300 MHz, DMSO), δ 16.3 and 20.7 (-CH$_2$C$_6$H(CH$_3$)$_4$); 48.2 (-NCH$_2$CHOCH$_2$-); 49.4 and 49.7 (-NCH$_2$CH$_2$N-); 52.1 (-NCH$_2$C$_6$H); 65.4 (CH$_2$CHOCH$_2$-); 71.0 (-OCHCH$_2$O-); 117.5, 117.6, 121.9, 122.1, 131.7, 132.1, 133.1, 133.8, 134.1,

134.3, 142.7 and 143.2. (Ar-$C$); 204.6 ($C$-Ag). m.p.: 105-107°C; ν$_{(CN)}$: 1667.2 cm$^{-1}$. Anal. Calcd for C$_{23}$H$_{26}$AgBrN$_2$O$_2$: C: 49.97; H: 5.07; N: 5.07. Found: C: 49.99; H: 5.09; N: 5.06.

## General procedure for the preparation of the ruthenium-NHC complexes, 2a-f

The ruthenium complexes (2a-f) have been prepared from NHC-silver complexes by transmetallation method [46]. To a mixture of Ag(I)-NHC complexes (2 mmol), Di-μ-chloro-bis[chloro(η$^6$-1-isopropyl-4-methylbenzene)ruthenium(II)] (1 mmol) and dichloromethane (30 mL) was stirred for 24 hours at room temperature in dark condition. The mixture was filtered through Celite and the solvent were evaporated under vacuum to afford the product as a red-brown solid. The crude product was recrystallized from dicloromethane:dietylether (1:3) at room temperature.

## Synthesis of dichloro[1-(2-methyl-1,4-benzodioxan)-3-benzylimidazolidin-2-ylidene](p-cymene)ruthenium(II), 2a

To a solution of bromo[1-(2-methyl-1,4-benzodioxan)-3-benzylimidazol-2-ylidene]silver(I) (0.14g., 0.28 mmol) in dichloromethane (30 mL), Di-μ-chloro-bis[chloro(η$^6$-1-isopropyl-4-methylbenzene)ruthenium(II)] (0.086 g, 0.14 mmol) was added. The reaction mixture was stirred for 24 hours at room temperature in dark condition. The reaction mixture was filtered through celite and the solvent were evaporated under vacuum to afford the product as a red-brown solid. The crude product was recrystallized from dicloromethane:dietylether (1:3) at room temperature. Yield: % 70 (0.12 g) *Analytical data for dichloro[1-(2-methyl-1,4-benzodioxan)-3-benzylimidazolidin-2-ylidene](p-cymene)ruthenium(II), 2a* $^1$H NMR (300 MHz, CDCl$_3$); δ 1.32 (d, 6H, $J$: 6.3 Hz Ru-C$_6$H$_4$CH(CH$_3$)$_2$); 2.18 (s, 3H, Ru-C$_6$H$_4$CH$_3$); 2.91-2.93 (m, 1H, Ru-C$_6$H$_4$CH(CH$_3$)$_2$); 3.48-3.53 (m, 4H, -NCH$_2$CH$_2$N-); 3.58-3.64 (m, 2H, -NCH$_2$CHO-); 3.94-3.98 (m, 1H, -CH$_2$CHOCH$_2$-); 4.28-4.38 (m, 2H, - OCHCH$_2$O-); 4.73 (s, 2H, -NCH$_2$C$_6$H$_5$); 5.35-5.37 (d, 2H, $J$: 5.8 Hz Ru-Ar-$H$); 5.47-5.49 (d, 2H, $J$: 5.6 Hz Ru-Ar-$H$); 6.86-6.90 (m, 4H, -OC$_6$H$_4$O- Ar-$H$); 7.28-7.39 (m, 5H, -CH$_2$C$_6$H$_5$ Ar-$H$). $^{13}$C NMR (300 MHz, CDCl$_3$); δ 18.9 (Ru-C$_6$H$_4$CH(CH$_3$)$_2$); 22.2 (Ru-C$_6$H$_4$CH(CH$_3$)$_2$); 30.7 (Ru-C$_6$H$_4$CH$_3$); 49.2 (-NCH$_2$CHOCH$_2$-); 50.3 and 52.6 (-NCH$_2$CH$_2$N-); 55.8 (-NCH$_2$C$_6$H$_5$); 66.1 (-CH$_2$CHOCH$_2$-); 74.6 (-OCHCH$_2$O-); 80.5, 83.4, 85.7, 86.8, 99.0 and 108.7. (Ru-Ar-$C$); 117.3, 117.4, 121.5, 121.7, 141.9 and 143.3. (-OC$_6$H$_4$O- Ar-$C$); 127.5, 127.7, 128.8, 129.0 and 137.0.(-NCH$_2$C$_6$H$_5$ Ar-$C$); 209.0 ($C$-Ru). m.p.: 112-114°C; ν$_{(CN)}$: 1492.5 cm$^{-1}$. Anal. Calcd for RuC$_{29}$H$_{34}$Cl$_2$N$_2$O$_2$: C: 56.62; H: 5.53; N: 4.56. Found: C: 56.60; H: 5.50; N: 4.54.

## Synthesis of dichloro[1-(2-methyl-1,4-benzodioxan)-3-(2-methylbenzyl)imidazolidin-2-ylidene](p-cymene)ruthenium(II), 2b

The synthesis of 2b was carried out in the same way as that described for 2a, but bromo[1-(2-methyl-1,4-benzodioxan)-3-(2-methylbenzyl)imidazol-2-ylidene]silver(I) (0.14 g., 0.28 mmol) was used instead of bromo[1-(2-methyl-1,4-benzodioxan)-3-benzylimidazol-2-ylidene]silver(I). Yield: % 64 (0.11 g) *Analytical data for dichloro[1-(2-methyl-1,4-benzodioxan)-3-(2-methylbenzyl)imidazolidin-2-ylidene](p-cymene)ruthenium(II), 2b* $^1$H NMR (300 MHz, CDCl$_3$); δ 1.32 (d, 6H, $J$: 6.8 Hz Ru-C$_6$H$_4$CH(CH$_3$)$_2$); 2.10 (s, 3H, Ru-C$_6$H$_4$CH$_3$); 2.34 (s, 3H, C$_6$H$_4$CH$_3$); 2.79-2.86 (m, 1H, Ru-C$_6$H$_4$CH(CH$_3$)$_2$); 3.46-3.50 (m, 4H, -NCH$_2$CH$_2$N-); 3.60-3.67 (m, 2H, -NCH$_2$CHO-); 3.81-3.84 (m, 1H,-CH$_2$CHOCH$_2$-); 4.25-4.40 (m, 2H, -OCHCH$_2$O-); 4.77 (s, 2H, -NCH$_2$C$_6$H$_4$); 5.30-5.32 (d, 2H, $J$: 5.7 Hz Ru-Ar-$H$); 5.40-5.42 (d, 2H, $J$: 5.4 Hz Ru-Ar-$H$); 6.85-6.90 (m, 4H,-OC$_6$H$_4$O-Ar-$H$); 7.25-7.29 (m,

4H, -CH$_2$C$_6$H$_4$(CH$_3$) Ar-*H*). $^{13}$C NMR ( 300 MHz, CDCl$_3$); δ 18.5 and 19.3 (C$_6$H$_4$CH(CH$_3$)$_2$); 21.9 (C$_6$H$_4$CH(CH$_3$)$_2$); 23.4 (C$_6$H$_4$CH$_3$); 30.7 (Ru-C$_6$H$_4$CH$_3$); 49.8 (-NCH$_2$CHOCH$_2$-); 50.5 (-NCH$_2$CH$_2$N-); 52.7 (-NCH$_2$C$_6$H$_5$); 66.1 (-CH$_2$CHOCH$_2$-); 74.7 (-OCHCH$_2$O-); 80.9, 81.1, 86.7, 96.8, 99.0. (Ru-Ar-*C* ); 117.4, 117.5, 121.5, 121.7, 142.0 and 143.4. (-OC$_6$H$_4$O- Ar-C); 127.2, 127.5, 130.9, 135.7 and 135.8.(-NCH$_2$C$_6$H$_5$Ar-C); 209.9 (*C*-Ru). m.p.: 203-205°C; $_{(CN)}$: 1493.2 cm$^{-1}$. Anal. Calcd for RuC$_{30}$H$_{36}$Cl$_2$N$_2$O$_2$: C: 57.27; H: 5.73; N: 4.45. Found: C: 57.25; H: 5.72; N: 4.43.

## Synthesis of dichloro[1-(2-methyl-1,4-benzodioxan)-3-(3-methylbenzyl)imidazolidin-2-ylidene](*p*-cymene)ruthenium(II), 2c

The synthesis of 2c was carried out in the same way as that described for 2a, but bromo[1-(2-methyl-1,4-benzodioxan)-3-(3-methylbenzyl)imidazol-2-ylidene]silver(I) (0.14 g., 0.28 mmol) was used instead of bromo[1-(2-methyl-1,4-benzodioxan)-3-benzylimidazol-2-ylidene]silver(I). Yield: % 75 (0.13 g) *Analytical data for dichloro[1-(2-methyl-1,4-benzodioxan)-3-(3-methylbenzyl)imidazolidin-2-ylidene](p-cymene)ruthenium(II), 2c* $^1$H NMR (300 MHz, CDCl$_3$); δ 1.31 (d, *J*: 5.8 Hz 6H, Ru-C$_6$H$_4$CH(CH$_3$)$_2$); 2.17 (s, 3H, Ru-C$_6$H$_4$CH$_3$); 2.36 (s, 3H, C$_6$H$_4$CH$_3$); 2.88-2.95 (m, 1H, Ru-C$_6$H$_4$CH(CH$_3$)$_2$); 3.37-3.44 ( m, 4H, -NCH$_2$CH$_2$N-); 3.47-3.50 (m, 2H, -NCH$_2$CHO-); 3.96-3.98 (m, 1H, -CH$_2$CHOCH$_2$-); 4.14-4.24 (m, 2H, -OCHCH$_2$O-); 4.26 (s, 2H, -NCH$_2$C$_6$H$_4$(CH$_3$)); 5.38-5.40 (d, 2H, *J*: 5.4 Hz Ru-Ar-*H*); 5.47-5.49 (d, 2H, *J*: 6.0 Hz Ru-Ar-*H*); 6.83-6.88 (m, 4H, -OC$_6$H$_4$O- Ar-*H*); 7.13-7.26 (m, 4H, -CH$_2$C$_6$H$_4$(CH$_3$) Ar-*H*). $^{13}$C NMR (300 MHz, CDCl$_3$), δ 18.8 (C$_6$H$_4$CH(CH$_3$)$_2$); 21.9 (Ru-C$_6$H$_4$CH(CH$_3$)$_2$); 23.3 (C$_6$H$_4$CH$_3$); 30.7 (Ru-C$_6$H$_4$CH$_3$); 48.0 (-NCH$_2$CHOCH$_2$-);49.1 and 51.6 (-NCH$_2$CH$_2$N-); 55.5 (-NCH$_2$C$_6$H$_4$(CH$_3$)); 66.1 (-CH$_2$CHOCH$_2$-); 73.0 (-OCHCH$_2$O-); 80.9, 81.5, 82.2, 86.7, 96.8 ve 99.0. (Ru-Ar-*C* ); 117.3, 117.5, 121.5, 121.8, 142.9 and 143.4 (-OC$_6$H$_4$O- Ar-C); 127.4, 127.9, 129.5, 129.9, 133.8 and 137.4. (-NCH$_2$C$_6$H$_5$Ar-C); 209.7 (*C*-Ru). m.p.: 204-206°C; ν$_{(CN)}$; 1493.3 cm$^{-1}$. Anal. Calcd for RuC$_{30}$H$_{36}$Cl$_2$N$_2$O$_2$: C: 57.27; H: 5.73; N: 4.45. Found: C: 57.24; H: 5.71; N: 4.44.

## Synthesis of dichloro[1-(2-methyl-1,4-benzodioxan)-3-(4-methylbenzyl)imidazolidin-2-ylidene](*p*-cymene)ruthenium(II), 2d

The synthesis of 2d was carried out in the same way as that described for 2a, but bromo[1-(2-methyl-1,4-benzodioxan)-3-(4-methylbenzyl)imidazol-2-ylidene]silver(I) (0.14 g., 0.28 mmol) was used instead of bromo[1-(2-methyl-1,4-benzodioxan)-3-benzylimidazol-2-ylidene]silver(I). Yield: % 69 (0.12 g) *Analytical data for dichloro[1-(2-methyl-1,4-benzodioxan)-3-(4-methylbenzyl)imidazolidin-2-ylidene](p-cymene)ruthenium(II), 2d* 1H NMR (300 MHz, CDCl$_3$) δ 1.29 (d, 6H, *J*: 6.6 Hz Ru-C$_6$H$_4$CH(CH$_3$)$_2$); 2.17 (s, 3H, Ru-C$_6$H$_4$CH$_3$); 2.36 (s, 3H, C$_6$H$_4$CH$_3$); 2.76-2.82 (m, 1H, Ru-C$_6$H$_4$CH(CH$_3$)$_2$); 3.37-3.43 ( m, 4H, -NCH$_2$CH$_2$N-); 3.56-3.63 (m, 2H, -NCH$_2$CHO-); 3.87-3.90 (m, 1H, -CH$_2$CHOCH$_2$-); 4.18-4.21 (m, 2H, -OCHCH$_2$O-); 4.26 (s, 2H, -NCH$_2$C$_6$H$_5$); 5.34-5.36 (d, 2H, *J*: 5.9 Hz Ru-Ar-*H*); 5.40-5.42 (d, 2H, *J*: 6.0 Hz Ru-Ar-*H*); 6.85-6.88 (m, 4H, -OC$_6$H$_4$O- Ar-*H*); 7.10-7.25 (m, 4H, -CH$_2$C$_6$H$_4$(CH$_3$) Ar-*H*). $^{13}$C NMR ( 300 MHz, CDCl$_3$), δ 18.9 and 19.0 (C$_6$H$_4$CH(CH$_3$)$_2$); 21.2 (Ru-C$_6$H$_4$CH(CH$_3$)$_2$); 22.3 (-C$_6$H$_4$CH$_3$); 30.7 (Ru-C$_6$H$_4$CH$_3$); 49.2 (-NCH$_2$CHOCH$_2$-); 55.5 (-NCH$_2$C$_6$H$_4$(CH$_3$)); 51.3 and 52.6 (-NCH$_2$CH$_2$N-); 65.9 (-CH$_2$CHOCH$_2$-); 73.2 (-OCHCH$_2$O-); 80.5, 81.3, 85.7, 96.8, 99.0 and 101.2. (Ru-Ar-*C* ); 117.3, 117.5, 121.5, 121.8, 143.1 and 143.4. (-OC$_6$H$_4$O- Ar-C); 127.5, 128.0, 129.3, 129.4, 133.8 and 137.4. (-NCH$_2$C$_6$H$_5$Ar-C); 209.7 (*C*-Ru). m.p.: 167-169°C; ν$_{(CN)}$; 1493.5 cm$^{-1}$. Anal. Calcd for RuC$_{30}$H$_{36}$Cl$_2$N$_2$O$_2$: 57.27; H: 5.73; N: 4.45. Found: C: 57.23; H: 5.70; N: 4.42.

## Synthesis of dichloro[1-(2-methyl-1,4-benzodioxan)-3-(2,4,6-trimethylbenzyl)imidazolidin-2-ylidene](*p*-cymene)ruthenium(II), 2e

The synthesis of 2e was carried out in the same way as that described for 2a, but bromo[1-(2-methyl-1,4-benzodioxan)-3-(2,4,6-trimethylbenzyl)imidazol-2-ylidene]silver(I) (15 g., 0.28 mmol) was used instead of bromo[1-(2-methyl-1,4-benzodioxan)-3-benzylimidazol-2-ylidene] silver(I). Yield: % 62 (0.17 g) *Analytical data for dichloro[1-(2-methyl-1,4-benzodioxan)-3-(2,4,6-trimethylbenzyl)imidazolidin-2-ylidene] (p-cymene)ruthenium(II), 2e* $^1$H NMR (300 MHz, CDCl$_3$); δ 1.31 (d, 6H, *J*: 6.4 Hz Ru-C$_6$H$_4$CH(CH$_3$)$_2$); 2.17 (s, 3H, Ru-C$_6$H$_4$CH$_3$); 2.27 and 2.28 (s, 9H, C$_6$H$_2$(CH$_3$)$_3$); 2.98-3.03 (m, 1H, Ru-C$_6$H$_4$CH(CH$_3$)$_2$); 3.32-3.39 (m, 4H, -NCH$_2$CH$_2$N-); 3.48-3.51 (m, 2H, -NCH$_2$CHO-); 3.78-3.85 (m, 1H, -CH$_2$CHOCH$_2$-); 4.04-4.11 (m, 2H, -OCHCH$_2$O-); 4.32 (s, 2H, -NCH$_2$C$_6$H$_2$); 5.37-5.39 (d, 2H, *J*: 5.9 Hz Ru-Ar-*H*); 5.46-5.48 (d, 2H, *J*: 6.0 Hz Ru-Ar-*H*); 6.83-6.87 (m, 4H, -OC$_6$H$_4$O- Ar-*H*); 7.28-7.30 (m, 2H, -CH$_2$C$_6$H$_2$(CH$_3$)$_3$ Ar-*H*). $^{13}$C NMR ( 300 MHz, CDCl$_3$); δ 18.7 (C$_6$H$_4$CH(CH$_3$)$_2$); 18.9 and 20.4 (C$_6$H$_2$(CH$_3$)$_3$); 21.0 (Ru-C$_6$H$_4$CH(CH$_3$)$_2$); 30.7 (Ru-C$_6$H$_4$CH$_3$); 45.3 (-NCH$_2$CHOCH$_2$-); 48.0 and 48.3 (-NCH$_2$CH$_2$N-); 52.7 (-NCH$_2$C$_6$H$_5$); 65.9 (-CH$_2$CHOCH$_2$-); 72.7 (-OCHCH$_2$O-); 80.6, 81.3, 86.2, 96.9, 99.8 and101.9. (Ru-Ar-*C*); 117.4, 117.5 121.4, 122.7, 138.0 and 143.4. (-OC$_6$H$_4$O- Ar-C); 126.6, 127.9, 129.0, 129.4, 137.4 and 137.6. (-NCH$_2$C$_6$H$_5$Ar-C); 209.2 (*C*-Ru). m.p.: 191-193°C; ν$_{(CN)}$: 1493.6 cm$^{-1}$. Anal. Calcd for RuC$_{32}$H$_{40}$Cl$_2$N$_2$O$_2$: C: 58.48; H: 6.09; N: 4.26. Found: C: 58.45; H: 6.07; N: 4.24.

## Synthesis of dichloro[1-(2-methyl-1,4-benzodioxan)-3-(2,3,5,6-tetramethylbenzyl)imidazolidin-2-ylidene](*p*-cymene)ruthenium(II), 2f

The synthesis of 2f was carried out in the same way as that described for 2a, but bromo[1-(2-methyl-1,4-benzodioxan)-3-(3,4,5,6-tetramethylbenzyl)imidazol-2-ylidene]silver(I) (16 g., 0.28 mmol) was used instead of bromo[1-(2-methyl-1,4-benzodioxan)-3-benzylimidazol-2-ylidene]silver(I). Yield: % 66 (0.12 g) *Analytical data for dichloro[1-(2-methyl-1,4-benzodioxan)-3-(2,3,5,6-tetramethylbenzyl) imidazolidin-2-ylidene](p-cymene)ruthenium(II), 2f* $^1$H NMR (300 MHz, CDCl$_3$); 1.34 (d, 6 H, *J*: 6.3 Hz Ru-C$_6$H$_4$CH(CH$_3$)$_2$); 2.18 (s, 3H, Ru-C$_6$H$_4$CH$_3$); 2.21 and 2.23 (s, 12H, C$_6$H(CH$_3$)$_4$); 2.94-2.98 (m, 1H, Ru-C$_6$H$_4$CH(CH$_3$)$_2$); 3.05-3.17 (m, 4H, -NCH$_2$CH$_2$N-); 3.39-3.48 (m, 2H, -NCH$_2$CHO-); 3.83-3.88 (m, 1H, -CH$_2$CHOCH$_2$-); 4.15-4.20 (m, 2H, -OCHCH$_2$O-); 4.52 (s, 2H, -NCH$_2$C$_6$H); 5.38-5.40 (d, 2H, *J*: 6.3 Hz Ru-Ar-*H*); 5.48-5.50 (d, 2H, *J*: 5.9 Hz Ru-Ar-*H*); 6.84-6.87 (m, 4H, -OC$_6$H$_4$O- Ar-*H*); 7.29 (s, 1H, -CH$_2$C$_6$H(CH$_3$)$_4$ Ar-*H*). $^{13}$C NMR (300 MHz, CDCl$_3$); δ 18.4 and 18.6 (C$_6$H$_4$CH(CH$_3$)$_2$); 19.0 and 20.5 (C$_6$H(CH$_3$)$_4$); 22.2 (Ru-C$_6$H$_4$CH(CH$_3$)$_2$); 30.6 (Ru-C$_6$H$_4$CH$_3$); 49.4 and 50.7 (-NCH$_2$CH$_2$N-); 66.0 (-CH$_2$CHOCH$_2$-); 52.3 (-NCH$_2$C$_6$H$_5$); 47.6 (-NCH$_2$CHOCH$_2$-); 73.1 (-OCHCH$_2$O-); 80.5, 81.3, 83.1, 96.8 and 101.2. ( Ru-Ar-*C* ); 117.3, 117.4, 121.7, 121.8 and 143.1. (-OC$_6$H$_4$O-Ar-C); 126.4, 127.2, 132.0, 133.5, 134.3 and 134.5. (-NCH$_2$C$_6$H$_5$Ar-C); 209.1 (*C*-Ru). m.p.: 191-193°C; ν$_{(CN)}$; 1494.3 cm$^{-1}$. Anal. Calcd for RuC$_{33}$H$_{42}$BrN$_2$O$_2$: C: 59.94; H: 6.85; N: 4.08. Found: C: 59.91; H: 6.83; N: 4.10.

## General method for the transfer hydrogenation of ketones

The catalytic hydrogen transfer reactions were carried out in a closed Schlenk flask under argon atmosphere. Substrate ketone (1 mmol), catalyst ruthenium(II)-NHC complex (2a-f) (0,01 mmol) and KOH (4 mmol) was heated to reflux in 10 mL of *i*-PrOH for 1 hr. The solvent was then removed under vacuum. At the conclusion of the reaction, the mixture was coled, extracted with ethylacetate/hexane (1:5), filtered through a pad of silica gel with copious washings, concentrated, and purified by flash chromatography on silica gel. The product distribution was determined by $^1$H NMR spectroscopy, GC and GC-MS.

## Results and Discussion

### Synthesis

The synthetic route for unsymmetrically 2-methyl-1,4-benzodioxan substituted Ag(I)-NHC complexes and their corresponding Ru(II)-NHC complexes described in this study is given in Scheme 1. The unsymmetrically substituted Ag(I)-NHC complexes 1a-f were prepared by stirring 1-(2-methyl-1,4-benzodioxan)-3-alkylimidazolidinium salts with 0.5 equivalents of Ag$_2$O in dichloromethane at room temperature for 24 hours. The Ag(I)-NHC complexes as off white solid in 85% to 91% yield. The silver complexes were soluble in halogenated solvent and insoluble in non-polar solvents. The complexes were characterized by spectroscopic techniques ($^1$H, $^{13}$C NMR, IR) and elemental analysis. $^1$H and $^{13}$C NMR spectra are consistent with the proposed formulate. In the $^1$H NMR and $^{13}$C NMR spectra in d-DMSO-d$_6$, loss of signals for the imidazolium protons (NCHN) (8.80-9.87 ppm) and imidazolium carbon (NCHN) at (157.9-159.1 ppm) showed the formation of the expected Ag(I)-NHC. The $^{13}$C NMR spectra, resonances of the carbene carbon atoms of complexes appeared in the range δ 204.6-205.7 ppm respectively for 1a-f. These signals are shifted downfield compared to the carbene precursors which further demonstrates the formation of expected Ag(I)-NHC. The IR data for Ag(I)-NHC complexes exhibit a characteristic v(C=N) band at 1666.8, 1645.9, 1661.5, 1662.6, 1665.9 and 1667.2 respectively, for 1a-f. The NMR and FT-IR values are similar to results of other Ag(I)-NHC complexes (Scheme 1). The 2-methyl-1,4-benzodioxan substituted Ru(II)-NHC complexes 2a-f were prepared from synthesized Ag(I)-NHC complexes via transmetallation method (Scheme 1). The air and moisture stable Ru(II)-NHC complexes were soluble in solvents such as chloroform and dichloromethane. The Ru(II)-NHC complexes 2a-f were prepared by stirring bromo[1-(2-methyl-1,4-benzodioxan)-3-alkylimidazol-2-ylidene]silver(I) with 0.5 equivalents of [RuCl$_2$(pcym)]$_2$ in dichloromethane at room temperature for 24 hours. The Ru(II)-NHC complexes as a red-brown solid in 64% to 75% yield. The Ru(II)-NHC complexes were soluble in halogenated solvent and insoluble in non-polar solvents. The Ru(II)-NHC complexes have been characterized by analytical and spectroscopic techniques. In the $^1$H NMR spectra, resonances for the isopropyl and methyl protons of the p-cymene group of complexes 2a-f in the range 1.29-1.34 (methyl of izopropyl group), 2.76-3.03 (single hydrogen of izopropyl group) and 2.14-2.18 ppm (p-methyl of p-cymene group) respectively showed the formation of the Ru(II)-NHC complexes. The $^{13}$C NMR spectra, resonances of the carbene carbon atoms of complexes appeared in the range δ 209.0-209.9 ppm respectively for 2a-f. These signals are shifted downfiel compared to corresponding Ag(I)-NHC complexes of Ru(II)-NHC complexes carbene carbons signal at the range 204.6-205.7 ppm respectively showed the formation of the expected Ru(II)-NHC complexes. The IR data for Ru(II)-NHC complexes exhibit a characteristic v(C=N) band at 1492.5, 1493.2, 1493.3, 1493.5, 1493.6 and 1494.3 respectively, for 2a-f. The NMR and FT-IR values are similar to results of other Ru(II)-NHC complexes.

### Structural characterization of 2a

Suitable crystal for X-ray crystallography to determine the molecular structure of 2a was obtained in a saturated dichloromethane solution with slow infusion of diethyl ether. The molecular structure of 2a is shown in Figure 1. Figures 1 and 2 were drawn using the PLATON

program. In the p-cymene ligand of the 2a in Figure 1, the arene ring (C21–C26) has a planar configuration [rms deviation=0.008 Å]. The C-C bond lengths within the p-cymene ring are similar, except from a shortening of the C24-C25 bond of 1.362(13) Å. The Ru-carbene distance in 2a is 2.052(8) Å. The distance of 1.690(3) Å between the centroid of the p-cymene ring and ruthenium is very close to that reported in other Ru$^{II}$ arene complexes [47,48]. The Ru-Cl1 and Ru-Cl2 bond lengths [2.4323(19) and 2.407(2) Å, respectively] are similar to those in other Ru$^{II}$ complexes [47,48]. In the crystal, molecules are connected by intermolecular C-H...Cl hydrogen bonds (Table 2, Figure 2), forming 3D network. H atoms were positioned geometrically and were refined using a riding model with U$_{iso}$(H)=1.2 or 1.5U$_{eq}$(C). The highest residual peak lies 1.22 Å from Ru1. The small proportion of reflections observed is a result of the rather poor quality of the very thin crystals obtained. Figure 2 shown a view of the crystal packing for 2a.

### Catalytic transfer hydrogenation of ketones

Ruthenium complexes have been used as active catalysts for transfer hydrogenation using 2-propanol as a hydrogen source. The reaction conditions for this transformation are economic, partly mild and environmentally benign friendly. 2-propanol using as a hydrogen source is a popular reactive solvent for the catalytic transfer hydrogenation since it is easy to handle and is relatively non-toxic, environmentally benign, and inexpensive. The volatile acetone by-product can also be easily removed. We have investigated and compared the catalytic properties of 2-methyl-1,4-benzodioxan substituted Ru(II)-NHC complexes 2a-f in the transfer hydrogenation of various methyl aryl ketones. The reduction of acetophenone with 2-propanol to 1-phenylethanol was chosen as a model reaction. The catalytic transfer hydrogenations of ketones were carried out using Ru(II)-NHC precatalyst (0.01 mmol), KOH (4 mmol) and substrate ketone (1.00 mmol) in 2-propanol at 80°C. The conversion was monitored by GC and NMR. It is well known that catalytic transfer hydrogenation is sensitive to the nature of the base. We surveyed K$_2$CO$_3$, Cs$_2$CO$_3$, NaOH, KOH, t-BuOK and NaOAc for the choice of base. The highest rate was observed when KOH was employed. A variety of ketones by 2-propanol were transformed to the corresponding secondary alcohols. Typical results are summerized in Table 3.

Is examined in general, all complexes 2a-f is seen to be reasonably active in hydrogen transfer reactions. Under the reaction conditions complex 2c turned out to be the active catalyst in comparison with 2a, 2b, 2d, 2e and 2f. The reduction of p-methoxyacetophenone with 2c was completed within 1 hr reaching 81%. In contrast, p-methoxacetophenone was reduced within 1 hr using 2a, 2b, 2d, 2e and 2f with 80, 70, 78, 74 and 68% conversion, respectively (Table 3).

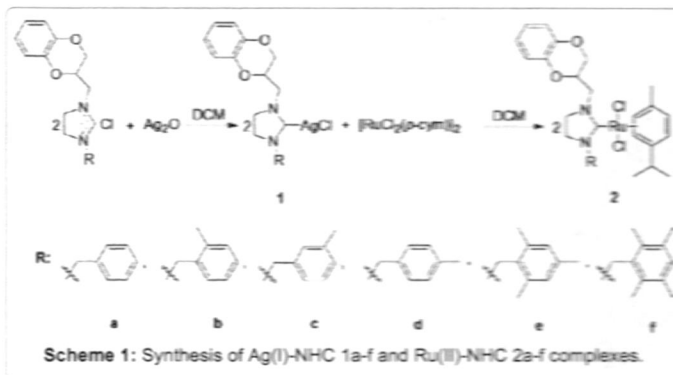

**Scheme 1:** Synthesis of Ag(I)-NHC 1a-f and Ru(II)-NHC 2a-f complexes.

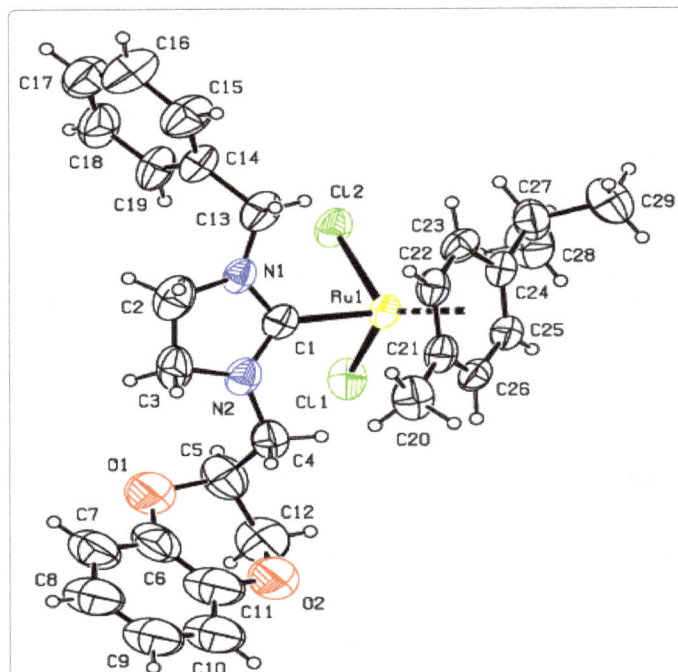

**Figure 1:** The molecular structure of 2a with the atom numbering scheme. Displacement ellipsoids for non-H atoms are drawn at the 30% probability level.

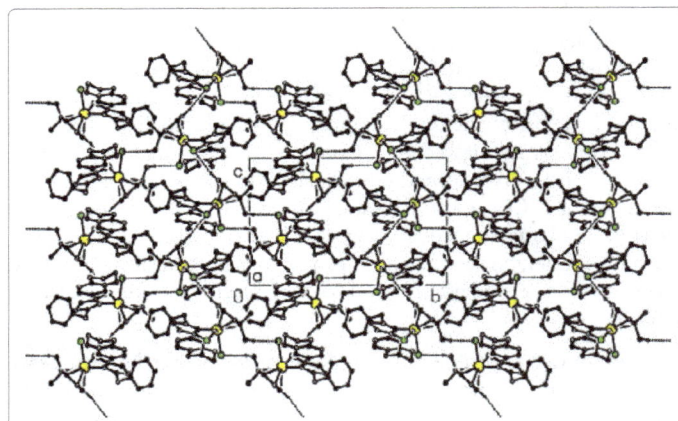

**Figure 2:** View of the hydrogen bonding and molecular packing of 2a along a-axis.

| | $D$-H | H...$A$ | D...$A$ | $D$-H...$A$ |
|---|---|---|---|---|
| C22-H22...Cl2$^i$ | 0.93 | 2.66 | 3.570 (7) | 168 |
| C28-H28C...Cl1$^{ii}$ | 0.96 | 2.81 | 3.769 (12) | 173 |

**Table 2:** Hydrogen-bond parameters for 2a (Å, °). Symmetry codes: (i) x, 1/2-y, -1/2+z; (ii) 1-x, 1-y, 2-z.

A variety of ketones were converted to be corresponding secondary alcohols. Typical results is illustrated in Table 3. Under those conditions *p*-methoxyacetophenone and *p*-fluoroacetophenone react neatly and in good yields with 2-propanole (Table 3). The existence of electron withdrawing (F) or electron donating (OCH₃) substituents on acetophenone (Table 3) has effect on the reduction of major of ketones to their corresponding alcohols. The more conversion of p-methoxyace-tophenone to secondary alcohole was obtained at a time 1 h (Table 3). The transformation of ketones with bulky substituents was not shown or mildly decreased. We tried this reaction with benzophenone at 1

| Entry | R | R$_1$ | Base | Complex | Yield (%) |
|---|---|---|---|---|---|
| 1 | H | -CH$_3$ | KOH | 2a | 75 |
| 2 | H | -CH$_3$ | KOH | 2b | 88 |
| 3 | H | -CH$_3$ | KOH | 2c | 70 |
| 4 | H | -CH$_3$ | KOH | 2d | 71 |
| 5 | H | -CH$_3$ | KOH | 2e | 70 |
| 6 | H | -CH$_3$ | KOH | 2f | 72 |
| 7 | MeO | -CH$_3$ | KOH | 2a | 80 |
| 8 | MeO | -CH$_3$ | KOH | 2b | 70 |
| 9 | MeO | -CH$_3$ | KOH | 2c | 81 |
| 10 | MeO | -CH$_3$ | KOH | 2d | 78 |
| 11 | MeO | -CH$_3$ | KOH | 2e | 74 |
| 12 | MeO | -CH$_3$ | KOH | 2f | 68 |
| 13 | F | -CH$_3$ | KOH | 2a | 82 |
| 14 | F | -CH$_3$ | KOH | 2b | 72 |
| 15 | F | -CH$_3$ | KOH | 2c | 86 |
| 16 | F | -CH$_3$ | KOH | 2d | 73 |
| 17 | F | -CH$_3$ | KOH | 2e | 87 |
| 18 | F | -CH$_3$ | KOH | 2f | 89 |
| 19$^b$ | H | -C$_6$H$_5$ | KOH | 2a | 94 |
| 20$^b$ | H | -C$_6$H$_5$ | KOH | 2b | 97 |
| 21$^b$ | H | -C$_6$H$_5$ | KOH | 2c | 99 |
| 22$^b$ | H | -C$_6$H$_5$ | KOH | 2d | 98 |
| 23$^b$ | H | -C$_6$H$_5$ | KOH | 2e | 98 |
| 24$^b$ | H | -C$_6$H$_5$ | KOH | 2f | 96 |

**Table 3:** Transfer hydrogenation of ketones catalyzed by 2a-f. $^a$Determined by GC-MS and yields are based on ketones. $^b$2h, 80°C.

hr. But, we have achieved low yields. Therefore, we have extended the duration of experiments for benzophenone to 2 hr. The benzophenone was reduced within 1 hr using 2a and 2b with 21% and 23% conversion, respectively. However, the yields lower than 2 hr, for example the reduction of benzophenone with 2a and 2b was completed within 94% and 97% respectively (Table 3, Scheme 1).

## Conclusions

As a result, we have described the preparation and catalytic activities of well-defined 2-methyl-1,4-benzodioxan substituted Ru(II)-NHC complexes 2a-f. Via the Ag(I)-NHC 1a-f transmetallation route, Ru(II)-NHC complexes were readily accessible and are effective catalyst precursors for the transfer hydrogenation of ketones. The catalytic activities of these six Ru(II)-NHC complexes have been examined for the transfer hydrogenation of ketones.

### Acknowledgements

This work was financially supported by İnönü University Research Fund (IUBAP 2011/25) and the Faculty of Arts and Sciences, Ondokuz Mayis University, Turkey, for the use of the Stoe *IPDS* 2 diffractometer (purchased under grant F.279 of the University Research Fund).

### References

1. Bantreil X, Broggiw J, Nolan SP (2009) N-Heterocyclic carbene containing complexes in catalysis. Annu Rep Prog Chem Sect B: Org Chem 105: 232-263.

2. Marion N, Nolan SP (2008) Well-defined N-heterocyclic carbenes-palladium(II) precatalysts for cross-coupling reactions. Acc Chem Res 41: 1440-1449.

3. Hahn FE, Jahnke MC (2008) Heterocyclic carbenes: synthesis and coordination chemistry. Angew Chem Int Ed Engl 47: 3122-3172.

4. Bourissou D, Guerret O, Gabbaï FP, Bertrand G (2000) Stable Carbenes. Chem Rev 100: 39-92.

5. Herrmann WA, Weskamp VP, Bohm WT (2001) Metal complexes of stable carbenes. Adv Organomet Chem 48: 1-355.

6. Gonzalez SD, Nolan SP (2005) 8 Carbene and transition metal-mediated transformations. Annu Rep Prog Chem B: Org Chem 101: 171-191.

7. Crabtree RH (2006) Some chelating C-donor ligands in hydrogen transfer and related catalysis. J Organomet Chem 691: 3146-3150.

8. Jafarpour L, Nolan SP (2001) Transition-Metal Systems Bearing a Nucleophilic Carbene Ancillary Ligand: From Thermochemistry to Catalysis. Adv Organomet Chem 46: 181-222.

9. Hu X, Castro-Rodriguez I, Olsen K, Meyer K (2004) Group 11 Metal Complexes of N-Heterocyclic Carbene Ligands:? Nature of the Metal Carbene Bond. Organometallics 23: 755-764.

10. Cavallo L, Correa A, Costabile C, Jacobsen HJ (2005) Steric and electronic effects in the bonding of N-heterocyclic ligands to transition metals. Organomet Chem 690: 5407-5413.

11. Süssner M, Plenio H (2005) pi-Face donor properties of N-heterocyclic carbenes. Chem Commun (Camb) : 5417-5419.

12. Christmann U, Vilar R (2005) Monoligated palladium species as catalysts in cross-coupling reactions. Angew Chem Int Ed Engl 44: 366-374.

13. Marion N, Navarro O, Mei J, Stevens ED, Scott NM, et al. (2006) Modified (NHC)Pd(allyl)Cl (NHC = N-heterocyclic carbene) complexes for room-temperature Suzuki-Miyaura and Buchwald-Hartwig reactions. J Am Chem Soc 128: 4101-4111.

14. Organ MG, Abdel-Hadi M, Avola S, Dubovyk I, Hadei N, et al. (2008) Pd-catalyzed aryl amination mediated by well defined, N-heterocyclic carbene (NHC)-Pd precatalysts, PEPPSI. Chemistry 14: 2443-2452.

15. Ulusoy M, Sahin O, Buyukgungor O, Cetinkaya B (2008) Imidazolium salicylaldimine frameworks for the preparation of tridentate N-heterocyclic carbene ligands. J Organomet Chem 693: 1895-1902.

16. Viciu MS, Navarro O, Germaneau RF, Kelly RA, Sommer W, et al. (2004) Synthetic and Structural Studies of (NHC)Pd(allyl)Cl Complexes (NHC=N-heterocyclic carbene). Organometallics 23: 1629-1635.

17. Jensen DR, Schultz MJ, Mueller JA, Sigman MS (2003) A well-defined complex for palladium-catalyzed aerobic oxidation of alcohols: design, synthesis, and mechanistic considerations. Angew Chem Int Ed Engl 42: 3810-3813.

18. Hartwig JF (2006) Discovery and understanding of transition-metal-catalyzed aromatic substitution reactions. Synlett 9: 1283-1294.

19. Bryliakov KP, Talsi EP (2008) Titanium-Salan-Catalyzed Asymmetric Oxidation of Sulfides and Kinetic Resolution of Sulfoxides with H2O2 as the Oxidant. Eur J Org Chem 19: 3369-3376.

20. Huynh HV, Han Y, Ho JHH, Tan GK (2006) Palladium(II) Complexes of a Sterically Bulky, Benzannulated N-Heterocyclic Carbene with Unusual Intramolecular C-H···Pd and Ccarbene···Br Interactions and Their Catalytic Activities. Organometallics 25: 3267-3274.

21. Garrison JC, Youngs WJ (2005) Ag(I) N-heterocyclic carbene complexes: synthesis, structure, and application. Chem Rev 105: 3978-4008.

22. Winkelmann O, Christian Näther, Lüning UJ (2008) Synthesis, structure and catalytic activity of a bimacrocylic NHC palladium allyl complex. Organomet Chem 693: 2784-2788.

23. Viciano M, Mas-Marzá E, Poyatos M, Sanaú M, Crabtree RH, et al. (2005) An N-heterocyclic carbene/iridium hydride complex from the oxidative addition of a ferrocenyl-bisimidazolium salt: implications for synthesis. Angew Chem Int Ed Engl 44: 444-447.

24. Xi Z, Zhang X, Chen W, Fu S, Wang D, et al. (2007) Synthesis and Structural Characterization of Nickel(II) Complexes Supported by Pyridine-Functionalized N-Heterocyclic Carbene Ligands and Their Catalytic Acitivities for Suzuki Coupling. Organometallics 26: 6636-6642.

25. Chiu PL, Lee HM (2005) Chemistry of the PCNHCP Ligand: Silver and Ruthenium Complexes, Facial/Meridional Coordination, and Catalytic Transfer Hydrogenation. Organometallics 24: 1692-1702.

26. Burling S, Field LD, Li HL, Messerle BA, Turner P, et al. (2003) Mononuclear Rhodium(I) Complexes with Chelating N-Heterocyclic Carbene Ligands Catalytic Activity for Intramolecular Hydroamination. Eur J Inorg Chem 17: 3179-3184.

27. Lin IJB, Vasam CS (2007) Preparation and application of N-heterocyclic carbene complexes of Ag(I). Coord Chem Rev 251: 642-670.

28. Hermann WA, Koher C, GooDen LJ, Artus GRJ (1996) Heterocyclic Carbenes: A High-Yielding Synthesis of Novel, Functionalized N-Heterocyclic Carbenes in Liquid Ammonia. Chem Eur J 2: 1627-1636.

29. Guerret O, Sole´S, Gornitzka H, Teichert M, Trinquier G, et al. (1997) 1,2,4-Triazole-3,5-diylidene: A Building Block for Organometallic Polymer Synthesis. J Am Chem Soc 119: 6668-6669.

30. Guerret O, Sole´S, Gornitzka H, Trinquier G, Bertrand G, et al. (2000) 1,2,4-Triazolium-5-ylidene and 1,2,4-triazol-3,5-diylidene as new ligands for transition metals. J Organomet Chem 600: 112-117.

31. Drozdzak R, Allaert B, Ledoux N, Dragutan I, Dragutan V, et al. (2005) Synthesis of Schiff Base-Ruthenium Complexes and Their Applications in Catalytic Processes. Adv Synth Catal 347: 1721-1743.

32. Selegue JP (2004) Metallacumulenes: from vinylidenes to metal polycarbides. Coord Chem Rev 248: 1543-1563.

33. Dragutan V, Dragutan I, Verpoort F (2005) Ruthenium Indenylidene Complexes. Platinum Metals Rev 49: 33-40.

34. Bruce MI (2004) Metal complexes containing cumulenylidene ligands. Coord Chem Rev 248: 1603-1625.

35. Cadierno V, Gamasa MP, Gimeno J (2004) Synthesis and reactivity of α,ß-unsaturated alkylidene and cumulenylidene Group 8 half-sandwich complexes. Coord Chem Rev 248: 1627-1657.

36. Dragutan I, Dragutan V (2006) Ruthenium Allenylidene Complexes. Platinum Metals Rev 50: 81-94.

37. Grubbs RH (2006) Olefin-metathesis catalysts for the preparation of molecules and materials (Nobel Lecture). Angew Chem Int Ed Engl 45: 3760-3765.

38. Dragutan V, Dragutan I, Balaban AT (2000) Metathesis Catalysed by the Platinum Group Metals. Platinum Metals Rev. 44: 58-66.

39. Perry MC, Burgess K (2003) Chiral N-heterocyclic carbene-transition metal complexes in asymmetric catalysis. Tetrahedron Assymetry 14: 951-961.

40. Dragutan I, Dragutan V, Delaude L, Demonceau A (2005) N-Heterocyclic carbenes as highly efficient ancillary ligands in homogeneous and immobilized metathesis ruthenium catalytic systems. ARKIVOC 10: 206-253.

41. Dragutan I, Dragutan V, Delaude L, Demonceau A (2007) NHC-Ru complexes-Friendly catalytic tools for manifold chemical transformations. Coordination Chem Rev 251: 765-794.

42. Gladiali S, Alberico E (2006) Asymmetric transfer hydrogenation: chiral ligands and applications. Chem Soc Rev 35: 226-236.

43. Stoe and Cie. X-AREA and X-RED32. Stoe and Cie, Darmstadt, Germany (2002).

44. Hübschle CB, Sheldrick GM, Dittrich B (2011) ShelXle: a Qt graphical user interface for SHELXL. J Appl Crystallogr 44: 1281-1284.

45. Farrugia LJ (2012) WinGX and ORTEP for Windows:an update. Appl Cryst 45: 849-854.

46. Wang HMJ, Lin IJB (1998) Facile Synthesis of Silver(I)-Carbene Complexes. Useful Carbene Transfer Agents. Organometallics 17: 972-975.

47. Arslan H, VanDerveer D, Yasar S, Özdemir I, Çetinkaya B, et al. (2007) Dichlorido 1-(2-methylbenzyl)-3-(2,4,6-trimethylbenzyl)imidazolidin-2-yl idene ruthenium(II). Acta Cryst E63: m1001-m1003.

48. Chandra M, Pandey DS, Puerta MC, Valerga P (2002) A p-cymene-ruthenium(II)-DMSO complex, [η⁶-C₁₀H₁₄)RuCl₂(DMSO)]. Acta Cryst E58: m28-m29.

# DFT Chemical Reactivity Analysis of Biological Molecules in the Presence of Silver Ion

**Linda-Lucila Landeros-Martinez, Erasmo Orrantia-Borunda and Norma Flores-Holguin***

*NANOCOSMOS Virtual Lab, Advanced Materials Research Center (CIMAV), Miguel de Cervantes 120, Complejo Industrial Chihuahua, Chihuahua 31190, México*

## Abstract

Silver ion oxidation process was studied in different biological molecules by Density Functional Theory; this by using the Becke three parameter Lee, Yang, and Parr functional and the Pople 6-31G (d) and Los Alamos LANL2DZ basis sets. The calculation was used to find the lowest energy molecular structure, molecular orbitals, and chemical reactivity parameters. Chemical hardness showed that the oxidation process begins in the purine and pyrimidine bases. The frontier orbitals electronic density distribution were analyzed in the biological molecule-silver ion complex. This distribution clearly revealed the transfer of electrons from highest occupied molecular orbital to lowest unoccupied molecular orbital indicating an oxidation process.

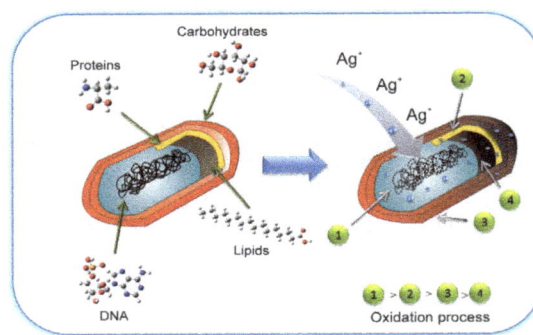

**Keywords:** Silver ion; DFT; Oxidation process

## Introduction

Because of its diverse properties silver is a versatile and safe agent with different uses. Silver ions had been used in biosensors [1], clothing [2], food industry [3-5], stainless steel coatings [6], beauty products [7], and in medical devices [8]. Silver is usually inert in its metallic form; however, it ionizes when in contact with skin moisture or fluids from a wound. As a result, it becomes highly reactive leading into an antibiotic behavior [9-12]. Studies of the inhibitory mechanism of silver ions on gram-positive and gram-negative bacteria have been reported [13,14], showing morphological changes in cytoplasm, cell membrane, and wall.

Currently, there is few information about how silver becomes bioactive. Some authors have reported that silver ions act when they penetrate the cell and get between purine and pyrimidine bases; thus, denaturing the deoxyribonucleic acid (DNA) [15]. Other authors indicate that bioactivity comes from deactivation of respiratory enzymes; this by forming complexes with the sulfur of the thiol group of cysteine [16,17]. It is also been reported that silver can be involved in catalytic oxidation reactions resulting from the formation of disulfide bonds (R-S-S-R). It catalyzes the reaction between oxygen molecules in the thiol groups. In such reaction, water is released as a by-product and two thiol groups are covalently joined through a disulfide bond [18]. The catalyzed form of silver in disulfide bonds might possibly change the three-dimensional structure of cell enzymes, and thus change their function. The effect of silver ions on bacteria can be difficult to understand. However, observation of morphological and structural changes can yield useful information for understanding the antibacterial effect and the inhibitory process of silver ions. Theoretical studies about the affinity of silver ions with DNA at a molecular level were performed to determine the interaction of silver ions with a cytosine and an adenosine basis, using ab initio calculations and Density Functional Theory (DFT) [19]. Another quantum chemical study, focused to shed light on the electronic and energetic properties of silver upon DNA adenosine and cytosine bases, was performed through DFT using the Becke three parameter Lee, Yang, and Parr functional (B3LYP) and the Minnesota family M06-L functionals [20]. According with the generated results with previous theoretical calculations and the experimental information mentioned above, it is considered imperative to extend the study of biological molecules in presence of silver ion, in an attempt to define the antibacterial mechanism, so, the aim of this research is the study of a silver ion antibacterial mechanism through an oxidation in the presence of biological molecules such as proteins, polysaccharides, lipids, and nucleic acids. In the particular case of biopolymers, they were analyzed according to their constituent monomers through a theoretical study; aiming to determine which parts of the bacterium cell react in the presence of the silver ion. To accomplish this goal, the calculation of the chemical reactivity parameters was done, among them: chemical potential (I), electron affinity (EA), electronegativity

**\*Corresponding author:** Norma Flores-Holguin, NANOCOSMOS Virtual Lab, Advanced Materials Research Center (CIMAV), Miguel de Cervantes 120, Complejo Industrial Chihuahua, Chihuahua 31190, México
E-mail: norma. lores@cimav.edu.mx

($\chi$), electrophilicity ($\omega$), chemical hardness ($\eta$), and electron donor potential ($\omega$-), which aid in understanding the oxidation process between silver ions and biological molecules. Also, highest occupied molecular orbital (HOMO) and lowest unoccupied molecular orbital (LUMO), which allow to observe electron transfer. All these parameters were determined through DFT.

## Materials and Methods

The studied biological molecules were selected considering them as representative of each group of the different kind of components of the bacterial cell structure: proteins, carbohydrates, lipids and DNA basis.

## Proteins

From the non-polar aliphatic R group, alanine was used, since it is abundant in living matter and it is the smallest chiral amino acid [21]. From the aromatic R group, the phenylalanine, since its structure contains an aromatic ring which allows to see its interaction with the silver ion. From the uncharged polar R group, cysteine and its structure contains sulphur and it has a high presence in peptides and proteins. It also has a reactive nature, and its inclusion of a thiol group has a preponderant role in the synthesis of proteins. Moreover, it is known that the thiol group interacts strongly with noble metals such as gold, silver, and cooper [22]. From the negatively charged R group, aspartic acid, since it contains dicarboxylic acids. Also, from the positively charged R group histidine was selected; it contains an imidazole functional group, and this feature will allow to evaluate interactions with the silver ion.

## Carbohydrates

Among monosaccharides, D-glucose was analyzed since it has dramatic effects on carbon metabolism regulation and it is responsible for cellular respiration regulation. Also, it has an effect on gluconeogenesis, making it capable of transporting and catabolizing sugars [23]. From polysaccharides, sucrose was analyzed because it has a high molecular weight and it is constituted by glucose and fructose. Also it is certainly involved in metabolic processes. Polysaccharides interfere in adenosine triphosphate (ATP) synthesis, and according to some researchers, silver ions inhibit oxidation of glycol, glucose and other molecules.

## Lipids

Lipids are fundamental structural components of cell membranes, they are little oxidizable molecules, and they serve as an energy reservoir for the cell. It is important to analyze the constituting elements of bacterial cell membranes for when there is an attack by foreign agents they become injured. The lipids analyzed in this research were palmitic acid, a saturated lipid, and palmitoleic acid, an unsaturated lipid. Palmitic acid promotes cell cycle progression, it accelerates cell proliferation, and it induces a transient and sequential activation of a series of kinases [24]. Palmitoleic acid is to be analyzed because it is biosynthesized through palmitic acid.

## DNA basis

In DNA, purine and pyrimidine bases (adenosine and cytosine) were analyzed. Adenosine monophosphate (AMP) was selected because it has an important role in energy metabolism. Through enzymatic reaction AMP forms bonds with other phosphate groups and plays an important role in incorporating amino acids into proteins [25]. Cytosine monophosphate (CMP) was selected since it is involved in the biosynthetic processes of phospholipids and uracil for carbohydrates [26]. There have been research papers that state silver ions bind to DNA transcription, while in the blocks they bond to the cells of the surface, thus disrupting bacterial respiration, and also they cause interference in ATP synthesis.

## Computational methods

Calculations were made using Density Functional Theory, DFT [27-29] to find the geometry in the minimum energy state in the gas phase of L-Alanine (Ala), L-Aspartic acid (Asp), L-Cysteine (Cys), L-Phenylalanine (Phe), L-Histidine (His), Adenosine, Cytosine, D-Glucose, Sucrose, palmitic acid, and palmitoleic acid, with functional B3LYP [30]. All calculations were performed with the Gaussian 09 software [31]. For carbon (C), Hydrogen (H), Oxygen (O), Nitrogen (N) and Phosphorus (P), the Pople 6-31G (d) [32] basis set was assigned, whereas Los Alamos LANL2DZ [33-35] basis set was elected for silver (Ag). The DFT approach has been successful in presenting a theoretical basis for chemical descriptors such as ionization potential (I), electron affinity (EA) which are defined by the energy difference $E_{(N)} - E_{(N-1)}$ and $E_{(N)} - E_{(N+1)}$, where N indicates the parent molecule, N-1 and N+1 correspond to the cation and anion radicals generated after electron transfer. Global hardness ($\eta$) [36], electronegativity ($\chi$) [36], electrophilicity ($\omega$) [37], indexes related to electronic structure and chemical reactivity; and the electron donor potential proposed by Gázquez et al. [38], which measures the capability to donate fractioned charges.

## Results and Discussion

The geometry optimization of the biological molecules in gas phase and frequency calculation were performed to make sure molecules were in their lowest energy level, Figure 1 shows the optimized geometries of the studied molecules. Condensed Fukui functions are mathematical expressions that define the sensitivity a molecular system has to experience changes in its electronic density, in different points of its structure. In a chemical reaction, a change in the number of electrons involves the addition or subtraction of at least one electron in the frontier orbitals [39]. Thus, calculating Fukui functions helps us determine the reactive sites of a molecule, based on the electronic density changes experienced by it during a reaction. The dual descriptor for nucleophilicity and electrophilicity was defined in terms of the variation of hardness with respect to the external potential; such dual descriptor was defined as the difference between nucleophilic and electrophilic Fukui functions, allowing to characterize both reactive behaviors [40]. The dual descriptor predicted the site reactivity induced by different donor and acceptor groups of the biological molecules. The condensed results of the Fukui indexes and dual descriptor showed which atoms are most susceptible to an electrophilic attack, see Table 1. These results were obtained with the Hirschfeld charge distribution [41]. The definition for these atoms was performed to establish the zone where the silver ion attraction was more likely to create the biological molecule-silver ion complex. Also, energy calculations were performed to obtain the most stable structure of the complex at a specific point, such as exemplified with the Ala-Ag+ amino acid. Figure 2. Qualitative chemical concepts such as electronegativity and hardness have been widely used in understanding various aspects of chemical reactivity. The theoretical basis for these concepts has been provided by DFT [42]. These reactivity indices are better appreciated in terms of the associated electronic structure of atoms and molecules such as electronegativity and hardness [38]. The obtained chemical reactivity parameters for the biological molecules are: ionization potential (I), which is defined as the energy needed to remove an electron from a molecule; electron affinity

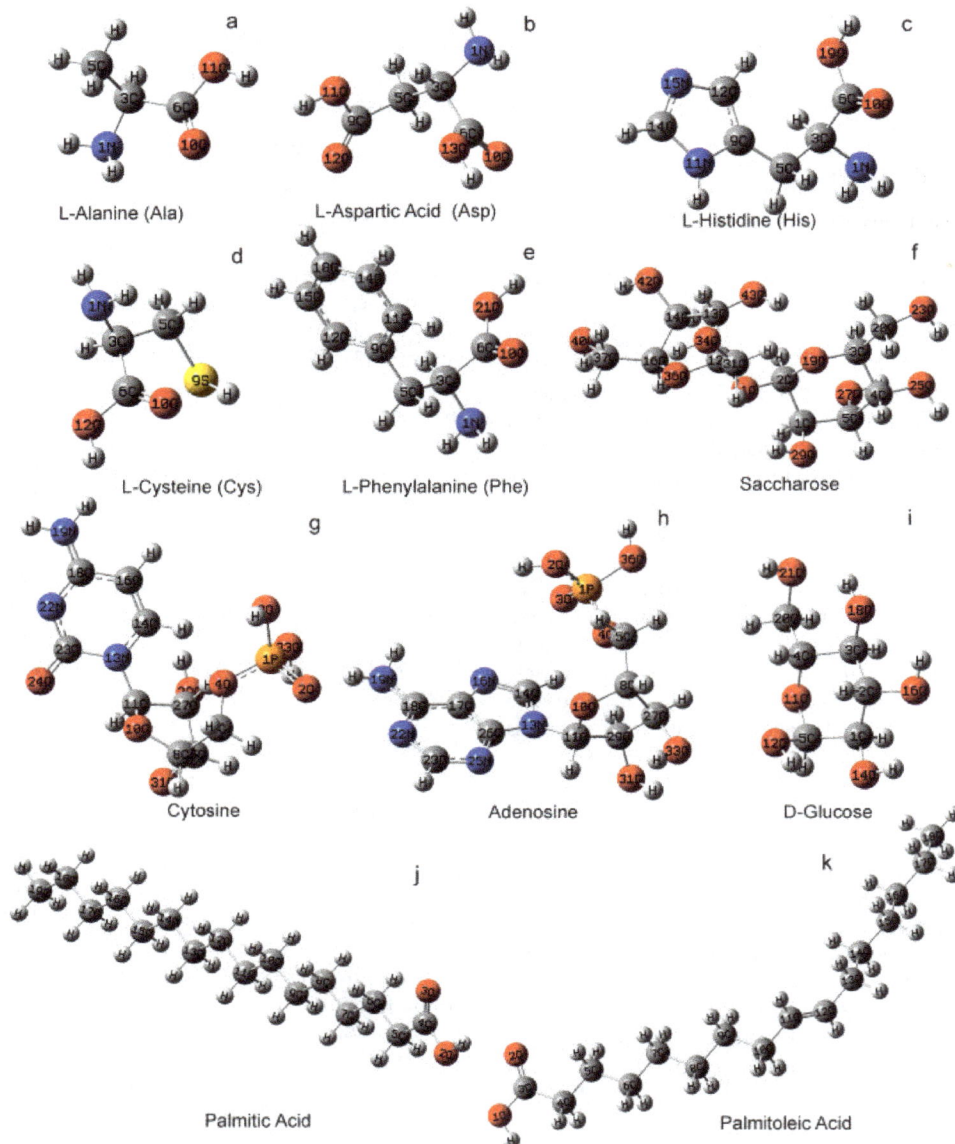

**Figure 1:** Optimized molecular structure of the minimum energy structure of biological molecules in the gas phase with the level of theory B3LYP/6-31G(d). (a) Ala, (b) Asp, (c) His, (d) Cys, (e) Phe, (f) Saccharose, (g) Cytosine, (h) Adenosine, (i) D-Glucose, (j) Palmitic Acid, (k) Palmitoleic Acid.

| Biological Molecules | $f_k^-$ | $f_k^2$ |
|:---:|:---:|:---:|
| Ala | N1 | N1 |
| Asp | N1 | N1 |
| Cys | S9 | S9 |
| Phe | N1 | N1 |
| Hys | C12 | C12 |
| Adenosine | N19 | N19 |
| Cytosine | O24 | O24 |
| D-Glucose | O16 | O12 |
| Saccharose | O43 | O43 |
| Palmitic Acid | O3 | C12 |
| Palmitoleic Acid | C11 | C11 |

**Table 1:** Electrophilic attack site and dual descriptor in biological molecules obtained with B3LYP/6-31G(d).

(EA), which measures the ability of a molecule to accept electrons and form anions; electronegativity ($\chi$), representing the tendency of atoms or molecules to attract electrons; electrophilicity ($\omega$), that gives an idea of the stabilization energy when the system acquires electrons from the environment up to saturation; and electron donor potential ($\omega$-). These reactivity information shows if a molecule is capable of donating charge. The reactivity parameters mentioned above were obtained using vertical approximation, in which the energy of the molecule in its anionic, cationic, and neutral states is calculated, keeping in mind the geometry of the fundamental state. Results are shown in Table 2. According with these results, the biological molecule Ala showed the highest ionization potential value, thus showing this amino acid would require the highest amount of energy to rip an electron off its structure, and therefore this amino acid will not oxidize easily in the presence of silver ions. On the other hand, Adenosine and Cytosine show the lowest

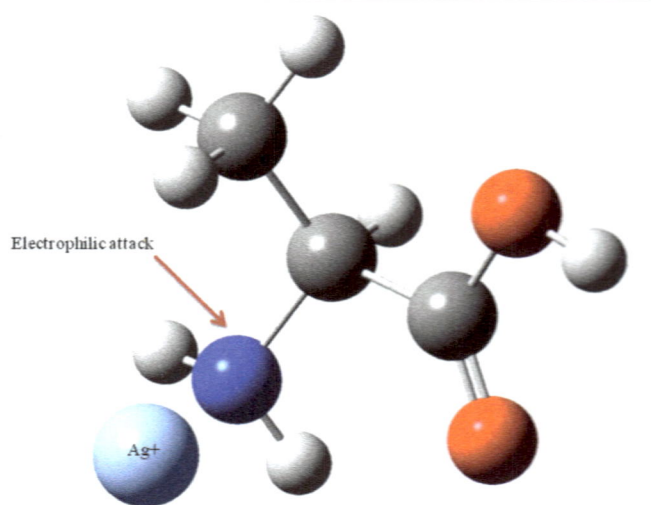

**Figure 2:** Approximation of silver ion in the site susceptible to electrophilic attack of Ala.

| Biological Molecules | I (eV) | AE (eV) | X (eV) | η(eV) | ω (eV) | ω⁻ (eV) |
|---|---|---|---|---|---|---|
| Ala | 9.27 | -2.76 | 3.26 | 6.02 | 0.88 | 3.26 |
| Asp | 9.14 | -1.96 | 7.71 | 5.55 | 1.16 | 3.65 |
| Cys | 8.71 | -2.11 | 3.30 | 5.41 | 1.01 | 3.33 |
| His | 8.18 | -2.14 | 3.02 | 5.16 | 0.88 | 3.04 |
| Phe | 8.18 | -1.70 | 3.24 | 4.94 | 1.06 | 3.30 |
| Adenosine | 8.06 | -0.95 | 3.56 | 4.51 | 1.41 | 3.75 |
| Cytosine | 7.98 | -1.04 | 3.47 | 4.51 | 1.34 | 3.64 |
| D-Glucose | 9.02 | -2.86 | 3.08 | 5.94 | 0.80 | 3.08 |
| Saccharose | 8.51 | -2.16 | 3.17 | 5.34 | 0.94 | 3.20 |
| Acid Palmitic | 8.87 | -2.57 | 3.15 | 5.72 | 0.87 | 3.16 |
| Acid Palmitoleic | 8.19 | -2.07 | 3.06 | 5.13 | 0.91 | 3.08 |

**Table 2:** Reactivity parameters of biological molecules calculated with B3LYP/6-31G(d).

ionization potential value, which indicates they will oxidize more easily in the presence of the silver ion [43]. There are some EA negative values because the energy of biological molecules is not absorbed, but released in the process of electron acceptance, namely it is required to supply energy in order to form the anion [43]. Biological molecules adenosine and cytosine are capable of donating electrons, getting oxidized more easily in contact with the silver ion. According to the electronic affinity results, biological molecules are more capable of donating electrons and therefore, more prone to oxidization. Agreeing to Dunning et al. [44], in order to obtain a reliable calculation of electronic affinity, it is necessary to use base complexes with high angular momentum scattered functions. In order to discard that the 6-31G (d) were the cause of the negative electron affinity results, the scattered function 6-31++G(d) was used on the Ala biological molecule. This calculation yield to -0.4479 eV EA result, a value that corroborates that even using a scattered function that allowed changing the angular momentum and the shape of the orbital, EA still results in a negative value. Regarding electronegativity, the highest value was 7.71 eV for Asp, this means that Asp presented the highest difficulty to be oxidized in the presence of the silver ion, since this amino acid tends to attract electrons more strongly. On the other hand, the purine and pyrimidine bases (adenosine and cytosine) show electronegativity values of 3.56 eV and 3.47 eV respectively, which indicates they can be oxidized more easily in the

presence of the silver ion. The chemical hardness values (η), one of the reactivity parameters considered to determine the oxidation process (namely a measure of the resistance of a system to transfer charge), are higher in D-Glucose and sucrose with 5.94 eV and 5.34 eV respectively. This indicates that D-Glucose and sucrose are not prone to yield charge. Adenosine and cytosine show the lowest η values (4.51 eV), which makes them the molecules more prone to yield electrons, thus being the molecules that are more easily oxidized in the presence of the silver ion. According to the results obtained, the reactivity order, expressed as the ease to be oxidized, as for value η is: DNA>Proteins>Carbohydrates>Lipids

Adenosine, cytosine and Asp show the highest (ω) value. About the electron donor potential, low values indicate an antioxidant behavior, thus, results show that lipids tend to stabilize the electron loss in the other parts of the bacterium. The analysis of the reactivity results confirm that the DNA is the effortlessness bacterium fragment to be oxidized. Figure 3 shows the proposed chemical reaction of the adenosine and cytosine basis interaction with silver ion through the heterocycles. This proposed is based in the Jeffrey et al. [45] work where they found that silver ions favor bonding strongly with heterocyclic bases and not with the phosphate groups. The association of silver with the purine base (adenosine) forms complexes via the N19, while with the pyrimidine bases (cytosine) the complex is formed by the O24. These sites agree with the electrophilic attack sites defined in this work. Also, the reaction with oxygen in the heterocycles is an association confirmed as well by the Jeffrey et al. in the cited paper above. A distribution analysis of the highest occupied molecular orbital (HOMO) and the lowest unoccupied molecular orbital (LUMO) of complexes of biological molecules-silver ion was performed in an attempt to observe the difference of this frontier orbitals distribution in presence of the silver ion. In all cases studied, the HOMO is localized over the structure of the amino acids, lipids, sugars, and purine and pyrimidine bases and the LUMO is found over the silver ion. If we also take into account that a global reaction called oxide reduction takes place (in every chemical reaction, the electrons one molecule loses, another must gain; in other words, one molecule is oxidized whereas the other one is reduced) [26], then we propose that the LUMO distribution over the silver ion, indicates an oxidation process, in which the biological molecules are oxidized and the silver is reduced. Figure 4 shows HOMO and LUMO orbitals for cytosine-Ag⁺ and adenosine-Ag⁺ complexes.

## Conclusions

The oxidation process of the silver ion upon the parts that constitute a bacterium were analyzed in this work, considering the analyzed amino acids as part of a protein, the purine and pyrimidine bases as part of DNA, D-glucose and sucrose as carbohydrates, and palmitic and palmitoleic acids as the lipids. The molecular characterization of the biological molecules includes the calculation of the molecular structure and chemical reactivity parameters, also the calculation of the complexes biological molecule-silver ion reactivity properties. Low chemical hardness values indicate which constitutive parts are more prone to yield electrons, thus generating an oxidation process. The results show that the process order is: DNA>Proteins>Carbohydrates>Lipids. In all the cases, the HOMO orbital is found in the biological molecules, whereas the LUMO orbital is found in the silver ion, indicating there is a HOMO to LUMO electron transfer, suggesting an oxidation process.

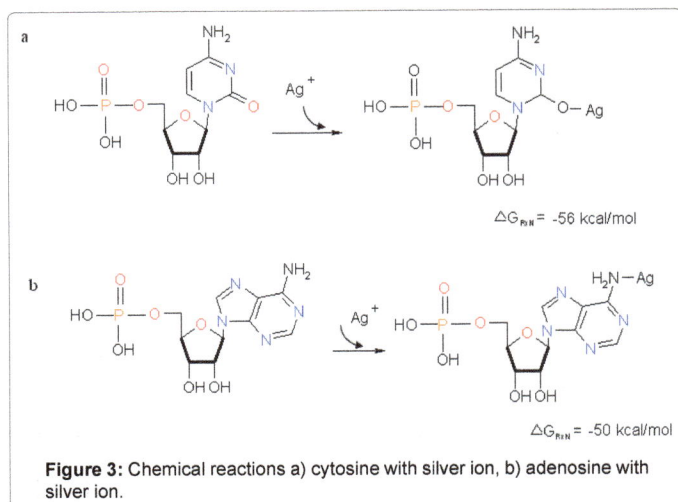

**Figure 3:** Chemical reactions a) cytosine with silver ion, b) adenosine with silver ion.

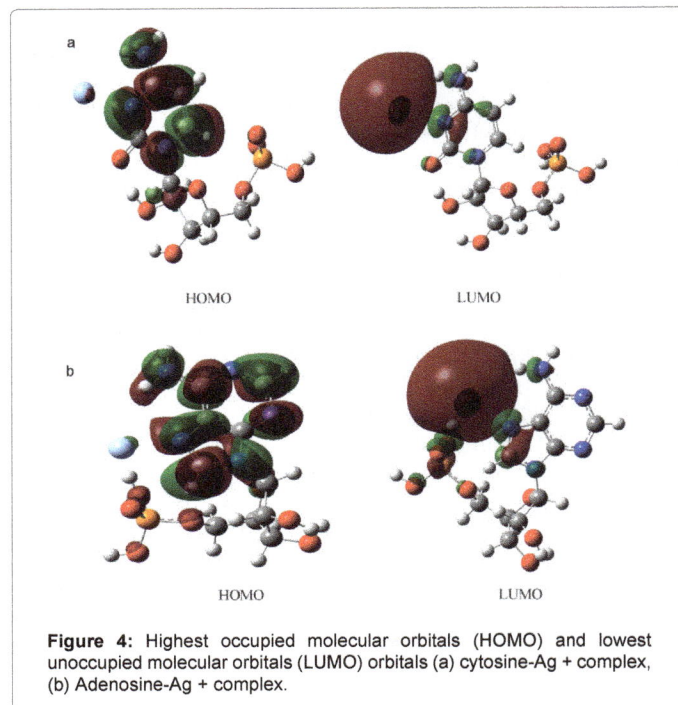

**Figure 4:** Highest occupied molecular orbitals (HOMO) and lowest unoccupied molecular orbitals (LUMO) orbitals (a) cytosine-Ag + complex, (b) Adenosine-Ag + complex.

## Acknowledgements

This work was supported by Consejo Nacional de Ciencia y Tecnología (CONACYT) and Centro de Investigación en Materiales Avanzados, S.C. (CIMAV). LLM gratefully acknowledge a fellowship from CONACYT. EOB is a researcher for CIMAV and CONACYT and NFH is a researcher of CIMAV and CONACYT.

## References

1. Tang L, Dong C, Ren J (2010) Highly sensitive homogenous immunoassay of cancer biomarker using silver nanoparticles enhanced fluorescence correlation spectroscopy. Talanta 81: 1560-1567.

2. Yuranova T, Rincon AG, Bozzi A, Parra S, Pulgarin C, et al. (2003) Antibacterial textiles prepared by RF-plasma and vacuum-UV mediated deposition of silver. Journal of Photochemistry and Photobiology A Chemistry 161: 27-34.

3. Chen X, Schluesener HJ (2008) Nanosilver: a nanoproduct in medical application. Toxicol Lett 176: 1-12.

4. Gupta A, Silver S (1998) Silver as a biocide: will resistance become a problem. Nat Biotechnol 16: 888.

5. Rai M, Yadav A, Gade A (2009) Silver nanoparticles as a new generation of antimicrobials. Biotechnol Adv 27: 76-83.

6. Kampmann Y, De Clerck E, Kohn S, Patchala DK, Langerock R, et al. (2008) Study on the antimicrobial effect of silver-containing inner liners in refrigerators. J Appl Microbiol 104: 1808-1814.

7. Kulthong K, Srisung S, Boonpavanitchakul K, Kangwansupamonkon W, Maniratanachote R (2010) Determination of silver nanoparticle release from antibacterial fabrics into artificial sweat. Part Fibre Toxicol 7: 8.

8. Tolaymat TM, El Badawy AM, Genaidy A, Scheckel KG, Luxton TP, et al. (2010) An evidence-based environmental perspective of manufactured silver nanoparticle in syntheses and applications: a systematic review and critical appraisal of peer-reviewed scientific papers. Sci Total Environ 408: 999-1006.

9. Russell AD, Hugo WB (1994) Antimicrobial activity and action of silver. Prog Med Chem 31: 351-370.

10. Magaña SM, Quintana P, Aguilar DH, Toledo JA, Ángeles-Chávez C, et al. (2008) Antibacterial activity of montmorillonites modified with silver. J Mol Catal A Chem 281: 192-199.

11. Lee SB, Otgonbayar U, Lee JH, Kim KM, Kim KN (2010) Silver ion-exchange sodium titanate and resulting effect on antibacterial efficacy. Surf Coat Technol 205: S172-S176.

12. Malachová K, Praus P, Rybková Z, Kozák O (2011) Antibacterial and antifungal activities of silver, copper, and zinc montmorillonites. Appl Clay Sci 53: 642-645.

13. Feng QL, Wu J, Chen GQ, Cui FZ, Kim TN, et al. (2000) A mechanistic study of the antibacterial effect of silver ions on Escherichia coli and Staphylococcus aureus. J Biomed Mater Res 52: 662-668.

14. Swathy JR, Sankar MU, Chaudhary A, Aigal S, Anshup, et al. (2014) Antimicrobial silver: an unprecedented anion effect. Sci Rep 4: 7161.

15. Klueh U, Wagner V, Kelly S, Johnson A, Bryers JD (2000) Efficacy of silver-coated fabric to prevent bacterial colonization and subsequent device-based biofilm formation. J Biomed Mater Res 53: 621-631.

16. Matsumura Y, Yoshikata K, Kunisaki S, Tsuchido T (2003) Mode of bactericidal action of silver zeolite and its comparison with that of silver nitrate. Appl Environ Microbiol 69: 4278-4281.

17. Gupta A, Maynes M, Silver S (1998) Effects of halides on plasmid-mediated silver resistance in Escherichia coli. Appl Environ Microbiol 64: 5042-5045.

18. Davies RL, Etris SF (1997) The development and functions of silver in water purification and disease control. Catalysis Today 36: 107-114.

19. Wu J, Fu Y, He Z, Han Y, Zheng L, et al. (2012) Growth mechanisms of fluorescent silver clusters regulated by polymorphic DNA templates: a DFT study. J Phys Chem B 116: 1655-1665.

20. Fortino M, Marino T, Russo N (2015) Theoretical study of silver-ion-mediated base pairs: the case of C-Ag-C and C-Ag-A systems. J Phys Chem A 119: 5153-5157.

21. Minkov VS, Yu AC, Boldyreva EV (2010) A study of the temperature effect on the IR spectra of crystalline aminoacids, dipeptids, and polyamino acids. VI. Alanine and DL-Alanine. Journal of structural Chemistry 51: 1052-1063.

22. Schreiber F (2000) Structure and growth of self-assembling monolayers. Prog Surf Sci 65: 151-257.

23. Rolland F, Winderickx J, Thevelein JM (2002) Glucose-sensing and -signalling mechanisms in yeast. FEMS Yeast Res 2: 183-201.

24. Wang X, Liu JZ, Hu JX, Wu H, Li YL, et al. (2011) ROS-activated p38 MAPK/ERK-Akt cascade plays a central role in palmitic acid-stimulated hepatocyte proliferation. Free Radic Biol Med 51: 539-551.

25. Haung H, Strik WO, Meyer W, Deibert K, Polzien P (1967) On the concentration changes of glucose, lactic acid, pyruvic acid and adenosine tri-,di- and monophosphoric acids (ATP, ADP, AMP) in the blood of the pulmonary artery before and after administration of theophylline ethylendiamine. Arzneimittelforschung 17: 1411-1414.

26. Pacheco LD (2001) Bioquímica estructural y aplicada a la medicina. Editorial IPN. México, DF.

27. Hohenberg P, Kohn W (1964) Inhomegeneous Electron Gas. Phys Rev 136: B864-B871.

28. Kohn W, Sham LJ (1965) Self-consistent equations including exchange and correlation effects. Phys Rev 140: 1133-1138.

29. Parr RG, Yang W (1989) Density-Functional Theory of Atoms and Molecules. Oxford University Press. New York, USA.

30. Becke AD (1993) Density functional thermochemistry. III. The role of exact exchange. J Chem Phys 98: 5648-5652.

31. Frisch MJ, Trucks GW, Schlegel HB, Scuseria GE, Robb MA, et al. (2009) Gaussian 09; Gaussian Inc., Wallingford CT, USA.

32. Rassolov VA, Ratner MA, Pople JA, Redfern PC, Curtiss LA (2001) 6-31G* Basis Set for Third-Row Atoms. J Comp Chem 22: 976-984.

33. Hay PJ, Wadt WR (1985) Ab initio effective core potentials for molecular calculations - potentials for the transition-metal atoms Sc to Hg. J Chem Phys 82: 270-283.

34. Wadt WR, Hay PJ (1985) Ab initio effective core potentials for molecular calculations - potentials for main group elements Na to Bi. J Chem Phys 82: 284-298.

35. Hay PJ, Wadt WR (1985) Ab initio effective core potentials for molecular calculations - potentials for K to Au including the outermost core orbitals. J Chem Phys 82: 299-310.

36. Pearson RG (1986) Absolute electronegativity and hardness correlated with molecular orbital theory. Proc Natl Acad Sci USA 83: 8440-8441.

37. Parr RG, Szentpaly LV, Liu S (1999) Electrophilicity Index. J Am Chem Soc 121: 1922-1924.

38. Gázquez JL (2008) Perspectives on the Density Functional Theory of Chemical Reactivity. J Mex Chem Soc 52: 3-10.

39. Fleming I (1976) Frontier Orbitals and Organic Chemical Reactions. Wiley, New York.

40. Morell C, Grand A, Toro-Labbé A (2005) New dual descriptor for chemical reactivity. J Phys Chem A 109: 205-212.

41. Hirschfeld FL (1977) Bonded-Atom Fragments for Describing Molecular Charge Densities. Theoret Chim Acta 44: 129-138.

42. Chattaraj S, Nath, Maiti B (2004) Reactivity descriptors. Computational medicinal chemistry for drug discovery. Editorial Marcel Dekker. New York.

43. Martínez A, Rodríguez-Gironés MA, Barbosa A, Costas M (2008) Donator acceptor map for carotenoids, melatonin and vitamins. J Phys Chem A 112: 9037-9042.

44. Dunning TH, Peterson KA, Mourik VT (2003) In Calculations of Electronic Affinities. A Roadmap in Dissociative Recombination of Molecular Ions whit Electrons. Guberman, SL, Eds, Kluver Academic Press: New York.

45. Petty JT, Zheng J, Hud NV, Dickson RM (2004) DNA-templated Ag nanocluster formation. J Am Chem Soc 126: 5207-5212.

# 3

# Binary Titanium (IV) Metal-organic Frameworks with Multidentate Ligands

**Balram Prasad Baranwal\*, Abhay Kumar Jain and Alok Kumar Singh**

*Coordination Chemistry Research Laboratory, Department of Chemistry, D.D.U. Gorakhpur University, Gorakhpur 273 009, India*

## Abstract

Some volatile binary metal-organic frameworks of titanium (IV), $[Ti(OOCR)_4]$ (where R = $C_{13}H_{27}$, $C_{15}H_{31}$, $C_{17}H_{35}$ or $C_{21}H_{43}$) were synthesized by the reaction of titanium tetrachloride and sodium salts of the straight chain fatty acids (prepared in situ) in 1:4 molar ratio. The isolated solid products were showing poor crystallinity and were characterized by their elemental analyses, molecular weight determinations, conductance, spectral (infrared, $^1$H NMR, $^{13}$C NMR, FAB mass and powder XRD) and TEM studies. Their monomeric nature was confirmed by molecular weight determinations and FAB mass spectra. Eight coordination number of titanium (IV) have been assigned in the isolated compounds. TEM indicated the particles are spherical in shape having ~200 nm diameter.

**Keywords:** Carboxylates; Metal-organic frameworks; Powder XRD; Titanium(IV)

## Introduction

Metal-organic frameworks (MOFs) of some transition metals have attracted a considerable attention in recent years [1-5] due to their wide range of applications. At present, various titanium-based materials have successfully been obtained in the form of nanotubes, nanofibers, nanowires, nanoflowers and nanocubes [6-8]. A key challenge for the industrial use of MOFs is to deliver them in a definite shape and size for their applications in many fields. There is an increasing trend in the research and development of $TiO_2$ coatings through metallo-organic chemical vapour deposition (MOCVD) for their applications as being semi conductor and chemically stable under different conditions. The evolution of metal–organic complex precursors of Ti, owing to the limitations of the alkoxide or other precursors including halides, is well documented in the literature [9,10]. Therefore, a good air and moisture stable, easy-to-handle and volatile titanium precursor is a prerequisite for MOCVD process [11]. One of the most active field in titanium(IV) chemistry is in the design of new compounds using different substituents having anticancer activity [12-14]. It has also been observed that only oxo- or di- or tri- substituted carboxylate derivatives of titanium(IV) are resulted after a number of trials to synthesize its tetracarboxylates [15-17].

Keeping in view of these objectives, some titanium(IV) complexes of electron- rich ligands have been recently synthesized in our laboratory, which are reported to be hydrolytically stable [18,19]. In this paper we report an easier method to synthesize titanium(IV) tetracarboxylates and solid products have been isolated. These compounds are volatile in nature as well as hydrolytically stable, having Ti-O-C linkage, a basic requirement for catalytic action. These have been characterized by different spectral studies to arrieveat their structure and coordination behaviour of the ligands.

## Experimental

### Materials and analytical methods

All the reactions were carried out under strictly anhydrous conditions. Glass apparatus with interchangeable quick fit joints were used throughout. Organic solvents (Qualigens) were dried and distilled before use by standard methods [20]. Carboxylic acids were used after distillation under reduced pressure (m.p. of myristic acid:

54°C, palmitic acid: 63°C, stearic acid: 70°C and behenic acid: 80°C). Titanium tetrachloride (BDH) was used as received. Titanium was estimated gravimetrically as $TiO_2$ [21].

### Physico-chemical measurements

Infrared spectra were recorded on a Perkin Elmer 1600 series FTIR spectrophotometer using KBr discs. $^1$H and $^{13}$C NMR spectra were recorded at 250.17 MHz on a Bruker DPX 250 NMR spectrometer in $CDCl_3$. FAB mass was done on a JEOL SX 102/ DA-6000 mass spectrometer using *m*-nitrobenzyl alcohol (NBA) as a matrix. Molecular weights were determined in a semi-micro ebulliometer (Gallenkamp) with a thermistor sensing device. Elemental analyses (C, H) were done on a Carlo-Erba 1108 elemental analyzer. Molar conductances were measured on century CC-601 digital conductivity meter at $10^{-2}$ -$10^{-3}$ molar solutions in benzene. Solid state conductance measurements were carried out with Keithley 6220 Precision current source and keithley 2182A Nanovoltmeter. Magnetic moment was measured on a Gouy balance using $Hg[Co(SCN)_4]$ as a calibrant. Powder XRD data were collected on a PW 1710 BASED diffractometer. The operating voltage of the instrument was 30 kV and the operating current was 20 mA. The intensity data were collected at room temperature over a $2\theta$ range of 5.025 - 79.925° with a continuous scan mode. Transmission electron microscopy (TEM) images were obtained on a Tecnai 30 G²S – Twin electron microscope with an accelerating voltage of 300 kV on the surface of a carbon coated copper grid.

### Synthesis of $[Ti (OOCC_{13}H_{27})_4]$ (1)

In myristic acid (2.88 g, 12.61 mmol), a solution of sodium isopropoxide prepared by dissolution of sodium (0.29 g, 12.61

---

**\*Corresponding author:** Balram Prasad Baranwal, Coordination Chemistry Research Laboratory, Department of Chemistry, D.D.U. Gorakhpur University, Gorakhpur 273009, India, E-mail: drbpbaranwal@gmail. com

mmol) in isopropanol (10 mL) was slowly added. The contents were refluxed for 2 h. To the sodium salt of myristic acid formed in situ, titanium tetrachloride (0.60 g, 3.16 mmol) in benzene (40 mL) was added dropwise with constant stirring. The mixture was stirred for 1 h followed by refluxing for 4 h. It resulted the formation of sodium chloride which was insoluble in the reaction mixture. This was removed by filtration using G 4 crucible. Excess solvent was removed under reduced pressure. The resulted solid was again dissolved in benzene in which trace amount of sodium chloride left was insoluble. Filtration and drying the solution *in vacuo* gave a light yellow solid. This gave negative test for chloride ions with silver nitrate. The compound thus obtained was further purified by distillation under reduced pressure (b.p. 220 °C at 5 mm pressure). Complexes **2** to **4** were synthesized analogously which were distilled under reduced pressure at 230 °C, 242 °C, 258 °C respectively and details of analytical as well as spectral results are given in Table 1.

## Results and Discussion

Titanium tetracarboxylates were synthesized by substitutions of chloride ions of titanium tetrachloride by sodium salts of long chain carboxylic acids in 1:4 molar ratio:

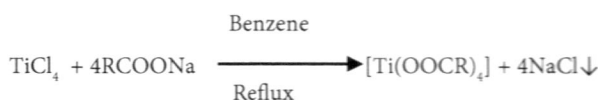

$$TiCl_4 + 4RCOONa \xrightarrow[\text{Reflux}]{\text{Benzene}} [Ti(OOCR)_4] + 4NaCl\downarrow$$

(Where R = $C_{13}H_{27}$; **1**, $C_{15}H_{31}$; **2**, $C_{17}H_{35}$; **3**, $C_{21}H_{43}$; **4**)

All the complexes were soluble in benzene, toluene, chloroform and dichloromethane. The molar conductances (at $10^{-2}$-$10^{-3}$ molar concentrations) in benzene were found 3 to 7 $\Omega^{-1}cm^2 mol^{-1}$ which indicated them to be non-electrolytes [22]. Solid state conductance measurements were done for all the complexes and the data were found in the range $1.1 \times 10^6$ - $2.3 \times 10^6$ $\Omega$ at 295 K using current $1 \times 10^{-8}$ A and voltage $1.4 \times 10^{-2}$ V. This clearly indicated them to show high resistance and we may say the complexes were behaving as insulators. Magnetic moment measurements indicated the diamagnetic nature for all the complexes which confirmed the absence of unpaired electrons in them. Elemental analyses were in good agreement with the calculated values (Table 1).

## Infrared spectra

In infrared spectra of all the complexes O–H stretching vibrations of carboxylic acid (at ~3400 $cm^{-1}$) were found absent. The bands at 1710 $cm^{-1}$ (CO stretching) and at 935 $cm^{-1}$ (OH deformation) of free carboxylic acids were also absent in the spectra. Two strong bands were observed at ~1590 $cm^{-1}$ and ~1465 $cm^{-1}$ which could be assigned to ($v_{asym}$OCO) (antisymmetric) and ($v_{sym}$OCO) (symmetric) vibrations of the carboxylate ions, respectively. The difference, $\Delta$ [$v_{asym}$OCO–$v_{sym}$OCO] was ~125 $cm^{-1}$ which indicated bidentate chelating nature of carboxylate ligands. This resulted in the formation of four symmetrical chelating rings. This enabled us to say that four carboxylate ions gave eight coordination number around Ti(IV) in these carboxylate complexes [23]. Analysis of the infrared bands (Table 1) also revealed that as the length of the alkyl group of carboxylic acids increased, there occurred shifts of $v_{asym}$(OCO) and $v_{sym}$(OCO) bands towards lower wave numbers. This effect can be explained by the influence of the type of alkyl group on the strength of the Ti-OOC interactions. The band observed around 470 $cm^{-1}$ could be assigned to Ti–O vibrations [23].

## Hydrolytic stability of the complexes

Titanium(IV) tetracarboxylates exhibited a high hydrolytic stability and were air stable for a longer time. This was tested by dissolving the complexes in benzene followed by adding 1 % water. After stirring the contents in open air for 12 h, no precipitate was apparently visible in the reaction mixture. Excess of solvent was removed *in vacuo* and no weigh loss was found. The analysis for titanium in this solid indicated to be titanium tetracarboxylate and the composition was not changed during the hydrolysis.

## ¹H NMR spectra

In the ¹H NMR spectra of **1–4**, no signal for –OH of free carboxylic acids ($\delta$ = 10.5 to 12 ppm) were observed indicating deprotonation of the acids. In the spectrum of **1**, a triplet appeared at $\delta$ = 0.90 ppm (12H) corresponded to methyl protons while a singlet corresponding to 80H of the 40-$CH_2$ groups was observed at $\delta$ = 1.26 ppm which could be interpreted for four myristate ions [-$OOCCH_2CH_2(CH_2)_{10}CH_3$] in the complex. The α- and β-$CH_2$ protons of four myristate ions was observed at $\delta$ = 2.48 ppm (8H), $\delta$ = 1.78 ppm (8H) respectively. Complexes **2, 3** and **4** showed a similar NMR pattern to **1** (Table 1).

## ¹³C NMR spectra

The ¹³C NMR spectra of **1–4** show signals corresponding to the carboxylato ligand. A signal at $\delta$ = 38.5 ppm corresponding to the α-carbon atom of methylene group and signals between $\delta$ 26.3 to 33.1 ppm corresponding to the carbon atoms of other methylene groups, and for –$CH_3$ carbon, a signal appears at $\delta$ = 15.0 ppm. Finally the signal assigned to the carbon atom of –COO group is observed at $\delta$ = 186.2 ppm.

Both ¹H and ¹³C NMR spectra suggested a similar nature of coordination for all the four carboxylate ions around titanium in the complexes.

## Molecular weight and FAB mass

Ebullioscopic method of molecular weight determinations showed that all the complexes were monomeric in refluxing benzene (Table

| Compound (Empirical formula) | Found (Calculated) | | | | | IR bands (cm⁻¹) | | | ¹H NMR (δ, ppm) |
|---|---|---|---|---|---|---|---|---|---|
| | Yield (Obtained) | C % | H % | Ti % | Molecular weight | $v_{asym}$OCO | $v_{sym}$OCO | Ti–O | |
| $C_{56}H_{108}O_8Ti$ **1** | 2.52 g, 83 % | 70.21 (70.24) | 11.41 (11.39) | 4.81 (4.99) | 963 (957) | 1592 | 1467 | 478 | 0.90 [t, 12H; (CH₃)₄] 1.26 [s, 80H; (-CH₂)₄₀] 1.78 [m, 8H; (β-CH₂)₄] 2.48 [t, 8H; (α-CH₂)₄] |
| $C_{64}H_{124}O_8Ti$ **2** | 3.23 g, 88% | 71.80 (71.87) | 11.68 (11.71) | 4.45 (4.48) | 1055 (1069) | 1589 | 1465 | 476 | 0.90 [t, 12H; (CH₃)₄] 1.26 [s, 96H; (-CH₂)₄₈] 1.78 [m, 8H; (β-CH₂)₄] 2.49 [t, 8H; (α-CH₂)₄] |
| $C_{72}H_{140}O_8Ti$ **3** | 3.85 g, 85 % | 73.18 (73.16) | 11.97 (11.96) | 4.01 (4.05) | 1198 (1182) | 1587 | 1462 | 473 | 0.90 [t, 12H; (CH₃)₄] 1.26 [s, 112H; (-CH₂)₅₆] 1.79 [m, 8H; (β -CH₂)₄] 2.50 [t, 8H; (α-CH₂)₄] |
| $C_{88}H_{172}O_8Ti$ **4** | 4.08 g, 81 % | 75.17 (75.14) | 12.28 (12.35) | 3.29 (3.40) | 1423 (1406) | 1581 | 1458 | 468 | 0.90 [t, 12H; (CH₃)₄] 1.27 [s, 144H; (-CH₂)₇₂] 1.80 [m, 8H; (β -CH₂)₄] 2.50 [t, 8H; (α-CH₂)₄] |

**Table 1:** Analytical and spectral data for titanium tetracarboxylates.

1). In FAB mass of **1** appearance of a peak at $m/z$ 959 corresponded to its monomeric nature. The peaks at $m/z$ 732, 501 and 274 showed the loss of one myristate ion at each stage. Therefore, at $m/z$ 274 $[Ti(OOCC_{13}H_{27})]^{3+}$ unit may be assigned (calculated $m/z$; 275). Some peaks on lower range may be due to the decomposed ions of indefinite compositions. Almost similar patterns were obtained in FAB mass of complexes **2**, **3** and **4**.

## Powder XRD and TEM Studies

The pattern and results of powder XRD suggested that the complexes showed poor crystallinity. Because of this single crystal XRD could not be done. Powder XRD were done for all the complexes and one spectrum for **2** along with its crystal data is given in Table 2 (Figure 1), which are comparable with titanium oxide oxalate hydroxide hydrate, both in diffraction intensity and position (JCPDS No. 48-1164). Particle size of the complexes was calculated by the standard Scherrer equation [24].

$$D = K\lambda / (\beta \cos\theta)$$

Where D is the particle size; K is a constant (= 0.94); $\lambda$ is X-ray wavelength ($\lambda$ = 1.5406 Å); $\theta$ is Bragg diffraction angle and $\beta$ is flex width which is converted into radian while calculation. The values obtained were in the range 180-195 nm.

TEM image for the complex **2** is given in Figure 2, which shows the primary particles are spherical in shape having ~200 nm average diameter of the particles.

## Conclusions

This communication demonstrates an easy method to synthesize titanium tetracarboxylates which have been isolated as volatile solids and are stable towards air and moisture. The stability and volatility of these metallo-organic titanium(IV) complexes were favoured by

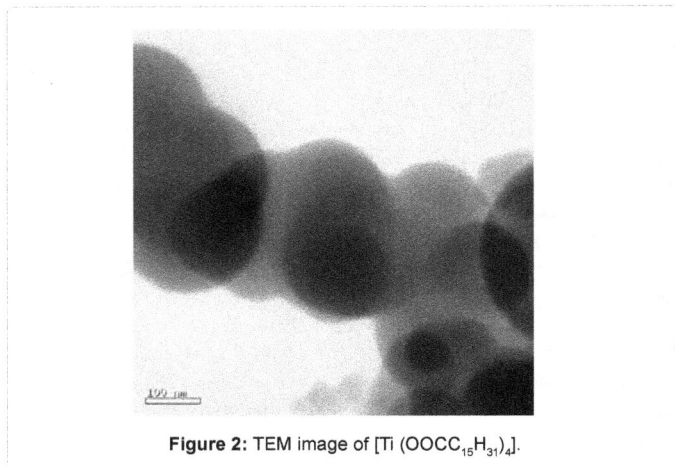

**Figure 2:** TEM image of [Ti $(OOCC_{15}H_{31})_4$].

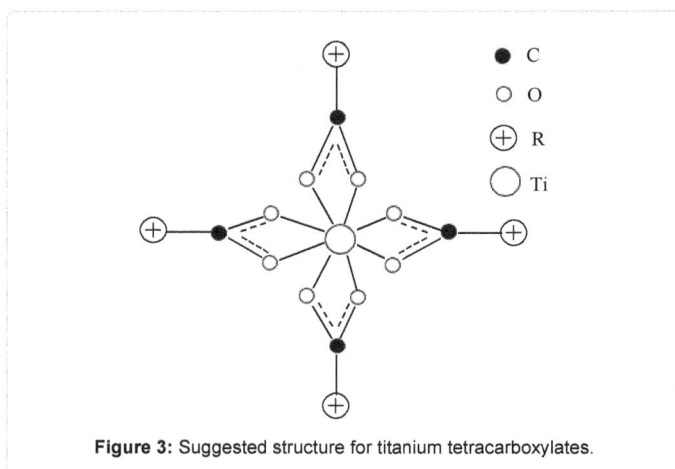

**Figure 3:** Suggested structure for titanium tetracarboxylates.

achieving a higher coordination number (eight). All the complexes exhibited ideal precursor behaviour and could be a potential candidate for the growth of $TiO_2$ thin film by MOCVD process at higher temperatures. The high solubility of these compounds in common organic solvents makes them suitable for liquid injection MOCVD process. Their physico-chemical characterization made us to conclude the bidentate chelating carboxylate ions around titanium(IV) giving coordination number eight as shown in Figure 3.

### Acknowledgements

Authors are thankful to the CSIR [No. 01(2293)/09/EMR-II] and UGC [No. 37-132/2009 (SR)], New Delhi for financial supports. They also thank CDRI, Lucknow for spectral and microanalysis.

| Peak No. | 2 Theta(°) | Flex width | d-value | | Intensity I/I$_o$ |
|---|---|---|---|---|---|
| 1 | 13.800 | 1.176 | 6.4117 | 56 | 9 |
| 2 | 17.800 | 0.941 | 4.9789 | 61 | 10 |
| 3 | 21.600 | 0.941 | 4.1108 | 635 | 100 |
| 4 | 24.000 | 0.941 | 3.7048 | 216 | 34 |
| 5 | 30.000 | 1.176 | 2.9761 | 52 | 9 |
| 6 | 41.000 | -- | 2.1995 | 71 | 12 |
| 7 | 41.600 | -- | 2.1692 | 81 | 13 |

**Table 2:** Powder XRD data of complex 2.

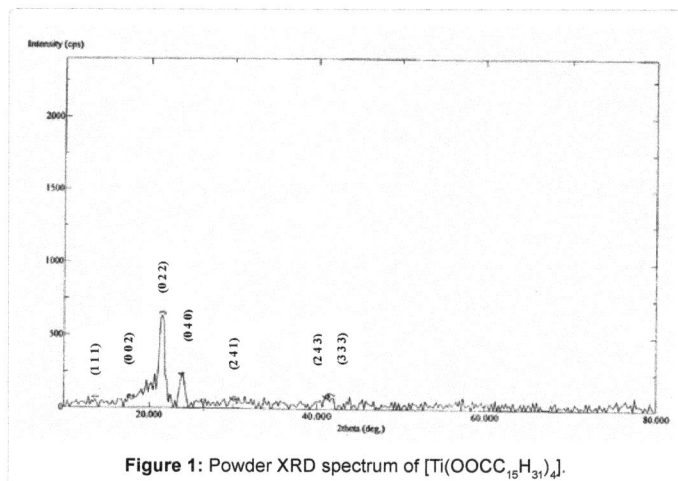

**Figure 1:** Powder XRD spectrum of [Ti$(OOCC_{15}H_{31})_4$].

### References

1. Trung TK, Trens P, Tanchoux N, Bourrelly S, Llewellyn PL, et al. (2008) Hydrocarbon adsorption in the flexible metal organic frameworks MIL-53(Al,Cr). J Am Chem Soc 130: 16926-16932.

2. Thallapally PK, Tian J, Kishan MR, Fernandez CA, Dalgarno SJ, et al. (2008) Flexible (breathing) interpenetrated metal-organic frameworks for CO$_2$ separation applications. J Am Chem Soc 130: 16842-16843.

3. Hartmann M, Kunz S, Himsl D, Tangermann O, Ernst S, et al. (2008) Adsorptive separation of isobutene and isobutane on Cu$_3$(BTC)$_2$. Langmuir 24: 8634-8642.

4. Henschel A, Gedrich K, Kraehnert R, Kaskel S (2008) Catalytic properties of MIL-101. Chem Commun 4192-4194.

5. Perry IV JJ, Perman JA, Zaworotko MJ (2009) Design and synthesis of metal-

organic frameworks using metal-organic polyhedra as supermolecular building blocks. Chem Soc Rev 38: 1400-1417.

6. Mancic LT, Marinkovic BA, Jardim PM, Milosevic OB, Rizzo F (2009) Precursor particle size as the key parameter for isothermal tuning of morphology from nanofibers to nanotubes in the Na2−xHxTinO2n+1 system through hydrothermal alkali treatment of rutile mineral sand. Cryst Growth Des 9: 2152-2158.

7. Zhao B, Chen F, Huang Q, Zhang J (2009) Brookite TiO$_2$ nanoflowers. Chem Commun 5115-5117.

8. Liu SJ, Wu XX, Hu B, Gong JY, Yu SH (2009) Novel anatase TiO2 boxes and tree-like structures assembled by hollow tubes: D,L-malic acid-assisted hydrothermal synthesis, growth mechanism, and photocatalytic properties. Cryst Growth Des 9: 1511–1518.

9. Jones AC, Leedham TJ, Wright PJ, Crosbie MJ, Fleeting KA, et al. (1998) Synthesis and characterisation of two novel titanium isopropoxides stabilised with a chelating alkoxide: their use in the liquid injection MOCVD of titanium dioxide thin films. J Mater Chem 8: 1773-1777.

10. Jones AC, Williams PA, Bickley JF, Steiner A, Davies HO, et al. (2001) Synthesis and crystal structures of two new titanium alkoxy–diolate complexes. Potential precursors for oxide ceramics. J Mater Chem 11: 1428-1433.

11. Woo K, Lee WI, Lee JS, Kang SO (2003) Novel titanium compounds for metal-organic chemical vapour deposition of titanium dioxide films with an ultrahigh deposition rate. Inorg Chem 42: 2378-2383.

12. Strohfeldt K, Tacke M (2008) Bioorganometallic fulvene-derived titanocene anti-cancer drugs. Chem Soc Rev 37: 1174-1187.

13. Tshuva EY, Peri D (2009) Modern cytotoxic titanium (IV) complexes; Insights on the enigmatic involvement of hydrolysis. Coord Chem Rev 253: 2098-2115.

14. Hartinger CG, Dyson PJ (2009) Bioorganometallic chemistry-from teaching paradigms to medicinal applications. Chem Soc Rev 38: 391-401.

15. Kapoor R, Bahl BK, Kapoor P (1986) Reactions of titanium(IV) chloride with carboxylic acids. Indian J Chem 25A: 271-274.

16. Alcock NW, Brown DA, Illson TF, Roe SM, Wallbridge MGH (1989) Preparation and reactions of some titanium(IV) carboxylate species. Polyhedron 8: 1846-1847.

17. Piszczek P, Richert M, Grodzicki A, Głowiak T, Wojtczak A (2005) Synthesis, crystal structures and spectroscopic characterization of [Ti$_4$O$_2$(OOCR)$_x$] (where R = Bu$^t$, CH$_2$Bu$^t$, C(CH$_3$)$_2$Et). Polyhedron 24: 663-670.

18. Baranwal BP, Fatma T, Singh AK, Varma A (2009) Nano-sized titanium(IV) ternary and quaternary complexes with electron-rich oxygen-based bidentate ligands. Inorg Chim Acta 362: 3461-3464.

19. Baranwal BP, Singh AK, Varma A (2011) Spectroscopic studies on some fluorescent mixed-ligand titanium(IV) complexes. Spectrochim Acta Part A 84: 125-129.

20. Armarego WLF, Perrin DD (1997) Purification of Laboratory Chemicals; 4th edition, Butterworth-Heinemann.

21. Jeffery GH, Bassett J, Mendham J, Denney RC (1997) Vogel's Textbook of Quantitative Inorganic Analysis, 5th edition, ELBS, England.

22. Geary WJ (1971) The use of conductivity measurements in organic solvents for the characterization of coordination compounds. Coord Chem Rev 7: 81-122.

23. Nakamoto K (1997) Infrared and Raman Spectra of Inorganic and Coordination Compounds Part B, 5th edition; Wiley: New York.

24. Quan CX, Bin LH, Bang GG (2005) Preparation of nanometer crystalline TiO$_2$ with high photo-catalytic activity by pyrolysis of titanyl organic compounds and photo-catalytic mechanism. Mater Chem Phys 91: 317-324.

# 4

# Chemical Constituents from the Polar Fraction of *Rubus suavissimus*

**Venkata Sai Prakash Chaturvedula\*, Rafael Ignacio San Miguel and Indra Prakash**

*The Coca-Cola Company, Organic Chemistry Department, Global Research and Development, One Coca-Cola Plaza, Atlanta, GA 30313, USA*

### Abstract

Systematic phytochemical study of the *n*-BuOH fraction of the aqueous extract of *Rubus suavissimus* resulted in the isolation of three diterpene glycosides namely rubusoside, suavioside-A and sugeroside; a phenolic glycoside quercetrin; and a lignan glycoside arctiin. The structures of the isolated compounds were characterized on the basis of extensive spectral data (1D and 2D NMR; and MS) and chemical studies which has not been reported earlier. This is the first report of the isolation of quercetrin and arctiin not only from the plant *R. suavissimus* but also from the genus *Rubus*. Also, herewith we are reporting the sweetness recognition threshold and sweetness enhancement effect of rubusoside, the major constituent of *R. suavissimus*.

**Keywords:** *Rubus suavissimus*, Rosaceae, Diterpene glycosides, Phenolic glycoside, Lignan glycoside, NMR, MS, rubusoside, sweetness recognition threshold, sweetness enhancement effect

## Introduction

*Rubus suavissimus* S. Lee belongs to *Rubus*, a large genus of flowering plants in the rose family, Rosaceae , subfamily Rosoideae. Raspberries, blackberries, and dewberries are common, widely distributed members of this genus. *R. suavissimus* is a perennial shrub grows widely grown in Guang-xi and Guang-dong, China [1]. The leaves of *R. suavissimus* are used to make beverage leaf tea by the local residents because of its intensely sweet flavor. It is generally known as tiancha in Chinese or Chinese sweet tea. Previous phytochemical studies of this plant mainly showed the presence of diterpene and triterpene glycosides as well as phenolic compounds [2-4]. The major constituent of this plant is the sweet diterpenoid glycoside rubusoside with an aglycone moiety belongs to the class of the diterpene, *ent*-13-hydroxykaur-16-en-19-oic

acid, known as steviol [5]. As a part of our research to discover natural sweeteners, we have recently reported several ditepene glycosides from *S. rebaudiana* [6-8] and triterpene glycosides from *Siraitia grosvenorii* [9].

In our continuing research to isolate natural compounds from various sweet taste plants collected from all over the World, we have isolated three diterpene glycosides rubusoside (**1**), suavioside-A (**2**) and sugeroside (**3**) as well as the phenolic and lignin glycosides namely quercetrin (**4**) and arctiin (**5**) respectively from the polar fraction of the aqueous extract of the leaves of *R. suavissimus* obtained from Chengdu Biopurify Phytochemicals Limited, China. This paper describes the isolation and structure elucidation of the isolated glycosides **1-5** (Figure 1) on the basis of extensive spectroscopic and chemical studies as well as in comparison of their physical and spectral properties reported from the literature. Also, we are herewith reporting the sweetness recognition threshold (SRT) and sweetness enhancement effect (SEE) of the predominant constituent of the plant, rubusoside (**1**).

## Results and Discussion

Compound **1** was isolated as a white powder and its molecular formula has been deduced as $C_{32}H_{50}O_{13}$ on the basis of its HRMS which showed $[M+NH_4]^+$ and $[M+Na]^+$ ions at *m/z* 660.3590 and 665.3136 respectively, and this was supported by the $^{13}C$ NMR spectral data. The $^1H$ NMR spectrum of **1** showed the presence of two methyl singlets at δ 1.24 and 1.26, two olefinic protons as singlets at δ 5.03 and 5.56 of an exocyclic double bond, nine methylene and two methine protons between δ 0.79-2.72 characteristic for the *ent*-kaurane diterpenoids isolated earlier from the plants belongs to the genus *Stevia* and *Rubus* [6-8,10,11]. The basic skeleton of kaurane diterpenoids was supported by COSY (H-1/H-2; H-2/H-3; H-5/H-6; H-6/H-7; H-9/H-11;

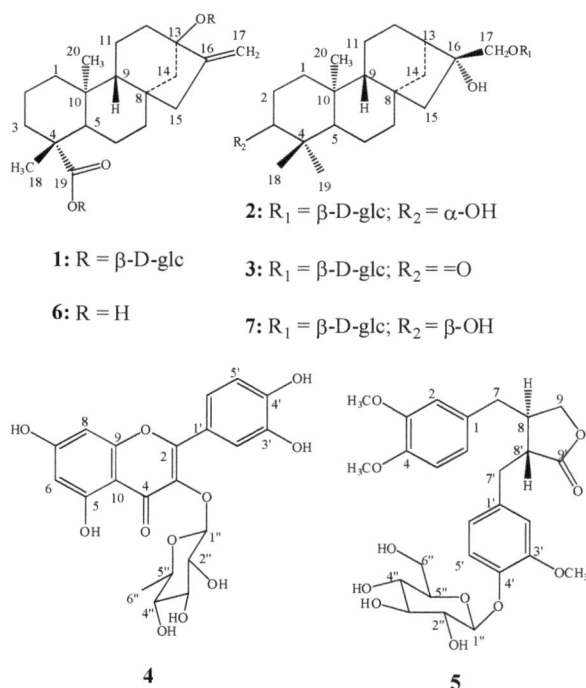

2: R₁ = β-D-glc; R₂ = α-OH

1: R = β-D-glc

3: R₁ = β-D-glc; R₂ = =O

6: R = H

7: R₁ = β-D-glc; R₂ = β-OH

4          5

**Figure 1:** Structures of **1-5** and other compounds.

**\*Corresponding author:** Venkata Sai Prakash Chaturvedula, The Coca-Cola Company, Organic Chemistry Department, Research and Technology, One Coca-Cola Plaza, Atlanta, GA 30313, USA, E-mail: vchaturvedula@na.ko.com

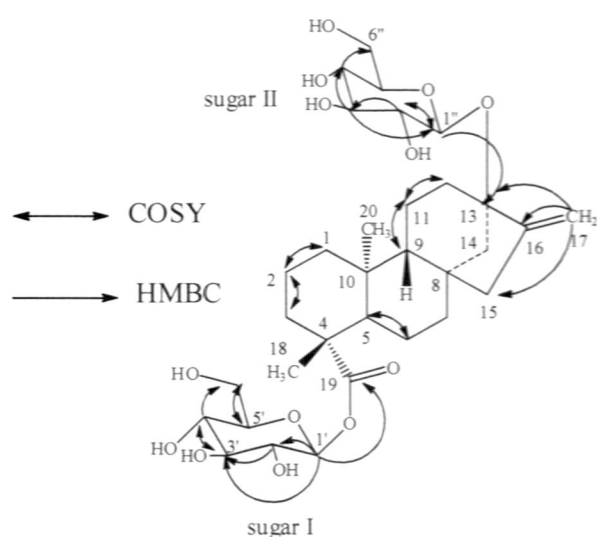

Figure 2: Key COSY and HMBC correlations of 1.

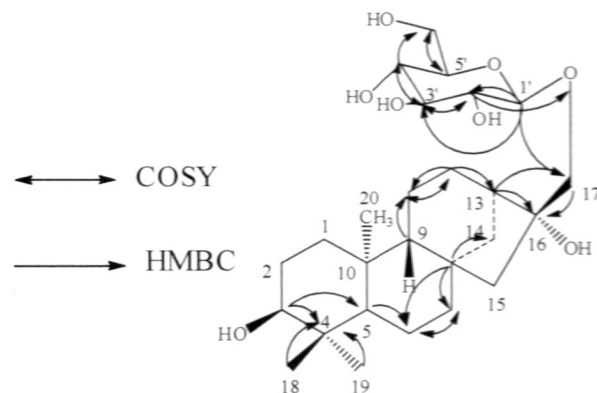

Figure 3: Key COSY and HMBC correlation of 2.

H-11/H-12) and HMBC (H-1/C-2, C-10; H-3/C-1, C-2, C-4, C-5, C-18, C-19; H-5/C-4, C-6, C-7, C-9, C-10, C-18, C-19, C-20; H-9/C-8, C-10, C-11, C-12, C-14, C-15; H-14/C-8, C-9, C-13, C-15, C-16 and H-17/C-13, C-15, C-16) correlations. The fragment ions observed at $m/z$ 481 and 319 in the positive ESI mode MS/MS spectrum of 1 indicating the presence of two hexose sugars in its structure. This was further supported by the $^{1}$H NMR spectrum of 1 which showed the presence of two anomeric protons at δ 5.13, and 6.15. Enzymatic hydrolysis of 1 furnished an aglycone which was identified as steviol (6) by comparison of $^{1}$H NMR [10] spectral data reported in the literature and co-TLC with standard compound. Acid hydrolysis of 1 afforded D-glucose that was identified by preparing the corresponding thiocarbamoyl-thiazolidine carboxylate derivative with L-cysteine methyl ester and O-tolyl isothiocyanate, and in comparison of its retention time with the standard sugars as described in the literature comparison [12]. The large coupling constants observed for the two anomeric protons at δ 5.13 ($J$=7.8 Hz), and 6.15 (d, $J$=8.5 Hz) suggested their β-orientation as reported for steviol glycosides [6-8].

The $^{13}$C NMR values for all the carbons were assigned on the basis of HSQC and HMBC correlations (Table 1 and Table 2). From COSY,

and HMBC correlations as shown in Figure 2, 1 was found to have a steviol aglycone moiety having a β-D-glucopyranosyl unit attached C-13 hydroxyl and another β-D-glucopyranosyl moiety in the form of an ester at C-19. This was supported by the HMBC correlations: H-1'/C-19, C-2', C-3'; and H-1"/C-13, C-2", C-3". A close comparison of the NMR spectral data of 1 with the reported literature values for rubusoside confirmed its structure.

The molecular formula of compound 2 was established as $C_{26}H_{44}O_{8}$ from its HRMS spectral data which showed $[M+NH_{4}]^{+}$ and $[M+Na]^{+}$ ions at $m/z$ 502.3327 and 507.2924 respectively. The $^{1}$H NMR spectrum of 2 showed the presence of three methyl singlets at δ 0.87, 1.02, and 1.17, eight methylene and two methine protons between δ 0.89-2.43, similar to 1. The $^{1}$H NMR of 2 also showed the presence of signal at δ 4.12 as a triplet like with $W_{1/2}$ = 2.6 Hz and a pair of doublets corresponding to an oxymethylene at δ 3.91 ($J$=10.6 Hz), 4.49 ($J$=10.2 Hz) and an anomeric proton as a doublet at δ 5.05. Acid hydrolysis of 2 afforded D-glucose that was identified by preparing its corresponding thiocarbamoyl-thiazolidine carboxylate derivative as in 1, and the coupling constant observed for the anomeric proton $J$=8.2 Hz suggested the β-orientation of the D-glucosyl unit. The $^{13}$C NMR spectrum of 2 showed the presence of nine oxygenated carbons between δ 63.2 and 107.2 of which six were assigned to the β-D-glucopyranosyl unit, leaving the assignment of the other three carbons.

| Position | 1 | 2 | 3 |
|---|---|---|---|
| 1 | 0.79 m, 1.68 m | 0.89 m, , 1.86 m | 1.25 m, 1.98 m |
| 2 | 1.41 m, 1.92 m | 1.34 m, 1.94 m | 1.82 dd (13.2, 12.2), 2.54 dd (12.0, 4.8) |
| 3 | 1.06 m, 2.34 d (12.4) | 4.12 t ($W_{1/2}$ 2.60) | - |
| 5 | 1.32 m | 1.57 m | 1.83 m |
| 6 | 1.42 m, 1.72 m | 1.42 m, 1.64 m | 1.40 m, 1.63 m |
| 7 | 1.32 m, 1.70 m | 1.47 m, 1.66 m | 1.45 m, 1.67 m |
| 9 | 0.94 m | 0.98 m | 0.97 m |
| 11 | 1.74 m | 1.65 m | 1.68 m |
| 12 | 1.76 m, 1.95 m | 1.54 m, 1.70 m | 1.51 m, 1.73 m |
| 13 | - | 2.02 m | 1.98 m |
| 14 | 2.24 m, 2.72 d (12.8) | 1.78 m, 2.41 m | 1.76 m, 2.43 m |
| 15 | 2.08 m, 2.52 m | 1.35 m, 1.82 m | 1.35 m, 1.82 m |
| 17 | 5.03 s, 5.56 s | 3.91 d (10.6), 4.49 d (10.2) | 3.95 d (10.4), 4.50 d (10.8) |
| 18 | 1.26 s | 0.87 s | 0.93 s |
| 19 | - | 1.02 s | 1.02 s |
| 20 | 1.24 s | 1.17 s | 1.10 s |
| 1' | 6.15 d (8.5) | 5.05 d (8.2) | 5.06 d (7.8) |
| 2' | 3.96 m | 4.04 m | 4.02 m |
| 3' | 4.08 m | 4.46 m | 4.45 m |
| 4' | 4.33 m | 4.25 m | 4.23 m |
| 5' | 4.21 m | 4.63 m | 4.64 m |
| 6' | 4.03 dd (4.2, 12.4), 4.41 dd (2.4, 9.2) | 3.61 dd (4.4, 12.2), 4.27 dd (2.5, 10.6) | 3.60 dd (4.4, 12.0), 4.24 dd (2.3, 10.2) |
| 1" | 5.13 d (7.8) | | |
| 2" | 4.19 m | | |
| 3" | 4.35 m | | |
| 4" | 4.23 m | | |
| 5" | 3.38 m | | |
| 6" | 4.22 dd (3.9, 12.2), 4.62 dd (2.2, 8.0) | | |

$^{a}$assignments made on the basis of COSY, HSQC and HMBC correlations; $^{b}$Chemical shift values are in δ (ppm); $^{c}$Coupling constants are in Hz.

Table 1: $^{1}$H NMR chemical shift values for 1–3 isolated from Rubus suavissimus recorded in $C_{5}D_{5}N$ $^{a-c}$.

| Position | 1 | 2 | 3 |
|---|---|---|---|
| 1 | 41.2 | 34.2 | 39.6 |
| 2 | 19.9 | 27.4 | 34.6 |
| 3 | 38.8 | 75.6 | 217.3 |
| 4 | 44.4 | 38.5 | 47.5 |
| 5 | 57.8 | 49.4 | 54.6 |
| 6 | 22.6 | 20.8 | 22.3 |
| 7 | 42.2 | 42.9 | 41.6 |
| 8 | 42.9 | 45.3 | 44.9 |
| 9 | 54.3 | 57.2 | 55.9 |
| 10 | 40.2 | 40.0 | 39.0 |
| 11 | 21.1 | 18.9 | 19.3 |
| 12 | 37.7 | 26.9 | 27.0 |
| 13 | 86.4 | 47.0 | 46.7 |
| 14 | 44.9 | 37.9 | 37.5 |
| 15 | 48.3 | 53.8 | 53.2 |
| 16 | 155.0 | 81.3 | 81.2 |
| 17 | 104.9 | 76.0 | 75.9 |
| 18 | 28.8 | 29.8 | 27.7 |
| 19 | 177.4 | 22.8 | 21.5 |
| 20 | 16.1 | 18.4 | 18.2 |
| 1' | 95.4 | 107.2 | 107.1 |
| 2' | 75.9 | 75.8 | 76.0 |
| 3' | 79.5 | 79.0 | 79.0 |
| 4' | 72.8 | 72.0 | 72.1 |
| 5' | 78.5 | 79.3 | 79.2 |
| 6' | 63.5 | 63.2 | 63.2 |
| 1" | 100.2 | | |
| 2" | 74.5 | | |
| 3" | 79.3 | | |
| 4" | 71.5 | | |
| 5" | 79.8 | | |
| 6" | 61.5 | | |

[a]assignments made on the basis of HSQC and HMBC correlations; [b] Chemical shift values are in δ (ppm)

**Table 2:** $^{13}C$ NMR chemical shift values for **1–3** isolated from *Rubus suavissimus* recorded in $C_5D_5N$ [a-b].

From HSQC and HMBC spectral data, it was found that the other three oxygenated carbons of **2** as one oxymethylene, one oxymethine and a tertiary hydroxyl resonating at δ 76.0, 75.6 and 81.3 respectively in its $^{13}C$ NMR spectrum. From the above spectral data and chemical studies, the structure was identified as an *ent*-kaurane diterpenenoid skeleton having a β-D-glucopyranosyl unit and three oxygenated carbons as mentioned above. A search from the literature indicated the presence of two compounds with the above functional groups namely *ent*-kaurane-3α, 16β, 17-triol-17-*O*-β-D-glucoside (suavioside A) and *ent*-kaurane-3β, 16β, 17-triol-17-*O*-β-D-glucoside (iwayoside A, **7**) isolated from *R. suavissimus* [13] and *Artemisia iwayomogi* [14] respectively. The key COSY and HMBC correlations as displayed in Figure 3 supported the basic skeleton of *ent*-kaurane-3, 16, 17-triol-17-*O*-β-D-glucoside. Since the spectral data for suavioside A and iwayoside A were reported in $CD_3OD$ and $C_5D_5N$ respectively and in order to compare the NMR values of **2** with the isolated compounds, its $^1H$ and $^{13}C$ NMR were also acquired in both the solvents. A close comparison of the NMR spectral data and optical rotation of **2** confirmed the structure as suavioside A.

Compound **3** was also obtained as a white amorphous powder and its molecular formula was inferred as $C_{26}H_{42}O_8$ from its HRMS spectral data that showed $[M+NH_4]^+$ and $[M+Na]^+$ ions at *m/z* 500.3220 and 505.2768 respectively, and this molecular composition was supported by the $^{13}C$ NMR spectral data. The $^1H$ NMR spectrum of **3** showed

the presence three methyl singlets, eight methylene and two methine protons, two protons corresponding to an oxymethylene group and an anomeric proton; identical to **2**. Acid hydrolysis confirmed the sugar and its configuration as D-glucose. A close comparison of the $^1H$ and $^{13}C$ NMR chemical shift values of **2** and **3** together with the ESI-MS data which has 2 amu difference, suggested the presence of a carbonyl group in **3** at C-3 position in place of an oxymethine proton in **2**. This was further supported by the presence of the peak observed in its $^{13}C$ NMR spectrum at δ 217.3, and the absence of a triplet like signal for the oxymethine proton at C-3 position and its corresponding carbon in the respective proton and carbon NMR spectral data of **3**. The above spectral and chemical data suggested its structure as **3**, which was reported as sugeroside earlier from *Ilex sugerokii* var. *brevipedunculata* [15] and *R. suavissimus* [13].

Compound **4** was obtained as yellow amorphous powder and its molecular formula was established as $C_{21}H_{20}O_{11}$ from its HRMS spectral data that showed $[M+H]^+$ and $[M+Na]^+$ ions at *m/z* 449.1075 and *m/z* 471.0892 respectively; this was supported by the $^{13}C$ NMR spectral data.

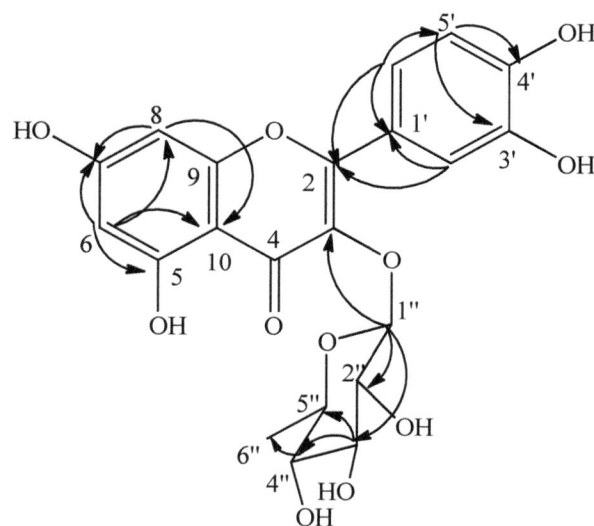

**Figure 4:** Key HMBC correlation of **4**.

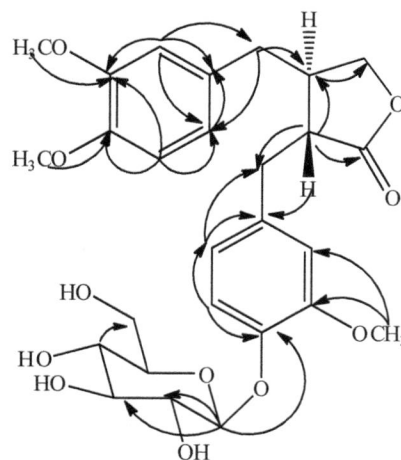

**Figure 5:** Key HMBC correlations of **5**.

| Position | 4 δ_H | 4 δ_C | 5 δ_H | 5 δ_C |
|---|---|---|---|---|
| 1 | | | | 133.0 |
| 2 | | 158.2 | 6.76 d (1.7) | 113.1 |
| 3 | | 136.5 | | 150.3 |
| 4 | | 179.6 | | 149.1 |
| 5 | | 163.5 | 6.93 d (7.2) | 114.7 |
| 6 | 6.65 d (1.6) | 95.0 | 6.72 dd (7.2, 1.9) | 121.7 |
| 7 | | 166.3 | 2.83 m | 38.4 |
| 8 | 6.72 d (1.8) | 100.2 | 2.52 m | 42.0 |
| 9 | | 158.7 | 3.93 dd (7.5, 8.9), 4.20 dd (7.3, 9.2) | 71.7 |
| 10 | | 105.9 | | |
| 1' | | 122.8 | | 132.1 |
| 2' | 8.03 d (1.7) | 117. | 7.02 d (1.9) | 113.6 |
| 3' | | 147.8 | | 150.1 |
| 4' | | 151.0 | | 147.4 |
| 5' | 7.31 d (8.1) | 116.9 | 7.58 d (7.5) | 116.7 |
| 6' | 7.73 dd (7.5, 1.8) | 122.6 | 6.88 dd (7.2, 1.6) | 122.7 |
| 7' | | | 3.07 m | 35.0 |
| 8' | | | 2.72 m | 47.1 |
| 9' | | | | 179.4 |
| 1" | 6.29 d (1.5) | 104.5 | 5.69 d (7.2) | 102.9 |
| 2" | 5.08 dd (4.3, 1.6) | 72.5 | 4.20 dd (7.2, 9.4) | 75.3 |
| 3" | 4.65 dd (9.3, 3.3) | 72.7 | 4.33 dd (9.2, 7.8) | 79.0 |
| 4" | 4.33 dd (9.8, 7.2) | 73.8 | 4.35 dd (9.4, 8.2) | 71.8 |
| 5" | 4.41 m | 73.0 | 4.12 ddd (9.4, 1.8, 7.9) | 79.3 |
| 6" | 1.49 d (6.0) | 18.1 | 4.40 dd (2.1, 12.4), 4.55 dd (4.2, 12.2) | 62.8 |
| OCH₃ | | | 3.75 | 56.4 |
| OCH₃ | | | 3.78 | 56.5 |
| OCH₃ | | | 3.80 | 56.5 |

[a]assignments made on the basis of COSY, HSQC and HMBC correlations; [b]Chemical shift values are in δ (ppm); [c]Coupling constants are in Hz

**Table 3:** $^1$H and $^{13}$C NMR chemical shift values for **4–5** isolated from *Rubus suavissimus* recorded in $C_5D_5N^{a-c}$.

The $^1$H NMR spectrum of **4** showed the presence of three meta coupled aromatic protons at δ 6.65 ($J$=1.6 Hz), 6.72 ($J$=1.8 Hz) and 8.03 ($J$=1.7 Hz); one ortho coupled aromatic proton at δ 7.31 ($J$=8.1 Hz); another ortho and meta coupled aromatic proton as doublet of doublets at δ 7.73 ($J$=7.5, 1.8 Hz); characteristic for the 3-substituted flavone. The $^1$H NMR of **4** also showed the presence of an anomeric proton as a doublet at δ 6.29 suggesting a sugar residue in its structure which was identified as L-rhamnosyl moiety on the basis of acid hydrolysis and by preparing the corresponding thiocarbamoyl-thiazolidine carboxylate derivative with L-cysteine methyl ester and O-tolyl isothiocyanate, and in comparison of its retention time with the standard sugars as described for compounds **1-3**. The presence of L-rhamnosyl moiety was further supported by the presence of the secondary methyl group as a doublet at δ 1.49 ($J$=6.0 Hz) as well as the absence of oxymethylene group at C-5 position of the sugar unit. The anomeric proton had a coupling constant of 1.5 Hz, similar to dulcosides A and B isolated from *S.rebaudiana* [16] confirming the α-orientation L-rhamnosyl moiety. The placement of the L-rhamnosyl moiety was identified at C-3 position on the basis of key HMBC correlations: H-2'/C-2, C-1', C-3' and H-1"/C-2, C-3, C-2" (Figure 4). Thus, based on the above spectral data the structure of **4** was assigned as quercetin-3-O-α-L-rhamnoside (quercetrin) consistent to the reported literature values [17,18].

Compound **5** was obtained as a hard gum and its molecular formula was inferred as $C_{27}H_{34}O_{11}$ from its HRMS spectral data that showed

[M+NH₄]⁺ and [M+Na]⁺ ions at *m/z* 552.2437 and *m/z* 557.1986 respectively. The $^1$H NMR spectrum of **5** showed the presence of six aromatic protons between δ 6.72-7.58; three methoxyl groups at δ 3.75, 3.78 and 3.80; two protons of an oxymethylene group at δ 3.93 and 4.20; two methylenes and two methines between δ 2.52-3.07; very similar to the lignan arctigenin. Acid hydrolysis of **5** furnished D-glucose and was identified as having β-orientation from the coupling constant of its anomeric proton appeared at δ 5.69 ($J$=7.2 Hz). From the above spectral data and chemical studies it was evident that the structure of **5** should contain a β-D-glucopyranosyl unit that has been attached to the aglycone moiety of arctigenin. From the HMBC correlations shown in Figure 5, the presence of β-D-glucopyranosyl unit was suggested at C-6' position unambiguously as in arctiin; confirmed its structure completely [19,20].

This is the first report of the isolation of quercetrin and arctiin not only from the plant *R. suavissimus* but also from the genus *Rubus*. Further, the detailed NMR characterizations have not been studied on some of the isolated glycosides and herewith we have assigned the entire proton and carbon values on the basis of COSY, HSQC and HMBC spectral data as well as confirmed the sugars and their configuration by hydrolysis studies. Also, we are reporting the $^1$H and $^{13}$C NMR data for all the isolated five compounds (1-5) in $C_5D_5N$ in this article.

## Experimental

### General

Melting points were measured using a SRS Optimelt MPA 100 instrument and are uncorrected. Optical rotations were recorded using a Rudolph Autopol V at 25 °C and NMR spectra were acquired on a Varian Unity Plus 600 MHz instrument using standard pulse sequences at ambient temperature. Chemical shifts are given in δ (ppm), and coupling constants are reported in Hz. HRMS data was generated with a Thermo LTQ Orbitrap Discovery mass spectrometer in the positive positive ion mode electrospray. Instrument was mass calibrated with a mixture of Ultramark 1621, MRFA [a peptide], and caffeine immediately prior to accurate mass measurements of the samples. Samples were diluted with water:acetonitrile:methanol (1:2:2) and prepared a stock solution of 50 ul concentration for each sample. Each sample (25 ul) was introduced via infusion using the onboard syringe pump at a flow injection rate of 120 ul/min. Low pressure chromatography was performed on a Biotage Flash system using a C-18 cartridge (40+ M, 35-70 μm). TLC was performed on Baker Si-C₁₈F plates and identification of the spots on the TLC plate was carried out by spraying 10% $H_2SO_4$ in EtOH and heating the plate at about 80°C. Analytical HPLC for sugar analysis was carried out with a Waters 600E multisolvent delivery system using a Phenomenex Luna C₁₈ non-chiral (150 x 4.6 mm, 5 μm) column.

### Plant material

The commercial sample consisting of the aqueous extract of the leaves of *R. suavissimus* was purchased from Chengdu Biopurify Phytochemicals, China. The plant material was identified by Professor Weiping He, Natural Plant Scientific Institute, Guangdong Ocean University, Guangxi, China and a voucher specimen is deposited at The Coca Cola Company, No. VSPC-3166-68.

### Isolation

The aqueous extract of the leaves of *R. suavissimus* (10g) was suspended in 100 ml water and extracted successively with *n*-hexane (3 x 100 ml), $CH_2Cl_2$ (3 x 100 ml) and *n*-BuOH (2 x 100 ml). The *n*-BuOH layer was concentrated under vacuum furnished a residue (2.5g) which was purified on a Biotage flash chromatography system using C-18 (100g) column (solvent system: gradient from 20-80 MeOH-water to 100% MeOH at 60 ml/min. detection at UV 210 nm) for 40 min. Fractions 13-20 (0.2g) were combined and further subjected to repeated flash chromatography purification with gradient from 40-80% MeOH-water at 30 ml/min for 30 min afforded suavioside A (2, $t$R, 18.6 min, 10.2 mg). Fractions 21-23 (1.4g) were combined and crystallized with MeOH furnished rubusoside (1, 1.1g). Purification of the combined fractions 24-27 (0.12g) using the gradient from 40-80% MeOH-water at 30 ml/min for 40 min furnished sugeroside (3, $t$R 16.4 min, 11.5 mg). Fractions 46-50 and 52-56 were combined to get residues 0.14 g and 0.11 g respectively, which on repeated purification using the gradient 60-90% MeOH-water at 30 ml/min for 40 min resulted quercetrin (4, $t$R 22.4 min, 6.8 mg), and arctiin (5, $t$R 19.6 min, 8.4 mg), respectively.

## Identification

*Rubusoside*: White powder, mp 177-179 °C [reported mp 178-181 °C]; $[\alpha]_D^{25}$ -37.20 (c 1.0, MeOH) [reported $[\alpha]_D^{18}$ -40.30 (c 0.8, MeOH)]; $^1$H NMR (600 MHz, C$_5$D$_5$N, δ ppm) and $^{13}$C NMR (150 MHz, C$_5$D$_5$N, δ ppm) spectroscopic data see Table 1 and Table 2; HRMS $m/z$ [M+NH$_4^+$] calcd for C$_{32}$H$_{54}$O$_{13}$N: 660.3595; found 660.3590 and [M+Na$^+$] calcd for C$_{32}$H$_{50}$O$_{13}$Na: 665.3159; found 665.3136.

*Enzymatic hydrolysis of 1.* A solution of 1 (500 µg) was dissolved in 5 ml of 0.1 M sodium acetate buffer, pH 4.5 and crude pectinase from *Aspergillus niger* (100 uL, Sigma-Aldrich, P2736) was added. The mixture was stirred at 50°C for 48 hr. The product precipitated out during the reaction was filtered and then crystallized from methanol (MeOH). The resulting steviol (6, 73 µg) was identical to an authentic sample by co-TLC and $^1$H NMR [10].

*Suavioside A*: White powder, $[\alpha]_D^{25}$ +45.62 (c 0.15, C$_5$H$_5$N) [reported $[\alpha]_D^{25}$ +47.21 (c 0.14, C$_5$H$_5$N)]; $^1$H NMR (600 MHz, C$_5$D$_5$N, δ ppm) and $^{13}$C NMR (150 MHz, C$_5$D$_5$N, δ ppm) spectroscopic data see Table 1 and Table 2; HRMS $m/z$ HRMS $m/z$ [M+NH$_4^+$] calcd for C$_{26}$H$_{48}$O$_8$N: 502.3380; found 502.3327 and [M+Na$^+$] calcd for C$_{26}$H$_{44}$O$_8$Na: 507.2934; found 507.2924.

*Sugeroside*: White powder, $[\alpha]_D^{25}$ -53.20 (c 0.35, C$_5$H$_5$N) [reported $[\alpha]_D^{23}$ -55.60 (c 0.36, C$_5$H$_5$N)]; $^1$H NMR (600 MHz, C$_5$D$_5$N, δ ppm) and $^{13}$C NMR (150 MHz, C$_5$D$_5$N, δ ppm) spectroscopic data see Table 1 and Table 2; HRMS $m/z$ [M+NH$_4^+$] calcd for C$_{26}$H$_{46}$O$_8$N: 500.3223; found 500.3220 [M+Na$^+$] calcd for C$_{26}$H$_{42}$O$_8$Na: 505.2777; found 505.2768.

*Determination of the configuration of sugars in 1-3:* Each compound 1-3 (500 µg) was hydrolyzed with 0.5 M HCl (0.5 mL) for 1.5 h. After cooling, the mixture was diluted with 5 ml water, passed through an Amberlite IRA400 column and the eluate was lyophilized. The residue was dissolved in pyridine (0.25 mL) and heated with L-cysteine methyl ester HCl (2.5 mg) at 60°C for 1.5 h, and then O-tolyl isothiocyanate (12.5 uL) was added to the mixture and heated at 60°C for an additional 1.5 h. The reaction mixture was analyzed by HPLC: column Phenomenex Luna C18, 150 x 4.6 mm (5 u); 25% acetonitrile-0.2% TFA water, 1 mL/min; UV detection at 250 nm. The sugar was identified as D-glucose in each experiment ($t$R, 12.26 to 12.42 min) [authentic samples, D-glucose ($t$R, 12.35) and L-glucose ($t$R, 11.12 min) [12].

*Quercetrin :* Yellow powder, mp 179.4-181.2 °C [reported mp 180-182 °C]; $^1$H NMR (600 MHz, C$_5$D$_5$N, δ ppm) and $^{13}$C NMR (150 MHz, C$_5$D$_5$N, δ ppm) spectroscopic data see Table 3; HRMS $m/z$ [M+H$^+$] calcd for C$_{21}$H$_{21}$O$_{11}$: 449.1084; found 449.1075 and [M+Na$^+$] calcd for C$_{21}$H$_{20}$O$_{11}$Na: $m/z$ 471.0903; found 471.0892.

*Arctiin :* Hard gum, $[\alpha]_D^{25}$ -52.60 (c 0.003, MeOH) [reported $[\alpha]_D^{23}$ -55.30 (c 0.0033, MeOH)]; $^1$H NMR (600 MHz, C$_5$D$_5$N, δ ppm) and $^{13}$C NMR (150 MHz, C$_5$D$_5$N, δ ppm) spectroscopic data see Table 3; HRMS $m/z$ [M+NH$_4^+$] calcd for C$_{27}$H$_{48}$O$_8$N: 552.2445; found 552.2437 and [M+Na$^+$] calcd for C$_{27}$H$_{34}$O$_{11}$Na: $m/z$ 557.1999; found 557.1986.

*Determination of the configuration of sugars in 4-5:* Each compound 4 and 5 (1 mg) was hydrolyzed with 2 M HCl (2 mL) for 18 h. After cooling, the mixture was diluted with 5 ml water, passed through an Amberlite IRA400 column, lyophilized and the residues obtained were converted to the corresponding thiocarbamoyl-thiazolidine carboxylate derivative with L-cysteine methyl ester and O-tolyl isothiocyanate as described above. The sugars were identified as L-rhamnose ($t$R, 21.32 min) and D-glucose ($t$R, 12.21 min) respectively from the hydrolysis experiments with 4 and 5 [authentic samples: D- rhamnose ($t$R, 11.73 min) and L- rhamnose ($t$R, 21.64 min); D-glucose ($t$R, 12.35) and L-glucose ($t$R, 11.12 min)] [12].

*Sweetness Recognition Threshold (SRT) measurement of 1:* The sweetness recognition threshold of 1 was measured by three experienced panelists in duplicated runs. All solutions were made in carbon-treated water and used at room temperature. Each of three subjects were asked to isosweet the random, different order blind samples against standard sugar solutions at 0.5%, 1.0% and 1.5% (w/v). The subjects were asked to focus on second sip of each sample and to rinse their mouths with water in between samples. The blind results indicated both duplicated runs yielded consistent results among samples at three different concentrations of 35, 50 and 65 ppm of 1 and the overall % sweetness equivalence (SE) averages were 0.59, 0.92 and 1.06, respectively. As a result, the SRT at 0.75% SE of sugar in water is estimated to be 42 ppm for 1. Similarly, the SRT of 1 in carbonated lemon-lime (LL) soda prototypes without sweetener was evaluated and determined to be 150 ppm.

*Sweetness Enhancement Effect (SEE) measurement of 1:* In order to find the SEE, test solutions of glucose, fructose and sucrose were prepared equivalent to 6% SE with carbon-treated water at room temperature at SRT of 1 in each carbohydrate solution. The same 6% SE solutions with the above three carbohydrates were prepared without 1. Experimental results indicated that 1 was found to have a slight, positive SEE (ca. 1% SE more) in glucose and fructose solutions whereas a relatively larger SEE in sucrose by ca. 2% SE compared to the solutions of glucose, fructose and sucrose without 1. Likewise, the SEE in carbonated lemon-lime (LL) soda prototypes was evaluated for sensory data in a trained, descriptive analysis of 10 descriptive analysis panelists by preparing 8% SE solution using high fructose corn syrup (HFCS), with and without 1 as described above. Each assessor evaluated all the beverage products in triplicate runs with 8 min break time between testing samples. Unsalted cracker, 0.75% saline solution and mineral grade water was used as a mouth rinse and refresher before testing each sample. Sensory data revealed that the 8% HFCS containing 1 had a SE of 9.2% which is 1.15 times as sweet as the prototype without 1.

## Conclusions

Five glycosides including three diterpene, a phenolic and a lignan were isolated from the commercial extract obtained from the leaves of *R. suavissimus* obtained from Chengdu Biopurify Phytochemicals Limited, China. The structures of all the isolated new compounds were identified as rubusoside (1), suavioside-A (2), sugeroside (3), quercetrin (4) and arctiin (5) on the basis of spectroscopic and chemical studies as well as by comparing their physical properties reported in the literature. This is the first report of the isolation of quercetrin and arctiin from *R. suavissimus* in nature. The complete ¹H and ¹³C NMR spectral assignments of all the isolated compounds are reported herewith in CD₅N₅ based on COSY, HSQC, HMBC, and MS/MS spectroscopic data as well as chemical studies. The sensory evaluation results demonstrated that the SRT of **1** in water and LL soda matrixes are 50 and 150 ppm respectively. Also, **1** showed ca.1% SSE in glucose and fructose, and ca.2% SSE in sucrose in aqueous solutions; whereas it showed 1.15 times SEE in LL soda prototypes at its SRT.

## Acknowledgements

We wish to thank Chengdu Biopurify Phytochemicals Limited, China for providing the *Rubus suavissimus* aqueous extract.

## References

1. Koh GY, Chou G, Liu Z (2009) Purification of a water extract of Chinese sweet tea plant (*Rubus suavissimus* S. Lee) by alcohol preparation. J Agric Food Chem 57: 5000-5006 and references cited therein.

2. Gao F, Chen F, Tanaka T, Kasai R, Seto T, Tanaka O (1985) 19α-hydroxyursane-type triterpene glucosyl esters from the roots of *Rubus suavissimus* S. Lee. Chem Pharm Bull 33: 37-40.

3. Wang J, Lu H (2007) Chemical constituents of *Rubus suavissimus* S. Lee. Zhong yao cai 30: 800-802.

4. Sugimoto N, Kikuchi H, Yamazaki T, Maitani T (2001) Polyphenolic constituents from the leaves of *Rubus suavissimus*. Nat Med (Tokyo, Jpn.) 55: 219.

5. Brandle JE, Starrratt AN, Gijen M (1998) *Stevia rebaudiana*: its agricultural, biological and chemical properties. Can J Plant Sci 78: 527-536.

6. Chaturvedula VSP, Rhea J, Milanowski D, Mocek U, Prakash I (2011) Two minor diterpene glycosides from the leaves of *Stevia rebaudiana*. Nat Prod Commun 6: 175-178.

7. Chaturvedula VSP, Prakash I (2011) A new diterpenoid glycoside from *Stevia rebaudiana*. Molecules 16: 2937-2943.

8. Chaturvedula VSP, Mani U, Prakash I (2011) Diterpene glycosides from *Stevia rebaudiana*. Molecules 16: 3552-3562.

9. Chaturvedula VSP, Prakash I (2011) Curcubitane glycosides from *Siraitia grosvenorii*. J Carb Chem 30: 16-26.

10. Ohtani K, Aikawa Y, Kasai R, Chou W, Yamasaki K, Tanaka O (1992) Minor diterpene glycosides from sweet leaves of *Rubus suavissimus*. Phytochemistry 31: 1553-1559.

11. Tanaka T, Kohda H, Tanaka O, Chen FH, Chou WH, Leu JL (1981) Rubusoside (β-D-glucosyl ester of 13-O-β-D-glucosyl-steviol), a sweet principle of *Rubus chingii Hu* (Rosaceae). Agric Biol Chem 45: 2165-2166.

12. Tanaka T, Nakashima T, Ueda T, Tomii K, Kouno I (2007) Facile discrimination of aldose enantiomers by reversed-phase HPLC. Chem. Pharm Bull 55: 899-901.

13. Hirono S, Chou WH, Kasai R, Tanaka O, Tada T (1990) Sweet and bitter diterpene glycosides from the leaves of *Rubus suavissimus*. Chem Pharm Bull 38: 1743-1744.

14. Liang YD, Yang SY, Kim JH, Lee YM, Kim YH (2010) A new diterpene glycoside from *Artemisia iwayomogi* Kitamura that enhances IL-2 secretion. Bull Korean Chem Soc 31: 2422-2423.

15. Ichikawa N, Ochi M, Kubota T (1973) Bitter principles of Aquifoliaceae. II. Structure of the bitter principles of *Ilex sugerokii var brevipedunculata and var longipedunculata*. Nippon Kagaku Kaishi 4: 785-793.

16. Kobayashi M, Horikawa S, Degrandi IH, Ueno J, Mitsuhashi H (1977) Dulcosides A and B, new diterpene glycosides from *Stevia rebaudiana*. Phytochemistry 16: 1405-1408.

17. Shen CJ, Chen CK, Lee SS. (2009) Polar constituents from *Sageretia thea* leaf characterized by HPLC-SPE-NMR assisted approaches. J Chin Chem Soc (Taipei, Taiwan) 56: 1002-1009.

18. Ma X, Tian W, Wu L, Cao X, Ito Y (2005) Isolation of querectin-3-O-L-rhamnoside from *Acer truncatum* Bunge by high-speed counter-current chromatography. J Chromatogr A 1070: 211-214.

19. Saklani A, Sahoo MR, Misra PD, Vishwakarma R (2011) *Saussura heteromalla* (D.Don) *Hand-Mazz.*: a new source of arctiin, arctigenin and chlorojanerin. Indian J Chem 50(B): 624-626.

20. Rahman MMA, Dewick PM, Jackson DE, Lucas JA (1990) Lignans of *Forsytha intermedia*. Phytochemistry 29: 1971-1980.

# 5

# Complexation of KBr₃ with Poly (Ethylene Glycol): Efficient Bromination of Aromatics Under Solvent-less Conditions

**Sanny Verma and Suman L. Jain***

*Chemical Sciences Division, CSIR-Indian Institute of Petroleum, Mohkampur, Dehradun-248005, India*

## Abstract

The complexation of $KBr_3$ with acyclic poly (ethylene glycol)-400 was carried out in which the chain of polyether suitably wrapped around the cation in a host-guest manner. The resulting complex was found to be an efficient brominating agent for the selective regioselective monobromination of various aromatic compounds in excellent yields under mild reaction conditions. In another protocol, bromination was carried out by using the PEG-embedded $KBr_3$ as a catalyst in the presence of hydrogen peroxide. The presence of hydrogen peroxide enhanced the reaction rates and provided selective bromination within very short reaction times.

**Keywords:** Host-guest complex; Bromination; Tribromide; PEG; Regioselective synthesis

## Introduction

After the synthesis of crown ethers and the discovery of their complexing properties toward alkali metal cations in 1967, the host-guest chemistry has developed rapidly and being used worldwide in various fields such as supra-molecular chemistry, biomimetic chemistry and materials science [1,2]. However, expensive nature and limited accessibility of the crown ethers make their utility limited. On the other hand, polyethylene glycols, referred as a "poor chemist's crown ethers" are well known to be inexpensive, non-volatile, non-toxic and biodegradable, are also have the tendency to bind with alkali cations as crown ethers. Owing to the easy availability and cost effective nature, PEG's, have widely been used in various applications [3,4]. Host-guest interactions of poly (ethylene glycols) to form inter polymer complexes through hydrogen bonding are well reported and have widely been used in drug-delivery systems [5,6]. However, host-guest complexation of PEG's with organic reagents for organic reactions is not known so far. Monobromination of activated aromatic compounds such as phenols, aromatic amines etc. is an important synthetic transformation as these compounds are valuable intermediates in a variety of well-known reactions such as Wurtz-type condensation reactions, hydrolysis reactions, formation of Grignard reagents, and many other useful syntheses [7,8]. Further, they tend to undergo multiple bromination when treated with elemental bromine under the usual bromination conditions [9]. Therefore selective bromination of these compounds to mono-brominated products is a challenging task. A variety of improved protocols by using expensive transition-metal based catalysts [10], alkali metal halides associated with $NaIO_4$ [11] or the combination of aqueous TBHP or $H_2O_2$ together with a hydrohalic acid [12] have also been reported for this transformation. However, these methods have certain limitations like use of expensive heavy transition metals, toxic/volatile chlorinated organic solvents and formation of polysubstituted and other side products. To encourage the development of sustainable synthetic methodologies, the use of eco-friendly reagents has become a subject of immense interest in present day chemistry [13,14]. In this regard, some ionic liquid tribromides (IL-Br₃) have been reported for the bromination of aromatic substrates and synthesis of bromo-esters from aromatic aldehydes [15]. The key drawbacks of these reagents are the use of expensive organic ammonium cations and difficult synthesis of these reagents. Ma et al. [16] studied the use of phosphonium based ionic reagents both polystyrene supported and ionic liquid based environmentally safe brominating agents for the bromination of unsaturated compounds. Le et al. [17] reported the regioselective monobromination of activated aromatics using 1-butyl-3-methylimidazolium tribromide ([Bmim]Br₃) under solvent free condition. Very recently Zolfigol et al. [18] reported the {[K.18-Crown-6]Br₃}n as a unique nanotube like structure as brominating agents. However, the expensive nature and limited accessibility of the crown ethers make this method of limited utility. This report led our interest towards utilizing the poly(ethylene)glycols as a cost-effective alternatives of expensive crown ethers for developing a new brominating agent. In continuation to our previous work on polyethylene glycols [19-23], we report herein a simple, economically affordable and efficient methodology for the selective mono-bromination of aromatic compounds under solvent-less conditions (Scheme 1).

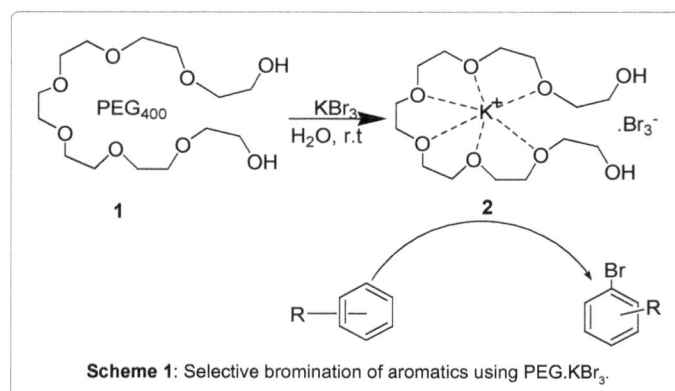

**Scheme 1**: Selective bromination of aromatics using PEG.KBr₃.

---

***Corresponding author:** Suman L. Jain, Chemical Sciences Division, CSIR-Indian Institute of Petroleum, Mohkampur, Dehradun-248005, India
E-mail: suman@iip.res.in

## Experimental

### Preparation of [K⁺PEG.Br₃⁻] (2)

In a round-bottomed flask (100 ml) bromine (68.6 mmol, 10.8 g, 3.5 ml) was added to a 34 % aq. solution of KBr (25 g; prepared by adding the KBr (9 g) in water (135 g) to prepare the solution of potassium tribromide. Polyethylene glycol (PEG$_{400}$; 27.4 g, 68.5 mmol) was added to the above solution and the mixture stirred for 5 h at room temperature. The dark orange-red colored solution was extracted with dichloromethane and the combined organic layer was dried over anhydrous sodium sulfate, concentrated under reduced pressure. The dark orange red viscous liquid was obtained in quantitative yield (44.5 g, 97 %). The loading of the KBr$_3$ in the prepared reagent 2 was found to be 1.2 mmol/g.

### General procedure for the bromination of aromatics

**Method A**: Stirred the mixture of the substrate (1 mmol) and reagent 2 (1 g, 1.2 mmol/g) at 30°C for the time as given in Table 1. At the end of the reaction (as analyzed by TLC), the reaction mixture was extracted with diethyl ether and the residual mixture was treated with bromine to regenerate the reagent 2. The combined organic layer was dried over anhydrous sodium sulfate and concentrated under reduced pressure. Conversion and selectivity of the products was determined by GC-MS and identity of the products was confirmed by comparing their physical constants, IR and NMR spectra with the authentic samples [24,25].

**Method B**: The reaction mixture containing substrate (1 mmol) and catalytic amount of reagent 2 (10 mol%, 0.1 g) was charged with hydrogen peroxide (1.2 mmol) and the resulting mixture was stirred until the reaction was completed (as analyzed by TLC). At the end of the reaction, the product was isolated by extraction with diethyl ether and the resulting residual layer was reused for bromination after adding the fresh substrate and hydrogen peroxide.

### Spectral data of selected compounds

*p*-Bromoaniline: (Table 1, entry 1): Brown solid; Mp (°C) [21] IR (KBr): 3472, 3379, 1615, 1489, 1282 cm⁻¹; ¹H NMR (300 MHz, CDCl₃): 3.6 (b, 2H), 6.6 (d, 2H, J=6.5 Hz), 7.3(d, 2H, J= 6.9). MS (m/z) 173.

*p*-Bromo-*o*-chloroaniline: (Table 1, entry 3): Liquid, IR: 3423, 3328, 3220, 2924, 1622, 1484 cm⁻¹; ¹H NMR (300 MHz, CDCl₃): 5.5 (b, 2H), 7.09 (m, 3H, J=6.8 Hz).

*o*-Bromo-*p*-toluidine: (Table 1, entry 5): Oil, IR: 3464, 3372, 3012, 2912, 1614 cm⁻¹; ¹H NMR (300 MHz, CDCl₃): 2.2 (s, 3H), 3.9 (b, 2H), 6.6 (d, 1H, J=0.8Hz), 6.9 (d, 1H, J=2.4 Hz), 7.2 (s, 1H, J=8.0Hz). MS m/z 187.

5-Bromo-2-aminopyridine (Table 1, entry 8): IR: 3402, 3186, 1650, 1569, 1485 cm⁻¹; ¹H NMR (300 MHz, DMSO): 6.8 (b, 2H, J=1 Hz), 8.2 (s, 2H,); ¹³C NMR (300 MHz, DMSO): 104.2, 155.2, 159.0, 161.7; MS m/z 175.

1-Bromo-2-napthol (Table 1, entry 11): ¹H NMR (300 MHz, CDCl₃): δ 5.7 (b, 1H), 7.4-7.6 (m, 4 H), 7.8-7.9 (m, 2H). MS: m/z 224.

1-Bromo-3-methoxy-2-napthol (Table 1, entry 12): ¹H NMR (300 MHz, CDCl₃): δ 3.4 (s, 3H), 5.6-5.7 (b, 1H), 7.3-7.5 (m, 4 H), 7.6 (s, 1H). ¹³C NMR (300 MHz, CDCl₃): 105.3, 119.0, 122.5, 126.1, 127.0, 133.4, 145.0. MS: m/z 254

| Entry | Substrate | Product | Method | Time /min | Conv[b] / Yield[c] / % |
|---|---|---|---|---|---|
| 1 | NH₂ (aniline) | NH₂, Br (para) | A / B | 30 / 8 | 98/95 / 98/96 |
| 2 | NH₂, Cl (para) | NH₂, Br, Cl | A / B | 25 / 10 | 96/93 / 97/95 |
| 3 | NH₂, Cl (ortho) | NH₂, Cl, Br | A / B | 30 / 10 | 97/94 / 98/96 |
| 4 | NH₂, NO₂ (para) | NH₂, Br, NO₂ | A / B | 35 / 10 | 94/88 / 95/90 |
| 5 | NH₂, Me | NH₂, Br, Me | A / B | 20 / 5 | 95/92 / 97/93 |
| 6 | NH₂, OMe | NH₂, Br, OMe | A / B | 25 / 10 | 99/95 / 99/94 |
| 7 | N(CH₃)₂ | N(CH₃)₂, Br | A / B | 25 / 5 | 100/97 / 100/96 |
| 8 | N, NH₂ (2-aminopyridine) | Br-N, NH₂ | A / B | 20 / 8 | 99/96 / 98/95 |
| 9 | OH (dimethylphenol) | OH, Br | A / B | 35 / 10 | 97/94 / 99/96 |
| 10 | OH (cresol) | OH, Br | A / B | 30 / 10 | 92 / 97 |
| 11 | naphthalene-OH | Br, naphthalene-OH | A / B | 30 / 8 | 94/90 / 92/85 |
| 12 | naphthalene-OH, OMe | Br, naphthalene-OH, OMe | A / B | 45 / 15 | 97/92 / 96/94 |

[a]Reaction conditions: substrate (1 mmol), [K⁺PEG.Br₃⁻] (1 g) at 30°C
[b]Conversion was determined by the GC-MS
[c]Isolated yields

**Table 1:** Bromination of aromatic compounds[a].

## Results and Discussion

For this purpose, we synthesized the potassium tribromide ($KBr_3$) by the addition of equimolar amounts of aq. solution KBr and $Br_2$ as following the literature procedure [26]. The prepared tribromide was stirred for 5 h in $PEG_{400}$ at 30°C, resulted in the formation of dark red-orange viscous oil which was dried under vacuum before use. The loading of the reagent 2 was found to be 1.2 mmol/g as determined by elemental analysis. The X-ray diffractogram of $[K^+PEG.Br_3^-]$ in the liquid phase at room temperature was not very clear and showed a broad low-angle peak between 19.5-22°. Unfortunately, we could not get any information regarding the structure of reagent 2. However, the prepared reagent was dark red in color and its red color was slowly lost as bromination occurred during the reaction. After completion of the bromination reaction, no bromine could be detected in the reaction mixture as analyzed by UV-Vis spectroscopy. Following the developed methodology, a variety of aromatic compounds were reacted with $[K^+PEG.Br_3^-]$ 2 (Method A) at room temperature under solvent free conditions. All the substrates were selectively converted to the corresponding monobrominated products in excellent yields. The results of these experiments are presented in Table 1. In contrast to the elemental bromine, the present method is advantageous as it is easy in handling and provides a highly efficacious approach for achieving the mono-brominated aromatic compounds selectively in higher yields. After completion of the reaction the brominated product was extracted with diethyl ether and the resulting residue containing [PEG.KBr] was treated with molecular bromine to regenerate the $[K^+PEG.Br_3^-]$.

In another protocol (Method B) the bromination of the aromatic compounds was taken place in the presence of aq. hydrogen peroxide by using $[K^+PEG.Br_3^-]$ 2 as catalyst. The results of these experiments are summarized in Table 1. After completion of the reaction, the product was separated by extraction with diethyl ether and the resulting residue could be used as such for the bromination reaction after addition of the fresh hydrogen peroxide and substrate. The presence of hydrogen peroxide enhanced the reaction rates significantly and bromination of the substrates was occurred within very short reaction times. The significant effect of hydrogen peroxide is probably due to the formation of hypobromous acid (HOBr) [27] in the system, which provides the active brominating species ($Br^+$) and allows the instant bromination (equation 1).

$$HBr + H_2O_2 \rightleftharpoons 2HOBr \qquad\qquad\qquad (eq. 1)$$

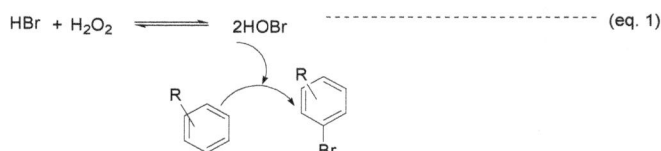

The develop method is advantageous in number of ways for example, the PEG-embedded $KBr_3$ reagent is readily formed by using the cost-effective reagents such as KBr, provides selective synthesis of mono-brominated products, reagent can easily be regenerated and recycled without loss in activity, solvent free conditions and moreover easy work-up of the products. Next, the effect of the various solvents was also studied for the bromination of aniline under described reaction conditions. Among the different solvents studied water was found to be better, whereas, solvent free condition was found to be best, making the method more advantageous in terms of product isolation as well as environmental viewpoints. The selectivity of the reaction was found to highly dependent upon the reaction temperature.

## Conclusion

In conclusion, we have developed a new, simple, cost effective and solvent-less methodology for the selective bromination of the aromatics by using the complex $[K^+PEG.Br_3^-]$ prepared by the $KBr_3$ with acyclic poly(ethylene)glycol-400 PEG, in which the chain of polyether suitably wrapped around the cation in a host-guest manner. This reagent is advantageous in a number of other respects such as it is readily formed, safer and easy in handling. Further, the reactions can be conducted at ambient temperatures and afforded high yields of the desired products.

### Acknowledgment

We kindly acknowledge the Director, IIP for his kind permission to publish these results and SV is thankful to CSIR, New Delhi for the research fellowship.

### References

1. Turro NJ (2005) Molecular structure as a blueprint for supramolecular structure chemistry in confined spaces. Proc Natl Acad Sci USA 102: 10766-10770.

2. Dodziuk H (2002) Introduction to Supramolecular Chemistry. Kluwer Academic Publisher, Dordrecht, The Netherlands.

3. Harris JM, Zalipsky ZS (1997) Poly(ethylene glycol) Chemistry and Biological Applications. American Chemical Society, Washington, DC, USA.

4. Chen J, Spear SK, Huddleston JG, Rogers RD (2005) Polyethylene glycol and solutions of polyethylene glycol as green reaction media. Green Chem 7: 64-82.

5. Xu DQ, Luo SP, Wang YF, Xia AB, Yue HD, et al. (2007) Organocatalysts wrapped around by poly(ethylene glycol)s (PEGs): a unique host-guest system for asymmetric Michael addition reactions. Chem Commun (Camb) 4393-4395.

6. Luo S, Zhang S, Wang Y, Xia A, Zhang G, et al. (2010) Complexes of ionic liquids with poly(ethylene glycol)s. J Org Chem 75: 1888-1891.

7. (1998) Ulmann's Encyclopedia of Industrial Chemistry. Wiley, Weinheim, Germany.

8. Beletskaya IP, Cheprakov AV (2000) The heck reaction as a sharpening stone of palladium catalysis. Chem Rev 100: 3009-3066.

9. Choudary BM, Someshwar T, Reddy CV, Kantam ML, Ratnam JK, et al. (2003) The first example of bromination of aromatic compounds with unprecedented atom economy using molecular bromine. Appl Catal A Gen 251: 397-409.

10. Larock RC (1989) Comprehensive Organic Transformation. VCH, New York, USA.

11. Dewkar GK, Narina SV, Sudalai A (2003) NaIO4-mediated selective oxidative halogenation of alkenes and aromatics using alkali metal halides. Org Lett 5: 4501-4504.

12. Podgorsek A, Stavber S, Zupan M, Iskra J (2009) Environmentally benign electrophilic and radical bromination 'on water': $H_2O_2$–HBr system versus N-bromosuccinimide. Tetrahedron 65: 4429-4439.

13. Ding K, Uozomi Y (2008) Handbook of Asymmetric Heterogeneous Catalysis Wiley. VCH Verlag, Weinheim, Germany.

14. Benaglia M (2009) Recoverable and Recyclable Catalysts. Wiley, VCH, Weinheim, Germany.

15. Bao W, Wang Z (2006) An effective synthesis of bromoesters from aromatic aldehydes using tribromide ionic liquid based on L-prolinol as reagent and reaction medium under mild conditions. Green Chem. 8: 1028-1033.

16. Ma K, Li S, Weiss RG (2008) Stereoselective bromination reactions using tridecylmethylphosphonium tribromide in a "stacked" reactor. Org Lett 10: 4155-4158.

17. Le ZG, Chen ZC, Hu Y, Zheng QG (2005) (Bmim)$Br_3$ as a new reagent for regioselective mono-bromination of activated aromatics under solvent-free conditions. Chin Chem Lett 16: 1007-1009.

18. Zolfigol MA, Chehardoli G, Salehzadeh S, Adams H, Ward MD (2007) $\{[K.18\text{-}Crown\text{-}6]Br_3\}_n$: A unique tribromide type and columnar nanotube-like structure for the oxidative coupling of thiols and bromination of some aromatic compounds. Tetrahedron Lett 48: 7969-7973.

19. Verma S, Jain SL, Sain B (2011) Poly(ethylene glycol) embedded potassium tribromide (PEG.KBr$_3$) as a recyclable catalyst for oxidation of alcohols. Ind Eng Chem Res 50: 5862-5865.

20. Verma S, Jain SL, Sain B (2011) PEG-embedded KBr(3): A recyclable catalyst for multicomponent coupling reaction for the efficient synthesis of functionalized piperidines. Beilstein J Org Chem 7: 1334-1341.

21. Jain SL, Singhal S, Sain B (2007) PEG-assisted solvent and catalyst free synthesis of 3,4-dihydropyrimidinones under mild reaction conditions. Green Chem 7: 740-741.

22. Verma S, Jain SL, Sain B (2010) PEG-embedded thiourea dioxide (PEG.TUD) as a novel organocatalyst for the highly efficient synthesis of 3,4-dihydropyrimidinones Tetrahedron Lett 51: 6897-6900.

23. Khatri PK, Jain SL, Sivakumar K LN, Sain B (2011) Polyethylene glycol clicked Co(II) Schiff base and its catalytic activity for the oxidative dehydrogenation of secondary amines. Org Biomol Chem 9: 3370-3374.

24. Grasselli JG, Ritcheg WM (1975) Atlas of Spectral Data and Physical Constants for Organic Compounds. (2ndedn). Cleveland, CRC Press, Ohio, USA.

25. Buckingham J, Donaghy SM (1982) Dictionary of Organic Compounds. (6thedn), Chapman and Hall, London.

26. DePriest RN (1990) Selective bromination of aromatic compounds using potassium tribromide. US Patent 4940807.

27. Ben-Daniel R, de Visser SP, Shaik S, Neumann R (2003) Electrophilic aromatic chlorination and haloperoxidation of chloride catalyzed by polyfluorinated alcohols: a new manifestation of template catalysis. J Am Chem Soc 125: 12116-12117.

# 6

# Antioxidant Mechanism of Active Ingredients Separated from *Eucalyptus globulus*

**N. M. El-Moein, E. A. Mahmoud and Emad A. Shalaby***

*Biochemistry Department, Faculty of Agriculture, Cairo University, Giza, Egypt, 12613*

## Abstract

The present study aimed to evaluate the antioxidant activity of petroleum ether, methanolic extracts and active ingredients separated from *Eucalyptus globulus* using three different antioxidant assays: 2,2 diphenyl picryl hydrazyl (DPPH), 2,2'- azino-bis [ethylbenzthiazoline-6-sulfonic acid] (ABTS) and β-carotene bleaching assay and identify the mode of action. The results revealed that, crude methanolic extract showed higher antioxidant activity against both DPPH and ABTS radicals than petroleum ether extract. The promising methanol soluble fraction of *Eucalyptus globulus* wood was fractionated on a silica gel column, using hexane, chloroform and ethyl acetate as the mobile phase to give three fractions (C1, C2 and C3), and both the antioxidant activity and chemical composition for raw and fractions were determined. One of the fractions isolated (C2) showed a remarkable antioxidant activity (EC$_{50}$ of 64.4 µg/ml, in comparison with 52.74 µg/ml for crude extracts) against ABTS radical method, and the chemical structures of separated active ingredients were identified using different spectroscopic methods such as 17-pentatricontene (C1), N,N-diphenyllauramide (C2) and O-benzyl-N-tert-butoxycarbonyl-D-serine (C3). Also, the mode of action of the promising fraction was determined.

**Keywords:** *Eucalyptus globulus*; Antioxidant activity; Chemical constituents; Mode of action

## Introduction

*Eucalyptus globulus* is one of the main forest species in Galicia, representing 27% of total wood volume. Eucalyptus wood is used mainly to produce cellulose pulp and, secondly, panels and boards. In both cases, eucalyptus bark is separated as a waste product and used as fuel. Yazaki and Hillis [1] detected ellagitannins, methyl and glycosyl derivatives of ellagic acid and free ellagic and gallic acids in methanolic extracts of the bark of various eucalyptus species. Gallotannins and cathechin were found in the tannins extracts obtained after acid hydrolysis of eucalyptus barks [2]. Methanolic extracts of *E. globulus* bark were characterized by an abundance of total phenols, polymeric proanthocyanidins and ellagitannins. Increasing interest in the replacement of synthetic antioxidants has led to the research into natural sources of antioxidants, especially in plant materials. Flavonoids and other polyphenols possess anti-tumoral, anti-allergic, anti-platelet, anti-ischemic, and anti-inflammatory activities, among others, and most of these effects are believed to be due to the antioxidant capacity [3]. Phenolic compounds in edible and non-edible plants have been reported to have antioxidant capacity. Several types of plant materials, such as vegetables, fruits, seeds, hulls, wood, bark, roots and leaves, spices and herbs, etc. have been examined as potential sources of antioxidant compounds. The antioxidant compounds from natural sources could be used to increase the stability of foods by preventing lipid peroxidation and also to protect oxidative damage in living systems by scavenging oxygen radicals. Natural antioxidants have been also proposed for use in topical pharmaceutical and cosmetic compositions [3]. The yield and antioxidant activity of natural extracts is dependent on the solvent used for extraction. Several procedures have been used [4].

*Eucalyptus* contains many chemical compounds that play several roles in the plant. These include defense against insect and vertebrate herbivores and protection against UV radiation and against cold stress. The best-known compounds are the terpenoids, which form most of the essential oil giving *Eucalyptus* foliage its characteristic smell. However, Eucalyptus is also a rich source of phenolic constituents such as tannins and simpler phenolics. Some of these have formed the basis of industries in the past. For example, tannins were extracted from *Eucalyptus astringens* and rutin from *Eucalyptus astringens* [5].

The synthetic antioxidants include butylated hydroxyanisole and butylated hidroxytoluene (BHA and BHT, respectively), propyl gallate (PG) and tertbutylhydroquinone (TBHQ). Their manufacture costs, the relative poor efficiency of natural tocopherols (also used as antioxidant agents) and the need of increased food additive safety give rise to a crescent demand on other natural and safe antioxidants sources. The search for cheap and widespread feedstock's for this purpose has led to the evaluation of residual materials, including several leaves, seeds and peels, generally considered as wastes [3,6]. The fibrous part of vegetal biomass can yield antioxidants after hydrolytic processing [7].

Extractions of phenolic compounds as antioxidants from eucalyptus (*Eucalyptus globulus*) bark were done by Vázquez et al. [8] who demonstrated the potential of eucalyptus bark as a source of antioxidant compounds.

This investigation was designed to identify the mechanism of active ingredients isolated from *Eucalyptus globulus* for their antioxidant activity, using three different antioxidant methods.

## Materials and Methods

### Materials

**Source of plant samples:** Sample of Eucalyptus (*Eucalyptus*

***Corresponding author:** Emad A. Shalaby, Biochemistry Department, Faculty of Agriculture, Cairo University, Giza, Egypt, E-mail: drshalaby@staff.cu.edu.eg

*Globulus* Family – Myrtaceae) was obtained from Experimental Station of Medicinal Plants, Faculty of Pharmacy, Cairo University, Giza.

**Chemicals:** All chemicals and reagents were obtained from Sigma chemical Co. (London, Lab. Poole), England (Cairo branch).

**Preparation of extracts:** Dried plant materials were pulverized using a mechanical grinder. 500 gr of powdered material was extracted with 1000 ml of petroleum ether (40-60) continuously for 6 hrs using the soxhlet apparatus. Then the residue of plant material extracted with 1000 ml of absolute methanol. Thereafter, the resulting petroleum ether and methanol extracts were reduced in rotary evaporator (40°C; $N_2$ stream), stored at 4°C until further use in the experiment.

## Methods

### Proximate analysis

**Determination of moisture:** The moisture content was determined according to the A.O.A.C. [9].

**Determination of crude proteins:** Total nitrogen (TN) was determined by the method of A.O.A.C. [9] using microkjeldahl method. Crude protein was calculated by multiplying TN by a factor of 6.25.

**Determination of total hydrolysable carbohydrate:** Total carbohydrate was estimated according to the method described by Dubois et al. [10].

**Determination of crude lipids:** Crude lipids were determined as described by A.O.A.C. [9] method using chloroform: methanol (2:1) as a solvent in soxhlet apparatus.

**Separation of fatty acids and unsaponofiable matter:** One gram of plant extracted lipids was saponified with methanolic KOH (30 ml, 1N) containing BHT (1 mg) at 60°C for 1 hr. under reflux. The unsaponifiable matter was extracted with ethyl ether, washed several times with distilled water and dried over anhydrous sodium sulfate and the solvent was evaporated.

**Separation of fatty acids:** The soap solution was acidified with sulfuric acid (5N) and the liberated fatty acids were extracted with ethyl ether, washed several times with distilled water then methylated with diazomethane ethereal solution [11].

**Identification of fatty acids:** Fatty Acids Methyl Esters (FAME) were analyzed by Gas Liquid Chromatography (GLC) according to Farag et al. [11] under the following conditions: column, a Thermo TR - FAME 70% cyanopropyl polysilphenylene - siloxane (30 m x 0.25 mm, film thickness 0.25 μm); detector, flame ionization; carrier gas, nitrogen; initial column temperature was 80°C and increased to 180°C at rate 4 C/min and hold at 140°C for 10 min; temperatures for injection (split ratio 1:100) and detector were 230 and 240°C, respectively. The fatty acids were identified by comparing their retention times with those of standard fatty acid methyl ester (purity 99% by GLC, sigma Co.).

**Identification of unsaponifiable matter:** The unsaponifiable compounds were identified by GLC using an instrument equipped with a Flame Ionization Detector (FID), a 30 m x 0.25 mm i.d. glass column. Thermo TR – 5MS (5% Phenyl Polysil Phenylene Siloxane). Initial column temperature was 70°C and increased to 280°C at 5°C/min and holded at 280°C for 20 min. injector and FID detector temperatures were 235 and 280°C, respectively and N2 was the carrier gas. The unsaponifiable compounds (hydrocarbon and sterols) were identified by comparing their retention times those of standard hydrocarbons

from C14 to C32 and some authentic sterols (Stigmasterol and β-sitosterol).

## Quantitative analysis of secondary metabolites

**Determination of total Glycosides:** Plant tissue was hydrolyzed with 2M HCl (Acid hydrolysis) at 100°C for 30 - 40 min. The cooled solution was extracted twice with ethyl acetate and the combined extracts were taken to dryness and the residue was taken up in a small volume of ethanol for spectrophotometer. The total sugars were determined as glucose with the phenol – sulfuric acid method according to Dubois et al. [10] using Jenway 1640 U.V/ Visible spectrophotometer for investigation.

**Determination of total Terpenes:** Total terpenes were estimated according to the method described by Ebrahimzadeh and Niknam [12].

One gram of dried plant sample was boiled with 15 ml of 40% ethyl alcohol for 4 hrs. A small amount of activated charcoal was added and the extract was filtered through Whatman filter paper No. 41 and the extract was completed to 50 ml (in a measuring flask) with distilled water.

10 ml of total extract was transferred to wide nick test tube, and then placed in an oven at 100°C in order to remove the water and after cooling; freshly prepared vanillin reagent (5 ml, 0.7% in 65% $H_2SO_4$) was added. The test tube was heated at 60 ± 1°C in a water bath for 1 hr, and then cooled in a crushed ice bath. The absorbance was determined spectrophotometrically using Jenway 1640 U.V/ Visible spectrophotometer after 1 min at 473 nm. A blank was prepared using distilled water instead of extract solution. Cholesterol was chosen as the standard

**Determination of total Alkaloids:** Total alkaloids contents were extracted according to Sabri et al. [13]. The alkaloid extract was dissolved in 2ml of chloroform, then 25ml of 0.02 N $H_2SO_4$ was added, The resulting solution was warmed to driven off the chloroform, cooled and titrated back the excess acid against 0.02 N NaOH solution, using methyl red as indicator. Each ml of $H_2SO_4$ (0.02N) was equivalent to 5.78 mg, of alkaloid.

**Determination of total polyphenols:** Total polyphenols were estimated according to the method described by Meda et al. [14]. Briefly, aliquots of 0.1 g sample extracts were dissolved in 1 ml ethanol. This solution (0.1 ml) was mixed with 2.8 ml of deionized water, 2 ml of 2% sodium carbonate and 0.1 ml of 50% folin-ciocalteu reagent. After incubation at room temperature for 30 min, the reaction mixture absorbance was measured at 750 nm against a deionized water blank using Jenway 1640 U.V/ Visible spectrophotometer. Ferulic acid was chosen as the standard.

### Determination of total flavonoids

A known weight (3 g) of dried plant materials was extracted with ethanol (10 ml, 70% v/v). The ethanolic extract was concentrated under vacuum to dryness, and the residue was dissolved in distilled water (10 ml). The aqueous solution was extracted with chloroform to remove the pigment and fatty materials. The defatted aqueous extract was evaporated to dryness and weighted. A known weight of the extract was placed in 10 ml volumetric flask. Then, 5 ml distilled water and 3 ml AlCl$_3$ (1:10, w/v) were added. After 6 min., 2 ml CH$_3$-COOK (1 M) was added and the total volume was made up to 10 ml with distilled water. The solution was well mixed and the absorbance was spectrophotometrically measured against a blank at 415 nm [15].

Quercetin was served as the standard compound for the preparation of calibration curve.

## Antioxidant activity

**DPPH method:** Quantitative measurement of radical scavenging properties was carried out according to Burits and Bucar [16]. The reaction mixture contained sample extracts 50, 100 μg/ml (or 80% MeOH as a blank) and 1 mL of a 0.002% (w/v) solution of DPPH in methanol. Butylated hydroxyl toluene (BHT) and Ascorbic acid were used for comparison or as a positive control. Discoloration was measured at 517 nm after incubation for 1, 15, 30 and 60 min. using Jenway 6305 U.V/ Visible spectrophotometer. Measurements were taken at least in triplicate. DPPH radical's concentration was calculated using the following equation:

DPPH scavenging effect (%) = Ao – A1 / Ao X100

Where, Ao was the absorbance of the control and A1 was the absorbance in the presence of the sample. The actual decrease in absorption induced by the test compounds was compared with the standards.

**β-Carotene-linoleic acid assay:** The antioxidant activity of the all extracts, based on coupled oxidation of β-carotene and Linoleic acid emulsion, was evaluated following the method of Taga et al. [17]. Briefly, 1ml of β-carotene (0.5 mg/ml) dissolved in chloroform was pipetted into a small round bottom flask. After removing the chloroform using a rotary evaporator under reduced pressure at low temperature (less than 30ºC), 10 mg of linoleic acid, 100 mg of Tween 40 and 50 ml of distilled water were added to the flask with vigorous shaking. Aliquots (1ml) of the prepared emulsion were transferred to a series of tubes each containing 50 and 100 μg/ml of extract or BHT and ascorbic acid as synthetic and natural standards. 3 ml of ethanol was finally added to them. A control sample was prepared exactly as before but without adding antioxidants. Each type of sample was prepared in triplicate. The test systems were placed in a shaking water bath at 50ºC for 60 hrs. the absorbance of each sample was read spectrophotometerically at 362 nm.

**ABTS radical method:** This assay was based on the ability of different substances to scavenge 2,2›- azino-bis (ethylbenzthiazoline-6-sulfonic acid (ABTS⁺) radical cation in comparison to a standard (BHT and Vit. C, 50 and 100 μg/ml). The radical cation was prepared by mixing a 7 mM ABTS stock solution with 2.45 mM potassium persulfate (1/1, v/v) and leaving the mixture for 4-16 h until the reaction was complete and the absorbance was stable. The ABTS⁺ solution was diluted with ethanol to an absorbance of 0.700 ± 0.05 at 734 nm for measurements. The photometric assay was conducted on 0.9 ml of ABTS⁺ solution and 0.1 ml of tested samples (at 50 and 100 μg/ml in MeOH solution) and mixed for 45 sec; measurements were taken at 734 nm after 1 min. The antioxidative activity of the tested samples was calculated by determined the decrease in absorbance at different concentrations by using the following equation:

E= ((Ac-At)/ Ac) x 100, where: At and Ac are the respective absorbance of tested samples and ABTS⁺[18].

## Separation of active gradient

25 gr of the *Eucalyptus globulus* crude Methanolic extract was fractionated using Liquid Chromatographic Column (50 x 1.5 cm, i.d packed with silica gel H (100 g). Gradient elution was carried out with petroleum ether, chloroform and their mixture with an increased polarity pattern (100% Petroleum ether to 100% chloroform), then with

chloroform, ethyl acetate and their mixture with an increased polarity pattern (100% chloroform to ethyl acetate 100%). Fractions (200 ml of each) were collected. Each was separated, evaporated under reduced pressure to dryness, redissolved in 5 ml of ethanol and monitored by TLC, using the solvent system Petroleum ether: ethyl acetate (70: 30, v/v and 90: 10, v/v). The TLC chromatogram was visualized under U.V. light at 365 nm and 245 nm before exposure to vaniline/sulfuric acid reagent.

All the sub-fractions were tested for their antioxidant activity by using ABTS and DPPH radical scavenging. The most potent fractions (C1, C2 and C3) were chosen for further identification using the chromatographic and spectroscopic methods as the following

**Gas chromatography–mass spectrometry:** GC–MS of National Research Center was used for identification the active groups in active gradients. Run time 2.02 min, low mass 49.97 m\z and high mass 700.00 m\z.

**FTIR:** JASCO FTIR spectra 460 plus, Japan was used for identification the active groups in active gradients.

**¹H-NMR:** The identification of compounds was confirmed by carried out H-NMR analysis using NMR Joel GIM, EX 270 (400 Hz).

## Statistical analysis

The present data was subjected to analysis of variance and the Least Significant Difference (L.S.D.) test was calculated to allow for a comparison between the average values of the studied factors.

## Results and Discussion

### Chemical composition

**Primary and secondary metabolites:** The chemical composition (primary and secondary metabolites) of eucalyptus bark is given in (Figure 1 and 2). The results indicated that, *Eucalyptus globulus* have high carbohydrate percentage reached to 20% as dry matter and less than lipids and protein content (30.0 and 25.0%) as illustrated in (Figure 1). These results are in agreement with those of Emara and Shalaby [19].

The total contents of alkaloids, terpenoids, glycosides and phenolic compounds of *Eucalyptus globulus* are shown in Figure 2. It could be noticed that the terpenoids content of sample have the maximum percentage (10.2%), With regard to Phenolic compounds content, it reaches to 5.0%. However, the glycoside and flavonoids content have the lowest value (0.2 and 0.05 % respectively). These results are in agreement with those of Mishra et al. [20] who found

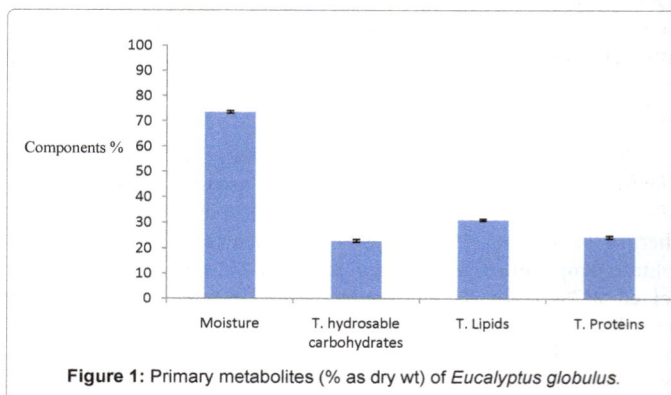

**Figure 1:** Primary metabolites (% as dry wt) of *Eucalyptus globulus*.

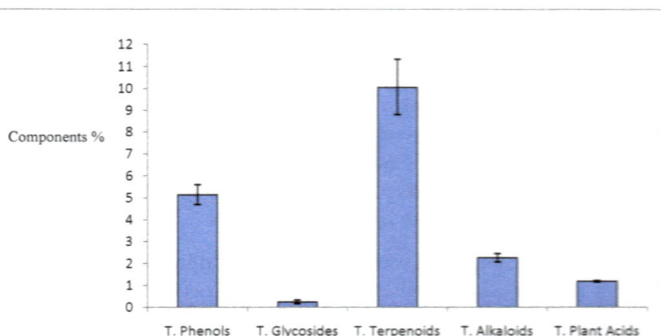

**Figure 2:** Secondary metabolites (% as dry wt) of *Eucalyptus globulus*.

that, Phytochemical screening of the *Eucalyptus globulus* showed the presence of flavonoids, terpenoids, saponins and reducing sugars.

**Fatty acids and hydrocarbons composition:** The relative percentages of total fatty acids are presented in Table 1. The total fatty acids patterns from *Eucalyptus globulus* shows relativity high level of saturated fatty acid compared with the unsaturated components (73.16 and 26.84 % respectively). Among saturated fatty acids Capric, Lauric, and palmetic acids, Lauric acid is the most common (29.6%). However, Palmitoleic acid has the highest amount of unsaturated fatty acids (10.13%).

The relative percentages of total unsaponifiable substances are presented in (Table 1). The levels of hydrocarbons $C_{20}$ and $C_{24}$ occur in large quantities among all the hydrocarbons isolated (11.14 and 34.86 % respectively). In addition, The *Eucalyptus globulus* have the high amount of sterol as hydrocarbon. From these sterols, β-sitosterol and stigmasterol have a percentage of 4.11 and 11.14%, respectively. These results are in agreement with those of Gutiérrez et al. [21], who studied the chemical composition of lipophilic Extractives from *Eucalyptus globulus* Labill. Wood and reported that, the main compounds identified included sterols, sterol esters, fatty acids, steroid ketones, hydrocarbons and triglycerides. Minor compounds such as fatty alcohols, mono- and diglycerides, waxes and tocopherols were also identified among the lipids from *E. globulus* wood.

## Antioxidant activity

Antioxidant efficiency of each extracts (methanol and petroleum ether) was carried out using three antioxidant methods. DPPH method was used as a principal antioxidant for fast test (H-donor method) and the other two methods were used to understand the mechanism of antioxidant activity.

The obtained results using the antioxidant bioassays (by different solvents) showed that, the antioxidant activity is concentration dependant. Tables 2 recorded the activity (using DPPH method) which increased by doubling the extract concentration (50 and 100 μg/ml). These results are in agreement with those of Mishra et al. [20].

The antioxidant activity against DPPH, ABTS and B-carotene bleaching methods of the methanolic extract is 75.6, 81.60 and 60.40% at 50 μg/ml respectively, and is increased to 90.24, 94.8 and 71.6% on doubling the extract concentration to 100 μg/ml respectively. However, petroleum ether extract give an antioxidant activity lower than methanolic extract at 50 and 100 μg/ml as shown in (Table 2) when also compared to BHT as a synthetic stander and ascorbic acid as a natural standard (ranged from 78 to 87%).

The obtained data clearly shows that the ABTS method recorded the highest antioxidant activity (94.8%), at a extract concentration of 100 μg/ml, which exceeds that of the standard BHT (84.6%), ascorbic acid (87%) and petroleum ether extract at the same concentration (22.84%). β-carotene bleaching antioxidant methods recorded much lower activities than those of DPPH and ABTS, as illustrated in (Table 2). These results are in agreement with those of Awika et al. [22] who found that, ABTS is a better choice than DPPH and is more sensitive than DPPH. The ABTS method has the extra flexibility in that it can be used at different pH levels (unlike DPPH, which is sensitive to acidic pH) and thus is useful when studying the effect of pH on antioxidant

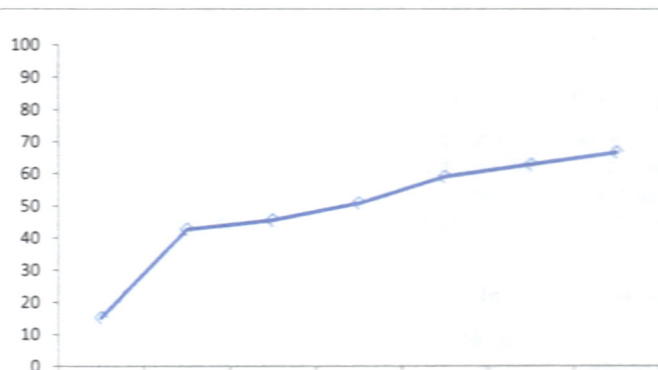

**Figure 3:** Dose response curve of methanolic extract of *Eucalyptus globulus* against DPPH radical.

17 pentatricontene

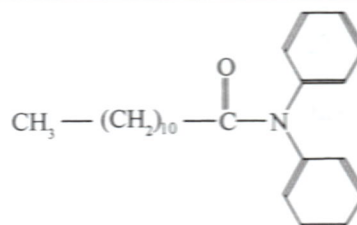

N,N-diphenyllauramide

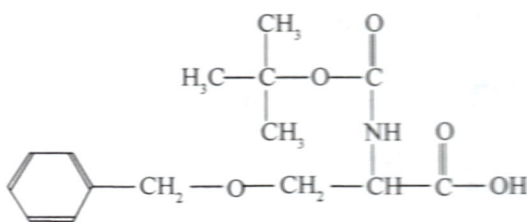

o-benzyl-N-tert butoxycarbonyl D serine

**Figure 4:** Suggested chemical structure of active principles separated from methanolic extract of *Eucalyptus globules*.

activity of various compounds. It is also useful for measuring antioxidant activities of samples extracted in acidic solvents. Additionally, ABTS is soluble in aqueous and organic solvents and is thus useful in assessing antioxidant activity of samples in different media. It is currently most commonly used in simulated serum ionic potential solution (pH 7.4 phosphate buffer containing 150 mM NaCl) (PBS). Another advantage of ABTS method is that samples reacted rapidly with ABTS in the aqueous buffer solution (PBS) reaching a steady state within 30 min. The DPPH reacted very slowly with the samples, approaching, but not reaching, the steady state after 8 hrs. This slow reaction was also observed when ABTS was reacted with samples in alcohol.

The antioxidant activity of methanol and petroleum ether extract correlated with the chemical constituents of *E. globulus* as shown in Figure 2, since phenolic compounds have been reported by Vazquez et al. [8] to bear high antioxidant activity.

The dose response curves of promising sample (Crude methanolic extract) was analyzed and the results are shown in Figure 3. It suggests that there is a positive correlation between the concentration of the sample (5-50 µg/ml) and the antioxidant activity against DPPH radical (12-70 %).

## Active ingredients structure and their antioxidant activity

Chromatographic and spectroscopic analysis of active compounds separated from methanol extract of *Eucalyptus globulus* suggest that, 17-pentatricontene (C1), N,N-diphenyllauramide (C2) and O-benzyl-N-tert-butoxycarbonyl-D-serine (C3) (Figure 4) were present with a molecular weight of 490, 351 and 295, respectively, and with a molecular formula of $C_{35}H_{70}$, $C_{24}H_{33}NO$ and $C_{15}H_{21}NO_5$, respectively. These compounds have high antioxidant activity (44.15, 77.59 and 49.00%, respectively against ABTS radical and 0.0, 22.42 and 12.18%, respectively against DPPH radical), compared to standard synthetic antioxidant BHT which has an activity of 78.85 and 84.60 % towards ABTS radical and DPPH radical, respectively.

This active compound was identified using different spectroscopic analysis methods as a terpenoid (C1) and amides (C2 and C3) (Figure 4) and was shown to exert potent antioxidant activity (Table 3) [23].

Our results also went parallelly with Lim et al. [24] who reported that dichloromethane fraction from methanol extract exhibited the strongest antioxidant activity (in red blood cell hemolysis and lipid peroxidation assays). Further fractionation by column chromatography, TLC, UV and IR showed that the separated four sub-fractions contain phenolic compounds and manifested potent antioxidant activities.

The mass spectrum of separated active ingredient (C1) indicates the presence of the following fragment ions: 490, 349,322, 279, 202, 169, 149, 111, 97 and 71 Dalton.

These results were confirmed by IR and $^1$H-NMR. The IR spectrum of active compound (C1) showed absorption at 2956 (-- CH aliphatic asymmetric stretch). 1630 (C=C), 1461 (Aliphatic $CH_2$ scissor for the methylene group).

| Hydrocarbons | | | Fatty acids | | |
|---|---|---|---|---|---|
| RR$_t$ | No. of carbons | Relative % | RR$_t$ | No. of carbons | Relative % |
| 0.73 | C14 | 8.52 | 0.40 | C8:0 | 3.8 |
| 0.76 | C16 | 0.18 | 0.51 | C10:0 | 20.54 |
| 0.78 | C18 | 0.59 | 0.62 | C12:0 | 29.6 |
| 0.79 | C19 | 3.12 | 1.0 | C16:0 | 19.1 |
| 0.83 | C20 | 11.14 | 1.10 | C16:1 | 10.13 |
| 0.86 | C21 | 1.18 | 1.81 | C18:2 | 6.99 |
| 0.97 | C22 | 2.39 | 2.05 | C18:3α | 8.74 |
| 1.0 | C24 | 34.86 | 2.43 | C18:3γ | 0.98 |
| 1.14 | C25 | 4.30 | % SFA | | 73.16 |
| 1.22 | C26 | 6.28 | % MUFA | | 10.13 |
| 1.27 | C27 | 3.67 | %PUFA | | 16.71 |
| 1.28 | C28 | 6.48 | | | |
| 1.30 | Stigmas sterol | 11.14 | | | |
| 1.31 | B-Sitosterol | 4.11 | | | |
| 1.33 | C32 | 1.99 | | | |

**Table 1:** The hydrocarbons and Fatty acids composition (Relative percentage) of *Eucalyptus globules*.

| Antioxidant methods | Pet. ether | | Methanol | | BHT | | Ascorbic acid | |
|---|---|---|---|---|---|---|---|---|
| | 50 µg/ml | 100 µg/ml | 50 µg/ml | 100 µg/ml | 50 µg/ml | 100 µg/ml | 50 µg/ml | 100 µg/ml |
| DPPH | 8.78 ± 3.42 | 17.7 ± 6.85 | 75.60 ± 9.79 | 90.24 ± 0.18 | 78.85 ± 2.18 | 84.60 ± 3.00 | 80.06 ± 12 | 87.63 ± 4.36 |
| ABTS | 15.89 ± 2.4 | 22.84 ± 1.9 | 81.60 ± 2.64 | 94.80 ± 5.72 | 80.2 ± 4.50 | 88.0 ± 2.61 | 84.61 ± 4.63 | 92.48 ± 4.0 |
| B-carotene bleaching | 14.85 ± 1.84 | 20.54 ± 2.00 | 60.40 ± 0.84 | 71.65 ± 4.73 | 26.7 ± 1.05 | 44.85 ± 4.51 | 45.31 ± 2.04 | 67.30 ± 6.45 |

**Table 2:** Antioxidant (%) activity of Methanol and pet. ether extracts of *Eucalyptus globulus* against three different antioxidant activities.

| Compounds | DPPH method | | ABTS method | |
|---|---|---|---|---|
| | 50 µg/ml | 100 µg/ml | 50 µg/ml | 100 µg/ml |
| C1 | 0.0 | 0.0 | 32.27 ± 3.48 | 44.15 ± 2.08 |
| C2 | 9.2 ± 0.67 | 22.42 ± 1.53 | 66.39 ± 2.06 | 77.59 ± 1.15 |
| C3 | 9.86 ± 0.68 | 12.18 ± 2.79 | 34.95 ± 2.52 | 49.00 ± 2.58 |

**Table 3:** Antioxidant activity (%) of pure compounds separated from methanolic extract of *Eucalyptus globules*.

The $^{1}$H-NMR data indicated that, the compound under study (C1) had the following types of protons; A multiplex signal at δ 3.372 ppm which is characteristic for unsaturated protons and the singlet signal at δ 1.280 ppm is characteristic for methylene group –CH$_2$- protons. Moreover, the singlet signals at δ 1.063-0.889 ppm is characteristic for methyl group–CH$_3$ protons.

The mass spectrum of separated active ingredient (C2) indicates the presence of the following fragment ions: 351, 305,255, 203, 191, 149, 119, 97, 81 and 71 Dalton.

These results were confirmed by IR and $^{1}$H-NMR. The IR spectrum of active compound (C2) showed absorption at 3423 (N-H), 2924 (-- CH aliphatic asymmetric stretch). 1735 (-- C=O absorption), 1600 (Aromatic nucleus), 1460 (Aliphatic CH$_2$ scissor for the methylene group).

The $^{1}$H-NMR data indicated that, the compound under study (C2) had the following types of protons; A multiplex signal at δ 7.260 ppm which is characteristic for aromatic protons, a singlet signal at δ 2.175 ppm which is characteristic for CH-C=O and a singlet signal at δ 1.640 ppm which is characteristic for methylene group –CH$_2$- protons. Moreover, the singlet signals at δ 1.063- 0.889 ppm is characteristic for methyl group–CH$_3$ protons.

The mass spectrum of separated active ingredient (C3) indicates the presence of the following fragment ions: 295, 282,246, 224, 211, 162, 149, 125, 104 and 77 Dalton.

These results were confirmed by IR and $^{1}$H-NMR. The IR spectrum of active compound (C2) showed an absorption at 3436 is characteristic for (O-H and N-H), 2927 for (-- CH aliphatic asymmetric stretch). 1727 for (-- C=O absorption), 1600-1500 for (Aromatic nucleus), 1460 for (Aliphatic CH$_2$ scissor for the methylene group) and (C-O, C-C and C-N in the finger print region).

The $^{1}$H-NMR data indicated that, the compound under study (C3) had the following types of protons: a multiplex signal at δ 7.264 ppm which is characteristic for aromatic protons, a singlet signal at δ 2.172 ppm which is characteristic for CH-C=O, a singlet signal at δ 2.44 ppm which is characteristic for methylene group –CH$_2$- protons beside aromatic group, and a singlet signal at δ 2.78 ppm which is characteristic for N-H protons. Moreover, the singlet signals at δ 1.256 ppm is characteristic for methyl group–CH$_3$ protons.

## Suggested mechanism

The antioxidant activity of the promising active compound especially C2 which was separated from *Eucalyptus globulus* against ABTS radical may be due to one of the following reasons:

First, it may be caused by the resonance phenomena of double bonds and lone pair atoms (N, O) in the chemical structure of the active compound. This structure may lead to radical formation at more than one sit e.g.: benzene ring, and formation of new covalent bond with ABTS radical and non-radical products [25].

Second, the presence of different electro-negative groups in the structure may lead to a less stability of different atoms (e.g.: methylene group) because these groups can attract electrons from methylene group and convert it to a radical or a carbonium ion. So, the activity of the active compound (C2) may be due to the reaction between methylene group radicals or hydrogen proton with ABTS radical [25].

## References

1. Yazaki Y, Hillis WE (1976) Polyphenol of *Eucalyptus globules*. Phytochemistry 15: 1180 -1181.

2. Fechtal PM, Riedl B (1991) Analyse de extraits tannants des ´ecorces des eucalyptus apr`es hydrolyse acide par la chromatographie en phase gaseuse coupl´ee avec la spectrometrie de masse (GC-MS). Holzforschung 45: 269-273.

3. Moure A, Cruz JM, Franco D, Dom-Inguez JM, Sineiro J, et al. (2001) Natural antioxidants from residual sources. Food Chem 72: 145–171.

4. Pokorny J, Korczak J (2001) Preparation of natural antioxidants. Antioxidants in Food: Practical Applications. CRC Press, Boca Raton, USA.

5. Lassak E, McCarthy T (1992) Australian Medicinal Plants. Melbourne: Methuen Australia.

6. Manthey JA, Grohmann K (2001) Phenols in citrus peel byproducts. Concentrations of hydroxycinnamates and polymethoxylated flavones in citrus peel molasses. J Agr Food Chem 49: 3268-3273.

7. González J, Cruz JM, Domínguez H, Parajó JC (2004) Production of antioxidants from *Eucalyptus globulus* wood by solvent extraction of hemicellulose hydrolysates. Food Chem 84: 243-251.

8. Vázquez G, Santos J, Sonia Freire M, Antorrena G, González-Álvarez J (2011) Extraction of antioxidants from eucalyptus (*Eucalyptus globulus*) bark . Wood Science and Technology 46: 443-457.

9. AOAC (2000) Official Methods of Analysis. Association of Official Analytical Chemist. USA.

10. Dubios M, Gilles KA, Hamilton JK, Rebers PA, Smith F (1956) Colorimetric Method for Determination of Sugars and Related Substances. Anal Chem 28: 350-356.

11. Farag RS, Badei AZMA, El-Baroty GSA (1989) Influence of thyme and clove essential oils on cottonseed oil oxidation. J Am Oil Chem Soc 66: 800-804.

12. Ebrahimzadeh, H, Niknam V (1998) A revised spectrophotometric method for determination of triterpenoid saponin. Indian Drugs 35: 379-381.

13. Sabri NN, El-Masry S, Khafagy SM (1973) Phytochemical investigation of *Hyoscyamus desertorum*. Planta Med 23: 4-9.

14. Meda A, Lamien CE, Romito M, Millogo J, Nacoulma OG (2005) Determination of the total phenolic, flavonoid and praline contents in Burkina Fasan honey, as well as their radical scavenging activity. Food Chem 91: 571-577.

15. Zhuang XP, Lu YY, Yang GS (1992) Extraction and determination of flavonoid in grinkgo. Chinese Herbal Medicine 23: 122-124.

16. Burits M, Bucar F (2000) Antioxidant activity of Nigella Sativa essential oil. Phytother Res 14: 323-328.

17. Taga MS, Miller EE, Pratt DE (1984) Chia seeds as a source of natural lipid antioxidants. J Am Oil Chem Soc 61: 928-931.

18. Re R, Pellegrini N, Proteggente A, Pannala A, Yang M, et al. (1999) Antioxidant activity applying an improved ABTS radical cation decolorization assay. Free Radic Biol Med 26: 1231-1237.

19. Emara KhS, Shalaby EA (2011) Seasonal variation of fixed and volatile oil percentage of four Eucalyptus spp. related to lamina anatomy. Afr J Plant Sci 5: 353-359.

20. Mishra AK, Sahu N, Mishra A, Ghosh AK, Jha S, et al. (2010) Phytochemical Screening and Antioxidant Activity of essential oil of Eucalyptus leaf. Pharmacognosy Journal 2: 21-24.

21. Gutiérrez A, del Río JC, González-Vila FG, Martín F (1999) Chemical Composition of Lipophilic Extractives from *Eucalyptus globulus* Labill. Wood. Holzforschung 53: 481-486.

22. Awika JM, Rooney LW, Ronald XW, Prior L, Cisneros-Zevallos L (2003) Screening Methods To Measure Antioxidant Activity of Sorghum (*Sorghum bicolor*) and Sorghum Products. J Agric Food Chem 51: 6657-6662.

23. Vázquez G, Fontenla E, Santos J, Freire MS, González-Álvarez J, et al. (2008) Antioxidant activity and phenolic content of chestnut (*Castanea sativa*) shell and eucalyptus (*Eucalyptus globulus*) bark extracts. Ind Crops Prod 28: 279-285.

24. Lim SN, Cheung PCK, Oai VEC, Ang PO (2002) Evaluation of Antioxidative Activity of Extracts from a Brown Seaweed, *Sargassum siliquastrum*. J Agric Food Chem 50: 3862-3866.

25. Shanab SMM, Shalaby EA, El-Fayoumy EA (2011) Enteromorpha Compressa Exhibits Potent Antioxidant Activity. J Biomed Biotechnol 726405: 1-11.

# 7

# Evaluation of Biofield Treatment on Atomic and Thermal Properties of Ethanol

**Mahendra Kumar Trivedi[1], Alice Branton[1], Dahryn Trivedi[1], Gopal Nayak[1], Omprakash Latiyal[2] and Snehasis Jana[2*]**

[1]*Trivedi Global Inc., 10624 S Eastern Avenue Suite A-969, Henderson, NV 89052, USA*
[2]*Trivedi Science Research Laboratory Pvt. Ltd., Hall-A, Chinar Mega Mall, Chinar Fortune City, Hoshangabad Rd., Bhopal, Madhya Pradesh, India*

### Abstract

Ethanol is a polar organic solvent, and frequently used as a fuel in automobile industries, principally as an additive with gasoline due to its higher octane rating. It is generally produced from biomass such as corn, sugar and some other agriculture products. In the present study, impact of biofield treatment on ethanol was evaluated with respect to its atomic and thermal properties. The ethanol sample was divided into two parts i.e., control and treatment. Control part was remained untreated. Treatment part was subjected to Mr. Trivedi's biofield treatment. Control and treated samples were characterized using Gas chromatography-mass Spectrometry (GC-MS), Differential scanning calorimetry (DSC), and High performance liquid chromatography (HPLC). GC-MS data revealed that isotopic abundance of $^{13}C$ i.e., $\delta^{13}C$ of treated ethanol was significantly changed from -199‰ upto 155‰ as compared to control. The DSC data exhibited that the latent heat of vaporization of treated ethanol was increased by 94.24% as compared to control, while no significant change was found in boiling point. Besides, HPLC data showed that retention time was 2.65 minutes in control, was increased to 2.76 minutes in treated ethanol sample. Thus, overall data suggest that biofield treatment has altered the atomic and thermal properties of ethanol.

**Keywords:** Biofield treatment; Ethanol; Gas Chromatography-Mass Spectrometry; Differential scanning calorimetry; High performance liquid chromatography

## Introduction

Ethanol or ethyl alcohol ($C_2H_5OH$) is a clear, volatile, colourless, and polar organic solvent. It is a source of energy being utilized with petrol/gasoline for vehicle fuel. Recently, the conventional fuel prices are increasing continuously, and further due to their limited natural resource; there is a huge demand to utilize the renewable biofuels produced from agriculture products. It is expected that, around 20% of petroleum will be replaced by another fuel in next 10 years [1]. Hence, the ethanol being a renewable source of energy, will play an important role in future. The ethanol fuel has very high octane rating than petrol, diesel, and gasoline. The high octane rating indicates the high fuel efficiency by mean of less premature combustion and prevents spark ignition. Ethanol contains 35% *w/w* oxygen, which assists gasoline to burn completely in ethanol-gasoline blend fuel. This complete combustion of ethanol-gasoline fuel reduces the gummy deposits in engines. Despite of all these positive benefits, the ethanol has less energy content than gasoline and petrol. From several decades ethanol has been produced as a by-product from sugar industries and is being used in beverage industry. Recently, ethanol has been synthesized from corn starch, wheat and other plant products. Zyakum et al. reported that carbon isotopic ratio $^{13}C/^{12}C$ of ethanol produced by fragmentation technique depends upon the substrate used, for instance, the ethanol produced using wheat as substrate was found lesser than cereals [2,3]. Besides, isotopic fractionation in a product can be changed through chemical reactions or a physicochemical process. The change in isotopic ratio $^{13}C/^{12}C$, affects the compound in two ways i.e., thermodynamic and kinetic isotopic. In kinetic isotope effect (KIE), different isotopes of same element are present in any compound, which have different bond length and bond strength. Furthermore, in a chemical reaction breaking of bonds plays a significant role, which ultimately determines the rate of a reaction. The thermodynamic isotopic effect concerns with the physicochemical properties such as heat of vaporization, boiling point, and vapor pressure. After considering of ethanol properties and fuel applications, authors wanted to investigate an approach that could be beneficial to modify the atomic and thermal properties of ethanol.

According to William Tiller, a physicist, proposed the existence of a new force related to human body, in addition to four well known fundamental forces of physics: gravitational force, strong force, weak force, and electromagnetic force [4]. A biophysicist Fritz-Albert Popp proposed that human physiology shows a high degree of order and stability due to their coherent dynamic states [5-8]. This emits the electromagnetic (EM) waves in form of bio-photons, which surrounds the human body and it is commonly known as biofield. Furthermore, a human has ability to harness the energy from environment/universe and it can transmit into any object (living or non-living) on the Globe. The object always receives the energy and responded into useful way, which known as biofield energy. This process is called biofield treatment. Mr. Trivedi's biofield treatment (The Trivedi Effect®) is well known to transform the characteristics in various fields such as material science [9-11], agriculture [12-14], microbiology [15-17], and biotechnology [18,19]. Biofield treatment has shown significant alteration in metals [20-22] and ceramics [23,24] with respect to their atomic and structural properties. Based on the excellent result obtained by biofield treatment in material science, the present study was undertaken to evaluate the impact of biofield treatment on atomic and thermal properties of ethanol.

## Experimental

The ethanol sample used in this experiment was procured from Sigma Aldrich, China. The ethanol sample was distributed into two

*Corresponding author: Snehasis Jana, Trivedi Science Research Laboratory Pvt. Ltd., Hall-A, Chinar Mega Mall, Chinar Fortune City, Hoshangabad Rd., Bhopal- 462026, Madhya Pradesh, India E-mail: publication@trivedisrl.com

equal parts, where one part was referred as control sample and another part was subjected to Mr. Trivedi's biofield treatment, which was considered as treated sample. The control and treated samples were characterized using Gas chromatography-mass spectrometry (GC-MS), Differential scanning calorimetry (DSC), and High performance liquid chromatography (HPLC) techniques.

## Gas Chromatography-Mass Spectrometry

The gas chromatography-mass spectroscopy (GC-MS) analysis was performed on Perkin Elmer/auto system XL with Turbo mass, USA, having detection limit up to 1 picogram. For GC-MS analysis the treated sample was further divided into three parts as T1, T2 and T3. The GC-MS data was obtained in the form of % abundance vs. mass to charge ratio (m/z), which is known as mass spectrum. The isotopic abundance of $^{13}$C was expressed by its deviation in treated ethanol sample as compared to control. Isotopic abundance of $^{13}$C i.e., $\delta^{13}$C was computed on thousand scale using equation as following:

$$\delta^{13}C\left(\text{‰}\right) = \frac{R_{Treated} - R_{control}}{R_{control}} \times 1000 \qquad (1)$$

Where, $R_{Treated}$ and $R_{Control}$ are the ratio of intensity at m/z=47 to m/z=46 in mass spectra of treated and control samples respectively.

## Differential scanning calorimetry

For thermal analysis, Differential Scanning Calorimeter (DSC) of Perkin Elmer/Pyris-1, USA, with a heating rate of 10°C/min and nitrogen flow of 5 mL/min was used. The boiling point and latent heat of vaporization of control and treated ethanol was recorded from their respective DSC curves. The percent change in boiling point and latent heat of vaporization was computed using following equations: Percent change in melting point was calculated using following equations:

$$\% \text{ change in boiling point} = \frac{\left[T_{Treated} - T_{Control}\right]}{T_{Control}} \times 100$$

Where, $T_{Control}$ and $T_{Treated}$ are the boiling point of control and treated samples, respectively.

Percent change in latent heat of vaporization was calculated using following equations:

$$\% \text{ change in Latent heat of vaporization} = \frac{\left[\Delta H_{Treated} - \Delta H_{Control}\right]}{\Delta H_{Control}} \times 100$$

Where, $\Delta H_{Control}$ and $\Delta H_{Treated}$ are the latent heat of vaporization of control and treated samples, respectively.

## High performance liquid chromatography

The HPLC analysis was performed on a Knauer High Performance Liquid Chromatograph (Berlin, Germany), which consists of a solvent delivery system Smartline Pump 1000 equipped with a UV 2600 detector. Chromatographic separation was achieved on a $C_{18}$ column (Eurospher 100) with a dimension of 250 × 4 mm and 5 µm particle size. The mobile phase used was methanol with a flow rate of 1 mL/min.

## Results and Discussion

## Gas Chromatography-Mass Spectrometry

The mass spectrum of control and treated samples (T1, T2 and T3) of ethanol are shown in Figure 1a-1c. In mass spectra of control and treated ethanol samples, different intensities were observed. Mass spectra showed that base peak at m/z=45 in control sample (Figure 1a),

whereas in treated samples the base peaks were found at m/z=31 in T1, T2, and T3 (Figure 1b-1c). Furthermore, the intense peaks with different mass to charge ratio (m/z), of possible molecular ions are illustrated in Table 1. It indicates that peak at m/z=31 and m/z=45 were due to $^{12}CH_2OH^+$ and $^{12}C_2H_4OH^+$, respectively in control and treated samples. Peaks at m/z=46 correspond to ethanol ion with carbon -12 ($^{12}CH_3$-$^{12}CH_2{}^{16}O^1H^+$), whereas the peak at m/z=47 corresponds to ethanol ion with carbon-13 i.e., $^{13}CH_3$-$^{12}CH_2OH^+$ or $^{12}CH_3$-$^{13}CH_2OH^+$. Computed result of isotopic ratio, $^{13}C/^{12}C$ from GC-MS spectra are presented in Table 2. Further, the isotopic abundance $\delta^{13}C$ computed using equation (1) is illustrated in Figure 2. It showed that the $\delta^{13}C$ was significantly increased by 5‰ and 155‰ (per 1000) in sample T1 and T3, respectively, while it was reduced by 199‰ in T2 as compared to control. This suggests that the $^{12}C$ atoms of T1 and T3, probably transformed into $^{13}C$ by capturing one neutron thereby increased $\delta^{13}C$. Whereas in T2, the $^{13}C$ atoms probably transformed into $^{12}C$ by emitting one neutron and reduced $\delta^{13}C$. This inter-conversion of $^{13}C$ and $^{12}C$ can be possible if a nuclear level reaction including the neutron and proton occurred after biofield treatment. Thus, it is assumed that biofield treatment possibly induced the nuclear level reactions, which may lead to alter the isotopic ratio $^{13}C/^{12}C$ in treated ethanol. Besides, it is reported that when lighter isotope ($^{12}C$) is substituted with heavier ($^{13}C$) in any compound or *vice versa*, it doesn't affect the nuclear charge and electronic structure, since $^{12}C$ and $^{13}C$ atoms are differentiated by one neutron, which is a neutral particle [25]. Thus the vibrational

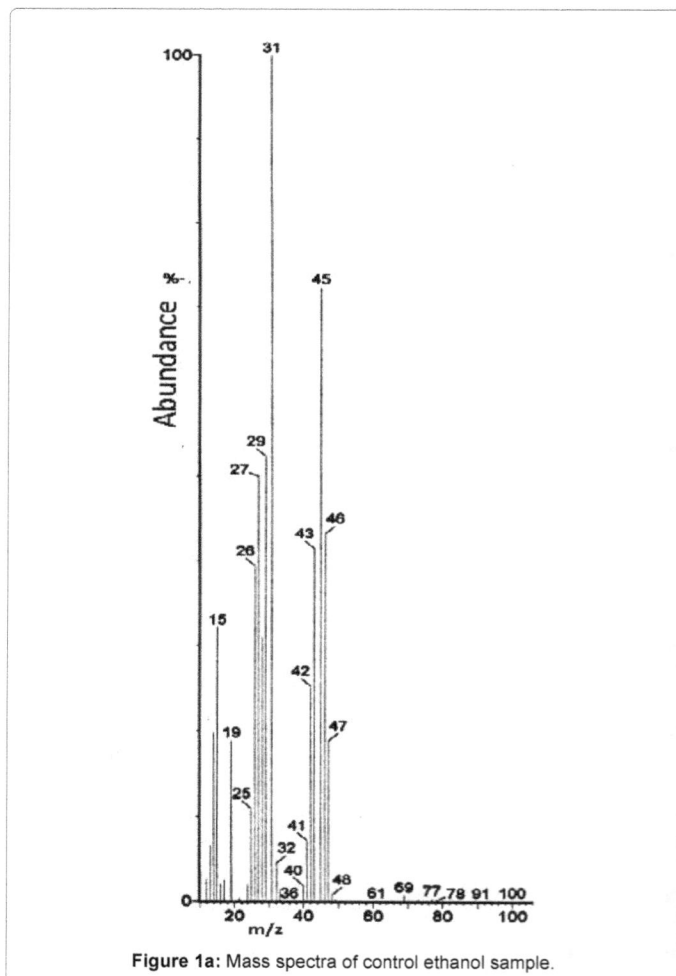

**Figure 1a:** Mass spectra of control ethanol sample.

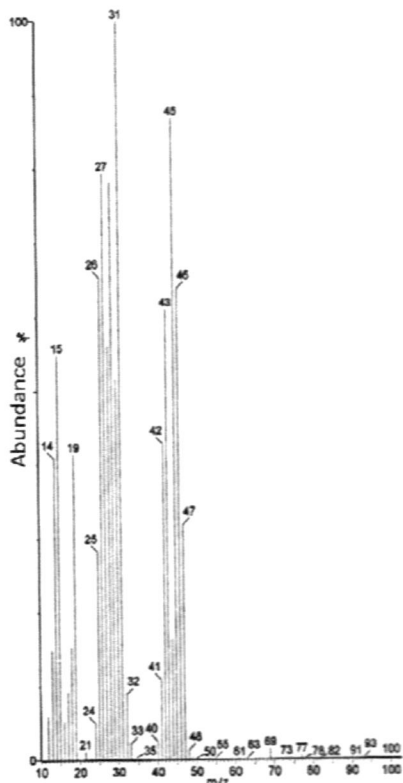

**Figure 1b:** Mass spectra of treated ethanol sample T1.

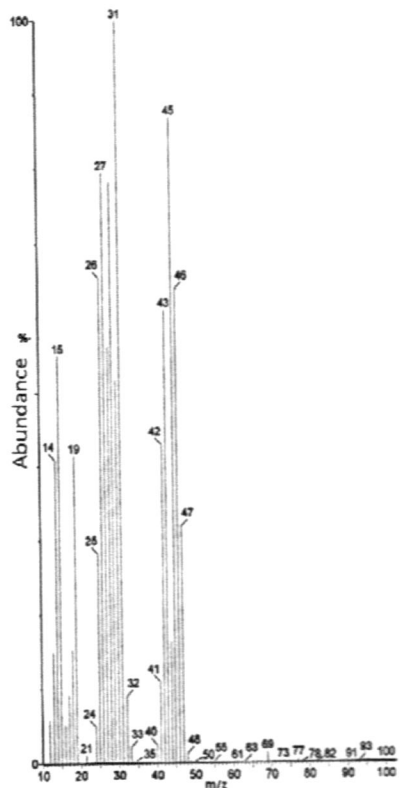

**Figure 1c:** Mass spectra of treated ethanol sample T2.

| Ratio m/z | Possible detected molecules |
|---|---|
| 29 | $^{12}C_2H_5^+$ |
| 31 | $^{12}CH_2OH^+$ |
| 45 | $^{12}C_2H_4OH^+$ |
| 46 | $^{12}CH_2-^{12}CH_2{}^{16}O^1H^+$ |
| 47 | $^{13}CH_3-^{12}CH_2OH^+$ or $^{12}CH_3-^{13}CH_2OH^+$ |
| 48 | $^{13}CH_3-^{13}CH_2OH^+$ or $^{13}CH_3-^{12}CH_2-^{17}OH^+$ |

**Table 1:** Identification of peaks in MS spectra of ethanol.

| Parameters | Control | T1 | T2 | T3 |
|---|---|---|---|---|
| Peak Intensity at m/z = 46 | 60.47 | 43.16 | 62.16 | 63.78 |
| Peak Intensity at m/z = 47 | 26.13 | 18.75 | 21.51 | 31.85 |
| Ratio of peak intensity at m/z = 47 to m/z = 46 | 0.4321 | 0.4344 | 0.3460 | 0.4993 |

**Table 2:** GC-MS isotopic abundance analysis result of ethanol.

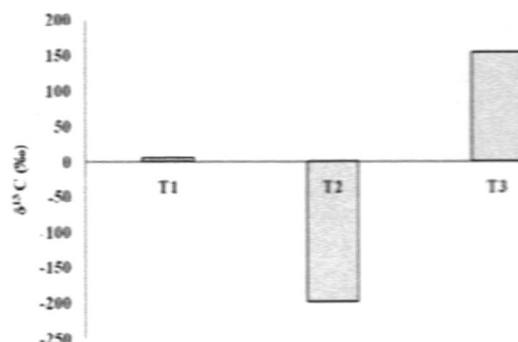

**Figure 2:** Effect of biofield treatment on isotopic abundance of $^{13}C$ ($\delta^{13}C$) of ethanol.

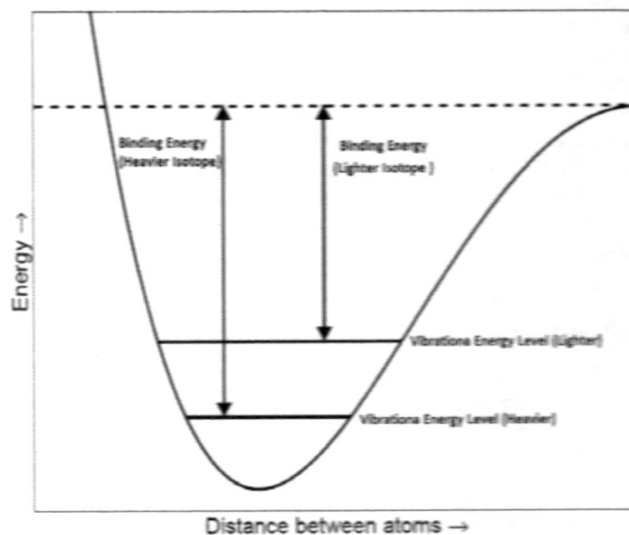

**Figure 3:** General vibrational potential energy curve.

potential energy curve remains same (Figure 3). Furthermore, the vibration energy level of a bond in a molecule can be represented by following equation [26].

$$E = \frac{(n+1)h\left(\dfrac{k}{\mu}\right)^{\frac{1}{2}}}{2\pi} \tag{2}$$

Where, n is quantum number (=0, 1, 2 …), h is Plank's constant, k is bond constant and $\mu$ is the reduced mass=$M_A \times M_B / (M_A + M_B)$

Where $M_A$ and $M_B$ are the atomic mass of atoms A and B respectively, which forms the bond A-B.

According to the equation (2) vibration energy level is inversely proportional to reduced mass. Furthermore, it is well known that the lower is the vibration energy level, higher the stability of molecules. Typically, in ethanol, many type of bonds are present such as $^{12}C$-$^{12}C$, $^{13}C$-$^{12}C$, $^{1}H$-$^{12}C$, $^{1}H$-$^{13}C$, $^{12}C$-$^{16}O$, $^{13}C$-$^{16}O$, and $^{16}O$-$^{1}H$ and reduced mass of these bond are illustrated in Table 3. It showed that reduced mass was higher in case of heavier isotope as compared to lighter. This indicates that ethanol with heavier isotope have low vibration energy level and more stability as compared to lighter ones. Furthermore, the higher stability of ethanol with heavier isotopes may lead to increase the enthalpy of reaction for combustion with oxygen. (Equation 3) This enthalpy of reaction is also known as heat of combustion of ethanol fuel.

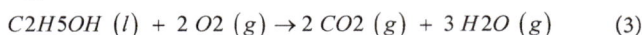

$$C2H5OH\ (l) + 2\ O2\ (g) \rightarrow 2\ CO2\ (g) + 3\ H2O\ (g) \qquad (3)$$

In case of treated ethanol, the amount of heavier isotopic $^{13}C$ atoms was significantly changed after biofield treatment. Thus, it is assumed that ethanol samples with higher isotopic abundance $\delta^{13}C$ (T1 and T3) might have higher stability and binding energy as compared to control [26]. In addition, the higher binding energy of treated ethanol may lead to increase the heat of combustion. On contrary, reverse might happen in treated T2. Thus, GC-MS data suggest that biofield treatment has significantly altered the isotopic abundance of $^{13}C$ in treated ethanol samples.

### Differential scanning calorimetry (DSC)

DSC was used for thermal analysis of control and treated ethanol samples. DSC curve of control and treated ethanol (T1) are shown in Figure 4a and 4b, respectively. Analysis result of boiling point and latent heat of vaporization ($\Delta H$) are presented in Table 4. Data showed that boiling temperature of ethanol was slightly reduced from 77.47°C (control) to 76.95°C after biofield treatment. Whereas, the latent heat of vaporization ($\Delta H$) was increased from 253 J/g (control) to 499.2 J/g in treated ethanol sample. It indicated that the $\Delta H$ of treated ethanol was significantly increased by 94.24% as compared to control. The increase in $\Delta H$ after biofield treatment could be due to higher intermolecular interaction in ethanol molecules in treated sample as compared to control. Furthermore, the increase in intermolecular interaction may enhance the thermal stability of ethanol after biofield treatment.

### High performance liquid chromatography (HPLC)

HPLC chromatogram of control and treated ethanol (T1) is shown in Figure 5a and 5b, respectively. The control ethanol showed a retention time ($R_t$) at 2.65 min, however after biofield treatment, it was shifted to 2.76 min. This increased (shift) of retention time can be attributed to reduced polarity of ethanol after treatment. Further, the decreased polarity after biofield treatment may be due to more interaction of treated ethanol with non-polar silica phase in C18 column that resulted into higher retention time. It is presumed that biofield treatment probably acting at atomic level, which reduced the electronegativity of oxygen atom present in ethanol and that might be responsible to decrease polarity. Due to this, hygroscopicity of treated ethanol might be reduced after biofield treatment. Thus, the reduced hygroscopic nature of treated ethanol might prevent the dilution of ethanol from moisture, which could maintain the purity of ethanol fuel with high energy content.

## Conclusion

In summary, the biofield treatment has significantly changed the isotopic abundance of $^{13}C$ and latent heat of vaporization in ethanol. The GC-MS data showed that biofield treatment has significantly changed the isotopic abundance of $^{13}C$ i.e., $\delta^{13}C$ from -199‰ upto 155‰ in treated ethanol as compared to control. It could be due to nuclear level transformation of $^{13}C$ and $^{12}C$, which probably induced through biofield treatment. Moreover, the higher $\delta^{13}C$ in treated ethanol may increase the stability of bonds, binding energy and heat of combustion. Besides, DSC data suggest that latent heat of vaporization of treated ethanol sample was increased by 94.24% as compared to

| Isotopes Bonds | Isotope type | Reduced mass $(M_A M_B /(M_A + M_B))$ |
|---|---|---|
| $^{12}C$-$^{12}C$ | Lighter | 6.00 |
| $^{13}C$-$^{12}C$ | Heavier | 6.24 |
| $^{1}H$-$^{12}C$ | Lighter | 0.923 |
| $^{1}H$-$^{13}C$ | Heavier | 0.928 |
| $^{12}C$-$^{16}O$ | Lighter | 6.85 |
| $^{13}C$-$^{16}O$ | Heavier | 7.17 |
| $^{16}O$-$^{1}H$ | Lighter | 0.94 |
| $^{16}O$-$^{2}H$ | Heavier | 1.77 |

**Table 3**: Possible isotopic bonds in ethanol.

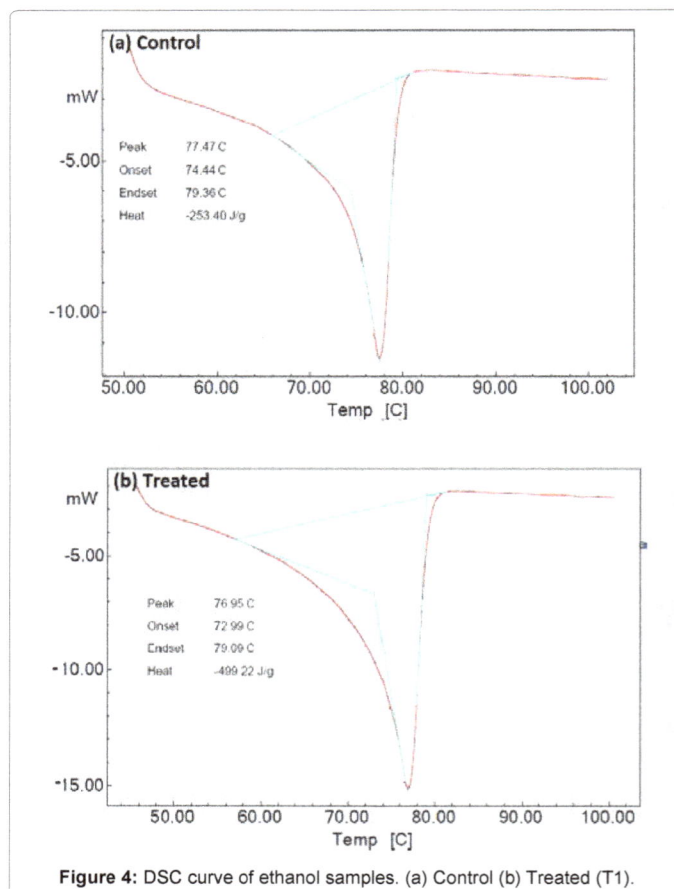

**Figure 4:** DSC curve of ethanol samples. (a) Control (b) Treated (T1).

| Parameters | Control | Treated (T1) | Percent Change |
|---|---|---|---|
| Boiling point (°C) | 77.47 | 76.95 | -0.67 |
| Latent heat of vaporization (J/g) | 253.4 | 492.22 | 94.24 |

**Table 4:** DSC analysis data of control and treated ethanol.

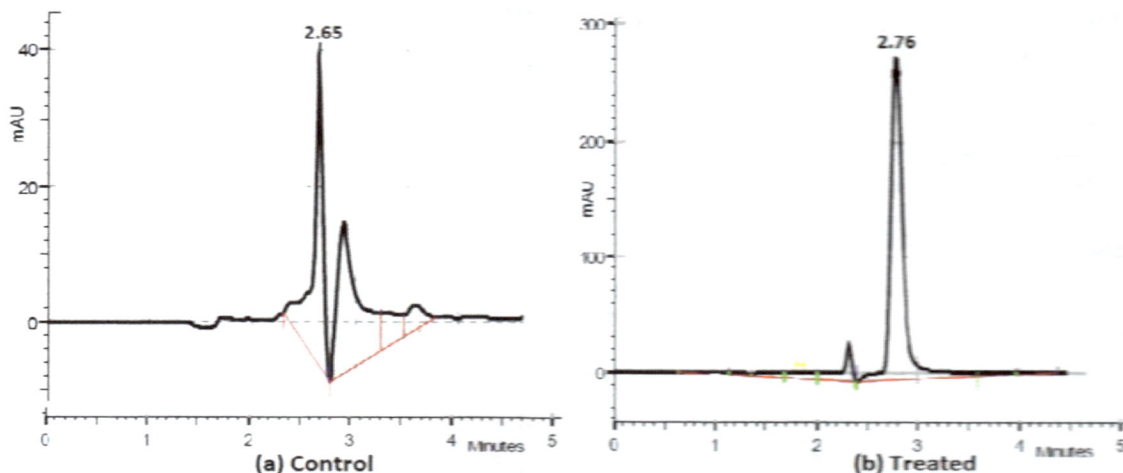

**Figure 5:** HPLC spectra of ethanol samples. (a) Control (b) Treated (T1).

control, which may be due to improved thermal stability of ethanol after biofield treatment. Nevertheless, the shift in retention time toward higher side in HPLC spectra of treated as compared to control revealed that the polarity of ethanol possibly reduced after biofield treatment, which may diminished the hygroscopic nature of ethanol. Therefore, the biofield treated ethanol with high energy content and lower hygroscopic nature could be utilized as a fuel in automobiles.

### Acknowledgements

We thank to all the staff of concern Laboratories, who supported us in conducting experiments. Authors also would like to thank Trivedi Science, Trivedi master wellness and Trivedi testimonials for their support during the work.

### References

1. Wen Z, Ignosh J, Arogo J (2009) Fuel ethanol. Virginia Tech 2009: 442-884.

2. Zyakun AM, Zakharchenko VN, Kudryavtseva AI, Peshenko VP, Mashkina LP et al. (2000) The use of 13C/12C isotope abundance ratio for characterization of the origin of ethyl alcohol. Appl Biochem Microbiol 36: 11-14.

3. Kirichenko EB, Zyakun AM, Bondar VA, Kirichenko AB, Bezruchko VV, et al. (1980) Ratio of stable carbon isotopes (13C/12C) in vegetative and generative organs of plants of the grass family. Doklady, botanical sciences. 250: 505-508.

4. Tiller WA (1997) Science and human transformation: subtle energies, intentionality and consciousness (1st edn). Pavior Publishing, Walnut Creek, California.

5. Popp FA, Chang JJ, Herzog A, Yan Z, Yan Y (2002) Evidence of non-classical (squeezed) light in biological systems. Phys Lett 293: 98-102.

6. Popp FA, Gu Q, Li KH (1994) Biophoton emission: Experimental background and theoretical approaches. Mod Phys Lett B 8: 21-22.

7. Popp FA (1992). Recent advances in biophoton research and its applications. World Scientific Publishing Co Pte Ltd.

8. Cohen S, Popp FA (2003) Biophoton emission of human body. See comment in PubMed Commons below Indian J Exp Biol 41: 440-445.

9. Trivedi MK, Tallapragada RM (2008) A transcendental to changing metal powder characteristics. Met Powder Rep 63: 22-28.

10. Trivedi MK, Tallapragada RM (2009) Effect of super consciousness external energy on atomic, crystalline and powder characteristics of carbon allotrope powders. Mater Res Innov 13: 473-480.

11. Dhabade VV, Tallapragada RM, Trivedi MK (2009) Effect of external energy on atomic, crystalline and powder characteristics of antimony and bismuth powders. Bull Mater Sci 32: 471-479.

12. Shinde V, Sances F, Patil S, Spence A (2012) Impact of biofield treatment on growth and yield of lettuce and tomato. Aust J Basic Appl Sci 6: 100-105.

13. Lenssen AW (2013) Biofield and fungicide seed treatment influences on soybean productivity, seed quality and weed community. Agricultural Journal 8: 138-143.

14. Sances F, Flora E, Patil S, Spence A, Shinde V, et al. (2013) Impact of biofield treatment on ginseng and organic blueberry yield. Agrivita J Agric Sci 35.

15. Trivedi MK, Patil S, Bhardwaj Y (2008) Impact of an external energy on Staphylococcus epidermis [ATCC-13518] in relation to antibiotic susceptibility and biochemical reactions - An experimental study. J Accord Integr Med 4: 230-235.

16. Trivedi MK, Patil S (2008) Impact of an external energy on Yersinia enterocolitica [ATCC -23715] in relation to antibiotic susceptibility and biochemical reactions: An experimental study. Internet J Alternat Med 6.

17. Trivedi MK, Patil S, Bhardwaj Y (2009) Impact of an external energy on Enterococcus faecalis [ATCC-51299] in relation to antibiotic susceptibility and biochemical reactions - An experimental study. J Accord Integr Med 5: 119-130.

18. Patil S, Nayak GB, Barve SS, Tembe RP, Khan RR, et al. (2012) Impact of biofield treatment on growth and anatomical characteristics of Pogostemon cablin (Benth.). Biotechnology 11: 154-162.

19. Altekar N, Nayak G (2015) Effect of biofield treatment on plant growth and adaptation. J Environ Health Sci 1: 1-9.

20. Trivedi MK, Patil S, Tallapragada RM (2012) Thought Intervention through bio field changing metal powder characteristics experiments on powder characteristics at a PM plant. Future Control and Automation LNEE173: 247-252.

21. Trivedi MK, Patil S, Tallapragada RM (2015) Effect of biofield treatment on the physical and thermal characteristics of aluminium powders. Ind Eng Manage 4: 151.

22. Trivedi MK, Patil S, Tallapragada RM (2013) Effect of biofield treatment on the physical and thermal characteristics of silicon, tin and lead powders. J Material Sci Eng 2: 125.

23. Trivedi MK, Patil S, Tallapragada RM (2013) Effect of biofield treatment on the physical and thermal characteristics of vanadium pentoxide powder. J Material Sci Eng S11: 001.

24. Trivedi MK, Patil S, Tallapragada RM (2014) Atomic, crystalline and powder characteristics of treated zirconia and silica powders. J Material Sci Eng 3: 144.

25. Wilson EB, Decius JC, Cross PC (1995) Molecular Vibrations. New York: Dover.

26. Califano S (1976). Vibrational States. New York: Wiley.

# 8

# Chemoselective Reduction of Nitroarenes to Aromatic Amines with Commercial Metallic Iron Powder in Water Under Mild Reaction Conditions

**Rajendra D Patil\* and Yoel Sassona**

*Casali Institute of Applied Chemistry, The Institute of Chemistry, The Hebrew University of Jerusalem, Jerusalem, Israel*

## Abstract

Novel, chemoselective, non-hazardous and mild reaction protocol was developed for an efficient reduction of nitroarenes to aromatic amines using "*iron activated water*". Water function as a terminal hydrogen source without any external catalyst, acid, salts or base and in the absence of a solvent. In the course of the reaction, zero valent iron was oxidized to magnetite.

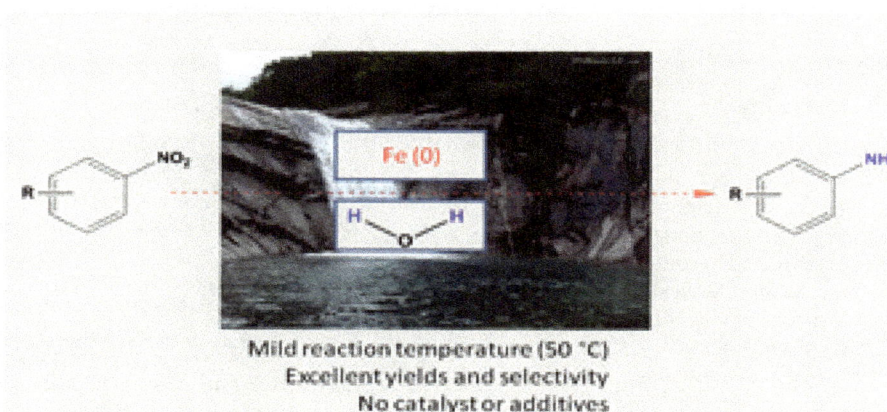

Mild reaction temperature (50 °C)
Excellent yields and selectivity
No catalyst or additives

**Keywords:** Water-iron; Hydrogenation; Reduction; Nitroarenes; Aromatic amines

## Introduction

The reduction of nitroarenes is an important process as the products, aromatic amines carrying chloro-, carbonyl, cyano, etc. groups are important intermediates in the synthesis of chemicals such as antioxidants, dyes, pigments, photographic, pharmaceutical and agricultural materials [1-3]. Several reviews and book chapters have covered the continuous progress in this field of reduction of nitroarenes to aromatic amines [4-7]. Aromatic amines can be produced from the corresponding nitroarenes by catalytic hydrogenation [8-15]. Catalytic hydrogenation is a clean and convenient method, but when other reducible groups are present in the molecule, it is difficult to reduce the nitro group selectively in a catalytic hydrogenation. Alternative to catalytic hydrogenation; the catalytic transfer hydrogenation (CTH) has been also used for nitro reductions in which alcohols, hydrocarbons, hydrazines, organic acids and their salts etc. are employed as hydrogen source with a wide range of metal-based catalysts [16-23]. The nitroarenes reduction using reducing metals such as zinc, tin and iron has been reported in presence of an acid [24,25] or salts [26,27]. However, notable disadvantages to these methods include high reaction temperatures, incompatibility of acid-sensitive functional groups, lack the chemoselectivity over other functional groups and reduction of aromatic nitro compounds often yield a mixture of products [28]. The notable applications of *in-situ* generated carbonic acids (from $CO_2$ and water) in conjunction with Fe/Zn as reducing metals have been recently demonstrated in the reduction of nitroarenes [29,30]. However necessary requirement of $CO_2$ pressure to activate iron through *in-situ*

generated carbonic acid (using water) and high reaction temperature (120°C) (to achieve high yields) were drawbacks of these methods [29,30]. An efficient catalytic reduction of water for generation of hydrogen is one of the most challenging transformations in chemistry [31]. Recently our research group has demonstrated that water activated through reducing metals (Zn, Mg) act as green source of hydrogen including for hydrogen transfer reactions [32,33]. Poliakoff and Boix [34] have tried the reduction of nitroarenes using a metallic reducing reagent directly in pure water at 250°C. The yield of aniline was only 10% using iron powder under the given reaction conditions. Wang *et al.* [35] applied nano-sized activated metallic iron powder for the reduction of nitroarenes directly in water, and good reaction yield can be achieved at 210°C. However, nano-sized metallic iron powder is expensive and also high reaction temperature is a limiting factor [35]. Ranu *et al.* [36] developed elegant method for reduction of nitroarenes

**\*Corresponding author:** Rajendra D Patil, Casali Institute of Applied Chemistry, The Institute of Chemistry, The Hebrew University of Jerusalem, Jerusalem 91904, Israel, E-mail: r_dpatil123@yahoo. co.in/ysasson@huji.ac.il

to aromatic amines using pre-synthesized iron nanoparticle. However, such pre-requisite synthesis of iron nanoparticle (using iron sulfate as iron precursor, sodium borohydride as reducing agent and citric acid as stabilizing agent) imposes additional chemicals and their cost, post processing problems and chemical wastes in the process. In above cases [34-36], an efficient reduction of nitroarenes using commercial metallic iron was in-effective. Moreover the search for new, mild, and selective reduction methods for nitro compounds to amines is still an active area of research. Herein, we are pleased to disclose a novel and mild reaction protocol for reduction of nitroarenes using cheap, non-hazardous, abundant, and eco-friendly "*water-iron*" pair as hydrogen donor (Scheme 1) without any external catalysts or additives. stoichiometric of the reaction can be formulated as follows (Scheme 1).

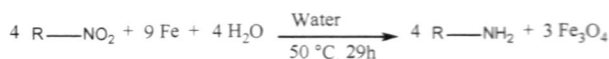

$$4 \ R-NO_2 + 9 \ Fe + 4 \ H_2O \xrightarrow[\text{50 °C, 29h}]{\text{Water}} 4 \ R-NH_2 + 3 \ Fe_3O_4$$

Scheme 1

## Experimental

### General

Chemicals were purchased from commercial firms [Nitroarenes purchased from Sigma Aldrich and iron powder (about 90 mesh) from BDH Chemicals] and used without further purification. Reactions were performed in a 30 Cm or 20 Cm pressure glass tube with closing cap. GC analyses were performed using Focus GC from Thermo Electron Corporation, equipped with low polarity ZB-1 column. GC-MS analyses were performed using Trace 1300 Gas Chromatograph model from Thermo Scientific, equipped with the Rxi-1ms (crossbond 100% dimethyl polysiloxane) column. 1H-NMR spectra were recorded with Bruker DRX-400 instrument in $CDCl_3$. XRD measurements were performed on a D8 Advance diffractometer (Bruker AXS, Karlsruhe, Germany).

### Experimental procedure for nitroarenes hydrogenations reactions

In a typical reaction; 0.23 g (4 mmol) iron powder, water (10 mL) and 0.14 g p-nitrotoluene (1 mmol) was placed in a pressure glass tube equipped with a magnetic stirrer. The tube was sealed and heated with stirring for 29 hours at 50°C. At the end of reaction, the reaction glass tube was allowed to come at room temperature. The product p-toludine was extracted with diethylether (15 × 3=45 mL) followed by filtration using Whatman paper, dried with magnesium sulphate and analyzed by GC. The GC analysis shows >99.9% of p-nitrotoluene conversion to p-toludine. The residue after solvent (diethylether) evaporation affords the desired p-toludine product of good purity (Isolated yield=90%). The isolated product was further characterized by NMR (¹H).

## Results and Discussion

In a typical example a mixture of 1 mmol p-nitrotoluene (0.14 g), iron powder (4 mmol, 0.22 g), and 10 ml of water was placed in a pressure glass tube equipped with a magnetic stirrer. The tube was sealed and heated with stirring at 50°C for 29 hours. After cooling the reaction mixture was found to contain 0.1 g of p-toludine and >99% yield (based on GC area) and 0.28 g (crude weight) of $Fe_3O_4$. p-Nitrotoluene was selected as model substrate for optimization study in the present work (Table 1). Initially reaction temperature was optimized while other reaction parameters [4 equivalent of iron, water (10 mL), 29 h] kept constant (Table 1). The reduction of p-nitrotoluene was carried out at

room temperature (RT) showed 64% conversion of p-nitrotoluene to p-toludine (Table 1, entry 1). In the next step reaction temperatures were increased from RT to 50°C (Table 1, entries 2-3); at 50°C (Table 1, entry 3) the quantitative and selective conversion (>99%) of p-nitrotoluene to p-toludine was observed. Further study for increase of reaction temperature upto 100°C reveals that, the reaction was selective below 80°C temperature (Table 1, entry 5) while at 100°C product selectivity was slightly decreased to 98% (Table 1, entry 6). It should be noted that the selective reduction of a nitro group to corresponding amine is a difficult task because reduction of aromatic nitro compounds often stops at an intermediate stage, producing hydroxylamines, hydrazines and azoarenes as side products [37]. Next various amount of iron powder from 4 equivalent to 2 equivalent (Table 1, entries 3, 8-9) were tested. The 4 equivalent of iron was sufficient for quantitative and selective reduction of p-nitrotoluene 1 to p-toludine 2 (Table 1, entry 3). Using 3.0 and 2.0 equivalent of iron, incomplete reduction of p-nitrotoluene was observed (Table 1, entries 8-9). In the complete absence of iron (Table 1, entry 11) or water (Table 1, entry 12), a neglible or no conversion of p-nitrotoluene to p-toludine was observed. The decrease of reaction time to less than 29 h resulted into incomplete p-nitrotoluene conversion (Table 1, entry 4). An effective stirring was found critical for the reaction to achieve quantitative conversion of nitroarenes. Finally 4 equivalent of iron, 50°C reaction temperature and 29 h reaction time set as optimum reaction parameters to achieve desired conversion and selectivity for nitro reduction under given conditions (Table 1, entry 3). Under the optimized reaction conditions (Table 1, entry 3), we performed hydrogenation of nitroarenes with diverse substituent groups. Importantly, the present reaction protocol was found to be a highly active and almost exclusively selective for the hydrogenation of substituted nitroarenes. Apart from p- nitrotoluene (Table 2, entry 1), other substituted nitrobenzenes having electron-donor or electron-withdrawing groups were also furnished with better to excellent yields (Table 2). Notably, halogen-substituted nitroarenes proceeded smoothly to the respective haloaromatic amines without any dehalogenation (Table 2, entries 4-6). Moreover, present reaction system also showed remarkable chemoselectivity in the hydrogenation of the challenging substrates bearing other easily reducible functional groups. The reducible functional groups in aromatic nitro substrates such as ether (Table 2, entries 7-8), nitrile (Table 2, entry 9), alkene (Table 2, entry 10), and ester (Table 2, entry 11) remained unaffected, thus giving the corresponding amines selectively. Moreover, bicyclic 2-nitronaphthalene also successfully reduced to corresponding 2-aminonaphthalene (Table 2, entry 12). We assert that the described system is composed of consecutive steps of hydrogen generation followed by hydrogenation rather than direct transfer hydrogenation (TH) from water to nitroarenes. This is supported by experimental observations. Under closed vessel conditions p-nitrotoluene was quantitatively converted to p-toludine (Table 1; entry 3). However, the reaction in an open tube showed only 41% conversion of p-nitrotoluene to p-toludine (Table 1, entry 12). In addition we found that iron is readily oxidized to magnetite even in the absence of p-nitrotoluene. The amount of hydrogen evolved from iron oxidation only with pure water was measured (for example; 2.9 mmol of hydrogen was evolved starting from 5 mmol of iron (0) powder under given conditions) [38]. It is well known fact that dissolved $CO_2$ may accelerate the water-iron reaction to generate the hydrogen [29,30]. To exclude the presence of $CO_2$ in water; we have used degassed distilled water (water was purged with nitrogen for 5 minutes before use) for p-nitrotoluene reduction under optimized reaction condition (Table 1, entry X). Here also reaction resulted into similar output (>99% conversion and selectivity) and rule out the role of dissolved $CO_2$ in the activation of water-iron

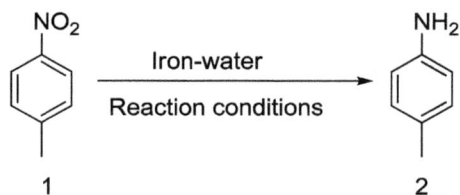

| Entry | Temp. (°C) | Conv. (%)[b] | Yield (%)[b] |
|---|---|---|---|
| 1 | RT | 64 | 64 |
| 2 | 40 | 93 | 93 |
| 3 | 50 | >99 | >99 |
| 4 | 50 | 94[c] | 94 |
| 5 | 80 | >99 | >99 |
| 6 | 100 | >99 | 98 |
| 7 | RT | >99[d] | >99 |
| 8 | 50 | 93[e] | 93 |
| 9 | 50 | 71[e] | 71 |
| 10 | 50 | 02[e] | 02 |
| 11 | 50 | --[f] | -- |
| 12 | 50 | 41[g] | 41 |

[a]Reaction conditions: p-nitrotoluene 1 (1 mmol), iron metal powder, water, 29 h
[b]Conversion and product 2 yield based on GC area.
[c]Reaction run for 27 h
[d]Reaction run for 60 h
[e]3 eqv., 2 eqv. and no iron used in the entries 8, 9 and 10 respectively
[f]No water used
[g]Open atmosphere reaction. NR=No reaction. RT=Room temperature

**Table 1:** Optimization of various reaction parameters[a].

to generate hydrogen. To detect the reaction intermediates we have performed GC-MS analysis of reaction mixture after 6 h. However p-toludine was only product observed. Based on the literature [36,37], we hypothesize that the reduction of aryl nitro compounds could have preceded via -N=O, -NHOH, -N=N(O)- as transient intermediates to provide the $NH_2$ product.

## Conclusion

In conclusion; novel, non-hazardous and mild process was developed for an efficient and chemoselective reduction of nitroarenes to aromatic amines using simple commerical metalic iron as reducing species and water as terminal hydrogen source. The straight forward operation, use of inexpensive and benign reagents as hydrogen donor (iron and water), high yields of aromatic amines and, above all, the unique chemoselectivity over a wide range of functional groups make this procedure an obvious choice for reduction of aromatic nitro compounds. No acid or base additives, salts were added neither salts generated during the current nitroarenes reduction. The end product magnetite generated in the reaction is useful commercial material and can be easily separable from reaction using external magnet. Therefore technicaly end aqueous effluent contains only pure water, and present system free from hazrordous waste generation. From the environmental point of view this novel methodology could be the economical, efficient

| Entry | Substrate | Product | Yield (%)[b,c] |
|---|---|---|---|
| 1 |  |  | 90 |
| 2 |  |  | 88 |
| 3 |  |  | 90 |
| 4 |  |  | 92 |
| 5 |  |  | 88 |
| 6 |  |  | 88 |
| 7 |  |  | 91 |
| 8 |  |  | 91 |
| 9 |  |  | 90 |
| 10 |  |  | 85 |
| 11 |  |  | 90 |
| 12 |  |  | 83 |

[a]Reaction conditions: Substrate (1-3 mmol), Iron powder (4 eqv.), water (10 mL), 50°C (oil bath), 29 h
[b]Conversion, yield and selectivity (for desired product to other products) was >99% (based on GC area) in all the entries
[c]Isolated yield

**Table 2:** Oxidation of various naphthalene derivatives[a].

and waste-free approch towards reduction of nitroarenes.

### Acknowledgments

We are thankful to Elad Meller for his help for GC-MS analysis.

### References

1. Ono N (2001) The Nitro Group in Organic Synthesis. Wiley-VCH: New York, USA.

2. Eller K, Henkes E, Rossbacher R, Hoke H (2000) Amines, Aliphatic. Ullmann's Encyclopedia of Industrial Chemistry.

3. Downing RS, Kunkeler PJ, Van Bekkum H (1997) Catalytic syntheses of aromatic amines. Catal Today 37: 121-136.

4. Dixon DJ, Morejon OP (2014) Recent Developments in the Reduction of Nitro and Nitroso Compounds. Comprehensive Organic Synthesis II. pp: 479-492.

5. Blaser HU, Steiner H, Studer, M (2009) Selective Catalytic Hydrogenation of Functionalized Nitroarenes: An Update. Chem Cat Chem 1: 210-221.

6. Tafesh AM, Weiguny J (1996) A Review of the Selective Catalytic Reduction of Aromatic Nitro Compounds into Aromatic Amines, Isocyanates, Carbamates, and Ureas Using CO. Chem Rev 96: 2035-2052.

7. Heathcock CH (1991) Comprehensive Organic Synthesis. Selectivity, Strategy and Efficiency in Modern Organic Chemistry 2. pp: 133-179.

8.  Lara P, Philippot K (2014) The Hydrogenation of Nitroarenes Mediated by Platinum nanoparticles: An Overview. Catal Sci Technol 4: 2445-2465.

9.  Kasparian AJ, Savarin C, Allgeier AM, Walker SD (2011) Selective catalytic hydrogenation of nitro groups in the presence of activated heteroaryl halides. J Org Chem 76: 9841-9844.

10. Corma A, Gonzalez-Arellano C, Iglesias M, Sanchez F (2009) Gold Complexes as Catalysts: Chemoselective Hydrogenation of Nitroarenes. Appl Catal A: Gen 356: 99-102.

11. Wua H, Zhuo L, He Q, Liao X, Shi B (2009) Heterogeneous Hydrogenation of Nitrobenzenes Over Recyclable Pd(0) Nanoparticle Catalysts Stabilized by Polyphenol-grafted Collagen Fibers. Appl Catal A: Gen 366: 44-56.

12. Takasaki M, Motoyama Y, Higashi K, Yoon SH, Mochida I, et al. (2008) Chemoselective hydrogenation of nitroarenes with carbon nanofiber-supported platinum and palladium nanoparticles. Org Lett 10: 1601-1604.

13. Corma A, Serna P, Concepción P, Calvino JJ (2008) Transforming nonselective into chemoselective metal catalysts for the hydrogenation of substituted nitroaromatics. J Am Chem Soc 130: 8748-8753.

14. Corma A, Serna P (2006) Chemoselective hydrogenation of nitro compounds with supported gold catalysts. Science 313: 332-334.

15. Chen Y, Wang C, Liu H, Qiu J, Bao X (2005) Ag/SiO2: a novel catalyst with high activity and selectivity for hydrogenation of chloronitrobenzenes. Chem Commun (Camb): 5298-5300.

16. Yang H, Cui X, Dai X, Deng Y, Shi F (2015) Carbon-catalysed reductive hydrogen atom transfer reactions. Nat Commun 6: 6478.

17. Jagadeesh RV, Natte K, Junge H, Beller M (2015) Nitrogen-Doped Graphene-Activated Iron-Oxide-Based Nanocatalysts for Selective Transfer Hydrogenation of Nitroarenes. ACS Catal 5: 1526-1529.

18. Rai RK, Mahata A, Mukhopadhyay S, Gupta S, Pei-Zhou L, et al. (2014) Room-Temperature Chemoselective Reduction of Nitro Groups Using Non-noble Metal Nanocatalysts in Water. Inorg Chem 53: 2904-2909.

19. Wienhöfer G, Sorribes I, Boddien A, Westerhaus F, Junge K, et al. (2011) General and selective iron-catalyzed transfer hydrogenation of nitroarenes without base. J Am Chem Soc 133: 12875-12879.

20. Sharma U, Verma PK, Kumar N, Kumar V, Bala M, et al. (2011) Phosphane-free green protocol for selective nitro reduction with an iron-based catalyst. Chemistry 17: 5903-5907.

21. Rajenahally VJ, Wienhofer G, Westerhaus FA, Surkus AE, Pohl MM, et al. (2011) Efficient and Highly Selective Iron-Catalyzed Reduction of Nitroarenes. Chem Comm 47: 10972-10974.

22. He L, Wang LC, Sun H, Ni J, Cao Y, et al. (2009) Efficient and selective room-temperature gold-catalyzed reduction of nitro compounds with CO and H(2)O as the hydrogen source. Angew Chem Int Ed Engl 48: 9538-9541.

23. Shi Q, Lu R, Jin K, Zhang A, Zhao D (2006) Simple and Eco-friendly Reduction of Nitroarenes to the Corresponding Aromatic Amines using Polymer-supported Hydrazine Hydrate over Iron oxide Hydroxide Catalyst. Green Chem 8: 868-870.

24. Coleman GH, McClosky SM, Suart FA (1955) Nitrosobenzene. Org Synth, Coll Vol 3. Wiley: New York, USA. pp: 668-670.

25. Hartman WW, Dickey JB, Stampfli JG (1943) 2,6-Dibromoquinone-4-chloroimide. Org Synth, Coll Vol 2; Wiley: New York, USA. pp: 175-178.

26. Chandrappa S, Vinaya K, Ramakrishnappa T, Rangappa KS (2010) An Efficient Method for Aryl Nitro Reduction and Cleavage of Azo Compounds Using Iron Powder/Calcium Chloride. Synlett 3019-3022.

27. Xiao ZP, Wang YC, Du GY, Wu J, Luo T, et al. (2010) Efficient Reducing System Based on Iron for Conversion of Nitroarenes to Anilines. Syn Commun 40: 661-665.

28. Wang L, Zhang Y (1999) Reduction of Aromatic Nitro Compounds Using Samarium Metal in the Presence of a Catalytic Amount of Iodine under Aqueous Media. Synlett 1065-1066.

29. Liu S, Wang Y, Jiang J, Jin Z (2009) The Selective Reduction of Nitroarenes to N-arylhydroxylamines using Zn in a CO2/H2O System. Green Chem 11: 1397-1400.

30. Gao G, Tao Y, Jiang J (2008) Environmentally Benign and Selective Reduction of Nitroarenes with Fe in Pressurized $CO_2$-$H_2O$ medium. Green Chem 10: 439-441.

31. Ismail AA, Bahnemann DW (2014) Photochemical Splitting of Water for Hydrogen Production by Photocatalysis: A review. Sol Energy Mater Sol Cells 128: 85-101.

32. Muhammad O, Sonavane SU, Sasson Y, Chidambaram M (2008) Palladium/Carbon Catalyzed Hydrogen Transfer Reactions using Magnesium/Water as Hydrogen Donor. Catal Lett 125: 46-51.

33. Mukhopadhyay S, Rothenberg G, Wiener H, Sasson Y (2000) Solid-solid Palladium-Catalysed Water Reduction with Zinc: Mechanisms of Hydrogen Generation and Direct Hydrogen Transfer Reactions. New J Chem 24: 305-308.

34. Boix C, Poliakoff MJ (1999) Selective Reductions of Nitroarenes to Anilines using Metallic Zinc in Near-critical Water. Chem Soc Perkin Trans 1: 1487-1490.

35. Wang L, Li PH, Wu ZT, Yan JC, Wang M, et al. (2003) Reduction of Nitroarenes to Aromatic Amines with Nanosized Activated Metallic Iron Powder in Water. Synthesis 13: 2001-2004.

36. Dey R, Mukherjee N, Ahammed S, Ranu BC (2012) Highly selective reduction of nitroarenes by iron(0) nanoparticles in water. Chem Commun (Camb) 48: 7982-7984.

37. de Noronha RG, Romão CC, Fernandes AC (2009) Highly chemo- and regioselective reduction of aromatic nitro compounds using the system silane/oxo-rhenium complexes. J Org Chem 74: 6960-6964.

38. Patil RD, Sasson Y (2015) Generation of Hydrogen from Zero-Valent Iron and Water: Catalytic Transfer Hydrogenation of Olefins in Presence of Pd/C. AsianJOC 4: 1258-1261.

# 9

# Formation of Spicules During the Long-term Cultivation of Primmorphs from the Freshwater Baikal Sponge *Lubomirskia baikalensis*

**L.I. Chernogor[1]\*, N.N. Denikina[1], S.I. Belikov[1] and A.V. Ereskovsky[2,3]**

[1]*Limnological Institute of the Siberian Branch of the Russian Academy of Sciences, Ulan-Batorskaya 3, Irkutsk 664033, Russia*
[2]*Department of Embryology, Faculty of Biology and Soils, Saint-Petersburg State University, Universitetskaja nab. 7/9, St. Petersburg 199034, Russia*
[3]*Centre d'Océanologie de Marseille, Station marine d'Endoume - CNRS UMR 6540-DIMAR, rue de la Batterie des Lions, 13007 Marseille, France*

## Abstract

Sponges (phylum Porifera) are phylogenetically ancient Metazoa that use silicon to form their skeletons. The process of biomineralization in sponges is one of the important problems being examined in the field of research focused on sponge biology. Primmorph cell culture is a convenient model for studying spiculogenesis. The aim of the present work was to produce a long-term primmorph culture from the freshwater Baikal sponge *Lubomirskia baikalensis* (class Demospongiae, order Haplosclerida and family Lubomirskiidae) in both natural Baikal water and artificial Baikal water to study the influence of silicate concentration on formation and growth of spicules in primmorphs. Silicate concentration plays an important role in formation and growth of spicules, as well as overabundance of silica leads to destruction of cell culture primmorphs. We also found that the composition of chemical elements (Si, O, C, and Na) varied along the length of growing spicules at cultivation in different media. The long-term culture of Baikal sponge primmorphs will be necessary for further investigations, and this system may serve as a powerful *in vitro* model to study spiculogenesis in Baikal siliceous sponges during the early stages of intracellular spicule formation to identify genes that affect biomineralization.

**Keywords:** Primmorphs, Spiculogenesis, Silica spicules, Microanalysis, Baikal sponges

## Introduction

Sponges (phylum Porifera) are the phylogenetically oldest Metazoa. Demosponges recorded before the end the Marinoan glaciation (635Myr ago) exist at present [1]. Processes of biomineralization and biosilification in sponges are the subject of active research and debate. Three sponge clades (class Demospongiae, order Hexactinellida) produce silica skeletons [2]. Their skeletons consist of spicules that form species-specific shapes. Their mineral components consist of silicon dioxide, silicon in amorphous opal-A and the bonded protein spongin (an analogue of collagen) [3,4]. The secretion of silica spicules in sponges is an intracellular process [3,5]. Many silica spicule-forming sponges have been characterized as containing and expressing silicatein genes [6-9]. The exact mechanism(s) of the action of silicatein remain undetermined, but the investigation of this compound is considered to be particularly important for the use of silica-based materials *in vitro* [10]. Formation of spicules is a genetically controlled process and highly relevant to nanobiotechnology [6,11-13].

It is very difficult to investigate the mechanism of spicule formation (spiculogenesis) at the cellular and molecular levels in adult sponges because spicules contain a wide variety of proteins that are involved in metabolism of silica and formation of spicules (biogenic silica). To avoid this complexity, the use of sponge cell cultures (primmorphs) has been suggested [14-16]. Primmorphs are 3D cell aggregates from sponge cells being a suitable model for studying spiculogenesis [8,17-21]. They are formed by dissociation of sponge cells and subsequent growth of aggregates. Previously, a primmorph system from *Suberites domuncula* was used to demonstrate that silicatein (a biosilica-synthesising enzyme) and silicase (a catabolic enzyme) are colocalized at the surfaces of growing spicules and in axial filaments [22].

The endemic freshwater Baikal sponge *Lubomirskia baikalensis* (Pallas 1776) was the subject of our investigation. These sponges are a good model for applied studies because *L. baikalensis* is the only Baikal sponge species that can easily be identified. Lake Baikal is situated in South-Eastern Siberia (51–56° N, 104–110° E), the world's largest (23,000 square kilometres), deepest (1,643 m), and oldest (> 24 million years [MiY]) freshwater body [23]. Sponges dominate the littoral zone of Lake Baikal, covering 47% of the available surfaces [24]. A percentage cover of 47% for sponges is unusual for a freshwater ecosystem and is difficult to compare with reported sponge biomass and occurrence for freshwater ecosystems [25,26]. Habitats of Baikal sponges differ considerably from those of other freshwater sponges because of hydrological and hydrochemical peculiarities of Lake Baikal: great depths, long ice period, low water temperature (10–12°C) in the upper layers in summer, high oxygen content, and low concentration of organic matter [27,28]. Moreover, Lake Baikal contains relatively low levels of nutrients, but the nutrients are stably balanced relative to the chemical composition ratios in phytoplankton [29]. Another peculiar characteristic of Lake Baikal is relatively high level of dissolved silicic acid. It is known, that the silicon concentration in Lake Baikal is 100 μM [27], whereas in marine coastal areas the mean silicon concentration at the surface is less than 3 μM [30] gradually increasing with depth [31]. The high content of silicic acid in Lake Baikal is caused by heavy influx of silicon from the rivers [32].

The main aim of this work was to study how the concentration of silicate influenced the formation and growth of spicules in primmorphs

**\*Corresponding author:** Lubov I. Chernogor, Limnological Institute of the Siberian Branch of Russian Academy of Sciences, 3 Ulan-Batorskaya Street, Irkutsk 664033, Russia, E-mail: lchernogor@ mail.ru

from the freshwater sponge *L. baikalensis* (as model system) in both natural Baikal water (NBW) and artificial Baikal water (ABW). The *L. baikalensis* primmorphs were cultivated in a previous study in both NBW and ABW for >10 months [33]. The prior investigations have indicated that neither 1.0% or 1.5% foetal bovine serum (FBS) (HyClone Laboratories, Inc.) nor classical Dulbecco's modified Eagle's medium (DMEM; Sigma) supplemented with antibiotics is necessary for the cultivation or growth of Baikal sponge primmorphs, as symbiotic green algae in cells of *L. baikalensis* and specific symbiotic microbial flora provide the freshwater sponge with necessary nutrition [34,35]. Moreover, a distinctive feature of Baikal sponges compared to other freshwater sponges is their ability to live symbiotically with various zoochlorellae and dinoflagellates [36,37]. Primmorphs are likely to grow due to photosynthetic products of their symbionts.

## Materials and Methods

### Sponge sampling

*L. baikalensis* (class Demospongiae, order Haplosclerida and family Lubomirskiidae) (Figure 1) has a branched shape and an encrusting base with erect (30–60 cm, up to 1 m high) dichotomous branches and rounded apices. Live specimens are a brilliant shade of green. The ectosomal skeleton consists of spicule tufts from the primary fibres. The skeleton consists of uniformly-spined megasclere oxeas (145–233 x 9–18 μm); microscleres are absent [38]. The *L. baikalensis* specimens were collected in Lake Baikal near Cape Listvenichny at a depth of 15 m (water temperature 3–4°C) using SCUBA in December, 2008 (minimal growth of diatoms). Apical parts of the specimens that were more than 30–40 cm high were collected. The samples were immediately placed in containers with Baikal water and ice and transported for 1.2 h at a constant water temperature (3–4°C) to Limnological Institute SB RAS (Irkutsk).

### Cultivation of primmorph as a model system

Primmorphs were obtained using a classical method of mechanical dissociation of cells [39]. A clean sponge was crushed, and the cell suspension obtained was subsequently filtered through a sterile 200-, 100-, and 29-μm-mesh nylon to eliminate pieces of skeleton and spicules of the maternal sponge. The cellular suspension was diluted to a concentration of 5×10⁶ cells/ml and used as the primary material for primmorph formation. The *L. baikalensis* primmorphs were cultivated as described previously [33] in both NBW and ABW at 3–6°C for >10 months, under conditions that mimicked the natural conditions as closely as possible.

NBW was taken from the depth of 500 m, passed through sterilizing filters, and treated with ultraviolet rays. ABW was prepared according to hydrochemical data on Lake Baikal water composition [27,28] excluding salts of silicon acid, the pH was adjusted to 7.9 like that in natural Baikal water [33]. All chemicals used were of reagent grade. The solution obtained was sterilized by filtration through 0.22-μm polycarbonate filters (Nalgene, Rochester, USA). Only plastic materials were used to prevent release of silicate from glassware.

Primmorphs (2112 pieces) were cultivated in 24-well plates (Nalge Nunc International) in NBW at 3–6°C and light intensity of 47 lx or 0.069 Wt with 12 h mode of day and night alteration. Real light intensity in Lake Baikal in the zone of sponge habitats is not exactly known. After 2 days, a portion (1440 pieces) of the primmorphs (1–2 mm in diameter) was transferred to separate wells of a 24-well plate with 2 ml ABW without silicate and cultivated for 6-8 weeks for adaptation to artificial conditions. The medium was replaced every day during the first week, then once a week, replacing only 50% of the medium. We tested the influence of silicate (Na₂SiO₄×7H₂O) on primmorph development and spicule growth by varying the silicate concentration between 70 and 120 μM, pH was adjusted to 7.9. The primmorphs were cultured in ABW without silicate. As a control, we cultivated primmorphs *in vitro* in NBW. Silicate concentrations in the wells permanently were controlled by spectrophotometric analysis during the experiments. The protocol was adapted from Strickland & Parsons [40]. Subsequently the extinction of standards and samples was measured with a spectrophotometer (Spectronic-20 Genesys, Spectronic Instruments, Rochester, USA) at a wavelength of 810 nm. A calibration curve was made from 0 to 100 μM.

### Spicule preparation and imaging

In our study, spicules totalling more than 1050 were measured in cell culture primmorphs *L. baikalensis*. Spicules were prepared on a single glass slide for each sample cultured either in ABW in the absence of silicate, ABW in the presence of 70 μM silicate and in NBW. A scalpel was used to cut 1-mm-thick sections that ran perpendicular to the surfaces of the specimens, which had been stored in 96% ethanol. Several drops of concentrated HNO₃ were applied to immerse primmorphs or sponge fragments, and the slides were heated slightly over a flame shaking to spread the liquid evenly [41]. Spicule samples were washed in distilled water (dH₂O) and air-dried. After mounting cleaned spicules on microscopic slides, we imaged them using an Axiovert 200 inverted light microscope (Zeiss, Germany). Spicule dimensions were measured by means of calibrated micrometer eyepieces with an amplification of 200× on an inverted light microscope (XDS-1B, COIC). Dynamics of formation and growth of spicules in primmorphs in different media was observed daily during a long period. We estimated a total of 10 view fields, using the methods described by Jones [42], spicule length was measured as the shortest distance between two ends, and spicule width was measured as cross-sectional thickness of the middle region of the spicule.

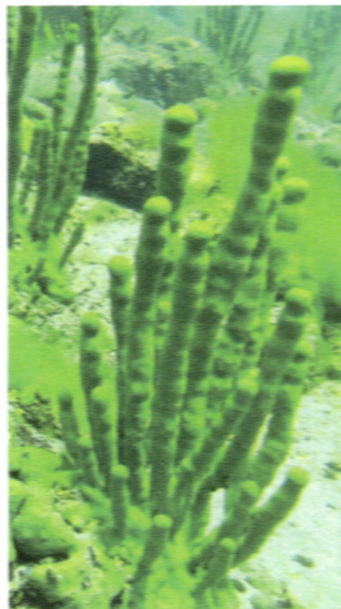

**Figure 1:** Underwater photo of *L. baikalensis in situ* from Lake Baikal at a depth of 15 m (Photo courtesy of I. Khanaev).

## SEM observation

Processes of spiculogenesis in primmorphs in different media (NBW, ABW, and ABW in the presence of 70 μM silicate) were examined by means of SEM. Subsequently, aliquots of primmorph (0.3 g) were analyzed for spicule formation and silicate content. The spicules were prepared from primmorph tissue samples that were first isolated with $HNO_3$, $H_2SO_4$, as described previously [43] with slight modifications (1:4 v/v, followed by n-butanol/water/SDS). Cleaned spicules (10 mg) were washed with distilled water and then with 96% ethanol at room temperature. Spicule separation was achieved by sedimentation for 5 minutes, and final separation by filtration through a sieve (29,10-μm mesh size). SEM images of the spicules was performed with an SEM 525M (Philips, Holland) and EVO 40 microanalyser microscope (Zeiss, Germany). The samples were mounted onto aluminium stubs, then sputtered with a 20-nm coat of a gold-palladium mixture and examined. We started to study 3-week juvenile spicules during three months. Some of the spicules were used for dimension measurements and some for observation by scanning electron microscopy (SEM).

## SEM X-ray microanalysis

Biosilica content and elemental composition of the cultivated spicules in ABW vs. NBW media were determined using a EVO 40 microanalyser microscope (Zeiss, Germany) for SEM X-ray microanalysis (A.V. Zhirmunsky Institute of Marine Biology FEB RAS, Vladivostok). Organic materials were cleared from the spicules, which were then mounted on platinum-coated aluminium stubs and examined using SEM with X-ray microanalysis (XMRA) to qualitatively assay the presence of Si, O, C and Na (i.e., the sum of these components was taken as 100%). SEM X-ray microanalysis uses the characteristic X-rays generated from an electron-bombarded sample to identify elemental constituents that comprise the sample [44]. This technique generates a spectrum in which the peaks correspond to specific X-ray lines, enabling the easy identification of elements present in the sample.

A 20-kV acceleration was used for the SEM-mode microanalysis. The gain rate was adjusted to 10000–20000 counts $s^{-1}$, and, for all of the received pulses at 20000 counts $s^{-1}$, the acquisition time was 100 s. The electron beam excitation was detected in a narrow window (10 $mm^2$). The crystal area in the detector was 10 $mm^2$. We used the 1.740-kV Si/K peak as a qualitative indicator of the presence of Si because it did not overlap with other peaks. Analyses were performed at 1105× magnification, producing spectra for small areas (3.0 and 1.0 μm) along the entire length of the growing spicules, including spicule tips and spines. We analyzed about 10–15 basic points in each spicule. The corresponding Si signals were examined in the spectra. Any significant differences in the Si content across the various compartments studied were assessed using the (INCA program). Moreover, peaks of the output pulses were used to map Si, O, C, and Na contents.

## Statistical analysis

Data on dimensional measurements of spicules were analyzed using Microsoft Excel. For spicules that were grown in various media we calculated length and diameter of spicules for three months using Microsoft Excel. Statistical analyses (descriptive statistics; Kolmogorov-Smirnov test on normal distribution; independent t-tests) were performed using WinSTAT for Excel.

## Results

### Spicule formation in *L. baikalensis* primmorph cell system

The *L. baikalensis* primmorph cultures in NBW, ABW and ABW

in the presence of 70 μM silicate were monitored for three months. Specimens were a brilliant shade of green. Diameter of primmorphs increased to 3–4 mm or more and continued to form spherical aggregates (Figure 2A,2B,2C). After three months in ABW with the presence of 70 μM silicate, the diameter of primmorphs increased to 4 mm or more, and we observed a smooth tissue-like surface with spicules that jutted out. In addition, we cultured primmorphs at a concentration of 120 μM silicate in ABW, the primmorphs become friable during the first 3 days and later died in culture.

The process of spicules formation appeared similar between NBW, ABW and ABW with the introduction of 70 μM silicate. Approximately on day 2, the spicules were squeezed out from the cell and into the extracellular space. Newly born spicules of usually curved shape were observed at this stage (Figure 2D,2E,2F). The spicules continued to increase in size after 12–14 days, reaching lengths of approximately 120 μm (Figure 2G,2H,2I). At the beginning of their formation, spicules were fleixble and curved and then straightened. We

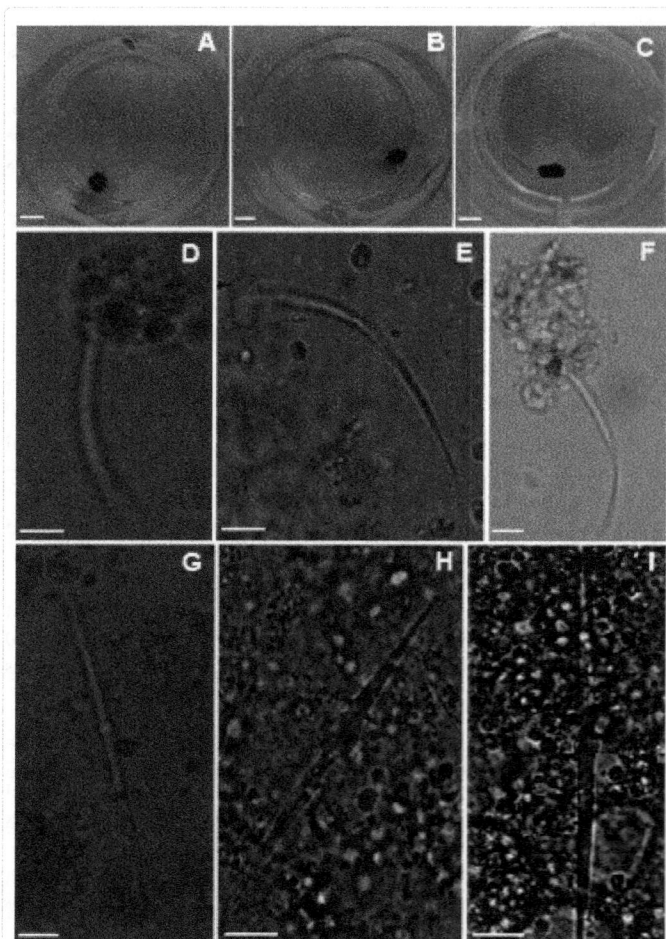

**Figure 2:** Light microscopy of live preparations. Dynamic formation of spicules *in vitro* in cell culture of the *L. baikalensis* primmorphs: (**A**) primmorphs cultivated for three months in NBW; (**B**) primmorphs cultivated for three months in ABW in the absence of silicate; (**C**) primmorphs cultivated for three months in ABW in the presence of 70 μM silicate; (**D**) newly formed spicule on 2 day in NBW; (**E**) formation of spicule on 2 day in ABW; (**F**) formation of spicule on 2 day in ABW in the presence of 70 μM silicate; (**G**) formation of spicule on 12 day in NBW; (**H**) formation of spicule on 12 day in ABW; (**I**) formation of spicule on 12 day in ABW in the presence of 70 μM silicate. Scale bars: (**A – C**) 0,5 mm; (**D – I**) 10 μm.

observed different stages of spicule development during three months of cultivation. Spicules, from newly born to mature, were mixed and distributed inside primmorphs for 2–3 months. During the initial stage of growth, spicules were smooth and flexible, without spines, and had an expansion bubble in the middle. Then spicules straightened, grew resembling the spicules of adult sponges. Spicules could be observed *in vivo* after 2–3 days; full-size spicules were present after 12–14 days of cultivation. At this point, growing spicules in primmorphs acquired sizes and shapes of adult spicules.

Using SEM, we found no significant differences between the spicules grown in different media. We also found no significant differences between the spicules cultivated in ABW with the introduction of 70 µM silicate (Figure 3A) and those cultivated in NBW or ABW without silicate (Figure 3B,3C). During the initial period of growth, spicules appeared smooth except for an expansion in the middle of spicules, called the "bubble".

The spicules observed in adult sponges were slightly different from those cultured from primmorphs. The spicules of adult sponge of *L. baikalensis* were slightly curved amphioxea covered with many spines (Figure 4); they were 150–210 µm in length and 8–15 µm in diameter [45,46]. High SEM magnification shows the spines at the surface of the spicules (Figure 4). In addition, the skeleton in the adult *L. baikalensis* sponges was of highly ordered arrangement of spicules.

Dynamics of spicule growth in primmorphs was observed for three months. During observations of primmorph cultivation in NBW, the number of spicules in the visual field during the first months was

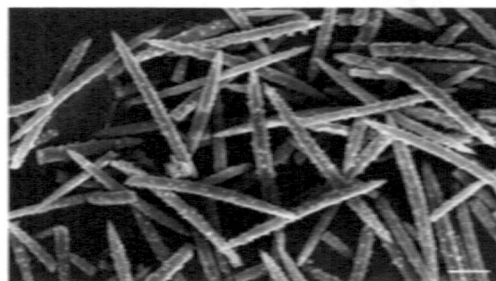

**Figure 4:** SEM image of adult spicules of *L. baikalensis*, this sample was used as a control. Scale bar: 20 µm.

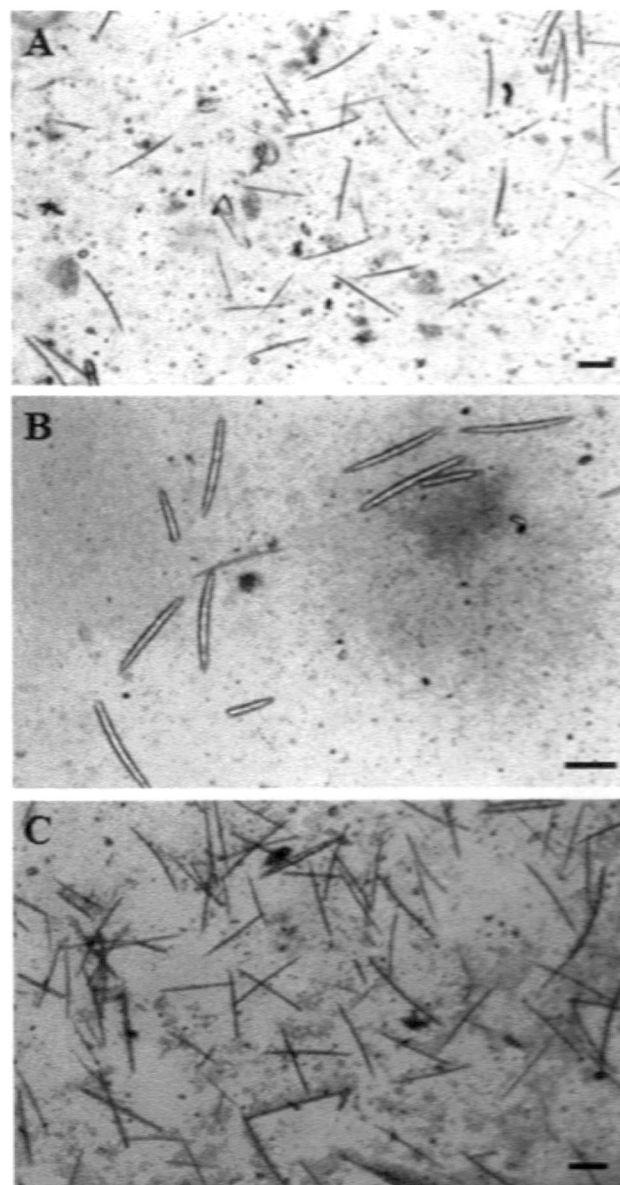

**Figure 5:** Light microscopy observations of spicules *L. baikalensis* primmorphs: A – growing spicules cultivated in NBW; B – growing spicules cultivated in ABW in the absence of silicate; C – growing spicules in ABW in the presence of 70 µM silicate. Scale bars: 60 µm.

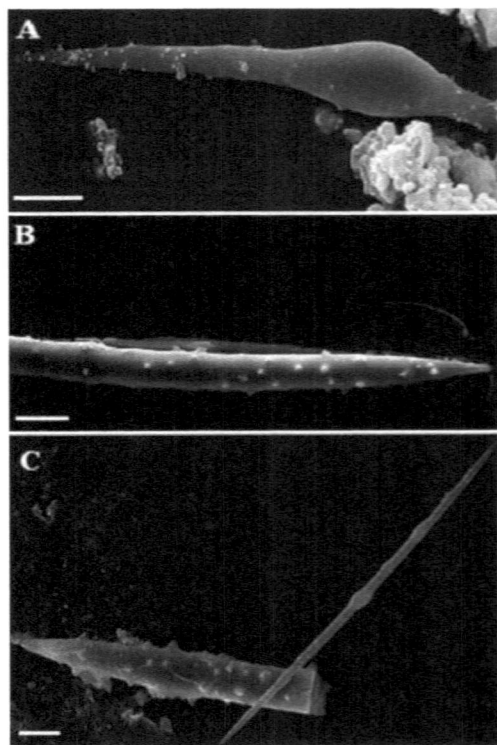

**Figure 3:** SEM image of spicules cultivated *in vitro* in primmorph of *L. baikalensis* during three months: (A) spicule cultivated in ABW in the presence of 70 µM silicate. The spicule is smooth with small spines and bubbles; (B) the spicule in NBW; (C) the spicule cultivated in ABW in the absence of silicate. Scale bars: 20 µm

determined to be 40–50 (Figure 5A) with a total of 10 view fields for each sample. Mean length and diameter of the spicules are given in Table 1. In the early growth the spicules lengths in the *L. baikalensis* primmorph cultures ranged from 90 to 200 µm to the end of three months. The diameter ranged from 4.0 to 7.5 µm. The mean diameter varied markedly at different developmental stages compared with the mean length. During the first month, spicules grew intensively forming new spicules.

In addition, in primmorphs that were cultivated *in vitro* in ABW without silicate, we observed insignificant spicule formation, the number of spicules in the visual field during the first months was determined to be 5–10 with a total of 10 view fields for each sample (Figure 5B). The mean spicule lengths continued to increase in size, reaching lengths of approximately 110–120 µm. The lengths of growing spicules in the *L. baikalensis* primmorph cultures ranged from 85 µm after 2 day of growth to 198.3 µm to three months (Table 1). The diameter ranged from 3.5 to 6.2 µm.

The highest number of spicules was formed in the primmorph at a concentration of 70 µM silicate in ABW (Figure 5C). The number of spicules in primmorphs was 100 or more per visual field during the initial months with a total of 10 view fields for each sample. Interestingly, that the growth of spicules is accelerates at adding concentration of 70 µM silicate in ABW. Lengths of the spicules in the *L. baikalensis* primmorph cultures ranged from 90 to 200 µm. The diameter ranged from 4.0 to 9.0 µm. In addition, we observed constant formation of new spicules during three months.

## X-ray microanalysis

We investigated qualitative elemental structure of growing spicules

during long-term cultivation in NBW, ABW or in ABW with the introduction of 70 µM silicate for three months. Full-size spicules formed to 3 weeks, they were flexible and brittle. Spicules that were analyzed are shown in Table 2. Mean lengths of spicules cultivated in NBW were approximately 120 µm for the first months and 193 µm to three months. The diameter ranged from 5.0 to 7.0 µm. Accumulations of silica along the lengths of the spicules were different (Figure 6A,6E,6F). General silicate contents of the spicules that were cultivated in NBW are indicated on the spectral diagram as percentages (Figure 6E). The composition of chemical elements varied along the length of spicules that were cultivated in NBW (Table 3). X-ray microanalysis indicated an increase in silica in the centers of spicules and low silica content on spines of spicules. As an example, we provide spectrum 8, which had a silicate content of up to 47.7% (Figure 6F). Oxygen content was high in spicules (Figure 6C). In addition, there was a small amount of Na (4–8%) at the ends of the spicules that decreased in the middles of spicules (1–3%).

We analyzed silica content in spicules cultured *in vitro* in ABW (Figure 7A,7E,7F). It was observed that the length and diameter of spicules were slightly smaller than of those cultured in NBW and ABW with 70 µM silicate. We recorded a more uniform distribution of silica along the spicules, but silicate contents increased towards the central part of spicules. The mean length of spicules ranged from 116.8 to 186.1 µm and the diameter ranged from 4.8–6.1 µm (Table 2). The composition of chemical elements (Si, O, C, and Na) varied along the length of spicules (Table 3). Spines of spicules contained 6.8 to 27 % silica. General silicate contents of spicules that were cultivated in ABW are given on the spectral diagram as percentages (Figure 7E). As an example, we provide spectrum 9, which had a silicate content of up to 25.9% (Figure 7F). Oxygen content was high (Figure 7C). In addition,

| | Natural Baikal water | | | Artificial Baikal water | | | Artificial Baikal water with 70µM silicate | | |
|---|---|---|---|---|---|---|---|---|---|
| Date (months) | 1 | 2 | 3 | 1 | 2 | 3 | 1 | 2 | 3 |
| **Developmental stage** | Length of spicules (µm) | | | | | | | | |
| Mean length (µm) | 121.7 | 168.1 | 193.7 | 117.3 | 159.9 | 187.4 | 121.1 | 169.7 | 193.5 |
| Standard deviation | ±16.7 | ±8.7 | ±6.1 | ±13.4 | ±16.6 | ±4.6 | ±16.9 | ±12.7 | ±3.6 |
| **Developmental stage** | Diameter of spicules (µm) | | | | | | | | |
| Mean diameter (µm) | 5.2 | 6.4 | 7.1 | 4.7 | 6.0 | 6.2 | 5.9 | 8.4 | 8.9 |
| Standard deviation | ±0.63 | ±0.28 | ±0.24 | ±0.68 | ±0.07 | ±0.09 | ±0.90 | ±0.42 | ±0.10 |

**Table 1:** Dimensional characteristics of growing spicules in the *L. baikalensis* primmorphs during three months using different methods of cultivation.

| | Natural Baikal water | | | Artificial Baikal water | | | Artificial Baikal water with 70µM silicate | | |
|---|---|---|---|---|---|---|---|---|---|
| Date (months) | 1 | 2 | 3 | 1 | 2 | 3 | 1 | 2 | 3 |
| Dimension | | | | | | | | | |
| Mean length (µm) | 120.5 | 167.8 | 193.0 | 116.8 | 160.5 | 186.2 | 121.8 | 169.0 | 193.5 |
| standard deviation | ±16.7 | ±8.7 | ±6.0 | ±13.5 | ±16.6 | ±4.63 | ±16.98 | ±12.73 | ±3.60 |
| Mean diameter (µm) | 5.0 | 6.4 | 7.0 | 4.5 | 5.9 | 6.3 | 6.0 | 8.5 | 9.0 |
| standard deviation | ±0.6 | ±0.2 | ±0.2 | ±0.68 | ±0.07 | ±0.09 | ±0.90 | ±0.42 | ±0.10 |
| Number of spicules observed by SEM | 28 | 27 | 28 | 25 | 24 | 22 | 28 | 29 | 28 |

**Table 2:** Mean dimensions of growing spicules in *L. baikalensis* primmorphs for X-ray microanalysis during three months using different methods of cultivation.

| Chemical elements | Natural Baikal water | | | Artificial Baikal water | | | Artificial Baikal water with 70µM silicate | | |
|---|---|---|---|---|---|---|---|---|---|
| | spines of sp. | ends of sp. | central part sp. | spines of sp. | ends of sp. | central part sp. | spines of sp. | ends of sp. | central part sp. |
| Silica (Si) | 9-10% | 22-33% | 27-47% | 6.8-27% | 9-17% | 33-35% | 22-23% | 30-35% | 36-47% |
| Oxygen (O) | 31-32% | 43-51% | 40-49% | 36-47% | 38-43% | 50-52% | 31-33% | 44-57% | 44-48% |
| Carbonate(C) | 56-58% | 16-35% | 27-38% | 24-56% | 37-51% | 13-14% | 43-46% | 20-24% | 10-17% |
| Natrium (Na) | 0.6-0.8% | 6.2-7.9% | 1.2-3.4% | 0.5-0.4% | 0.79-1.7% | 0.8-1.5% | 0.3-0.8% | 0.3-0.6% | 6.5-7.0% |

sp. – spicules

**Table 3:** Composition of chemical elements along the length of growing spicules in primmorphs of *L baikalensis* using different methods of cultivation.

there was a small Na amount – 0.7% at the ends of the spicules which increased to 1.5% in the centers.

Table 2 demonstrates results of spicules cultivated in ABW with 70 µM silicate. The length of spicules of primmorphs *L. baikalensis* ranged from 121.5 to 192.8 µm and the diameter ranged from 6.0–9.0 µm. Relatively large amounts of silicate were also observed during long-term primmorph cultivation in ABW with 70 µM silicate, as indicated on the spectra as percentages (Figure 8A,8E,8F). The increase in silicate was recorded in the central regions of the spicules (Table 3). In the bubble, 44% silicate, 43% oxygen, and 12% carbonate were observed. The silicate content decreased towards the spicule ends and spines. However, the content of oxygen and carbonate increased in spicule spines relative to the ends (Table 3). As an example, we provide spectrum 7, in which the silicate content was 43.5% (Figure 8F). In addition, at the ends of the growing spicules there was 6.5% of Na, which decreased to 0.5% in the centers. Thus, during long-term primmorph cultivation in ABW, silicate levels gradually increased in growing spicules, resulting in 44–50% silicate, 38–47% oxygen, and 10–15% carbonate. The composition of chemical elements (Si, O, C, and Na) varied along the length of the growing spicules at cultivation in NBW, ABW and ABW with 70 µM silicate during three months.

## Discussion

We developed a primmorph culture from the endemic Baikal sponge *L. baikalensis* as a model to study biosilica formation. To obtain

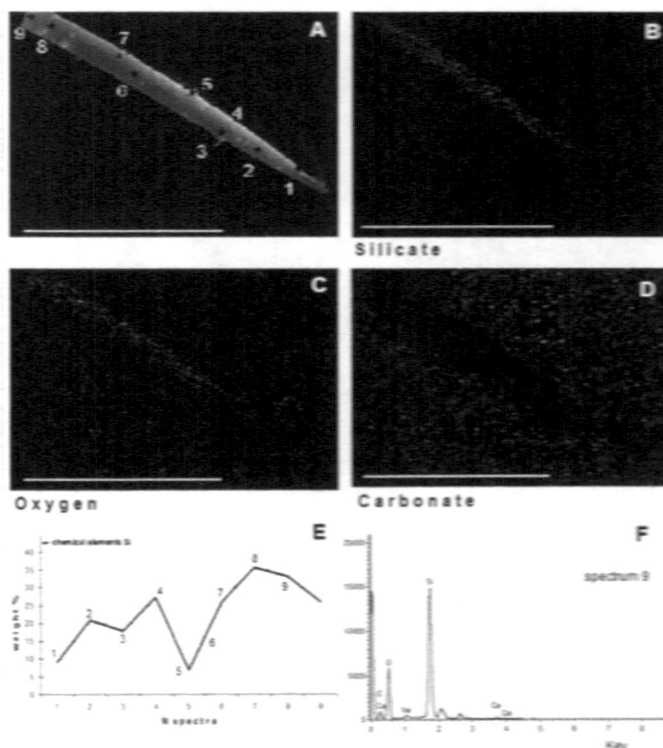

Figure 7: SEM image of spicules in primmorphs of *L. baikalensis* cultivated in ABW during three months. XMRA spectra obtained for freshwater spicules: (A) points spectra; (B) silica contents; (C) oxygen contents; (D) carbonate contents analyzed in the wall of the spicules. (E) general silica contents of spicules that were cultivated in ABW, as indicated on the spectra as percentages; (F) as an example, we provide spectrum 9, in which the silica content was 25.9%. Scale bars: 20 µm.

a viable long-term primmorph culture, we aimed to create cultivation conditions that resembled natural conditions as closely as possible. Primmorphs can grow based on the sustenance of photosynthetic products of their symbionts [33,36,37]. We observed a good survival of primmorphs under different culture conditions.

Here, we report spicule formation in *L. baikalensis* primmorphs that were grown in various media. Concentration of silicate influenced the formation and growth of spicules in primmorphs from the freshwater sponge *L. baikalensis*. We indicated that spicules are formed and grow in primmorphs during *in vitro* cultivation in NBW. The NBW has enough silicate for siliceous Baikal sponges, to build their spicule skeletons. Intensive formation of spicules was also observed in the primmorphs in the presence of 70 µM silicate in ABW. Similar results were observed in primmorphs of marine sponges in the presence of 60 µM silicate [7,47]. However, at a concentration of 120 µM of silicate primmorphs became friable during the first 3 days and later died in culture. Higher concentrations are likely to be toxic, e.g. 120 µM, and inhibit cultivation of primmorphs. Polymerization of silicic acid is a process which, if it is neither constrained nor controlled, is highly cytotoxic [48]. It is known that silica concentration in Lake Baikal is 100 µM [27]. Moreover, Lake Baikal contains relatively low levels of nutrients, but nutrients are stably balanced relative to chemical composition ratios in phytoplankton [29].

We indicated that spicules are formed in primmorphs and grow during *in vitro* cultivation in ABW without silicate introduction.

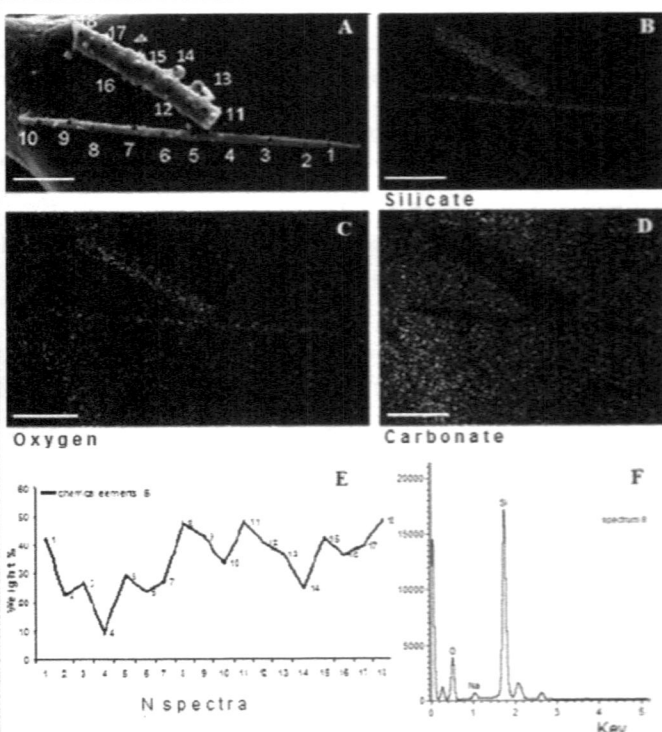

Figure 6: SEM image of spicules in primmorphs of *L. baikalensis* cultivated in NBW during three months with the points of spectra (numbers) — (A); (B – D) X-ray microanalysis (XMRA) spectra obtained for freshwater spicules. (B) silica contents; (C) oxygen contents; (D) carbonate contents analyzed in the wall of spicules. (E) general silica contents of the spicules that were cultivated in NBW are indicated on the spectral diagram as percentages; (F) as an example, we provide spectrum 8, which had a silica content of up to 47.7%. Scale bars: 30 µm.

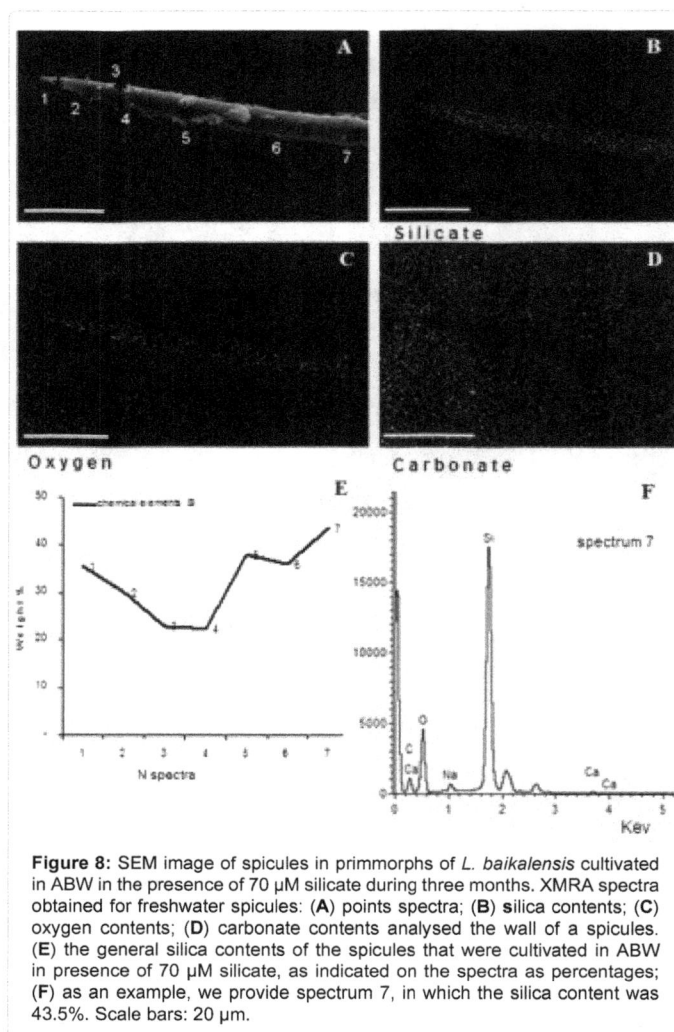

**Figure 8:** SEM image of spicules in primmorphs of *L. baikalensis* cultivated in ABW in the presence of 70 µM silicate during three months. XMRA spectra obtained for freshwater spicules: (**A**) points spectra; (**B**) silica contents; (**C**) oxygen contents; (**D**) carbonate contents analysed the wall of a spicules. (**E**) the general silica contents of the spicules that were cultivated in ABW in presence of 70 µM silicate, as indicated on the spectra as percentages; (**F**) as an example, we provide spectrum 7, in which the silica content was 43.5%. Scale bars: 20 µm.

different media we observed intensive growth in length and diameter of spicules during three months. We also observed some (minor) differences in length and diameter of the spicules at cultivation in ABW without silicate introduction.

The SEM X-ray microanalysis data provided some information on the process of biosilica deposition during the cell cultivation of *L. baikalensis* primmorphs. Distributions of silica along the lengths of the spicules were significantly different in all of the specimens. The highest concentration of silica was observed at the bubble (the central parts of the young spicules contained up to 50% silica), and minimal amount of silica was observed at the spicule ends and spines (Table 3). We also found a great number of spicules with the "bubble" in the primmorphs cultured in ABW and ABW with the introduction of 70 µM silicate. It is likely that silica is synthesized in the middle of the spicule, where the thickening occurs. However, at cultivation in ABW without silicate introduction we observed a more uniform distribution of silica on the spicules. Probably the lack of silicon influences the growth of spicules. Moreover, accumulation of silica in spicules occurs on the intracellular molecular level [43,49]. For example, in the marine sponge *Crambe crambe*, TEM X-ray microanalysis has indicated that the extracellular space between sclerocyte and growing spicule contains 50–65% Si [52]. This intracellular process is related to polycondensation of silicic acid that is mediated by enzyme silicatein [43,49]. Moreover, we observed robust maintenance of O (41%) and C (36%) and slight maintenance of Na (8%) at the ends of spicules in the samples. These concentrations are reduced to 0.5–1% in the central region of spicules, confirming the presence of biogenic silica in growing spicules of *L. baikalensis* primmorphs. This distribution of elements is probably related to metabolism of biosilica in the primmorph cells. Based on the data discussed here, Si, O, C, and Na are the most important elements for construction of spicules. All of these elements were involved in the basic structure of siliceous spicules in primmorphs of Baikal sponges. SEM X-ray microanalysis allowed us to provide X-ray spectra and to demonstrate accumulation of biogenic silica in cultured freshwater spicules in various media. Silica plays an important role in formation of spicules, the overabundance of silica leads to destruction of cell culture primmorphs. A tendency for gradually increasing Si-levels is observed during long-term spicule growth in cultures. The fact that the model culture of *L. baikalensis* primmorphs lives for a long time under these conditions makes it possible to study genes and proteins participating in the formation of silicon spicules. In addition, the study adds important knowledge to understanding of silica deposition in spicules of primmorphs.

Overall, the Lubomirskia primmorph system is shown to be a good model for studying silica biomineralization due to biomass growth independent of external nutrients based on photosynthetic symbionts that present in primmorphs. Primmorphs cultivation *in vitro* for a long period of time will allow the creation of a live controlled model system under experimentally controlled conditions in the absence of any additional organic component and are a good model system that offers controlled experimental conditions to understand a mechanism of silica transport from the environment into sponge cells of Baikal sponges (i.e., the mechanism of spicule biosynthesis). Primmorph system described in this work can be considered as a powerful novel model system to study basic mechanisms of silicatein expression (an enzyme responsible for spicule silicification) for the identification and genetic determination of the proteins involved in the process biomineralization, and dimensional changes of spicules during developmental process of individual primmorphs. On the basis of this work we will be able to carry out further research of some

Earlier, the *L. baikalensis* primmorphs were cultivated for >10 months in ABW, the composition of which was close to the Lake Baikal water [27,28] without silicate [33]. Pure primmorph culture after dissociation of sponge cells is difficult to obtain as there are traces of diatoms and spicules from adult sponge. We observed by light microscopy that cells in the primmorphs decomposed diatoms and spicules from adult sponges; it is likely to receive biogenic silica for formation of new spicules. Moreover, siliceous spicules are known to contain a definite axial filament and their synthesis is under genetic control [12,43,49]. It is known that siliceous spicules of sponges are formed intracellularly in sclerocytes and during their growth they are surrounded by a membrane called a silicalemma [3]. Spicule synthesis is rapid; for example, spicules of the freshwater sponge *Ephydatia fluviatilis* grow at a rate of 5 µm/h [50,51]. It is known that spicules of Baikal sponges are formed in two stages [39]. At the first stage of the intracellular formation a thin spicule (fiber) is formed. At the second stage a spicule goes out from of cell into the extracellular space and grows to the required thickness [39]. The production of spicules occurs inside specialized cells, the sclerocytes, where silica is deposited in an organized way [3]. Overall spiculogenesis of freshwater *L. baikalensis* primmorphs is also rapid. On the second day, they are squeezed out from sclerocytes into the intercellular space and continue to lengthen to full-size spicules on 12–14 days of cultivation. At cultivation in

processes and mechanisms occurring in primmorphs and their cells. In addition, these methods will allow the investigation of the early stages of intracellular spicule formation.

## Acknowledgements

This work was supported by the following grants: RFBR 07-04-00103a, 09-04-00337; the European Commission 6th Framework Programme (project: Research, Technological Development and Demonstration, NMP4-CT-2006-031541); the Integration Project SB RAS 38-2009; and the European Marie Curie Mobility Programme (fellowship to A. Ereskovsky, MIF1-CT-2006-040065-980066). We thank D. Fomin for technical assistance with the scanning electron microscopy and microanalysis. The authors are indebted to the Microscopy Centre Service Staff of the A.V. Zhirmunsky Institute of Marine Biology, FEB RAS, Vladivostok and the Microscopy Centre of the Limnological Institute of the Siberian Branch of the Russian Academy of Sciences, Irkutsk.

## References

1.  Love GD, Grosjean E, Stalvies C, Fike DA, Grotzinger JP, et al. (2009) Fossil steroids record the appearance of Demospongiae during the Cryogenian period. Nature 457: 718-721.

2.  Uriz MJ (2006) Mineral skeletogenesis in sponges. Can J Zool 84: 322-356.

3.  Simpson TL (1984) The Cell Biology of sponges. Springer-Verlag, New York, NY.

4.  De La Rocha CL (2003) Silicon isotope fractionation by marine sponges and the reconstruction of the silicon isotope composition of ancient deep water. Geology 31: 423-426.

5.  Uriz MJ, Turon X, Becerro MA, Agell G (2003) Siliceous spicules and skeleton frameworks in sponges: origin, diversity, ultrastructural patterns, and biological functions. Microsc Res Tech 62: 279-299.

6.  Morse DE (1999) Silicon biotechnology: harnessing biological silica production to construct new materials. Trends Biotechnology 17: 230-232.

7.  Krasko A, Lorenz B, Batel R, Schröder HC, Müller IM, et al. (2000) Expression of silicatein and collagen genes in the marine sponge Suberites domuncula is controlled by silicate and myotrophin. Eur J Biochem 267: 4878-4887.

8.  Müller WEG, Krasko A, Pennec GL, Schröder HC (2003) Biochemistry and cell biology of silica formation in sponges. Microsc Res Tech 62: 368-377.

9.  Mohri K, Nakatsukasa M, Masuda Y, Agata K, Funayama N (2008) Toward Understanding the morphogenesis of siliceous spicules in freshwater sponge: differential mRNA expression of spicule-type-specific silicatein genes in Ephydatia fluviatilis. Dev Dyn 237: 3024-3039.

10. Kozhemyako VB, Veremeichik GN, Shkryl YN, Kovalchuk SN, Krasokhin VB, et al. (2010) Silicatein Genes in Spicule-Forming and Nonspicule-forming Pacific Demosponges. Mar Biotechnol 12: 403-409.

11. Cha JN, Shimizu K, Zhou Y, Christianssen SC, Chmelka BF, et al. (1999) Silicatein filaments and subunits from a marine sponge direct the polymerization of silica and silicones in vitro. Proc Natl Acad Sci U S A 96: 361-365.

12. Müller WEG, Krasko A, Pennec GL, Steffen R, Ammar MSA, et al. (2003b) Molecular mechanism of spicule formation in the demosponge Suberites domuncula: silicatein - collagen - myotrophin. Prog Mol Subcell Biol 33: 195-221.

13. Ehrlich H, Worch H (2007) Sponges as natural composites: from biomimetic potential to development of new biomaterials. Museu National, Rio de Janeiro, Brasil 303-312.

14. Custodio MR, Prokic I, Steffen R, Koziol C, Borojevic R, et al. (1998) Primmorphs generated from dissociated cells of the sponge Suberites domuncula: a model system for studies of cell proliferation and cell death. Mech Ageing Dev 105: 45-59.

15. Koziol C, Borojevic R, Steffen R, Müller WE (1998) Sponges (Porifera) model systems to study the shift from immortal to senescent somatic cells: the telomerase activity in somatic cells. Mech Ageing Dev 100: 107-120.

16. Müller WE, Wiens M, Batel R, Steffen R, Schroder HC, et al. (1999) Establishment of a primary cell culture from a sponge: primmorphs from Suberites domuncula. Mar Ecol Prog Ser 178: 205-219.

17. Müller WEG, Rothenberger M, Boreiko A, Tremel W, Reiber A, et al. (2005) Formation of siliceous spicules in the marine demosponge Suberites domuncula. Cell Tissue Res 321: 285-297.

18. Zhang W, Zhang X, Cao X, Xu J, Zhao Q, et al. (2003) Optimizing the formation of in vitro sponge primmorph from Chinese sponge Stylotella agminata (Ridley). J Biotechnol 100: 161-168.

19. Zhang X, Cao X, Zhang W, Yu X, Jin M (2003) Primmorphs from archaeocytes - dominant cell population of the sponge hymeniacidon perleve: improved cell proliferation and spiculogensis. Biotechnol Bioeng 84: 583-590.

20. Cao X, Fu W, Yu X, Zhang W (2007) Dynamics of spicule evolution in marine sponge Hymeniacidon perlevis during in vitro cell culture and the seasonal development in field. Cell Tissue Res 329: 595-608.

21. Cao X, Yu X, Zhang W (2007) Comparison of Spiculogenesis in in Vitro ADCP-Primmorph and Explants Culture of Marine Sponge Hymeniacidon perleve with 3-TMOSPU Supplementation. Biotechnol Prog 23: 707-714.

22. Eckert C, Schroder HC, Brandt D, Perovic-Ottstadt S, Müller WE (2006) Histochemical and electron microscopic analysis of spiculogenesis in the Demosponge Suberites domuncula. J Histochem Cytochem 54: 1031–1040.

23. Belikov SI, Kaluzhnaya OV, Schröder HC, Krasko A, Müller IM, et al. (2005) Expression of silicatein in spicules from the Baikalian sponge Lubomirskia baicalensis. Cell Biol Int 29: 943-951.

24. Pile AJ, Patterson M.R, Savarese M, Chernykh VI, Fialkov VA (1997) Trophic effects of sponge feeding within Lake Baikal's littoral zone. Sponge abundance, diet, feeding efficiency, and carbon flux. Limnol Oceanogr 42: 178-184.

25. Frost TM, De Nagy GS, Gilbert JJ (1982) Population dynamics and standing biomass of the freshwater sponge Spongilla lacustris. Ecology 63: 1203-1210.

26. Bailey RC, Day KE, Norris RH, Reynoldson TB (1995) Macroinvertebrate community structure and sediment bioassay results from nearshore areas of North American Great Lakes. J Great Lakes Res 21: 42-52.

27. Falkner KK, Measures CI, Herbelin SE, Edmond JM (1991) The major and minor element geochemistry of Lake Baikal. Limnol Oceanogr 36: 413-423.

28. Grachev MA (2002) About a modern condition of ecological system of Lake Baikal. Publishing House The Siberian Branch of the Russian Academy of Science, Novosibirsk.

29. Genkai-Kato M, Sekino T, Yoshida T, Miyasaka H, Khodzher TV, et al. (2002) Nutritional diagnosis of phytoplankton in Lake Baikal. Ecological Research 17: 135–142.

30. Maldonado M, Carmona MC, Uriz MJ, Cruzado A (1999) Decline in Mesozoic reef-building sponges explained by silicon limitations. Nature 401: 785-788.

31. Tréguer P, Nelson DM, van Bennekom AJ, DeMaster DJ, Leynaert A, et al. (1995) The silica balance in the world ocean: a reestimate. Science 268: 375-379.

32. Bukharov AA, Fialkov VA (2001) Baikal in numbers. Baikal Museum, Irkutsk 72.

33. Chernogor LI, Denikina NN, Belikov SI, Ereskovsky AV (2011) Long-term cultivation of primmorphs from freshwater Baikal sponges Lubomirskia baikalensis. Mar Biotechnol 13: 782-792.

34. Efremova SM (1981) The structure and embryogenesis of the Baikal sponge Lubomirskia baicalensis (Pallas) and relations of Lubomirskiidae with other sponges. In: Morphogenesis in sponges. Korotkhova, G. P. (eds), Leningrad University, Leningrad. 33: 93-107.

35. Latyshev NA, Zhukova NV, Efremova SM, Imbs AB, Glysina OI (1992) Effect of habitat on participation of symbionts in formation of the fatty acid pool of freshwater sponges of Lake Baikal. Comparative Biochemistry & Physiology 102: 961-965.

36. Sand-Jensen K, Pedersen MF (1994) Photosynthesis by symbiotic algae in the freshwater sponge, Spongilla lacustris. Limnol Oceanogr 39: 551-561.

37. Bil K, Titlyanov E, Berner T, Fomina I, Muscatine L (1999) Some Aspects of the Physiology and Biochemistry of Lubomirska baikalensis, a Sponge from Lake Baikal Containing Symbiotic Algae. Symbiosis 26: 179-191.

38. Manconi R, Pronzato R (2002) Suborder Spongillina subord. nov.: Freshwater Sponges. In: Hooper JNA, Van Soest RWM (eds), Systema Porifera: A Guide to the Classification of Sponges. New York, Kluwer Academic /Plenum Publishers 1: 921-1020.

39. Rezvoj PD (1936) Freshwater sponges Fam. Spongillidae & Lubomirskiidae, Vol. 2. Fauna URSS Academie Sciences URSS, Moscow.

40. Strickland JDH, Parsons TR (1972) A practical handbook of seawater analysis. Queen's Printer, Ottawa, Canada.

41. Kelly-Borges M, Pomponi S (1992) The simple fool's guide to sponge taxonomy. Harbor Branch Oceanographic Institution, Fort Pierce, Florida.

42. Jones WC (1984) Spicule dimensions as taxonomic criteria in the identification of haplosclerid sponges from the shores of Anglesey. Zoological Journal of the Linnean Society 80: 239-259.

43. Shimizu K, Cha J, Stucky GD, Morse DE (1998) Silicatein alpha: cathepsin L-like protein in sponge biosilica. Proceeding of the National Academy of Sciences USA 95: 6234-6238.

44. Goldstein JI, Newbury DE, Echlin PE, Joy DC (1992) Scanning Electron Microscopy and X-ray Microanalysis, 2nd ed., Plenum Press, New York.

45. Dybowski W (1880) Studien über die Spongien des russischen Reiches mit besonderer Berücksichtigung der Spongien-Fauna des Baikal-Sees. Mém Acad Sci St. Petersburg sér. 27: 1-71.

46. Masuda Y, (2009) Studies on the taxonomy and distribution of freshwater sponges in the Lake Baikal. Prog Mol Subcell Biol 47: 81-110.

47. Krasko A, Schröder HC, Batel R, Grebenjuk VA, Steffen R, et al. (2002) Iron induces proliferation and morphogenesis in primmorphs from the marine sponge Suberites domuncula. DNA Cell Biol 21: 67-80.

48. Iler RK (1979) The chemistry of silica. Wiley, New York.

49. Weaver JC, Pietrasanta LI, Hedin N, Chmelka BF, Hansma PK, et al. (2003) Nanostructural features of demosponge biosilica. J Struct Biol 144: 271–281.

50. Weissenfels N, Landschoff HW (1977) Bau und Funktion des Süßwasserschwammes E. fluviatilis L. (Porifera). IV. Die Entwicklung der monaxialen SiO2-Nadeln in Sandwich-Kulturen. Zool Jahrb Abt Anat 98: 355-371.

51. Weissenfels N (1989) Biologie und Mikroskopische Anatomie der Süßwasserschwämme (Spongillidae). Gustav Fischer Verlag, Stuttgart.

52. Uriz MJ, Turon X, Becerro MA (2000) Silica deposition in Demospongiae: spiculogenesis in Crambe crambe. Cell Tissue Res 301: 299-309.

# 10

# Characterisation of Physical, Spectral and Thermal Properties of Biofield treated Resorcinol

**Mahendra Kumar Trivedi[1], Alice Branton[1], Dahryn Trivedi[1], Gopal Nayak[1], Ragini Singh[2] and Snehasis Jana[2]\***

[1]*Trivedi Global Inc., 10624 S Eastern Avenue Suite A-969, Henderson, NV 89052, USA*
[2]*Trivedi Science Research Laboratory Pvt. Ltd., Hall-A, Chinar Mega Mall, Chinar Fortune City, Hoshangabad Rd., Bhopal, Madhya Pradesh, India*

## Abstract

Resorcinol is widely used in manufacturing of several drugs and pharmaceutical products that are mainly used for topical ailments. The main objective of this study is to use an alternative strategy i.e., biofield treatment to alter the physical, spectral and thermal properties of resorcinol. The resorcinol sample was divided in two groups, which served as control and treated group. The treated group was given biofield treatment and both groups i.e., control and treated were analysed using X-ray diffraction (XRD), Fourier transform-infrared (FT-IR) spectroscopy, UV-Visible (UV-Vis) spectroscopy, Differential scanning calorimetry (DSC) and Thermogravimetric analysis (TGA). The results showed a significant decrease in crystallite size of treated sample i.e., 104.7 nm as compared to control (139.6 nm). The FT-IR and UV-Vis spectra of treated sample did not show any change with respect to control. Besides, thermal analysis data showed 42% decrease in latent heat of fusion. The onset temperature of volatilization and temperature at which maximum volatilization happened was also decreased by 16% and 12.86%, respectively. The significant decrease in crystallite size may help to improve the spreadability and hence bioavailability of resorcinol in topical formulations. Also increase in volatilization temperature might increase the rate of reaction of resorcinol when used as intermediate. Hence, biofield treatment may alter the physical and thermal properties of resorcinol and make it more suitable for use in pharmaceutical industry.

**Keywords:** Resorcinol; Biofield energy treatment; X-Ray diffraction; Fourier transform infrared spectroscopy; Ultraviolet-Visible spectroscopy; Differential scanning calorimetry; Thermogravimetric analysis

## Abbreviations

XRD: X-Ray Diffraction; FT-IR: Fourier Transform Infrared; DSC: Differential Scanning Calorimetry; TGA: Thermogravimetric Analysis; DTG: Derivative Thermogravimetry; NCCAM: National Centre for Complementary and Alternative Medicine

## Introduction

Resorcinol is a dihydric phenol having the hydroxyl group at 1 and 3 positions in the benzene ring [1]. It occurs naturally in argan oil as main natural phenol. It is white crystalline powder having a faint odour and bitter-sweet taste [2]. It is used as a chemical intermediate in manufacturing of pharmaceuticals, dyestuffs and fungicides such as *p*-aminosalicylic acid, hexylresorcinol and light screening agents for protecting plastics from UV lights [3,4]. It is used in the formulation of several pharmaceuticals such as acne creams, hair dyes, anti-dandruff shampoos, and sun tan lotions. It also possesses various therapeutic uses such as topical antipruritic and antiseptic. It is used to treat seborrheic dermatitis, psoriasis, corns, warts and eczema. It is effective in the treatment of several dermatological problems due to its antibacterial, antifungal and keratolytic effects [5,6]. Resorcinol solution in ethyl alcohol (Jessner's solution) is used in chemical peeling [7], and it has special medical use as biological glue (gelatin-resorcinol-formaldehyde glue) in cardiovascular surgery [8,9]. Despite its wide pharmaceutical applications, some side effects are also associated with it, for instance, mild skin irritation, skin redness, etc. It is also hygroscopic i.e., absorb moisture from the air and turns pink on exposure to air or light [10]. By conceiving the usefulness of resorcinol, the present study was attempted to investigate an alternative way that can improve the physical and thermal properties of resorcinol. In recent years, biofield treatment was proved to be an alternative method that has an impact on various properties of living organisms and non-living materials in a cost effective manner. It is already demonstrated that energy can neither be created nor be destroyed, but it can be transferred through various processes such as thermal, chemical, kinetic, nuclear, etc. [11-13]. Similarly, electrical current exists inside the human body in the form of vibratory energy particles like ions, protons, and electrons and they generate a magnetic field in the human body [14,15]. This electromagnetic field of the human body is known as biofield, and energy associated with this field is known as biofield energy [16,17]. The human beings are infused with this precise form of energy, and it provides regulatory and communications functions within the organism [18,19]. The health of living organisms can be affected by balancing this energy from the environment through natural exchange process [20]. National Centre for Complementary and Alternative Medicine (NCCAM), which is part of the National Institute of Health (NIH) places biofield therapy (putative energy fields) as subcategory of energy medicine among complementary and alternative medicines. The healing therapy is also considered under this category [21,22]. Thus, the human has the ability to harness the energy from environment or universe and can transmit it to any living or non-living object. This process is termed as biofield treatment. Mr. Trivedi's unique biofield treatment (The Trivedi Effect®) is well known and significantly studied in different fields such as microbiology [23-25], agriculture [26-28], and biotechnology [29,30]. Recently, it was reported that biofield treatment has changed the atomic, crystalline and powder characteristics as well as spectroscopic

**\*Corresponding author:** Snehasis Jana, Trivedi Science Research Laboratory Pvt. Ltd., Hall-A, Chinar Mega Mall, Chinar Fortune City, Hoshangabad Rd., Bhopal-462026, Madhya Pradesh, India
E-mail: publication@trivedisrl.com

characters of different materials. Moreover, alteration in physical, thermal and chemical properties were also reported in materials like antimony, bismuth and ceramic oxide [31,32]. Hence, based on above results the current study was designed to determine the impact of biofield treatment on physical, spectral and thermal properties of resorcinol.

## Materials and Methods

### Study design

Resorcinol was procured from Loba Chemie Pvt. Ltd., India. The sample was divided into two parts and referred as control and treatment. The treatment sample in sealed pack was handed over to Mr. Trivedi for biofield treatment under standard laboratory conditions. Mr. Trivedi provided the treatment through his energy transmission process to the treatment group without touching the sample. The biofield treated sample was returned in the same sealed condition for further characterization using XRD, FT-IR, UV-Vis, DSC, and TGA techniques. For determination of FT-IR and UV-Vis spectroscopic characters, the treated sample was divided into two groups i.e., T1 and T2. Both treated groups were analysed for their spectral characteristics using FT-IR and UV-Vis spectroscopy as compared to control resorcinol sample.

### X-ray diffraction (XRD) study

XRD analysis was carried out on Phillips, Holland PW 1710 X-ray diffractometer system. The X-ray generator was equipped with a copper anode with nickel filter operating at 35 kV and 20 mA. The wavelength of radiation used by the XRD system was 1.54056 Å. The XRD spectra were acquired over the 2θ range of 10°-99.99° at 0.02° interval with a measurement time of 0.5 second per 2θ intervals. The data obtained were in the form of a chart of 2θ vs. intensity and a detailed table containing peak intensity counts, d value (Å), peak width (θ°), and relative intensity (%).

The average size of crystallite (G) was calculated from the Scherrer equation [33] with the method based on the width of the diffraction patterns obtained in the X-ray reflected crystalline region.

$$G=k\lambda/(bCos\theta)$$

Where, k is the equipment constant (0.94), λ is the X-ray wavelength (0.154 nm), B in radians is the full-width at half of the peaks and θ the corresponding Bragg angle.

Percent change in crystallite size was calculated using the following equation:

$$\text{Percent change in crystallite size}=[(G_t-G_c)/G_c] \times 100$$

Where, $G_c$ and $G_t$ are crystallite size of control and treated powder samples, respectively [34].

### Fourier transform-infrared (FT-IR) spectroscopic characterization

The powdered sample was mixed in spectroscopic grade KBr in an agate mortar and pressed into pellets with a hydraulic press. FT-IR spectra were recorded on Shimadzu's Fourier transform infrared spectrometer (Japan). FT-IR spectra are generated by the absorption of electromagnetic radiation in the frequency range 4000-400 cm$^{-1}$. The FT-IR spectroscopic analysis of resorcinol (control, T1 and T2) was carried out to evaluate the impact of biofield treatment at atomic and molecular level like bond strength, stability, rigidity of structure etc. [35].

### UV-Visible spectroscopic analysis

The UV-Vis spectral analysis was measured using Shimadzu UV-2400 PC series spectrophotometer. It involves the absorption of electromagnetic radiation from 200-400 nm range and subsequent excitation of electrons to higher energy states. It is equipped with 1 cm quartz cell and a slit width of 2.0 nm. The UV-Vis spectra of resorcinol were recorded in methanol solution at ambient temperature. This analysis was performed to evaluate the effect of biofield treatment on the structural property of resorcinol sample. The UV-Vis spectroscopy gives the preliminary information related to the skeleton of chemical structure and possible arrangement of functional groups. With UV-Vis spectroscopy, it is possible to investigate electron transfers between orbitals or bands of atoms, ions and molecules existing in the gaseous, liquid and solid phase [35].

### Differential scanning calorimetry (DSC) study

Differential scanning calorimeter (DSC) of Perkin Elmer/Pyris-1 was used to study the melting temperature and latent heat of fusion (ΔH). The DSC curves were recorded under air atmosphere (5 mL/min) and a heating rate of 10°C/min in the temperature range of 50°C to 350°C. An empty pan sealed with cover pan was used as a reference sample. Melting temperature and latent heat of fusion were obtained from the DSC curve.

Percent change in latent heat of fusion was calculated [36] using following equations to observe the difference in thermal properties of treated resorcinol sample as compared to control:

$$\% \text{ change in latent heat of fusion} = \frac{\left[\Delta H_{Treated} - \Delta H_{Control}\right]}{\Delta H_{Control}} \times 100$$

Where, $\Delta H_{Control}$ and $\Delta H_{Treated}$ are the latent heat of fusion of control and treated samples, respectively.

### Thermogravimetric analysis/Derivative thermogravimetry (TGA/DTG)

Thermal stability of control and treated samples of resorcinol was analysed by using Mettler Toledo simultaneous Thermogravimetric analyser (TGA/DTG). The samples were heated from room temperature to 400°C with a heating rate of 5°C/min under air atmosphere. From TGA curve, onset temperature $T_{onset}$ (temperature at which sample start losing weight) and from DTG curve, $T_{max}$ (temperature at which sample lost its maximum weight) were observed [37].

Percent change in $T_{max}$ was calculated using following equation:

$$\% \text{ change in } T_{max} = [(T_{max, treated} - T_{max, control})/ T_{max, control}] \times 100$$

Where, $T_{max, control}$ and $T_{max, treated}$ are the temperature at which sample lost its maximum weight due to volatilization in control and treated sample, respectively. Similarly, the percent change in onset temperature at which sample start losing weight was also calculated.

## Results and Discussion

### X-ray diffraction

X-ray diffraction analysis was conducted to study the crystalline nature of the control and treated samples of resorcinol. XRD diffractogram of control and treated samples of resorcinol are shown in Figure 1 and results are given in Table 1. The XRD diffractogram of control resorcinol showed intense crystalline peaks at 2θ equal to 18.04°, 18.18°, 19.11°, 19.68°, 19.93°, and 20.08°. The intense peaks indicated the crystalline nature of resorcinol. The XRD diffractogram

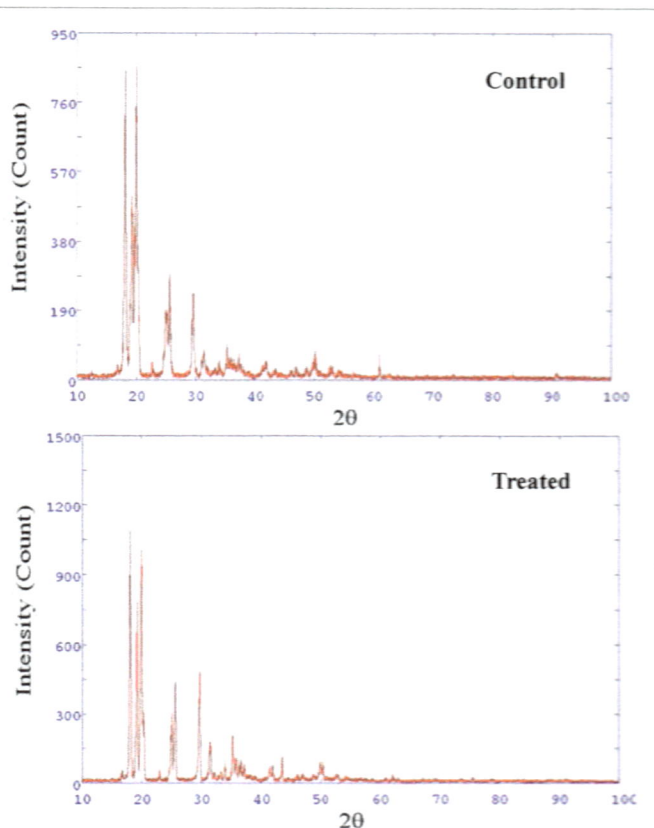

Figure 1: X-ray diffractogram (XRD) of control and treated samples of resorcinol.

| Parameter | Control | Treated |
|---|---|---|
| Volume of unit cell × 10$^{-23}$ (cm$^3$) | 57.162 | 57.190 |
| Crystallite size (nm) | 139.60 | 104.70 |

Table 1: XRD analysis of control and treated samples of resorcinol.

of treated resorcinol showed the crystalline peaks at 2θ equal to 18.04°, 18.18°, 19.19°, 20.04° and 29.73°. The peaks in treated sample showed high intensity as compared to control that indicated that crystallinity of treated resorcinol sample increased along the corresponding plane as compared to the control. It is presumed that biofield energy may be absorbed by the treated resorcinol molecules that may lead to form a symmetrical crystalline long range pattern that further results in increasing the symmetry of resorcinol molecules. Besides, the crystallite size was found to be 139.6 nm in control sample whereas, it was reduced to 104.7 nm in treated resorcinol. The crystallite size was reduced by 25% in treated resorcinol as compared to control. Other parameters like the volume of unit cell and molecular weight showed very slight change (0.05%) as compared to control sample. The effect of biofield treatment on crystallite size was also reported previously [37,38]. It is hypothesized that biofield treatment might produce the energy that causes the fracturing of grains into subgrains hence; the crystallite size was decreased in treated sample as compared to control. As resorcinol is used in many topical formulations, the decrease in crystallite size may improve its spreadability over the skin that further affects its bioavailability [39]. Hence, the treated resorcinol with decreased crystallite size may improve its bioavailability when used in topical formulations.

Spectroscopic studies

## FT-IR analysis

The FT-IR spectra of control and treated (T1 and T2) samples are shown in Figure 2. The spectra showed characteristic vibrational frequencies as follows:

**Carbon-Hydrogen vibrations:** The aromatic structure of resorcinol showed the presence of C-H stretching vibrations in the region 3100-3000 cm$^{-1}$ which was the characteristic region. The frequency of C-H stretching was overlapped with O-H stretching frequencies in all three samples, i.e., control, T1 and T2. The C-H in-plane bending vibrations were observed at 1379 cm$^{-1}$ in control and T1 sample whereas, at 1381 cm$^{-1}$ in T2 sample. The C-H out-of-plane bending vibrations appeared at 773 cm$^{-1}$ in control and T1 sample whereas, at 777 cm$^{-1}$ in T2 sample.

**Oxygen-Hydrogen vibrations:** In the present study, the O-H stretching vibration was observed at 3257-3207 cm$^{-1}$ in control sample whereas at 3263-3200 cm$^{-1}$ in T1 and 3281-3072 cm$^{-1}$ in T2 sample. Generally the O-H band were appeared at frequency range 3600-3300 cm$^{-1}$; however, broadening of the peak may occur in the presence of H-bonded O-H stretching. Hydrogen bonding may shift the peaks to lower frequencies as it was seen in FT-IR spectra of control and treated

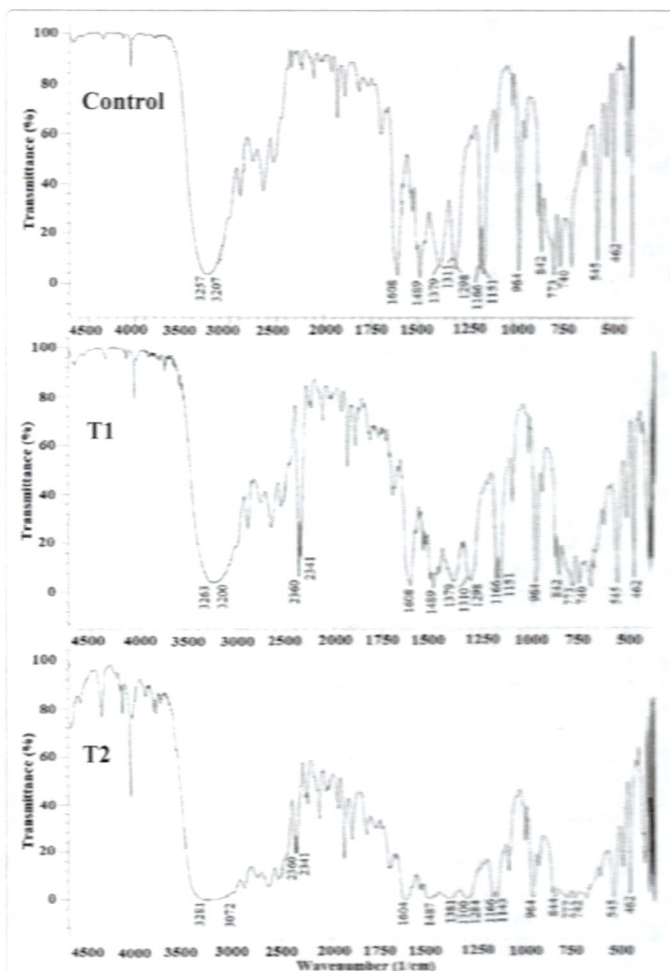

Figure 2: FT-IR spectra of control and treated (T1 and T2) samples of resorcinol.

samples of resorcinol. Hence, it confirmed the presence of H-bonding on resorcinol sample.

### C-OH group vibration

The most important peaks due to C-OH stretching mode were appeared at 1311-1298 cm$^{-1}$ and 1166-1151 cm$^{-1}$ as doublet peak in the control sample. In T1 sample, the peaks were appeared at 1310-1298 cm$^{-1}$ and 1166-1151 cm$^{-1}$ whereas; in T2 sample the peaks were appeared at 1300-1284 cm$^{-1}$ and 1166-1143 cm$^{-1}$. The C-OH bending peak was appeared at 462 cm$^{-1}$ in all three samples i.e., control, T1, and T2.

### Ring vibration

The fundamental vibrational modes of C-C stretching generally occurred in the region of 1600-1400 cm$^{-1}$. In the present study, the peaks observed at 1608 and 1489 cm$^{-1}$ in control and T1 sample were assigned to C-C stretching vibrations. Whereas, in T2, these peaks were appeared at 1604 and 1487 cm$^{-1}$. Another peak due to ring vibration was appeared at 545 cm$^{-1}$ in all three samples, i.e., control, T1, and T2. The other important peaks were appeared at 842 and 740 cm$^{-1}$ due to *meta* di-substituted ring in control and T1 sample. Whereas, the same peaks were appeared at 844 and 742 cm$^{-1}$ in T2 sample. The overall analysis was supported by literature data [4] and showed that there was no significant difference between observed frequencies of control and treated (T1 and T2) samples. Hence, it showed that biofield treatment might not induce any significant change at bonding level.

## UV-Vis spectroscopic analysis

The UV spectra of control and treated samples (T1 and T2) of resorcinol are shown in Figure 3. The UV spectrum of control sample showed absorption peaks at $\lambda_{max}$ equal to 205, 275 and 281 nm and was well supported by literature [40]. The absorbance peaks were appeared at the same wavelength in treated samples. In T1 sample, the peaks were found at $\lambda_{max}$ equal to 204, 275 and 281 nm and in T2 sample, they were appeared at $\lambda_{max}$ equal to 205, 275 and 281 nm. It showed that no change was found in UV spectroscopic properties, i.e., related to structure skeleton, functional groups or energy for electron transfers between orbitals or bands of atoms of treated resorcinol as compared to control.

## Thermal studies

**DSC analysis:** DSC was used to determine the latent heat of fusion ($\Delta$H) and melting temperature in control and treated samples of resorcinol. The DSC analysis results of control and treated samples of resorcinol are presented in Table 2. In a solid, the amount of energy required to change the phase from solid to liquid is known as the latent heat of fusion. The result showed that $\Delta$H was decreased from 179.77 J/g (control) to 103.47 J/g in treated resorcinol. It indicated that $\Delta$H was decreased by 42.45% in treated sample as compared to control. It was previously reported that resorcinol molecules possess rigid structure but as the temperature increases, this rigidity breaks down. The molecules rearrange into a hydrocarbon resembling structure and achieve lower van der walls interactions [41]. Hence, it is hypothesized that biofield treatment might produce the energy. This energy probably causes deformation of hydroxyl bond in treated resorcinol, and it needs less energy in the form of $\Delta$H to undergo the process of melting. Previously, our group reported that biofield treatment has altered $\Delta$H in lead and tin powder [42]. Moreover, the melting temperature of treated (112.56°C) sample showed very slight change with respect to control (111.18°C) resorcinol sample.

**TGA/DTG analysis:** TGA/DTG of control and biofield treated samples are summarized in Table 2. TGA thermogram (Figure 4) showed that control resorcinol sample started losing weight around 200°C (onset) and stopped around 246°C (end set) which could be due to volatilization of resorcinol [43]. However, the treated resorcinol started losing weight around 168°C (onset) and terminated around 215°C (end set). It indicated that onset temperature of treated resorcinol was decreased by 16% as compared to control. Besides, DTG

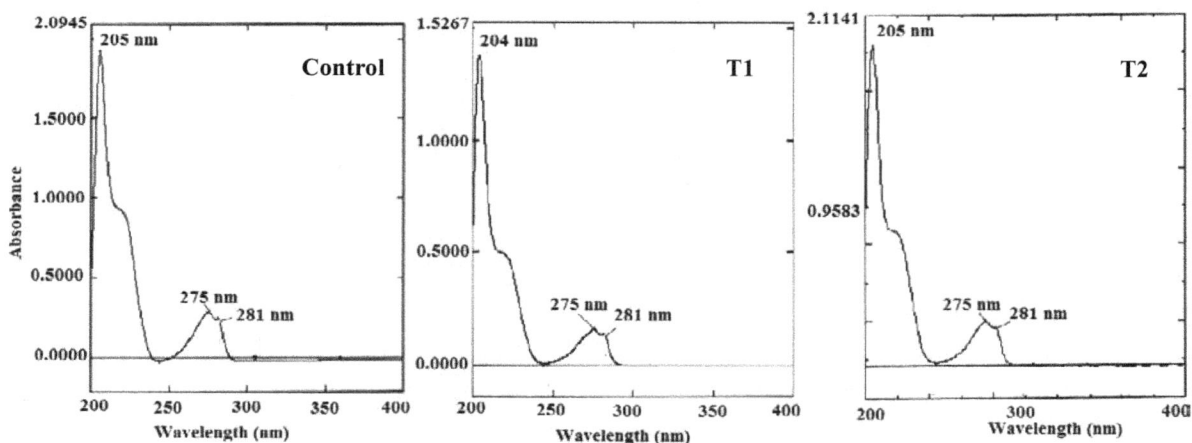

**Figure 3:** UV-Vis spectra of control and treated (T1 and T2) samples of resorcinol.

| Parameter | Control | Treated |
|---|---|---|
| Latent heat of fusion $\Delta$H (J/g) | 179.77 | 103.47 |
| Melting point (°C) | 111.18 | 112.56 |
| Onset temperature (°C) | 200 | 168 |
| $T_{max}$ (°C) | 217.11 | 189.2 |

**Table 2:** Thermal analysis of control and treated samples of resorcinol. $T_{max}$: Temperature at which maximum weight loss occur.

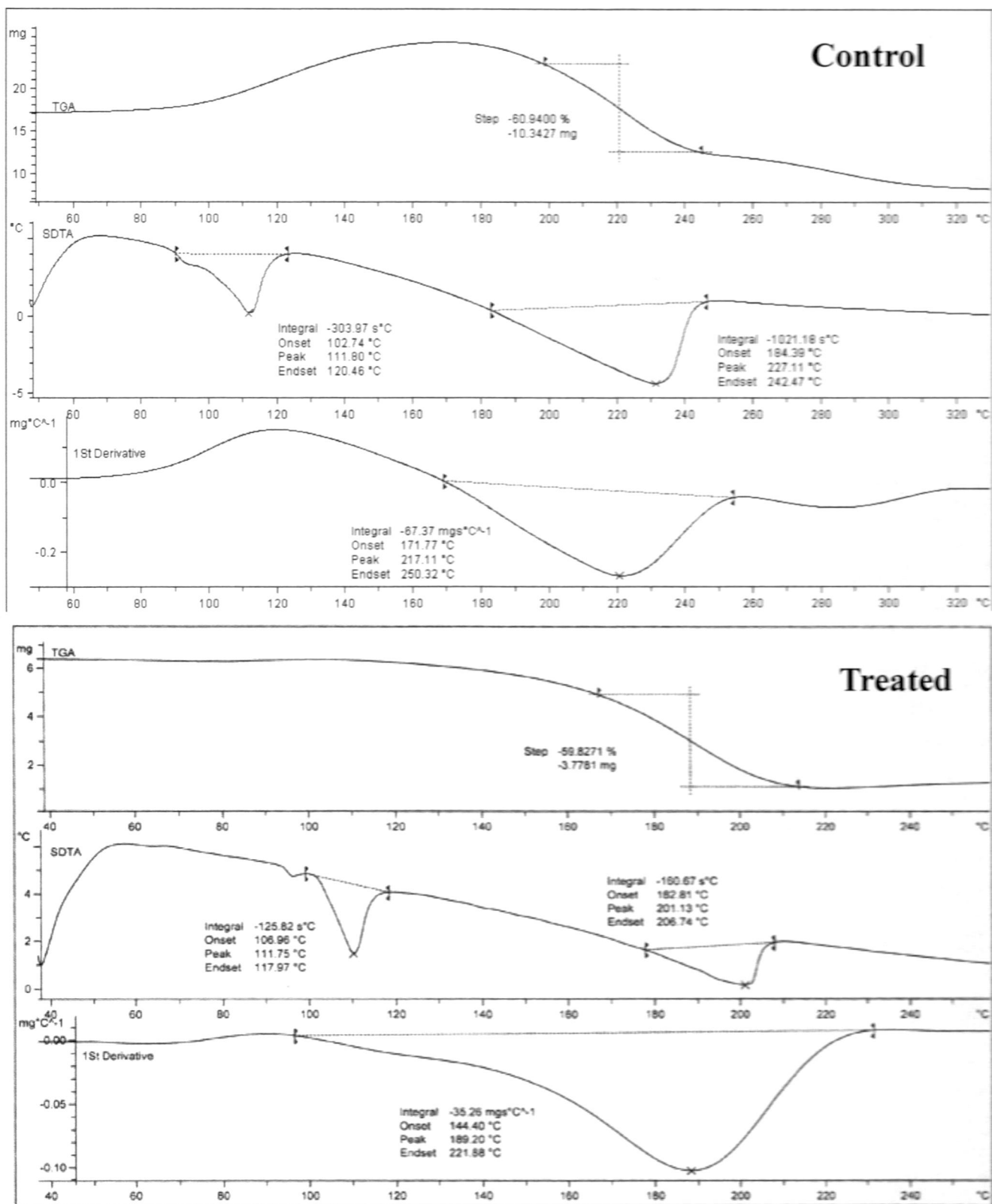

**Figure 4:** TGA/DTG thermogram of control and treated samples of resorcinol.

thermogram data showed that $T_{max}$ in control sample was 217.11°C and in treated sample, it was found at 189.2°C. It showed that $T_{max}$ was decreased by 12.86% in treated sample as compared to control. Furthermore, the reduction in onset temperature and $T_{max}$ in treated resorcinol with respect to control sample may be correlated with the increase in volatilization of treated resorcinol after biofield treatment. A possible reason for this reduction is that biofield energy might cause some alteration in internal energy which probably resulted into earlier volatilization of treated resorcinol sample as compared to control. Also, decrease in volatilization temperature indicated that resorcinol molecules change their phase from liquid to gas at low temperature, which may results in fasten the rate of those reactions where resorcinol can be used as an intermediate in synthesis [44]. Hence, overall observations suggest that biofield treated resorcinol can be used to enhance the reaction kinetics and yield of the end product.

## Conclusion

The XRD results showed the decrease in crystallite size (25%) in treated sample as compared to the control that may occur due to biofield treatment that probably produces the energy which leads to fracturing of grains into subgrains. The reduced crystallite size of treated resorcinol sample may be used to improve its bioavailability in topical preparations. The DSC analysis of treated sample showed 42.45% decrease in ΔH value as compared to control, which probably occurred due to deformation of hydroxyl bond in treated sample. The biofield treatment might affect the structure rigidity of resorcinol and hence reduced the latent heat of fusion. TGA/DTG analysis revealed that onset temperature of volatilization and $T_{max}$ were decreased by 16% and 12.86%, respectively. This reduction in volatilization temperature of treated sample might be helpful for resorcinol to be used as a chemical intermediate in the synthesis of various pharmaceuticals. Hence, above study concluded that biofield treatment might alter the physical and thermal properties of resorcinol that could make it more useful in pharmaceutical industries by increasing the bioavailability and reaction kinetics.

### Acknowledgements

The authors would like to acknowledge the whole team of Sophisticated Analytical Instrument Facility (SAIF), Nagpur and MGV Pharmacy College, Nashik for providing the instrumental facility. We are very grateful to Trivedi Science, Trivedi Master Wellness and Trivedi Testimonials for their support in this research work.

### References

1. Durairaj RB (2005) Resorcinol: Chemistry, technology and applications. Springer-Verlag, Berlin, Germany.

2. Charrouf Z, Guillaume D (2007) Phenols and polyphenols from Argania spinosa. Am J Food Technol 2: 679-683.

3. Schmiedel KW, Decker D (2000) Resorcinol. Ullmann's encyclopedia of industrial chemistry. Wiley-VCH, Weinheim.

4. Dressler H (1994) Resorcinol, its uses and derivatives. Springer, NewYork, USA.

5. Hahn S (2006) Resorcinol. Concise international chemical assessment document. WHO Press, Geneva, Switzerland.

6. Karam PG (1993) 50% resorcinol peel. Int J Dermatol 32: 569-574.

7. Hernandez-Perez E, Jaurez-Arce V (2000) Gross and microscopic findings with a combination of Jessner solution plus 53% resorcinol paste in chemical peels. J Cosmet Surg 17: 85-89.

8. Bachet J, Guilmet D (1999) The use of biological glue in aortic surgery. Cardiol Clin 17: 779-796.

9. Kazui T, Washiyama N, Bashar AH, Terada H, Suzuki K, et al. (2001) Role of biologic glue repair of proximal aortic dissection in the development of early and midterm redissection of the aortic root. Ann Thorac Surg 72: 509-514.

10. Gile TJ (2004) Safety never takes a holiday. Clin Leadersh Manag Rev 18: 342-348.

11. Becker RO, Selden G (1985) The body electric: Electromagnetism and the foundation of life, New York City, William Morrow and Company.

12. Barnes RB (1963) Thermography of the human body. Science 140: 870-877.

13. Born M (1971) The Born-Einstein Letters. (1st edn), Walker and Company, New York.

14. Prakash S, Chowdhury AR, Gupta A (2015) Monitoring the human health by measuring the biofield "aura": An overview. IJAER 10: 27637-27641.

15. Einstein A (1905) Does the inertia of a body depend upon its energy-content? Ann Phys 18: 639-641.

16. Rivera-Ruiz M, Cajavilca C, Varon J (2008) Einthoven's string galvanometer: the first electrocardiograph. Tex Heart Inst J 35: 174-178.

17. Rubik B (2002) The biofield hypothesis: its biophysical basis and role in medicine. J Altern Complement Med 8: 703-717.

18. Garland SN, Valentine D, Desai K, Langer C, Evans T, et al. (2013) Complementary and alternative medicine use and benefit finding among cancer patients. J Altern Complement Med 19: 876-881.

19. Peck SD (1998) The efficacy of therapeutic touch for improving functional ability in elders with degenerative arthritis. Nurs Sci Q 11: 123-132.

20. Saad M, Medeiros RD (2012) Distant healing by the supposed vital energy-Scientific bases. Complementary Therapies for the Contemporary Healthcare. U.S.

21. Thomas AH (2012) Hidden in plain sight: The simple link between relativity and quantum mechanics. Swansea, UK.

22. NIH, National Center for Complementary and Alternative Medicine. CAM Basics. Publication 347. [October 2, 2008]. Available at: http://nccam.nih.gov/health/whatiscam/

23. Trivedi MK, Patil S, Shettigar H, Gangwar M, Jana S (2015) An effect of biofield treatment on multidrug-resistant Burkholderia cepacia: A multihost pathogen. J Trop Dis 3: 167.

24. Trivedi MK, Patil S, Shettigar H, Gangwar M, Jana S (2015) Antimicrobial sensitivity pattern of Pseudomonas fluorescens after biofield treatment. J Infect Dis Ther 3: 222.

25. Trivedi MK, Patil S, Shettigar H, Bairwa K, Jana S, et al. (2015) Phenotypic and biotypic characterization of Klebsiella oxytoca: An impact of biofield treatment. J Microb Biochem Technol 7: 203-206.

26. Shinde V, Sances F, Patil S, Spence A (2012) Impact of biofield treatment on growth and yield of lettuce and tomato. Aust J Basic Appl Sci 6: 100-105.

27. Sances F, Flora E, Patil S, Spence A, Shinde V, et al. (2013) Impact of biofield treatment on ginseng and organic blueberry yield. Agrivita J Agric Sci 35: 22-29.

28. Lenssen AW (2013) Biofield and fungicide seed treatment influences on soybean productivity, seed quality and weed community. Agricultural Journal 8: 138-143.

29. Nayak G, Altekar N (2015) Effect of biofield treatment on plant growth and adaptation. J Environ Health Sci 1: 1-9.

30. Patil SA, Nayak GB, Barve SS, Tembe RP, Khan RR, et al. (2012) Impact of biofield treatment on growth and anatomical characteristics of Pogostemon cablin (Benth.). Biotechnology 11: 154-162.

31. Dabhade VV, Tallapragada RR, Trivedi MK (2009) Effect of external energy on atomic, crystalline and powder characteristics of antimony and bismuth powders. Bull Mater Sci 32: 471-479.

32. Trivedi MK, Nayak G, Patil S, Tallapragada RM, Latiyal O, et al. (2015) Studies of the atomic and crystalline characteristics of ceramic oxide nano powders after bio field treatment. Ind Eng Manage 4: 161.

33. Alexander L, Klug HP (1950) Determination of crystallite size with the x-ray spectrometer. J Appl Phys 21: 137-142.

34. http://shodhganga.inflibnet.ac.in:8080/jspui/bitstream/10603/7080/16/16_chapter%207.pdf

35. Pavia DL, Lampman GM, Kriz GS (2001) Introduction to spectroscopy. (3rd edn), Thomson Learning, Singapore.

36. http://academic.brooklyn.cuny.edu/physics/sobel/CoreExpts/CalorLatent.html

37. Trivedi MK, Patil S, Tallapragada RM (2015) Effect of biofield treatment on the physical and thermal characteristics of aluminium powders. Ind Eng Manage 4: 151.

38. Trivedi MK, Patil S, Tallapragada RM (2012) Thought intervention through bio field changing metal powder characteristics experiments on powder characteristics at a PM plant. Future Control and Automation LNEE 173: 247-252.

39. Niazi SK (2009) Handbook of Pharmaceutical Manufacturing Formulations: Semisolid Products. (2nd edn), CRC Press.

40. Lide DR, Milne GWA (1994) Handbook of Data on Organic Compounds. (3rd edn), CRC Press, USA.

41. Robertson JM, Ubbelohde AR (1938) A new form of resorcinol II. Thermodynamic properties in relation to structure. Proc Royal Soc London Ser A 167: 136-147.

42. Trivedi MK, Patil S, Tallapragada RM (2013) Effect of biofield treatment on the physical and thermal characteristics of silicon, tin and lead powders. J Material Sci Eng 2: 125.

43. O'Neil MJ (2013) The Merck Index - An encyclopedia of chemicals, drugs, and biologicals. Royal Society of Chemistry, Cambridge, UK.

44. Espenson JH (1995) Chemical kinetics and reaction mechanisms. (2nd edn), Mcgraw-Hill, U.S.

# 11

# 3',5'-Dibromo-2',4'-dihydroxy Substituted Chalcones: Synthesis and *in vitro* Trypanocidal Evaluation

**K. L. Ameta[1]\*, Nitu S. Rathore[1], Biresh Kumar[1], Edith S. Maalaga M[2], Manuela Veraastegui[2] and Robert H. Gilman[3]**

[1]*Department of Chemistry, FASC, Mody Institute of Technology & Science, Lakshmangarh-332311, Rajasthan, India*
[2]*Department of Microbiology, Faculty of Science and Philosophy, Universidad Peruana Cayetano Heredia, Lima, Peru*
[3]*Department of International Health, Bloomberg School of Public Health, Johns Hopkins University, Baltimore, Maryland-21205, USA*

## Abstract

A new series of 3', 5'-dibromo-2', 4'-dihydroxy substituted chalcones was synthesized and evaluated for their inhibitory effect against *Trypanosoma cruzi* (Chagas disease). Some of these compounds 3c, 3g and 3m showed 85.53, 85.03 and 83.34 *invitro* percentage growth of inhibition respectively; while compounds 3b, 3e, 3i and 3l showed 71.23, 71.95, 67.53 and 68.88 percentage growth of inhibition respectively with nifurtimox and benznidazole as reference drugs. 3l was the compound with a good anti-trypanocidal activity, the lower cytotoxicity, higher therapeutic index (14.5), and was the best candidate in comparison with the others. The structures of the newly synthesized compounds (3a-t) were determined by elemental analysis, FTIR, $^1$HNMR, $^{13}$CNMR and mass spectroscopic studies.

**Keywords:** Chalcones; *Trypanosoma cruzi*; Chagas disease; Inhibition

## Introduction

Currently, an infectious disease crisis of global proportion is threatening hard-won gains in health and life expectancy. Infectious diseases are the world's largest killer of children and young adults. Chagas disease is one of them which are caused by the protozoan parasite *T. cruzi*. It is a major cause of illness, morbidity, long-term disability, and death in Latin America. This disease is the third largest parasitic disease burden in the world and an estimated 10 million people are infected with this disease worldwide, mostly in Latin America [1]. In Latin America, infection with *T. cruzi* is responsible for Chagas disease, which is the leading cause of heart disease [2]. Despite the alarming health, economic and social consequences of this parasite infection and the limited existing drug therapy (nifurtimox and benznidazole) suffer from a combination of drawbacks including poor efficacy and serious side effects. Therefore, there is an urgent need for new chemotherapeutic agents with novel mechanisms of action [3-6]. Chalcones are a diverse group of compounds that can be synthesized or obtained from natural sources. This type compounds is 1, 3-diaryl-2-propen-1-ones and belong to the flavonoid family. These compounds are small molecules that exert various biological activities [7-10]. Moreover, they provide an opportunity for chemist to synthesize a wide variety of bioactive heterocycles [11-14] due to the presence of α, β-unsaturated carbonyl functionality. However, the search for an efficient synthesis for chalcones remains a challenging task.

The most widely used classical method for synthesizing chalcones is the Claisen-Schmidt condensation or through microwave irradiation [15-17] and ultrasonic irradiation because of their rapidity and improvement in yields [18,19]. In a continuation of our earlier endeavour [20-22] to design and synthesize novel bioactive heterocycles, and to consider the biological and medicinal importance of chalcones, herein, we reported a new series of chalcones via conventional as well as non-conventional microwave irradiation method and their trypanocidal evaluation.

## Results and Discussion

### Chemistry

3',5'-dibromo-2',4'-dihydroxyacetophenone (1) was treated with substituted aromatic aldehydes (2) (Note: From the structure, they are not aromatic aldehydes) to give substituted chalcones (3a-t) as shown in Scheme 1, with 74-84% yields. The structures of the newly synthesized compounds (3a-t) were determined on the basis of analytical and spectroscopic data. Thus, FTIR spectrum showed bands at 1688-1722 (C=O), 1642-1654 (-CH=CH), $^1$H NMR spectrum revealed the presence of a doublet at δ 7.48-7.98 corresponding to α Hydrogen and doublet at δ 8.11-8.42 corresponding to β Hydrogen and $^{13}$C NMR spectrum revealed the presence of Cα group (122.96-131.94), Cβ group (133.36-155.64), C=O group (187.87-192.40) ppm. All these newly synthesized compounds were evaluated for their *in vitro* trypanocidal evaluation using nifurtimox and benznidazole as reference drugs.

### Biological evaluation

The results of percentage growth of inhibition are summarized in (Table 1). The compounds 3a, 3d, 3f, 3h, 3j, 3k, 3n, 3o, 3r, 3s, 3t don't have trypanocidal activity but there are nine active compounds that do have trypanocidal activity, which were evaluated for their IC$_{50}$ and their cytotoxicity. Table 2 shows the compounds' IC$_{50}$, cytotoxicity, therapeutic index, and the values of the compounds with more anti trypanocidal activity. When the range of therapeutic index is short, testing *in vivo* is not recommended; we found that compounds 3b, 3c, 3e, 3g, 3i, 3l, 3m, 3p, and 3q have a good IC$_{50}$ and their cytotoxicity is low, which means that these compounds could be used *in vivo* studies. 3l was the compound with a good anti- trypanocidal activity, the lower cytotoxicity, higher therapeutic index 14.5, and is the best candidate in comparison with the others. This compound has a therapeutic index lower than that of benznidazole and nifurtimox.

**\*Corresponding author:** K. L. Ameta, Department of Chemistry, Faculty of Arts Science and Commerce (FASC), MITS University, Lakshmangarh, Sikar, Rajasthan-332311 India
E-mail: klameta77@hotmail.com, klameta.fasc@mitsuniversity.ac.in

## Experimental section

**General:** All melting points (m.p.) were determined in open capillaries on Veego (VMP – PM) melting point apparatus and are uncorrected. The purity of the compounds was routinely checked by thin layer chromatography (TLC) with Silica Gel-G (Merck). The instruments used for spectroscopic data are the IR spectrophotometer Brucker Alpha-Zn-Se, $^1$HNMR and $^{13}$CNMR (CDCl$_3$) on 500 MHz FT-NMR spectrometer Bruker AV III, GC-MS (EI-MS fragment) performed on JEOL GC Mate spectrometer and elemental analysis was carried out on a Carlo Erba 1108 analyzer within ± 0.5% of the theoretical values. Column chromatography was performed on silica

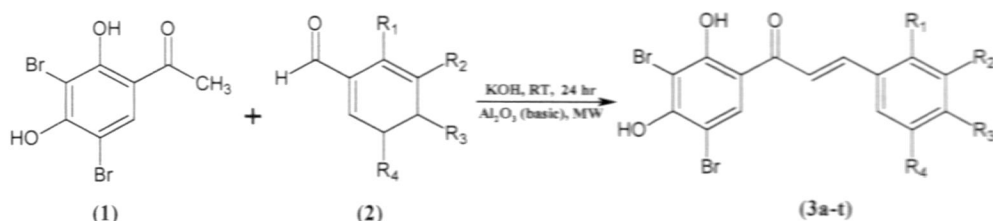

| Entry | 3a | 3b | 3c | 3d | 3e | 3f | 3g | 3h | 3i | 3j | 3k | 3l | 3m | 3n | 3o | 3p | 3q | 3r | 3s | 3t |
|---|---|---|---|---|---|---|---|---|---|---|---|---|---|---|---|---|---|---|---|---|
| R$_1$ | H | Cl | H | H | Cl | F | H | CH$_3$ | H | H | H | H | H | H | OH | H | H | H | H | 2-furyl |
| R$_2$ | H | H | Cl | H | H | H | H | H | H | NO$_2$ | H | H | H | OH | H | OCH$_3$ | OCH$_3$ | OCH$_3$ | OCH$_3$ | |
| R$_3$ | H | H | H | Cl | Cl | H | F | H | CH$_3$ | H | OH | OCH$_3$ | Br | OCH$_3$ | H | OCH$_3$ | OCH$_3$ | OH | OH | |
| R$_4$ | H | H | H | H | H | H | H | H | H | H | H | H | H | H | Br | H | OCH$_3$ | H | Br | |

**Scheme 1:** Synthesis of the title compounds (3a-t).

| Entry | Concentration Used | % Growth of Inhibition | Entry | Concentration Used | % Growth of Inhibition |
|---|---|---|---|---|---|
| 3a | 10 ug/mL | 9.84 | 3l | 10 ug/mL | 68.88 |
| 3b | 10 ug/mL | 71.23 | 3m | 10 ug/mL | 83.34 |
| 3c | 10 ug/mL | 85.53 | 3n | 10 ug/mL | 33.14 |
| 3d | 10 ug/mL | 29.51 | 3o | 10 ug/mL | 9.58 |
| 3e | 10 ug/mL | 71.95 | 3p | 10 ug/mL | 54.22 |
| 3f | 10 ug/mL | 25.57 | 3q | 10 ug/mL | 51.41 |
| 3g | 10 ug/mL | 85.03 | 3r | 10 ug/mL | 15.41 |
| 3h | 10 ug/mL | 13.06 | 3s | 10 ug/mL | 10.40 |
| 3i | 10 ug/mL | 67.53 | 3t | 10 ug/mL | 5.80 |
| 3j | 10 ug/mL | 19.34 | Nifurtimox | 10 ug/mL | 68.50 |
| 3k | 10 ug/mL | 23.32 | Benznidazole | 10 ug/mL | 86.77 |

*Each value is the mean of three experiments

**Table 1:** Biological evaluation of synthesized chalcones against *Trypanosoma cruzi*; % Growth inhibition.

| Entry | Concentration used | % Growth of Inhibition | IC$_{50}$[a] (ug/mL) | Cytotoxicity[b] (ug/mL) | T. I.[c] |
|---|---|---|---|---|---|
| 3b | 10 ug/mL | 71.23 | 6.04 | 32 | 5.3 |
| 3c | 10 ug/mL | 85.53 | 4.55 | 10 | 2.2 |
| 3e | 10 ug/mL | 71.95 | 5.53 | 32 | 5.8 |
| 3g | 10 ug/mL | 85.03 | 5.24 | 25 | 4.8 |
| 3i | 10 ug/mL | 67.53 | 5.48 | 30 | 5.5 |
| 3l | 10 ug/mL | 68.88 | 2.07 | 14 | 14.5 |
| 3m | 10 ug/mL | 83.34 | 5.42 | 30 | 2.6 |
| 3p | 10 ug/mL | 54.22 | 4.54 | 28 | 6.2 |
| 3q | 10 ug/mL | 51.41 | 9.23 | 25 | 2.7 |
| Nifurtimox | 10 ug/mL | 68.50 | 0.47 | 27 | 57.44 |
| Benznidazole | 10 ug/mL | 86.77 | 0.81 | >50 | >62 |

*Each value is the mean of three experiments
* [a] IC$_{50}$: concentration that produces 50% inhibitory effect, [b]cytotoxicity TD: dose required to produce of 50% cell LLCMK2, [c] TI: therapeutic index: IC$_{50\ LLC-MK2}$/IC$_{50,\ T\ cruzi}$

**Table 2:** Biological evaluation of active samples against *Trypanosoma cruzi*: IC$_{50}$, Cytotoxicity & therapeutic index.

gel (Merck, 60-120 mesh). Microwave assisted reaction was carried out on a commercially modified MW synthesis system model CATA-R, operating 700W, generating 2450 MHz frequency.

### Synthetic procedures for (3a-t):

- #### Conventional solution phase method

A mixture of 3',5'-dibromo-2',4'-dihydroxyacetophenone 1 (0.01 mol) and substituted aromatic aldehydes 2 (Note: From the structure, they are not aromatic aldehydes ) (0.01 mol) was stirred in 30 mL ethanol and then 15mL 40% KOH solution was added to it. The mixture was kept overnight at room temperature and then it was poured into crushed ice and acidified with HCl. The solid was obtained by filtering and then it was crystallized from ethanol and offered the analytical samples of (3a-t).

- #### Non-conventional solid phase method

To a solution of 3',5'-dibromo-2',4'-dihydroxyacetophenone 1 (0.01 mol) and substituted aromatic aldehyde 2 (Note: From the structure, they are not aromatic aldehydes ) (0.01 mol) in 1 mL DMF placed in 100 mL borosil flask, was added 4 g basic alumina. The mixture was uniformly mixed with glass rod and dried to remove the solvent under air. Adsorbed material was irradiated inside a microwave oven for 4-6 min. at medium power level (700 W). After completion of the reaction (monitored by TLC), the reaction mixture was cooled at room temperature and the product was extracted with dichloromethane (2×20 mL). Removal of the solvent and subsequent recrystallization from ethanol afforded analytical samples of (3a-t). The synthesis of title compounds is shown in Scheme 1 and the comparison of reaction times and yields of compounds (3a-t) under microwave and classical method are showed in Table 3.

**Spectroscopic data of the synthesized compounds are shown below:**

### (2E)-1-(3', 5'-dibromo-2', 4'-dihydroxyphenyl)-3-phenylprop-2-en-1-one

| Entry | Reaction time | | Yield (%) | |
|-------|---------------|---------------|-----|-----------|
| | MW (hrs or min?) | Classical (hrs) | MW | Classical |
| 3a | 7 | 17 | 76 | 72 |
| 3b | 8 | 18 | 79 | 69 |
| 3c | 6 | 18 | 77 | 67 |
| 3d | 7 | 19 | 80 | 68 |
| 3e | 8 | 18 | 81 | 65 |
| 3f | 9 | 17 | 83 | 66 |
| 3g | 6 | 20 | 80 | 70 |
| 3h | 8 | 22 | 77 | 64 |
| 3i | 8 | 21 | 81 | 65 |
| 3j | 7 | 19 | 80 | 68 |
| 3k | 7 | 20 | 80 | 66 |
| 3l | 7 | 20 | 84 | 67 |
| 3m | 8 | 19 | 80 | 68 |
| 3n | 9 | 21 | 82 | 66 |
| 3o | 9 | 20 | 79 | 65 |
| 3p | 7 | 19 | 84 | 70 |
| 3q | 8 | 21 | 81 | 69 |
| 3r | 7 | 22 | 77 | 67 |
| 3s | 6.9 | 22 | 78 | 68 |
| 3t | 7 | 20 | 74 | 61 |

**Table 3:** Comparison of reaction time and yield of synthesized chalcones under MW and classical method.

**(3a).** Yellow solid; Yield (?) m.p. 128-129°C. IR (KBr, cm$^{-1}$): 3350 (Ar-OH), 3071, 3009 (Ar-H), 1688 (-C=O), 1642 (-CH=CH), 862 (C-Br). $^1$HNMR (500 MHz, CDCl$_3$) δ: 7.20-7.40 (m, 5H, Ar-H), 7.51 (d, αH, J=15.4Hz), 8.32 (d, βH, J=15.4Hz), 8.10 (s, 1H, Ar-H), 11.10 (s, 2H, 2 x Ar-OH). $^{13}$CNMR (125 MHz, CDCl$_3$) δ: 79.55, 99.01, 103.62, 116.20, 123.96 (Cα), 126.14, 129.07, 130.48, 132.34, 133.37, 139.84 (Cβ), 160.37, 161.70, 190.95 (C=O). MS: m/z 397.80 (M$^+$). Calcd. for C$_{15}$H$_{10}$Br$_2$O$_3$: 45.25, H, 2.51%. Found: C, 45.30, H, 2.46%.

### (2E)-3-(2-chlorophenyl)-1-(3', 5'-dibromo-2', 4'-dihydroxyphenyl) prop-2-en-1-one

**(3b).** Orange solid; Yield (?) m.p. 120-121°C. IR (KBr, cm$^{-1}$): 3393 (Ar-OH), 3061, 3006 (Ar-H), 1693 (-C=O), 1645 (-CH=CH), 862 (C-Br), 753 (C-Cl). $^1$HNMR (500 MHz, CDCl$_3$) δ: 7.28-7.61 (m, 4H, Ar-H), 7.78 (d, αH, J=15.6Hz), 8.13 (s, 1H, Ar-H), 8.37 (d, βH, J=15.6Hz), 11.05 (s, 2H, 2 x Ar-OH). $^{13}$CNMR (125MHz, CDCl$_3$) δ: 79.55, 97.91, 105.92, 116.20, 124.89 (Cα), 127.04, 130.30, 131.03, 132.72, 133.38, 140.06 (Cβ), 160.97, 188.15 (C=O). MS: m/z 432.30 (M$^+$). Calcd. for C$_{15}$H$_9$Br$_2$O$_3$Cl: C, 41.64, H, 2.08%. Found: C, 41.69, H, 2.13%.

### (2E)-3-(3-chlorophenyl)-1-(3', 5'-dibromo-2', 4'-dihydroxyphenyl) prop-2-en-1-one

**(3c).** Yellow solid; Yield (?) m.p. 141-142°C. IR (KBr, cm$^{-1}$): 3368 (Ar-OH), 3071, 3009 (Ar-H), 1695 (-C=O), 1642 (-CH=CH), 862 (C-Br), 758 (C-Cl). $^1$HNMR (500 MHz, CDCl$_3$) δ: 7.22-7.42 (m, 4H, Ar-H), 7.48 (d, αH, J=15.6Hz), 10.72 (s, 2H, 2 x Ar-OH), 8.06 (s, 1H, Ar-H), 8.34 (d, βH, J=15.6Hz). $^{13}$C-NMR (125 MHz, CDCl$_3$): = 81.55, 97.97, 106.12, 116.43, 124.81 (Cα), 128.54, 130.30, 133.79, 134.03, 138.07 (Cβ), 148.22, 161.93, 187.87 (C=O). MS: m/z 432.30 (M$^+$). Calcd. for C$_{15}$H$_9$Br$_2$O$_3$Cl: C, 41.64, H, 2.08%. Found: C, 41.59, H, 2.12%.

### (2E)-3-(4-chlorophenyl)-1-(3', 5'-dibromo-2', 4'-dihydroxyphenyl) prop-2-en-1-one

**(3d).** Brown solid; Yield (?) m.p. 129-130°C. IR (KBr, cm$^{-1}$): 3365 (Ar-OH), 3054, 3011 (Ar-H), 1701 (-C=O), 1648 (-CH=CH), 862 (C-Br), 753 (C-Cl). $^1$H-NMR (500 MHz, CDCl$_3$) δ: 7.26-7.64 (m, 4H, Ar-H), 7.89 (d, αH, J=15.6Hz), 10.89 (s, 2H, 2 x Ar-OH), 8.08 (s, 1H, Ar-H), 8.40 (d, βH, J=15.6Hz). $^{13}$C NMR (125 MHz, CDCl$_3$) δ: 88.37, 94.61, 104.43, 115.19, 124.89 (Cα), 128.54, 130.71, 133.38, 136.19, 140.06 (Cβ), 149.21, 163.93, 188.15 (C=O). MS: m/z 432.30 (M$^+$). Calcd. for C$_{15}$H$_9$Br$_2$O$_3$Cl: C, 41.64, H, 2.08%. Found: C, 41.59, H, 2.12%.

### (2E)-1-(3', 5'-dibromo-2', 4'-dihydroxyphenyl)-3-(2, 4-dichlorophenyl) prop-2-en-1-one

**(3e).** Orange solid; Yield (?). m.p. 124-125°C. IR (KBr, cm$^{-1}$): 3394 (Ar-OH), 3081, 3005 (Ar-H), 1699 (-C=O), 1652 (-CH=CH), 862 (C-Br), 783,754 (C-Cl). $^1$HNMR (500 MHz, CDCl$_3$) δ: 7.36-7.69 (m, 3H, Ar-H), 7.75 (d, αH, J=15.6Hz), 8.08 (s, 1H, Ar-H), 8.37 (d, βH, J=15.6Hz), 11.05 (s, 2H, 2 x Ar-OH). $^{13}$CNMR (125 MHz, CDCl$_3$) δ: 87.19, 96.27, 108.76, 115.19, 125.40 (Cα), 128.04, 131.93, 133.08, 135.40, 143.72 (Cβ), 149.21, 163.93, 188.22 (C=O). MS: m/z 466.80 (M$^+$). Calcd. for C$_{15}$H$_8$Br$_2$O$_3$Cl$_2$: C, 38.56, H, 1.71%. Found: C, 38.62, H, 1.76%.

### (2E)-1-(3', 5'-dibromo-2', 4'-dihydroxyphenyl)-3-(2-fluorophenyl) prop-2-en-1-one

**(3f).** Brown solid; Yield (?), m.p. 102-103°C. IR (KBr, cm$^{-1}$): 3410 (Ar-OH), 3092, 3009 (Ar-H), 1710 (-C=O), 1651 (-CH=CH), 862 (C-Br), 1256 (C-F). $^1$H-NMR (500 MHz, CDCl$_3$) δ: 7.28-7.52 (m, 4H, Ar-H), 7.82 (d, αH, J=15.6Hz), 10.95 (s, 2H, 2 x Ar-OH), 8.06 (s, 1H,

Ar-H), 8.25 (d, βH, J=15.6Hz). $^{13}$C-NMR (125 MHz, CDCl$_3$) δ: 79.73, 98.98, 103.44, 115.23, 127.82 (Cα), 129.86, 130.92, 133.67, 145.26 (Cβ), 149.29, 155.66, 163.88, 188.65 (C=O). MS: m/z 415.80 (M$^+$). Calcd. for C$_{15}$H$_9$Br$_2$O$_3$F: C, 43.29, H, 2.16%. Found: C, 43.34, H, 2.11%.

### (2E)-1-(3', 5'-dibromo-2', 4'-dihydroxyphenyl)-3-(4-fluorophenyl) prop-2-en-1-one

(3g). Brown solid; Yield (?), m.p. 158-159°C. IR (KBr, cm$^{-1}$): 3372 (Ar-OH), 3089, 3005 (Ar-H), 1705 (-C=O), 1652 (-CH=CH), 862 (C-Br), 1261 (C-F). $^1$H-NMR (500 MHz, CDCl$_3$) δ: 7.26-7.61 (m, 4H, Ar-H), 7.89 (d, αH, J=15.6Hz), 10.90 (s, 2H, 2 x Ar-OH), 8.09 (s, 1H, Ar-H), 8.34 (d, βH, J=15.6Hz). $^{13}$C-NMR (125 MHz, CDCl$_3$) δ: 81.23, 99.09, 103.49, 116.29, 127.83 (Cα), 130.86, 132.29, 133.37, 145.96 (Cβ), 155.66, 158.35, 163.57, 190.95 (C=O). MS: m/z 415.80 (M$^+$). Calcd. for C$_{15}$H$_9$Br$_2$O$_3$F: C, 43.29, H, 2.16%. Found: C, 43.24, H, 2.22%.

### (2E)-1-(3', 5'-dibromo-2', 4'-dihydroxyphenyl)-3-(2-methylphenyl) prop-2-en-1-one

(3h). Yellow solid; Yield (?), m.p. 119-120°C. IR (KBr, cm$^{-1}$): 3395 (Ar-OH), 3289 (CH$_3$), 3093, 3007 (Ar-H), 1710 (-C=O), 1649 (-CH=CH), 862 (C-Br). $^1$H-NMR (500 MHz, CDCl$_3$) δ: 2.63 (3H, s, -CH$_3$), 7.27-7.74 (m, 4H, Ar-H), 7.89 (d, αH, J=15.6Hz), 8.11 (s, 1H, Ar-H), 8.42 (d, βH, J=15.6Hz), 11.08 (s, 2H, 2 x Ar-OH). $^{13}$C-NMR (125 MHz, CDCl$_3$) δ: 21.21 (CH$_3$), 80.33, 99.47, 103.24, 117.86, 125.98 (Cα), 130.15, 132.28, 134.84, 146.70 (Cβ), 155.61, 158.57, 161.63, 189.11 (C=O). MS: m/z 411.80 (M$^+$). Calcd. for C$_{16}$H$_{12}$Br$_2$O$_3$: C, 46.62, H, 2.91%. Found: C, 46.55, H, 2.96%.

### (2E)-1-(3', 5'-dibromo-2', 4'-dihydroxyphenyl)-3-(4-methylphenyl) prop-2-en-1-one

(3i). Yellow solid; Yield (?), m.p. 117-118°C. IR (KBr, cm$^{-1}$): 3391 (Ar-OH), 3283 (CH$_3$), 3091, 3005 (Ar-H), 1699 (-C=O), 1645 (-CH=CH), 862 (C-Br). $^1$H-NMR (500 MHz, CDCl$_3$) δ: 2.67 (3H, s, -CH$_3$), 7.29-7.79 (m, 4H, Ar-H), 7.91 (d, αH, J=15.6Hz), 10.92 (s, 2H, 2 x Ar-OH), 8.07 (s, 1H, Ar-H), 8.39 (d, βH, J=15.6Hz). $^{13}$C-NMR (125 MHz, CDCl$_3$) δ: 21.65 (CH$_3$), 81.41, 99.66, 105.41, 117.81, 127.36 (Cα), 131.56, 133.36, 138.84, 147.43 (Cβ), 155.82, 160.37, 161.65, 191.16 (C=O) ppm. MS: m/z 411.80 (M$^+$). Calcd. for C$_{16}$H$_{12}$Br$_2$O$_3$: C, 46.62, H, 2.91%. Found: C, 46.67, H, 2.87%.

### (2E)-1-(3', 5'-dibromo-2', 4'-dihydroxyphenyl)-3-(3-nitrophenyl) prop-2-en-1-one

(3j). Orange solid; Yield (?), m.p. 159-160°C. IR (KBr, cm$^{-1}$): 3389 (Ar-OH), 3091, 3011 (Ar-H), 1715 (-C=O), 1654 (-CH=CH), 1531 (Asy Ar-NO$_2$), 1350 (Sym Ar-NO$_2$), 862 (C-Br). $^1$H-NMR (500 MHz, CDCl$_3$) δ: 7.25-7.78 (m, 4H, Ar-H), 7.97 (d, αH, J=15.6Hz), 10.90 (s, 2H, 2 x Ar-OH), 8.01 (s, 1H, Ar-H), 8.11 (d, βH, J=15.6Hz). $^{13}$C-NMR (125 MHz, CDCl$_3$) δ: 83.47, 99.61, 103.98, 116.87, 124.46 (Cα), 130.14, 133.36, 138.84, 144.25 (Cβ), 148.22, 152.11, 161.22, 186.79 (C=O). MS: m/z 442.80 (M$^+$). Calcd. for C$_{15}$H$_9$Br$_2$O$_5$N: C, 40.65, H, 2.03, N, 3.16%. Found: C, 40.59, H, 2.08, N, 3.12%.

### (2E)-1-(3', 5'-dibromo-2', 4'-dihydroxyphenyl)-3-(4-hydroxyphenyl) prop-2-en-1-one

(3k). Brown solid;Yiled (?), m.p. 101-102°C. IR (KBr, cm$^{-1}$): 3395 (Ar-OH), 3087, 3007 (Ar-H), 1710 (-C=O), 1648 (-CH=CH), 862 (C-Br). $^1$HNMR (500 MHz, CDCl$_3$) δ: 7.32-7.83 (m, 4H, Ar-H), 7.98 (d, αH, J=15.6Hz), 8.12 (s, 1H, Ar-H), 8.19 (d, βH, J=15.6Hz), 11.02 (s, 2H, 2 x Ar-OH), 11.50 (s, 1H, Ar-OH). $^{13}$CNMR (125 MHz, CDCl$_3$) δ: 80.20, 99.41, 115.25, 116.13, 123.88 (Cα), 131.09, 133.38, 155.64

(Cβ), 160.35, 162.81, 190.92 (C=O). MS: m/z 413.80 (M$^+$). Calcd. for C$_{15}$H$_{10}$Br$_2$O$_4$: C, 43.50, H, 2.42%. Found: C, 43.55, H, 2.37%.

### (2E)-1-(3', 5'-dibromo-2', 4'-dihydroxyphenyl)-3-(4-methoxyphenyl) prop-2-en-1-one

(3l). Red solid; Yiled (?). m.p. 159-160°C. IR (KBr, cm$^{-1}$): 3402 (Ar-OH), 3089, 3007 (Ar-H), 2830 (OCH$_3$), 1715 (-C=O), 1651 (-CH=CH), 862 (C-Br). $^1$HNMR (500 MHz, CDCl$_3$) δ: 3.86 (s, 3H, OCH$_3$), 7.00-7.48 (m, 4H, Ar-H), 7.84 (d, αH, J=16.0Hz), 10.82 (s, 2H, 2 x Ar-OH), 8.08 (s, 1H, Ar-H), 8.17 (d, βH, J=16.0Hz). $^{13}$CNMR (125 MHz, CDCl$_3$) δ: 56.29 (OCH$_3$), 80.25, 99.84, 115.05, 116.04, 123.87 (Cα), 129.31, 131.71, 133.40, 138.26 (Cβ), 156.20, 160.35, 161.77, 190.77 (C=O). MS: m/z 427.80 (M$^+$). Calcd. for C$_{16}$H$_{12}$Br$_2$O$_4$: C, 44.88, H, 2.80%. Found: C, 44.83, H, 2.74%.

### (2E)-3-(4-bromophenyl)-1-(3', 5'-dibromo-2', 4'-dihydroxyphenyl) prop-2-en-1-one

(3m). Orange solid; Yield (?). m.p. 130-131°C. IR (KBr, cm$^{-1}$), 3418 (Ar-OH), 3087, 3009 (Ar-H), 1718 (-C=O), 1652 (-CH=CH), 862 (C-Br). $^1$HNMR (500 MHz, CDCl$_3$) δ: 7.04-7.58 (m, 4H, Ar-H), 7.88 (d, αH, J=15.6Hz), 10.93 (s, 2H, 2 x Ar-OH), 8.06 (s, 1H, Ar-H), 8.14 (d, βH, J=15.6Hz). $^{13}$CNMR (125 MHz, CDCl$_3$) δ: 77.28, 99.57, 115.51, 118.13, 124.14 (Cα), 129.75, 132.19, 133.04, 136.16 (Cβ), 153.66, 160.35, 161.97, 188.79 (C=O). MS: m/z 476.70 (M$^+$). Calcd. for C$_{15}$H$_9$Br$_3$O$_3$: C, 37.76, H, 1.89%. Found: C, 37.70, H, 1.93%.

### (2E)-1-(3', 5'-dibromo-2', 4'-dihydroxyphenyl)-3-(3-hydroxy-4-methoxyphenyl) prop-2-en-1-one

(3n). Orange solid; Yield (?). m.p. 152-153°C. IR (KBr, cm$^{-1}$) :3398 (Ar-OH), 3087, 3011 (Ar-H), 2830 (OCH$_3$), 1720 (-C=O), 1649 (-CH=CH), 862 (C-Br). $^1$HNMR (500 MHz, CDCl$_3$) δ: 3.87 (s, 3H, OCH$_3$), 7.11-7.61 (m, 4H, Ar-H), 7.91 (d, αH, J=15.6Hz), 8.00 (s, 1H, Ar-H), 8.11 (d, βH, J=15.6Hz), 11.13 (s, 2H, 2 x Ar-OH). $^{13}$CNMR (125 MHz, CDCl$_3$) δ: 56.39 (OCH$_3$), 80.25, 99.84, 116.04, 119.56, 122.96 (Cα), 129.87, 132.12, 133.16, 145.08 (Cβ), 153.66, 160.36, 161.68, 192.40 (C=O). MS: m/z 443.80 (M$^+$). Calcd. for C$_{16}$H$_{12}$Br$_2$O$_5$: C, 43.26, H, 2.70%. Found: C, 43.32, H, 2.65%.

### (2E)-3-(5-bromo-2-hydroxyphenyl)-1-(3', 5'-dibromo-2', 4'-dihydroxyphenyl) prop-2-en-1-one

(3o). Brown solid; Yield (?). m.p. 121-122°C. IR (KBr, cm$^{-1}$): 3432 (Ar-OH), 3089, 3009 (Ar-H), 1721 (-C=O), 1652 (-CH=CH), 862 (C-Br). $^1$HNMR (500 MHz, CDCl$_3$) δ: 7.03-7.68 (m, 4H, Ar-H), 7.94 (d, αH, J=15.6Hz), 10.78 (s, 2H, 2 x Ar-OH), 8.01 (s, 1H, Ar-H), 8.16 (d, βH, J=15.6Hz). $^{13}$C-NMR (125 MHz, CDCl$_3$) δ: 77.27, 99.03, 116.20, 119.64, 125.75 (Cα), 129.81, 132.29, 133.37, 135.85 (Cβ), 155.73, 158.25, 160.36, 190.71 (C=O). MS: m/z 492.70 (M$^+$). Calcd. for C$_{15}$H$_9$Br$_3$O$_4$: C, 36.53, H, 1.83%. Found: C, 36.58, H, 1.76%.

### (2E)-1-(3', 5'-dibromo-2', 4'-dihydroxyphenyl)-3-(3, 4-dimethoxyphenyl) prop-2-en-1-one

(3p). Brown solid; Yield (?). m.p. 151-152°C. IR (KBr, cm$^{-1}$): 3436 (Ar-OH), 3091, 3009 (Ar-H), 2832 (OCH$_3$), 1699 (-C=O), 1649 (-CH=CH), 862 (C-Br). $^1$HNMR (500 MHz, CDCl$_3$) δ: 3.93 (s, 6H, 2 x OCH$_3$), 6.99-7.48 (m, 3H, Ar-H), 7.98 (d, αH, J=15.6Hz), 10.97 (s, 2H, 2 x Ar-OH), 8.02 (s, 1H, Ar-H), 8.14 (d, βH, J=15.6Hz). $^{13}$C-NMR (125 MHz, CDCl$_3$) δ: 56.77 (OCH$_3$), 60.81, 99.89, 105.78, 109.64, 115.05, 116.15, 126.51 (Cα), 129.31, 132.41, 132.97, 135.68 (Cβ), 153.13, 160.35, 162.17, 189.21 (C=O). MS: m/z 457.80 (M$^+$). Calcd. for C$_{17}$H$_{14}$Br$_2$O$_5$: C, 44.56, H, 3.06%. Found: C, 44.51, H, 3.11%.

**(2E)-1-(3', 5'-dibromo-2', 4'-dihydroxyphenyl)-3-(3,4,5-trimethoxyphenyl) prop-2-en-1-one**

**(3q).**Orange solid; Yield (?) m.p. 128-129°C. IR (KBr, cm⁻¹): 3438 (Ar-OH), 3091, 3009 (Ar-H), 2832 (OCH₃), 1699 (-C=O), 1649 (-CH=CH), 862 (C-Br). ¹HNMR (500 MHz, CDCl₃) δ: 3.87 (s, 9H, 3 x OCH₃), 6.97-7.53 (m, 2H, Ar-H), 7.98 (d, αH, J=15.6Hz), 8.05 (s, 1H, Ar-H), 8.16 (d, βH, J=15.6Hz), 11.05 (s, 2H, 2 x Ar-OH). ¹³C-NMR (125 MHz, CDCl₃) δ: 56.17 (OCH₃), 56.45, 60.89, 99.81, 105.16, 110.64, 113.38, 120.56, 126.57 (Cα), 128.09, 131.31, 134.63 (Cβ), 155.13, 160.35, 162.17, 191.12 (C=O). MS: m/z 487.80 (M⁺). Calcd. for C₁₈H₁₆Br₂O₆: C, 44.28, H, 3.28% Found: C, 44.32, H, 3.23%.

**(2E)-1-(3', 5'-dibromo-2', 4'-dihydroxyphenyl)-3-(4-hydroxy-3-methoxyphenyl) prop-2-en-1-one**

**(3r).**Yellow solid; Yiled (?) m.p. 110-111°C. IR (KBr, cm⁻¹): 3438 (Ar-OH), 3089, 3009 (Ar-H), 2830 (OCH₃), 1718 (-C=O), 1651 (-CH=CH), 862 (C-Br). ¹HNMR (500 MHz, CDCl₃) δ: 3.89 (s, 3H, OCH₃), 7.09-7.58 (m, 4H, Ar-H), 7.96 (d, αH, J=15.6Hz), 10.88 (s, 2H, 2 x Ar-OH), 8.02 (s, 1H, Ar-H), 8.13 (d, βH, J=15.6Hz). ¹³CNMR (125 MHz, CDCl₃) δ: 56.17 (OCH₃), 99.40, 108.76, 110.42, 114.38, 119.31, 124.24 (Cα), 127.54, 129.61, 133.38 (Cβ), 155.36, 160.34, 161.58, 190.91 (C=O). MS: m/z 443.80 (M⁺). Calcd. for C₁₆H₁₂Br₂O₅: C, 43.26, H, 2.70%. Found: C, 43.21, H, 2.75%.

**(2E)-3-(3-bromo-4-hydroxy-5-methoxyphenyl)-1-(3', 5'-dibromo-2', 4'-dihydroxyphenyl) prop-2-en-1-one**

**(3s).** Brown solid; Yield (?) m.p. 99-100°C. IR (KBr, cm⁻¹): 3435 (Ar-OH), 3092, 3005 (Ar-H), 2833 (OCH₃), 1720 (-C=O), 1652 (-CH=CH), 862 (C-Br). ¹HNMR (500 MHz, CDCl₃) δ: 3.94 (s, 3H, OCH₃), 7.03-7.63 (m, 4H, Ar-H), 7.89 (d, αH, J=16.0Hz), 8.03 (s, 1H, Ar-H), 8.13 (d, βH, J=16.0Hz), 10.95 (s, 1H, Ar-OH), 11.27 (s, 2H, 2 x Ar-OH). ¹³CNMR (125MHz, CDCl₃) δ: 56.53 (OCH₃), 77.28, 99.11, 108.21, 110.42, 115.14, 119.31, 122.56, 129.85 (Cα), 130.09, 130.47, 133.37 (Cβ), 155.87, 160.37, 162.61, 189.67 (C=O). MS: m/z 526.70 (M⁺). Calcd. for C₁₆H₁₁Br₃O₅: C, 37.21, H, 2.09%. Found: C, 37.26, H, 2.04%.

**(2E)-1-(3', 5'-dibromo-2', 4'-dihydroxyphenyl)-3-(furan-2-yl) prop-2-en-1-one**

**(3t).** Yellow solid; Yield (?) m.p. 132-133°C. IR (KBr, cm⁻¹):3435 (Ar-OH), 3087, 3009 (Ar-H), 1722 (-C=O), 1649 (-CH=CH), 862 (C-Br). ¹HNMR (500 MHz, CDCl₃) δ: 7.09-7.59 (m, 3H, Ar-H), 7.91 (d, αH, J=16.0Hz), 8.00 (s, 1H, Ar-H), 8.16 (d, βH, J=16.0Hz), 11.09 (s, 2H, 2 x Ar-OH). ¹³CNMR (125 MHz, CDCl₃) δ: 99.42, 113.10, 117.96, 131.94 (Cα), 132.29, 133.36 (Cβ), 151.31, 155.63, 160.36, 162.64, 190.74 (C=O). MS: m/z 387.80 (M⁺). Calcd. for C₁₃H₈Br₂O₄: C, 40.22, H, 2.06%. Found: C, 40.16, H, 2.11%.

## Biological evaluation: materials and method

**The parasite:** *T. cruzi* (Tulahuen C4) transfected with β-galactosidase (Lac Z) gene was obtained from Institute of Scientific Research and Advanced High Technology services - Panama (AIP). The strain was maintained in monolayer Vero cells (African Green Monkey cells line (ATCC/CCL-81)) in complete RPMI 1640 medium without phenol red (Sigma company, St. Louis MO modified - R8755), supplemented with 10% heat inactivated fetal bovine serum. All cultures and assays were conducted at 37 °C under an atmosphere of 5% CO₂ 95% air mixture.

*Invitro* **trypanocidal evaluation:** The anti-trypanocidal activity was evaluated by the colorimetric method based on reducing of the substrate chlorophenolred-β-ᴅ-galactopyranoside (CPRG) for β-galactosidase resulting from the expression of the gene for *T. cruzi* (Tulahuen C4) [22]. The assay was realised in 96 wells plates containing monolayer VERO cells which were infected with 5×10⁴ trypomastigotes (Tulahuen C4). We grew the parasite using VERO cells that were infected with *T. cruzi* trypomastigotes. The parasite is in the trypomastigote stage before it infects the cells. Once it infects the Vero cells, it enters an amastigotes stage and begins to reproduce as amastigotes. When it is released from the cells, it returns to the original trypomastigote stage to infect new cells. All the active compounds showed anti-trypanocidal activity, passed through a second test for determining the inhibitory concentration of 50% growth of the parasites (IC₅₀). These compounds were evaluated at 10, 2, 0.4, 0.8 and 0.16 ug/mL and incubated for 5 days at 37°C, relative humidity 95% and 5% CO₂.

The intensity of colour resulting from the cleavage of CPRG by *T. cruzi* (Tulahuen C4) β-galactosidase was measured at 570 nm using a reader boards VersaMax Micro™ microplate reader. The IC₅₀ of the compound was calculated by logarithmic regression of the values of OD obtained, compared with the untreated control. Those samples showing IC₅₀ values <50ug/mL, have been further tested for cytotoxicity evaluation. Nifurtimox (Bayer) was used as a control at concentrations of 0.1, 1 and 10ug/mL. Negative Control is comprised of 50uL of a solution containing DMSO, equivalent to the DMSO contained in samples (working dilution).

**Cytotoxicity assay:** Active Compounds were screened for cytotoxicity against VERO cell line, at a maximum concentration of 50μg / mL. Briefly, Vero cell line were seeded into 96-well plate at a total concentration of 12 x10⁴ cells/well in 100 uL of RPMI-1640 media without phenol red with 10% FBS. Cells were allowed to attach for 24 hrs. The wells were incubated with five decreasing concentrations, diluted in RPMI 1640 modified media or RPMI 1640 modified media alone, used as control. After 72 hrs, a colorimetric MTT assay was performed. Wells were incubated for 4 hrs with 5mg/mL of the tetrazolium salt MTT (3 - [4.5 dimethylthiazol-2-y1] -2, 5-diphenyl tetrazolium bromide/ Aldrich company, St. Louis MO). The sobrenadant were removed and cells lysed with 100% isopropanol. The absorbance was measured using an ELISA microplate reader (VersaMax Micro™ microplate reader) at 570nm. Tetrazolium salts are cleaved to formazan by mitochondrial enzymes in viable cells. Therefore, an increase in the OD reading, as a result of production of formazan, indirectly measures cell viability. The GI₅₀ value was defined as the concentration of test sample resulting in a 50% reduction of absorbance as compared with untreated controls that received a serial dilution of the solvent in which the test samples were dissolved, and was determined by linear regression analysis.

## Conclusions

Two noteworthy features are apparent from our study project on the synthesis of small molecules of medicinal interest. Firstly, a novel series of substituted chalcones has been synthesized and it is concluded from Table 2 that classical procedure is tedious, time consuming, low yield and requiring an appreciable amount of solvent as compare to the environmentally benign synthetic procedure utilizing microwave irradiation (MWI) under solvent free conditions, over inorganic solid support. Secondly, it was observed from the results obtained by the trypanocidal evaluation that compounds: 3b, 3c, 3e, 3g, 3i, 3l, 3m, 3p, and 3q have a good IC₅₀ and their cytotoxicity is low, this means that these compounds could be used in future *in vivo* studies. 3l was the compound with a good anti- trypanocidal activity, the lower cytotoxicity, higher therapeutic index 14.5, and is the best candidate in comparison with the others. This compound has a therapeutic index lower than that of benznidazole and nifurtimox. Our results demonstrate the potential of these compounds as a new class of small molecule inhibitors of *T. cruzi*. The further biologically assay of the tested compounds gives an idea about the possible development for a new encouraging framework in this field that may lead to the discovery of potent trypanocidal drug.

## Acknowledgements

The authors are thankful to Prof. B. L. Verma, M.L.S. University and Dr. Sunil Jhakoria, Dean, FASC, MITS University for their constant encouragement during this work. Authors are also thankful to the Head, Sophisticated Analytical Instrument Facility, Indian Institute of Technology, Madras for spectral analysis.

## References

1. (2010) Chagas disease (American trypanosomiasis) fact sheet (revised in June 2010). Wkly Epidemiol Rec 85: 334-336.

2. WHO: African trypanosomiasis or sleeping sickness (2001) World Health Org. Fact Sheet 259.

3. Fujii N, Mallari JP, Hansell EJ, Mackey Z, Doyle P, et al. (2005) Discovery of potent thiosemicarbazone inhibitors of rhodesain and cruzain. Bioorg Med Chem Lett 15: 121-123.

4. Siles R, Chen SE, Zhou M, Pinney KG, Trawick ML (2006) Design, synthesis and biochemical evaluation of novel cruzain inhibitors with potential application in the treatment of Chagas' disease. Bioorg Med Chem Lett 16: 4405-4409.

5. Guido RVC, Oliva G, Montanari CA, Andricopulo AD (2008) Structural Basis for Selective Inhibition of Trypanosomatid Glyceraldehyde-3-Phosphate Dehydrogenase: Molecular Docking and 3D QSAR Studies. J Chem Inf Model 48: 918-929.

6. Borchhardt DM, Mascarello A, Chiaradia LD, Nunes RJ, Oliva G, et al. (2010) Biochemical Evaluation of a Series of Synthetic Chalcone and Hydrazide Derivatives as Novel Inhibitors of Cruzain from Trypanosoma cruzi. J Brazilian Chem Soc 21: 142-150.

7. Herencia F, Ferrandiz ML, Ubeda A, Dominguez JN, Charris JE, et al. (1998) Synthesis and anti-inflammatory activity of chalcone derivatives. Bioorg Med Chem Lett 8: 1169-1174.

8. Wu X, Wilairat P, Go ML (2002) Antimalarial activity of ferrocenyl chalcones. Bioorg Med Chem Lett 12: 2299-2302.

9. Cheng JH, Hung CF, Yang SC, Wang JP, Won SJ, et al. (2008) Synthesis and cytotoxic, anti-inflammatory, and anti-oxidant activities of 2′,5′-dialkoxylchalcones as cancer chemopreventive agents. Bioorg Med Chem 16: 7270-7276.

10. Lin YM, Zhou Y, Flavin MT, Zhou LM, Nie W, et al. (2002) Chalcones and flavonoids as anti-tuberculosis agents. Bioorg Med Chem 10: 2795-2802.

11. Vegas S, Diaz JA, Darias V, Sanchez Mateo CC, Albertos LM (1998) Antidepressant activity of new hetero [2, 1] benzothiazepine derivatives. Pharmazie 52: 130-134.

12. El-Deeb IM, Lee SH (2010) Design and synthesis of new anticancer pyrimidines with multiple-kinase inhibitory effect. Bioorg Med Chem 18: 3860-3874.

13. Basaif SA, Faidallah HM, Hassan SY (1997) Synthesis and biological activity of some new pyrazoline and pyrazole derivatives. JKAU Sci 9: 83-90.

14. Jiang B, Yang CG, Xiong WN, Wang J (2001) Synthesis and cytotoxicity evaluation of novel indolylpyrimidines and indolylpyrazines as potential antitumor agents. Bioorg Med Chem 9: 1149-1154.

15. Ameta KL, Kumar Biresh, Rathore Nitu S (2011) Microwave induced synthesis of some novel substituted 1, 3-diarylpropenones and their antimicrobial activity. E-J Chem 8: 665-670.

16. Rao SS, Gahlot US, Dulawat SS, Vyas R, Ameta KL, et al. (2003) Microwave induced improved synthesis and anti-bacterial activity of some chalcones and their 1-acyl-3,5-diaryl-2-pyrazolines. Afinidad 60: 271-276.

17. Gahlot US, Rao SS, Dulawat SS, Ameta KL, Verma BL (2003) A facile one-pot microwave assisted conversion of 3′-5′-dibromo/ diiodo-4′-hydroxy substitutes chalcones to 2-substituted -4,6-diaryl pyrimidines using S-benzylisothiouronium chloride (SBT) and their antibacterial activities. Afinidad 60: 558-562.

18. Ameta KL, Rathore Nitu S, Kumar Biresh (2011) Synthesis of some novel chalcones and their facile one-pot conversion to 2-aminobenzene-1, 3-dicarbonitriles using malononitrile. An Univ Bucuresti Chimie 20: 15-24.

19. Ameta KL, Rathore Nitu S, Kumar Biresh (2012) Synthesis and in vitro anti breast cancer activity of some novel 1, 5-benzothiazepine derivatives. J Serbian Chem Soc 77: 725-731.

20. Ameta KL, Kumar Biresh, Rathore Nitu S, Verma BL (2012) Facile synthesis of some novel 2-substituted-4,6-diarylpyrimidines using 4′-hydroxy-3′,5′-dinitrochalcones and S-benzylthiouronium chloride. Org Commun 5: 1-11.

21. Srivastava YK, Ameta KL, Verma BL (2002) Synthesis of some 3N-substituted flavones. Indian J Heterocyclic Chem 11(4): 279-282.

22. Buckner FS, Verlinde CL, La Flamme AC, Van Voorhis WC (1996) Efficient technique for screening drugs for activity against Trypanosoma cruzi using parasites expressing beta-galactosidase. Antimicrob Agents Chemother 40: 2592-2597.

# 12

# *In Vitro* Synthesis of Ten Starches by Potato Starch-Synthase and Starch-Branching-Enzyme Giving Different Ratios of Amylopectin and Amylose

Mukerjea R, Sheets RL, Gray AN and Robyt JF*

*Laboratory of Carbohydrate Chemistry and Enzymology, Department of Biochemistry, Biophysics and Molecular Biology, Iowa State University, Ames, IA 50011, USA*

## Abstract

The objectives of this study was the use of two highly purified enzymes, potato Starch-Synthase (SS) and Starch-Branching-Enzyme (SBE) to obtain *in vitro* the syntheses of ten starches, with different ratios of amylopectin and amylose. The amylose was first synthesized by the reaction of SS with ADP [14C] Glc to give ten identical amyloses; the amylopectin was then synthesized by the action of SBE to give ten starches with different amounts of amylopectin. Two of the 14C-amyloses were reacted with 1.0 mIU of purified Starch-Branching-Enzyme (SBE) for 75 and 130 sec, respectively. In a second synthesis, 3 of the amyloses were autoclaved and reacted with 1.0 mIU for 15, 45, and 75 sec. In the third synthesis, 5 amyloses were reacted with different amounts of SBE from 0.50 to 0.01 mIU for 130 sec each. The 10 synthesized starches were separated into amylopectin and amylose fractions. The amylopectin and amylose ratios ranged from 99.9% amylopectin and 0.1% amylose to 10% amylopectin and 90% amylose. The results show that 10 different starches were synthesized, using only two enzymes, SS and SBE. No primers were involved. Glycogen and debranching enzymes were also not involved.

**Keywords:** Starch-Synthase; Starch-Branching-Enzyme; *In vitro* synthesis of starch; ADPGlc; Amylose; Amylopectin; Ten kinds of starches

## Introduction

Starch usually is composed of a mixture of two polysaccharides: amylose and amylopectin. Amylose is a α-1, 4-linked D-glucopyranose polymer and amylopectin has a number of α-1,4-linked D-glucopyranose chains, joined to other α-1,4-linked D-glucopyranose chains by 5-6 % α-1,6 branch glycosidic linkages [1]. Most starches contain 75-80 % amylopectin with 25-20 % amylose, respectively. There are, however, some starches that contain 100% amylopectin (waxy varieties) and some that are high amylose starches, containing 50-80% amylose and 50-20% amylopectin, respectively. Amylopectins are branched 5-6 %, giving average branch chain lengths of 20-16 D-glucopyranose units, [1] respectively. Using pulse and chase reactions of highly purified Starch-Synthase (SS) with ADP-[14C] Glc and nonlabeled ADPGlc, respectively, we recently have shown that starch chains are biosynthesized de novo from the reducing-end, without a primer requirement [2]. The synthesis starts by forming two D-glucopyranosyl covalent intermediates at the active-site of SS; the C-4-OH group of one of the glucopyranosyl intermediates then makes an attack on the C-1-carbon of the other glucopyranosyl intermediate to transfer the glucose and make a α-1,4-glycosidic linkage. The free catalytic group then attacks another ADPGlc unit and forms a D-glucopyranosyl-intermediate. The synthesis then proceeds processively, going back-and-forth between the covalent D-glucopyranosyl-C-4-OH groups and the growing D-glucopyranose-chains. The D-glucopyranosyl units are transferred to the reducing-ends of the growing chains. The synthesis is called a "two catalytic-site insertion mechanism" and results in the processive addition of D-glucopyranose from ADPGlc to the reducing-ends of growing starch chains [2] (Figure 1) During the pulse reaction, the reducing-ends become labeled with [14]C-D-glucose and that when the pulsed-polysaccharides are isolated, reduced with $NaBH_4$, and acid hydrolyzed, [14]C-D-glucitol is obtained; and in the chase reaction, the [14]C-D-glucose is chased from the reducing-ends into the chain, and in reduction and acid hydrolysis, the [14]C-D-glucitol is decreased, because nonlabeled D-glucose units are being added to the reducing-ends in the chase. If the synthesis had been to the nonreducing-ends of primers, no D-glucitol would ever have been formed in the pulsed

reactions and therefore decreased in the chase reactions. In the present study, the *in vitro* biosynthesis of starch was performed by using only two highly purified potato starch enzymes: Starch-Synthase (SS) and Starch-Branching-Enzyme (SBE). The amounts of synthesized amylose was approximately the same for each reaction and ten different kinds of starches were obtained by the conversion of different amounts of amylose into amylopectin by the reaction of SBE with the amyloses for different lengths of time or by different amounts of SBE.

## Experimental

### Materials

Adenosine 5'-diphospho-α-D-glucose (ADPGlc); dithiothreitol (DTT); polyvinyl alcohol 50K; PPO and PoPoP were obtained from Sigma-Aldrich Chem. Co. (St. Louis, MO. USA). ADP-[14C]Glc (333 mCi/mmol) was obtained from American Radiolabeled Chemicals, Inc. 101 Arc Drive, St. Louis, MO 63147, USA. Amylose was obtained by fractionation of potato starch as given in section, Separation of Amylose from Amylopectin, in the Methods Section [3], or obtained from Sigma-Aldrich Chem. Co. Starch-Synthase (SS) used was Fraction 23 [3] a highly purified enzyme, with a specific activity of 944 mIU/mg (1.0 IU is an International Unit that incorporates 1.0 μmole of D-glucose into starch/min) and was shown to be primer-free [4]. Starch-Branching-Enzyme (SBE) was Fraction 4 [3]. It had a specific activity of 198 mIU/mg, where 1.0 IU equal 1.0 μM of α-1,6 branch linkages formed/min and also was shown to be primer-free, along with

*Corresponding author: John F Robyt, Laboratory of Carbohydrate Chemistry and Enzymology, Department of Biochemistry, Biophysics and Molecular Biology, Iowa State University, Ames, IA 50011, USA
E-mail: jrobyt@iastate.edu

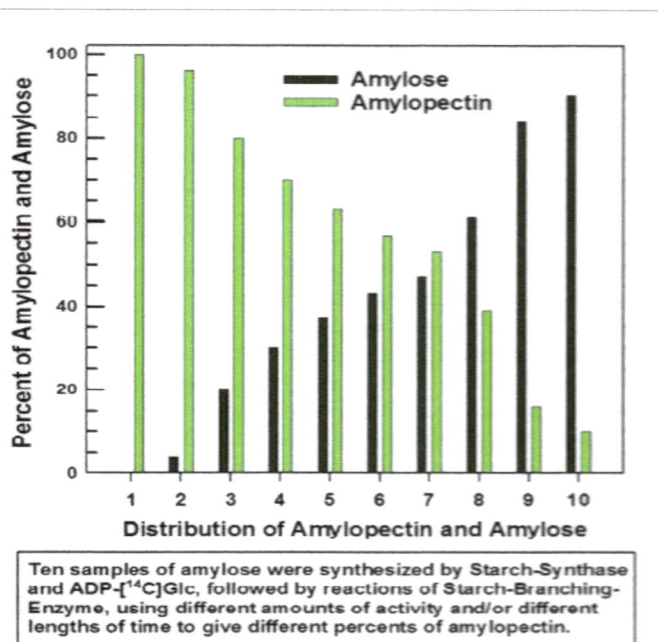

Ten samples of amylose were synthesized by Starch-Synthase and ADP-[$^{14}$C]Glc, followed by reactions of Starch-Branching-Enzyme, using different amounts of activity and/or different lengths of time to give different percents of amylopectin.

**Figure 1:** Distributions of amylopectin and amylose in the ten in vitro synthesized starches by highly purified potato Starch-Synthase and Starch-Branching-Enzyme.

the Standard Buffer [2]. The Standard Buffer was pH 8.4 and contained 10 mM glycine, 2 mM EDTA, 0.04% (w/v) polyvinyl alcohol 50K, and 1 mM DTT. Liquid Scintillation Cocktail was prepared, containing 5.0 g PPO and 0.1 g PoPoP in 1 L of toluene (Table 1).

## Methods

**Assay for starch-synthase (SS):** The SS assay reagent contained (75 µL) Standard Buffer (pH 8.4) with 20 mM (0.05 µCi) ADP-[$^{14}$C] Glc; 25 µL of SS samples were added to the assay reagent and incubated at 37°C for 30 min; 25 µL aliquots were then taken in triplicate and each added to 1.5 cm$^2$ Whatman 3MM paper, which is immediately added to 100 mL of MeOH, with stirring for 10 min, to stop the reaction and precipitate the synthesized $^{14}$C-amylose onto the paper, and wash away ADP-[$^{14}$C] Glc, and buffer material. The papers are washed 2-more times with 100 mL of MeOH and then dried and counted in 10 mL of toluene scintillation cocktail. A control blank is prepared by adding the Standard Buffer and the ADP-[$^{14}$C] Glc substrate in the same volume as the reaction digest, with Standard Buffer substituted for the SS enzyme sample. 1.0 mIU=1.0 nmole of D-glucose incorporated into amylose/min.

**Assay for starch-branching-enzyme (SBE):** Starch-Branching-Enzyme (SBE) (360 µL) in Standard Buffer (pH 8.4) is added to 200 µL (3 mg/mL) amylose and incubated at 37°C for 60 min; the reaction was stopped by placing the digest in a boiling water bath for 5 min and then cooled and centrifuged to remove denatured protein. The supernatant was removed and 2-10 µL of potassium triiodide reagent (2 mg iodine and 20 mg potassium iodide) was added to the digest supernatant to the point of just giving a slight blue triiodide color, and oxidize the DTT in the buffer. This was followed by a micro-titration (Hamilton Syringe) with 0.15 mM of silver nitrate, until the blue color just disappeared; the resulting silver iodide was removed by centrifugation. The pH was then made 5.2 by the addition of 10 µL of 40 mM pyridine-acetate buffer,

containing 200 units of isoamylase (where 1 unit=1.0 nmole of a-(1→6) bonds hydrolyzed/min), and the isoamylase reaction was allowed to proceed for 30 min at 37°C. Three 100-µL aliquots were added to the wells of a microplate, containing 100 µL of copper bicinchoninate reagent, and the reducing value was determined, using maltose (20-200 µg/mL), as a standard [4]. The number of nmoles of a-(1→6) branched linkages formed/min by SBE was calculated from the reducing value obtained from the isoamylase reaction. A control blank was prepared by adding the Standard Buffer with the substrate, substituting water for the enzyme sample, and then carrying it through the isoamylase reaction, giving the exact reaction conditions used for the samples. One unit of SBE = 1.0 nmol of a-(1→6) branch linkages formed/min.

**Separation of Amylose from Amylopectin in the synthesized starches [5]:** Potato starch (1.0 g in 100 mL) was added to each of the enzyme digests as a carrier when the reaction was over. The pH was adjusted to 7.0 and stirred for 30 min at 21-22°C; 22 mL of 1-butanol was added and stirred for 24 hr at 22°C to precipitate the amylose/1-butanol complexes and were then centrifuged at 20K rpm for 30 min, washed with 20 mL of water; and the amylose was then suspended in 40 mL of water, and the 1-butanol was removed by heating to boiling with the addition of water to maintain the volume. The amylose was then precipitated by the addition of 3 volumes of ethanol and placed at 4°C for 24 hr. The amylopectin that remained in the supernatant from the precipitation of the amylose by butanol-1 was precipitated by the addition of 3 volumes of ethanol and placed at 4°C for 24 hr. The amylose and amylopectin were centrifuged and their pellets were washed with 40 mL of acetone/water, 80/20, (v/v), and then treated with 100 mL of acetone, and dried under vacuo for 24 hr.

## *In vitro* synthesis of starch by SS reacting with ADP-[$^{14}$C] Glc and SBE reacting with the synthesized amylose to give amylopectin

**Synthesis type I:** Two reactions were performed. Both reactions had 1.05 mL containing 400 mIU of SS. The reactions were initiated by the addition of 30 µL of (25.4 mg, 0.1 µCi) ADP-[$^{14}$C]Glc and allowed to go for 135 min to give 1.2 Conversion Periods (CP), where 1.0 CP is the theoretical length of time for the complete reaction of ADP-[$^{14}$C] Glc. The two reactions were initiated by the addition of 30 µL of SBE (1.0 mIU) to each of the two amylose reactions. The first reaction was allowed to go for 75 sec and the second reaction for 130 sec. The reactions were stopped by the addition of 0.1 M trifluoroacetic acid (TFA) to give a pH of 2.0, and 1.0 g of carrier potato starch was added to each reaction, and the amyloses and amylopectins were separated as described under separation of Amylose from Amylopectin in the synthesized starches [5]. The amounts (20-100 mg) of synthesized amyloses and amylopectins were obtained by Liquid Scintillation Spectrometry. All of the pipetting for 100 µL and less for Synthesis Type I was done, using Hamilton Syringes to give the highest accuracy.

**Synthesis type II:** Three samples of amylose were synthesized identical to the amyloses in Type I. The reactions were allowed to go for 135 min, 1.2 Conversion Periods, where 1 CP is the theoretical length of time to give complete reaction of ADP-[$^{14}$C] Glc. The digests were then precipitated with 3 volumes of ethanol, which was washed 5-times with 80/20 acetone water, and dissolved in 0.6 mL of water by autoclaving at 121°C, followed by making the volume 1.05 mL by adding Standard pH 8.4 Buffer. The branching reactions were then started by the addition of 30 µL (1.0 mIU) of SBE to the three amylose samples and the reactions were allowed to go 15, 45, and 75 sec, respectively. The reactions were stopped by adding 0.1 M TFA to give pH 2, and 1.0 g of carrier potato

| Starch Sample Number | Synthesis Type[a] | Percent Amylopectin Synthesized[b,c] | Percent Amylose Synthesized[b] | Amount of SBE used mIU[c,d] | Time of SBE Reaction Sec[d] |
|---|---|---|---|---|---|
| 1 | I | 99.9 | 0.1 | 1.0 | 130 |
| 2 | III | 96.0 | 4.0 | 0.5 | 130 |
| 3 | III | 80.0 | 20.0 | 0.3 | 130 |
| 4 | II | 70.0 | 30.0 | 1.0 | 75 |
| 5 | II | 63.0 | 37.0 | 1.0 | 45 |
| 6 | III | 57.0 | 43.0 | 0.2 | 130 |
| 7 | I | 53.0 | 47.0 | 1.0 | 75 |
| 8 | III | 39.0 | 61.0 | 0.1 | 130 |
| 9 | II | 16.0 | 84.0 | 1.0 | 15 |
| 10 | III | 10.0 | 90.0 | 0.01 | 130 |

[a]For each of the types of reactions, 400 mIU (1.0 mIU = 1.0 nmole of $\alpha$-1,4 linkages synthesized/min) of Starch-Synthase was first allowed to go for 135 min or 1.2 CP, where 1.0 CP is the theoretical length of time necessary to convert 100% of the ADPGlc into an identical amount of amylose for each of the ten syntheses.

[b]Presented in decreasing amounts of amylopectin and increasing amounts of amylose, respectively.

[c]The resulting amounts of amylopectin and amylose synthesized is dependent on the amount of Starch-Branching-Enzyme (1.0 to 0.01 mIU) and the length of time (130 sec to 15 sec) that it was allowed to react.

[d]1.0 mIU = 1.0 nmole of $\alpha$-1,6 branch-linkages synthesized per min.

**Table 1:** Percent distribution of amylopectin and amylose in ten in vitro synthesized starches, obtained by the reaction of potato Starch-Synthase and Starch-Branching-Enzyme in different amounts or for different lengths of time.

starch was added to each reaction, and the amyloses and amylopectins were separated as described under Methods. The amounts (20-100 mg) of synthesized amyloses and amylopectins were obtained by Liquid Scintillation Spectrometry. All of the pipetting for 100 μL and less for Synthesis Type II was done, using Hamilton Syringes to give the highest accuracy.

**Synthesis type III:** Five amyloses were synthesized as in Type I. Varying amounts (0.5 mIU, 0.3 mIU, 0.2 mIU, 0.1 mIU, and 0.01 mIU) of SBE were added and the reactions were allowed to go for 130 sec. They were stopped by the addition of 0.1 M of TFA to give pH 2, followed by the addition of 1.0 g of carrier potato starch to each reaction, and the amyloses and amylopectins were separated as described under Separation of Amylose from Amylopectin in the synthesized starches [5]. The amounts (20-100 mg) of synthesized amyloses and amylopectins were determined by Liquid Scintillation Spectrometry. All of the pipetting for 100 μL or less for Synthesis Type III was done, using Hamilton Syringes to give the highest accuracy.

## Results and Discussion

Three types of syntheses were performed. All three start with the same amounts of amylose, synthesized by highly purified potato SS to give [14]C-labeled amylose. The results of the different percentages of amylopectin and amylose for the ten syntheses are given in (Table 1). It should be noted that the 75 sec reaction of SBE in the Type I synthesis gave different amounts of amylopectin than did the 75 sec reaction of SBE in Type II synthesis. Type I synthesis gave 53% amylopectin, whereas Type II synthesis only gave 16% amylopectin. In the Type II synthesis, the SS had been completely denatured by precipitating and autoclaving before its addition. The SS in the Type I synthesis was not autoclaved and SS activity remained after SBE had been added and amylose continued to be synthesized during the branching reaction by SBE to give much larger amounts of amylopectin by some synergistic mechanism. In (Table 1), the percent distribution for each synthesized

amylopectin and amylose are given. The synthesized starches had (a) "very high amylopectin" (99.9% amylopectin and 0.1% amylose); (b) "normal" amounts of amylopectin (80% amylopectin and 20% amylose) and (70% amylopectin and 30% amylose); (c) intermediate amounts of amylopectin and amylose (57% amylopectin and 43% amylose) and (53% amylopectin and 47% amylose); (d) "high amylose starch" (39% amylopectin and 61% amylose); and (e) "very high amylose starch" (16% amylopectin and 84% amylose) and (10% amylopectin and 90% amylose). These results show that various kinds of starches, containing different amounts of amylopectin and amylose, have been synthesized, *in vitro*, by only two highly purified enzymes, potato SS and SBE, in which the SBE was allowed to react with the same amounts of amylose for different lengths of time or with different amounts of activity for a constant length of time. Previous studies of starch biosynthesis [6-10] involved various amounts of primers, phosphorylase, and glycogen. During the course of these studies, Hanes found that if he started with $\alpha$-Glc-1-P and starch, without any phosphate, some glucose units were added to the nonreducing-ends of the starch chains, giving synthesis. This, in essence, was the origin of the requirement for a primer in polysaccharide biosynthesis. It was found shortly thereafter [7], however, that only a few (2-3) glucose units were added to the nonreducing-ends of the starch chains. In plants and animals, it also was found that the concentration ratio of Pi to $\alpha$-Glc-1-P was 20-40 folds higher *in vivo*, and phosphorylase was exclusively a degradative enzyme rather than a synthetic enzyme [7-10]. Hanes' observation [6], however, set the stage for the primer concept for polysaccharide biosynthesis and many investigators made futile searches for the putative primers for polysaccharide biosynthesis. One such study was by Koepsell et al. [11], who added isomaltose and maltose to digests of sucrose and Leuconostoc mesenteroides B-512F dextransucrase, and found that these reactions formed isomaltodextrins and maltosyl-isomaltodextrins, with maltose at the reducing-ends, respectively. Both also inhibited the biosynthesis of dextran and were therefore not primers. Likewise, Mukerjea and Robyt [12] found that maltose, maltotriose, and maltododecaose (G12) also inhibited starch biosynthesis by maize, wheat, and rice starch granules, very similar to the findings of Koepsell et al. [11] for the biosynthesis of dextran by dextransucrase.In the early 1960s, Leloir et al. [13-15] also found that active Starch-Synthase, was entrapped inside starch granules, and when reacted with adenosine diphospho-[14]C-glucose (ADP-[[14]C]Glc), [14]C-starch chains were synthesized in the granule. The [14]C-labeled starch was dissolved in buffer and reacted with the exo-acting ß-amylase that hydrolyzes starch from the nonreducing-ends, to give [14]C-labeled maltose. From this result, it has been widely assumed that the starch was being biosynthesized by the addition of [14]C-labeled glucose to the nonreducing-ends of primers. Later, Mukerjea and Robyt, [12] suggested that the Leloir et al. [15] conclusion that the formation of [14]C-maltose by ß-amylase action on their synthesized [14]C-starch, which suggested that the addition of glucose was to the nonreducing-ends of primers, was not necessarily correct: if the starch chains had been synthesized de novo by the addition of D-glucose to the reducing-ends, without the involvement of a primer, the entire starch chains would have been labeled with [14]C-D-glucose and ß-amylase would also have produced 14C-maltose.In 1948, Wolf et al. [16] assumed that sweet corn starch apparently originated from globules of polysaccharide on the surface of sweet corn. These globules were assumed to be glycogen and, therefore, the starch arose from glycogen. The globules, however, were never shown to be glycogen and could very well have been pre-amylopectin. Nevertheless, from their observation, it was proposed that glycogen was an intermediate in the biosynthesis of amylopectin and amylose. In 1951 Hobson et al. logically indicated, however, that amylose had to be synthesized first and then part of the amylose converted into amylopectin [17]. However, in 1958 Erlander [18], stated that cumulative evidence indicated that glycogen was the precursor for the formation of the linear amylose chains by the action of isoamylase or debranching-enzyme with the glycogen to give amylose. The released chains, however, that would be released would be quite short, ~10-12 D-glucose units, and did not come close to the number average degree of polymerization (dpn) values obtained for several amyloses: the smallest known was from amylomaize-VII, a high amylose starch, in which the amylose has a dpn of ~400 glucose units; then amylomaize starch, with 20% amylose, has a $dp_n$ of ~800 glucose units; potato starch, with 25% amylose, has $dp_n$ of ~1,000 glucose units; and wheat starch, with 25% amylose, has a $dp_n$ of ~4,000 D-glucose units. In 1990, Baba et al. [19] found that glycogen was the best primer for starch biosynthesis. This observation furthered the hypothesis that starch arose from glycogen. In 1996, Ball et al. [20] proposed that amylopectin was synthesized from glycogen. Glycogen is branched 10-12%, having an average of 8-10 glucose units per chain. They proposed that glycogen was debranched to give 5-6% branching and the chains were then elongated at the nonreducing-ends by Starch-Synthase to give an average of 20-25 glucose units in amylopectin. In 1998, Ball et al. [21] further hypothesized that the amylose, that was initially released from the glycogen was also elongated by Starch-Synthase addition of several hundred glucose units to the nonreducing-ends to give high-molecular weight amyloses. Ball and Morell [22] proposed that glycogen was a precursor for the formation of starch. There are a number of problems with these hypotheses. One problem is that experimental attempts to elongate linear maltodextrin chains and branched maltodextrin chains from the nonreducing-ends by Starch-Synthase only gives the addition of a few (1-3) glucose units to the nonreducing-ends of linear and branched putative primers [23,24] Mukerjea and Robyt [12] further found, a compounding result for Starch-Synthesis, in which different concentrations of the putative maltodextrin primers (maltose, maltotriose, and maltododecaose) inhibited starch biosynthesis, instead of stimulating it, as would have been expected for primers. Using pulse and chase techniques with ADP-[14C] Glc, Mukerjea and Robyt [2] recently showed that Starch-Synthase synthesizes linear amylose chains, de novo, by the addition of D-glucose units to the reducing-ends of growing amylose chains and not to the nonreducing-ends of a primer. Further, Robyt et al. [25] showed that the need for a primer in starch biosynthesis had been perpetuated for over 70-years by a primer myth that arose from the use of Tris-type buffers (Tris, Bicine, and Tricine), which they have recently shown to be potent inhibitors for Starch-Synthase [25]. It was then shown that the inhibition can be partially reversed, giving ~10% Starch-Synthase activity for Tris and Bicine, but 0% for Tricine buffers by the addition of the putative primers that actually are activators that release a small percentage of the ADPGlc substrate that is complexed with the Tris-buffers, producing inhibition [25]. Along with the carbohydrate putative primers, a high concentration of a non-carbohydrate, 500 mM Na-citrate, had also been found by Pollock and Preiss to give about the same degree (10%) activation of SS in the presence Tris-type buffers as did 10 mg/mL glycogen [26]. Na-Citrate is not a carbohydrate and its structure could, therefore, not be acting as a primer. Thus, it was shown that these so-called primers (glycogen, starch, maltodextrins, and Na-Citrate) were not primers, but activators that partially reversed the Tris-buffer inhibition [25]. In the present study, the in vitro biosynthesis of ten different kinds of starches, having different percent ratios of amylopectin and amylose, were obtained using only two highly purified potato enzymes, Starch-Synthase and Starch-Branching-Enzyme. The amounts of amylopectin and amylose synthesized are given in Table 1.

The ratios range from 99.9% amylopectin and 0.1% amylose to 10% amylopectin and 90% amylose. All ten starches were biosynthesized in this study by first synthesizing identical amounts of amylose by Starch-Synthase and ADP-[14C] Glc, and then having different amounts or different lengths of time of reaction for Starch-Branching-Enzyme to synthesize the a-1,6-branch linkages with the amylose chains, to give 10 different kinds of 14C-labeled starches, with different percentages of amylopectin and amylose see Table 1. The amounts of amylose and amylopectin are also given in Figure 1 so that a comparison can be readily made between the ten synthesized starches. It should be noted that the relative activity of Starch-Branching-Enzyme must be considerably less than the activity of Starch-Synthase, as amylopectin is only branched 5-6%, although the amount of amylopectin is usually considerably larger than the amount of amylose; however, the amount of branching enzyme activity must then be considerably less than the amount of starch synthesizing activity to obtain a starch with significant amounts of amylose e.g., 20-25%.Starch-Debranching-Enzyme(s) that supposedly would react with glycogen and amylopectin were not present in the reactions, and the putative primers (glycogen, maltodextrins, or pre-formed starch) were also not present. The earlier hypotheses put forward by Ball et al. [20,21], in which it was proposed that glycogen was a precursor for the biosynthesis of both amylose and amylopectin, by the action of a Debranching-Enzyme, followed by the elongation of the chains of the partially debranched glycogen and the debranched amylose chains by Starch-Synthase to give amylopectin and an extensively elongated debranched amylose chains also by Starch-Synthase shows that the Ball et al. [20,21] hypotheses are impossible and do not occur because (i) Debranching-Enzyme(s) are not involved in the synthesis and have not been shown to be involved in any study and certainly was not involved in the present study, and (ii) it has been shown that Starch-Synthase also does not elongate starch chains from the nonreducing-ends [2], as is required by the Ball et al. hypotheses [21,22]. Starch-Synthase synthesizes linear, a-1,4 linked chains, de novo, processively from the reducing-end, as shown by Mukerjea and Robyt [2]. The present study, thus, shows that glycogen is not involved in starch biosynthesis as was proposed by Ball and Morell [22], and linear a-1,4-glucose chains are not synthesized from the nonreducing-end of a primer, as has been hypothesized by Ball et al. [21] and many others.While the in vitro biosynthesis is not identical to the in vivo biosynthesis that occurs in chloroplasts and amyloplasts, this study closely mimics the in vivo biosynthesis, capable of giving different kinds of starches. An important point here is that if this synthesis, to give ten different kinds of starches, with different ratios of amylopectin and amylose, can occur in vitro, using only two enzymes (highly purified Starch-Synthase and Starch-Branching-Enzyme), it certainly should be able to also occur in vivo in chloroplasts and amyloplasts. The study, further, shows that glycogen, Debranching-Enzyme(s), and amylose elongation from the nonreducing-ends, are not involved in starch biosynthesis, and the synthesis only requires two kinds of enzymes, Starch-Synthase(s) that synthesize amylose chains, de novo, from the reducing-end, and Starch-Branching-Enzyme(s) that synthesize a-1,6-linked branches to give amylopectin from the amylose in different proportions to produce different kinds of starches. What the study does not show is how the Starch-Branching-Enzyme is regulated and controlled in vivo to give the different ratios of amylopectin and amylose. In this study, it was controlled by first synthesizing amylose and then adding small amounts of Branching-Enzyme for varying lengths of time to convert a portion of the amylose into amylopectin. The major synthesis was performed by Starch-Synthase synthesis (400 mIU of SS reacting with 400 nmoles of ADP-[14C] Glc reaction for 135 min) to synthesize 14C-amylose; and the amylopectin fraction was then synthesized by the addition of 1.0-0.01 mIU of Starch-Branching-Enzyme for 130 sec to 15 sec with the $^{14}$C-amyloses to give the different percentages of the amylopectin fractions. The action of the Branching-Enzyme is, thus, only a fraction of the Starch-Synthase reaction, as amylopectin is only branched to the extent of 5-6% and, therefore, only a relatively small amount of the branching enzyme activity is required to obtain the varying amounts of the amylopectin fractions, to give the different ratios of amylopectin and amylose (Table 1 for the individual number of SBE units and number of seconds that they reacted to obtain the 10 starches). Schwall et al. [27] published a paper in 2000, in which a very high-amylose, potato starch was obtained by simultaneously inhibiting two SBE, A and B isoforms, to give about 1% of the wild-type activities. This project and result further indicates that it is SBE that controls the amount of amylopectin and not a Debranching-Enzyme action on glycogen, followed by extension of the nonreducing-ends of the resulting debranched glycogen and short amylose chains by Starch-Synthase.

## References

1. Whistler RL (1965) Starch: Chemistry and Technology Chap. 1, pp. 1-8 in, Vol. I Whistler RL, Paschall EF, BeMiller JN, Roberts HJ, eds., Academic Press, New York; and/or Robyt JF (2008) Properties and Occurrence of Starch in General Properties, Occurrence, and Preparation of Carbohydrates, Chap. 1.2, Glycoscience II, 2nd Ed : 72-73, (Fraser-Reid B, Tatsuta K, Thiem J, eds.) Springer-Verlag, Berlin, Heidelberg.

2. Mukerjea R, Robyt JF (2012) De novo biosynthesis of starch chains without a primer and the mechanism for its biosynthesis by potato starch-synthase. Carbohydr Res 352: 137-142.

3. Mukerjea Ru, Falconer DJ, Yoon SH, Robyt JF (2010) Large-scale isolation, fractionation, and purification of soluble starch synthesizing enzymes: Starch-synthase and branching enzyme from potato tubers. Carbohydr. Res 345: 1555-1563.

4. Fox JD, Robyt JF (1991) Miniaturization of three carbohydrate analyses using a microsample plate reader. Anal Biochem 195: 93-96.

5. Gilbert L, Gilbert GA, Spragg SP (1964) Separation of amylose and amylopectin fractions of starch, in Methods Carbohydr. Chem Starch, Whistler RL, Smith RJ, Be Miller JN, eds: 25-27, Academic Press, New York, Vol. IV

6. Hanes CS (1940) The reversible formation of starch from glucose-1-phosphate catalyzed by potato phosphorylase. Proc Roy Soc Bull 129: 174-208.

7. Green DE, Stumpf PK (1942) Starch Phosphorylase from Potato. J Biol Chem 142: 355-366.

8. Trevelyan WE, Mann PF, Harrison JS (1952) The phosphorylase reaction. I. Equilibrium constant: principles and preliminary survey. Arch Biochem Biophys 39: 419-439.

9. Ewart MH, Siminovitch D, Briggs DR (1954) Studies on the Chemistry of the Living Bark of the Black Locust in Relation to its Frost Hardiness. VIII. Possible Enzymatic Processes Involved in Starch-Sucrose Interconversions. Plant Physiol 29: 407-413.

10. Liu TT, Shannon JC (1981) Measurement of Metabolites Associated with Nonaqueously Isolated Starch Granules from Immature Zea mays L. Endosperm. Plant Physiol 67: 525-529.

11. koepsell HJ, Tsuchiya HM, Hellman NN, Kazenko A, Hoffman CA, et al. (1953) Enzymatic synthesis of dextran; acceptor specificity and chain initiation. J Biol Chem 200: 793-801.

12. Mukerjea Ru, Robyt JF (2005) Starch biosynthesis: the primer nonreducing-end mechanism versus the nonprimer reducing-end two-site insertion mechanism. Carbohydr Res 340: 245-255.

13. Rongine De Fekete MA, Leloir LF, Cardini CE (1960) Mechanism of starch biosynthesis. Nature 187: 918-919.

14. Recondo E, Leloir LF (1961) Adenosine diphosphate glucose and starch synthesis. Biochem Biophys Res Commun 6: 85-88.

15. Leloir LF, De Fekete MA, Cardini CE (1961) Starch and oligosaccharide synthesis from uridine diphosphate glucose. J Biol Chem 236: 636-641.

16. Wolf MJ, MacMasters MM, Hubbard JE, Rist CE (1948) Comparison of corn starches at various stages of kernel maturity. Cereal Chem. 25: 312-325.

17. Hobson PN, Whelan WJ, Peat S (1950) A 'de-branching' enzyme in bean and potato. Biochem J 47: xxxix.

18. Erlander SR (1958) A proposed mechanism for the synthesis of starch from glycogen. Enzymologia 19: 273-283.

19. Baba T, Noro M, Hiroto M, Arai Y (1990) Properties of primer dependent starch synthesis catalyzed by starch synthase from potato tubers. Phytochemistry 29: 719-723.

20. Ball S, Guan HP, James M, Myers A, Keeling P, et al. (1996) From glycogen to amylopectin: a model for the biogenesis of the plant starch granule. Cell 86: 349-352.

21. Van de Wal M, D'Hulst C, Vincken JP, Buléon A, Visser R, et al. (1998) Amylose is synthesized in vitro by extension of and cleavage from amylopectin. J Biol Chem 273: 22232-22240.

22. Ball SG, Morell MK (2003) From bacterial glycogen to starch: understanding the biogenesis of the plant starch granule. Annu Rev Plant Biol 54: 207-233.

23. Damager I, Denyer K, Motawia MS, Møller BL, Blennow A (2001) The action of starch synthase II on 6'''-alpha-maltotriosyl-maltohexaose comprising the branch point of amylopectin. Eur J Biochem 268: 4878-4884.

24. Damager I, Olsen CE, Blennow A, Denyer K, Møller BL, et al. (2003) Chemical synthesis of methyl 6'-alpha-maltosyl-alpha-maltotrioside and its use for investigation of the action of starch synthase II. Carbohydr Res 338: 189-197.

25. Mukerjea R, McIntyre AP, Robyt JF (2012) Potent inhibition of starch-synthase by Tris-type buffers is responsible for the perpetuation of the primer myth for starch biosynthesis. Carbohydr Res 355: 28-34.

26. Pollock C, Preiss J (1980) The citrate-stimulated starch synthase of starchy maize kernels: purification and properties. Arch Biochem Biophys 204: 578-588.

27. Schwall GP, Safford R, Westcott RJ, Jeffcoat R, Tayal A, et al. (2000) Production of very-high-amylose potato starch by inhibition of SBE A and B. Nat Biotechnol 18: 551-554.

# 13

# Characterization of Thermal and Physical properties of Biofield Treated Acrylamide and 2-Chloroacetamide

**Mahendra KT[1], Shrikant P[1], Rakesh KM[1], and Snehasis J[2]**

[1]Trivedi Global Inc., 10624 S Eastern Avenue Suite A-969, Henderson, NV 89052, USA
[2]Trivedi Science Research Laboratory Pvt. Ltd., Hall-A, Chinar Mega Mall, Chinar Fortune City,Hoshangabad Rd., Bhopal- 462026, Madhya Pradesh, India

## Abstract

Acrylamide (AM) and 2-chloroacetamide (CA) are widely used in diverse applications such as biomedical, drug delivery, waste water treatment, and heavy metal ion removal. The objective of this study was to evaluate the influence of biofield treatment on physical and thermal properties of amide group containing compounds (AM and CA). The study was performed in two groups (control and treated). The control group remained as untreated, and biofield treatment was given to treated group. The control and treated compounds were characterized by X-ray diffraction (XRD), differential scanning calorimetry (DSC), thermogravimetric analysis (TGA) and surface area analysis. XRD of treated AM showed decrease in intensity of peaks as compared to control sample. However, the treated AM showed increase in volume of unit cell (0.16%) and molecular weight (0.16%) as compared to control. The crystallite size was decreased by 33.34% in treated AM as compared to control Whereas, the XRD diffractogram of treated CA showed increase in intensity of crystalline peaks as compared to control. The percentage volume of unit cell (-1.92%) and molecular weight (-1.92%) of treated CA were decreased as compared to control. However, significant increase in crystallite size (129.79%) was observed in treated CA as compared to control. DSC of treated AM showed increase in melting temperature as compared to control sample. Similarly, the treated CA also showed increase in melting temperature with respect to control. Latent heat of fusion ($\Delta H$) was significantly changed in treated AM and CA as compared to control samples. TGA showed increase in thermal stability of treated AM and CA which was evidenced by increase in thermal decomposition temperature ($T_{max}$) as compared to control. Surface area analysis of treated AM showed increase (31.6%) in surface area as compared to control. However, a decrease (30.9%) in surface area was noticed in treated CA as compared to control. Study results suggest that biofield treatment has significant impact on the physical and thermal properties of AM and CA.

**Keywords:** Acrylamide; 2-chloroacetamide; Biofield treatment; X-ray diffraction; Differential scanning calorimetry; Thermogravimetric analysis; Surface area analysis

**Abbreviation:** XRD: X-Ray Diffraction; DSC: Differential Scanning Calorimetry; TGA: Thermogravimetric Analysis; AM: Acrylamide; CA: 2-Chloroacetamide

## Introduction

Acrylamide (AM) is a monomer used for synthesis of polyacrylamide that is commonly utilized for biomedical, pharmaceutical and wastewater treatment applications. AM based polymer was introduced as a support matrix for electrophoresis in 1959 [1]. Generally, the AM based gels are produced using bifunctional cross linker such as N-methylene bisacrylamide [2]. AM based polymers are widely used in areas such as enzyme immobilization [3,4], carrier for delivery of drugs and bioactive compounds [5-9], stimuli responsive materials [10,11] and in non-absorbable soft tissue fillers used for body contouring in reconstructive surgery [12]. On the other hand 2-chloroacetamide (CA) is a chlorinated organic compound used as herbicide and preservative [13]. The amide group of CA has better selectivity towards mercury binding, hence it has been commonly used for heavy metal ion removal applications [14]. The structural and conformational nature of CA have been previously studied by many researchers [15,16]. However, the AM and CA have toxicity problems which need to be addressed in order to make it useful for other areas such as biomedical and novel drug development. Biofield treatment has been recently used as a strategy for changing the atomic and physical properties of various materials [17-20].

Bioelectrography dates back to 1770 when German scientist George Christopher Lichtenberg observed light coming out from subjects in electrical fields [21]. Researchers have showed that short lived electrical events or action potential exist in several type of mammalian cells such as neurons, muscle cells, and endocrine cells [22]. For instance the cells present in central nervous system of human body communicate with each another by means of electrical signals that travel along the nerve processes. Therefore, it was hypothesized that biofield exists around the human body and the evidence was found using Electromyography, Electrocardiography and electroencephalogram [23]. Human has the ability to harness the energy from environment or universe and can transmit into any living or nonliving object around the Globe. The object(s) always receive the energy and responds in a useful way that is called biofield energy and the process is known as biofield treatment. Mr. Mahendra K. Trivedi is known to transform the characteristics of various living and non-living things using his biofield treatment. The biofield treatment had significantly improved the production and quality of various agricultural products [24-27]. Biofield treatment has shown excellent results in improving antimicrobial susceptibility, and alteration of biochemical reactions, as well as induced alterations in characteristics of pathogenic microbes [28-30]. The biofield treatment had also caused an increase in growth and anatomical characteristics

*Corresponding author: Snehasis J, Trivedi Science Research Laboratory Pvt. Ltd., Hall-A, Chinar Mega Mall, Chinar Fortune City, Hoshangabad Rd., Bhopal- 462026 Madhya Pradesh, India, E-mail: publication@trivedisrl.com

of an herb *Pogostemon cablin* that is commonly used in perfumes, in incense/insect repellents, and alternative medicine [31]. By conceiving the above research outcome of biofield treatment, an attempt was made here to study the influence of biofield on physical and thermal properties of AM and CA.

## Materials and Methods

Acrylamide (AM) and 2-chloroacetamide (CA) were procured from S D Fine Chemicals Pvt., Ltd., India. Each sample was divided into two parts; one was kept as a control, while other was subjected to Mr. Trivedi's biofield treatment and coded as treated sample. The treatment group in sealed pack was handed over to Mr. Trivedi for biofield treatment under standard laboratory condition. Mr. Trivedi provided the treatment through his energy transmission process to the treated group without touching the sample. The biofield treated samples were returned in the similar sealed condition for characterization using XRD, DSC, TGA and surface area analysis techniques.

## Characterization

**X-ray diffraction (XRD) study:** XRD analysis was carried out on Phillips, Holland PW 1710 X-ray diffractometer system, which had a copper anode with nickel filter. The radiation of wavelength used by the XRD system was 1.54056 Å. The data obtained from this XRD were in the form of a chart of 2θ vs. intensity and a detailed table containing peak intensity counts, d value (Å), peak width (θ°), relative intensity (%) etc. Additionally, PowderX software was used to calculate unit cell volume.

The crystallite size (G) was calculated by using formula:

$G=k\lambda/(bCos\theta)$

Here, λ is the wavelength of radiation used and k is the equipment constant (=0.94). However, the percentage change in all parameters such as, unit cell volume and percent change in crystallite size was calculated using the following equation:

Percent change in unit cell volume=$[(V_t-V_c)/V_c] \times 100$

Where, $V_c$ and $V_t$ are the unit cell volume of control and treated powder samples respectively

Percent change in crystallite size=$[(G_t-G_c)/G_c] \times 100$

Where, $G_c$ and $G_t$ are crystallite size of control and treated powder samples respectively.

The molecular weight of atom was calculated using following equation:

Molecular weight=number of protons × weight of a proton + number of neutrons x weight of a neutron + number of electrons x weight of an electron.

Molecular weight in g/Mol was calculated from the weights of all atoms in a molecule multiplied by the Avogadro number ($6.023 \times 10^{23}$). The percent change in molecular weight was calculated using the following equation:

Percent change in molecular weight=$[(M_t-M_c)/M_c] \times 100$

Where, $M_c$ and $M_t$ are molecular weight of control and treated powder sample respectively.

**Differential scanning calorimetry (DSC) study:** The control and treated samples (AM and CA) were analyzed by using a Pyris-6 Perkin Elmer DSC on a heating rate of 10°C/min under air atmosphere and air was flushed at flow rate of 5 mL/min.

**Thermogravimetric analysis-differential thermal analysis (TGA-DTA):** Thermal stability of control and treated samples (AM and CA) were analyzed by using Metller Toledo simultaneous TGA and Differential thermal analyzer (DTA). The samples were heated from room temperature to 400°C with a heating rate of 5°C/min under air atmosphere.

**Surface area analysis:** Surface area of AM and CA were characterized by surface area analyzer, SMART SORB 90 BET using ASTM D 5604 method which had a detection range of 0.2-1000 m²/g.

## Results and Discussion

### X-ray diffraction

X-ray diffraction study was conducted to study the crystalline nature of the control and treated AM. XRD diffractogram of control AM showed (Figure 1) intense crystalline peaks at 2θ equals to 11.90°, 19.34°, 19.55°, 24.00°, 24.51°, 28.27°, 28.53°, 36.38° and 49.21°. These peaks showed the crystalline nature of AM. Whereas, the treated AM showed decreased intensity of the XRD peaks. The XRD diffractogram showed crystalline peaks at 2θ equals to 11.75°, 11.99°, 19.17°, 19.32°, 23.69°, 23.97°, 24.14°, 28.50° and 36.05°. The decrease in intensity of crystalline planes after biofield treatment of AM may be due to disturbed regular pattern of the atoms. The volume of unit cell, crystallite size and change in molecular weight were computed from the XRD diffractogram using Powder X software. The results are presented in Figure 2. An increase (0.16%) in volume of unit cell was observed in treated AM as compared to control. The treated AM showed 0.16% increase in molecular weight as compared to control. It is hypothesized that biofield energy possibly acted on treated AM crystals at nuclear level and altered the number of proton and neutrons as compared to control, which may led to increase the molecular weight. The treated AM (46.20 nm) showed (Table 1) decrease in crystallite size as compared to control (69.31 nm) and percentage decrease in crystallite size was 33.34% (Figure 3). The existence of severe lattice strains was evidenced by the change in unit cell volume. Thus, it is assumed that presence of these internal strains might be a reason for fracturing the grains into sub grains which leads to decrease in crystallite size of treated sample. The XRD diffractogram of control and treated CA are presented in Figure 4. The XRD diffractogram of control CA showed occurrence of intense crystalline peaks at 2θ equals to 26.87°, 33.09°, 35.32° and 35.45°. This clearly showed the crystalline nature of the control sample. However, the treated CA showed the increase in intensity of the XRD peaks. The treated sample showed (Figure 4) XRD peaks at 17.64°, 27.09°, 33.24°,

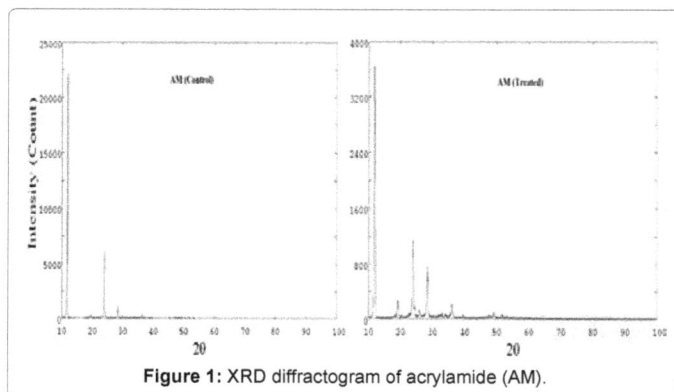

**Figure 1:** XRD diffractogram of acrylamide (AM).

35.52°, 35.63° and 44.78°. The significant increase in intensity of the XRD peaks revealed the enhanced crystalline nature of the sample. It is presumed that biofield energy may be absorbed by the CA crystals, which led to formation of more symmetrical crystalline long-range pattern; that caused increase in crystallinity. The volume of unit cell of treated CA was decreased with respect to control sample and the percentage decrease in volume of unit cell was -1.92%. The molecular weight of treated CA was reduced by 1.92% as compared to control and this may be due to decrease in number of protons and neutrons after biofield treatment (Figure 2). However, the treated CA showed (Table 1) a significant increase in crystallite size (108.60 nm) as compared to control CA (47.26 nm). The percentage change in crystallite size was 129.79%. This showed (Figure 3) the significant impact of biofield treatment on increasing the crystallite size of treated CA with respect to control. It was previously reported that crystallite size increases with elevation in temperature or thermal energy. Hence, it is assumed that biofield energy may reduce the thermodynamically driving force which automatically causes decrease in nucleus densities and raises the crystallite size [32,33].

## DSC studies

The DSC thermogram of control and treated AM are shown in Figure 5. The DSC thermogram of control AM showed the presence of sharp endothermic peak at 86°C which was due to melting temperature of the control. However, the treated AM showed (Figure 5) an increase in melting temperature peak (88°C) as compared to control. It indicated that biofield treated AM had better thermal stability as compared to control. DSC thermogram of control and treated CA are depicted in Figure 6. The DSC graph of the control sample showed sharp endothermic inflexion at 114°C which was responsible for melting of the compound. DSC thermogram of control CA showed another two endothermic peaks at 135°C and 168°C. Katayama performed X-ray investigation on CA and found that there are two forms of compound in crystalline state, one being more stable than other [34]. Hence, it is presumed that two endothermic peaks in control CA may be represented to two crystalline polymorphs present in the compound. Whereas, the treated CA showed (Figure 6) an increase in melting temperature and it was observed at 116°C. It confirmed that biofield treatment may induced regular pattern in the CA atoms and hence increase in thermal stability. It was reported that γ radiation treatment increased the thermal stability of poly (3-hexadecythiophene). Therefore, it is presumed that biofield treatment may cause conformational changes and crosslinking in AM and CA compounds which led to increase in thermal stability [35]. Latent heat of fusion (ΔH) was calculated from the DSC data and results are depicted in Table 2. The control AM showed a ΔH of 231.96 J/g and it was decreased to 156.83 J/g in treated

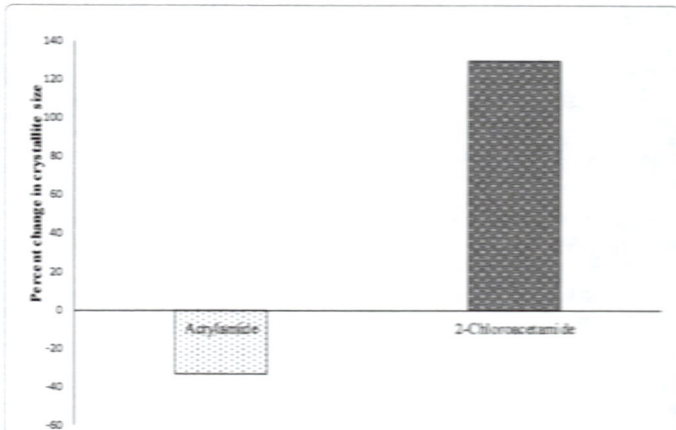

Figure 3: Percent change in crystallite size ('G' x 10⁻⁹ nm) of treated acrylamide (AM) and 2-chloroacetamide (CA).

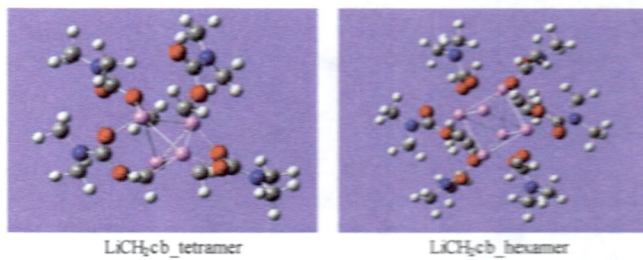

Figure 4: XRD diffractogram of 2-chloroacetamide (CA).

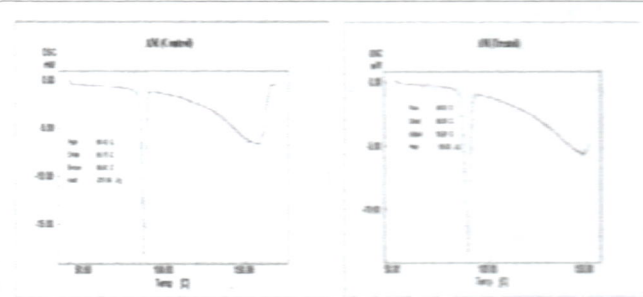

Figure 5: DSC thermogram of acrylamide (AM).

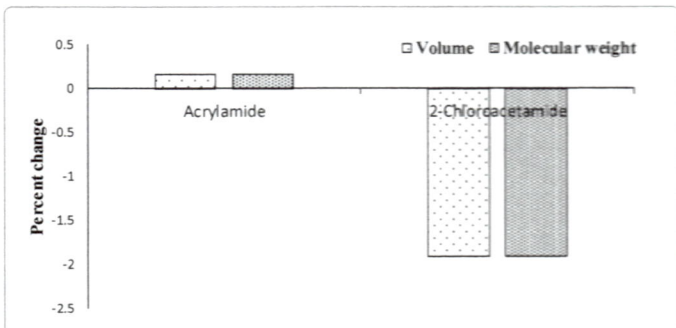

Figure 2: Percent change in volume of unit cell (10⁻²³ cm³) and molecular weight of treated acrylamide (AM) and 2-chloroacetamide (CA).

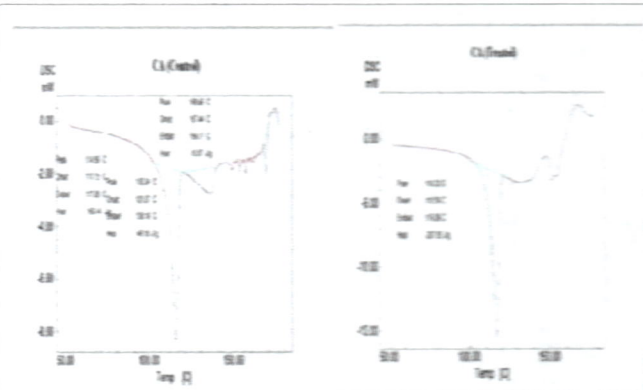

Figure 6: DSC thermogram of 2-chloroacetamide (CA).

| Compound Characteristics | Sample type | Acrylamide | 2-Chloroacetamide |
|---|---|---|---|
| Volume of unit cell ($10^{-23}$ cm³) | Control | 423.4 | 392.13 |
| | Treated | 424.09 | 384.61 |
| Crystallite Size 'G' x $10^{-9}$ m | Control | 69.31 | 47.26 |
| | Treated | 46.20 | 108.60 |
| Molecular Weight (g/mol) | Control | 72.93 | 94.77 |
| | Treated | 73.05 | 92.95 |

**Table 1:** XRD data (volume of unit cell, crystallite size and molecular weight) of acrylamide (AM) and 2-chloroacetamide (CA).

| Sample | Control ($\Delta$H J/g) | Treated ($\Delta$H J/g) | % Change in $\Delta$H |
|---|---|---|---|
| Acrylamide | -231.96 | -156.83 | -32.39 |
| 2-Chloroacetamide | -169.44 | -207.55 | 22.49 |

**Table 2:** Latent heat of fusion ($\Delta$H) of control and treated compounds (Acrylamide and 2-Chloroacetamide).

AM. The $\Delta$H was changed by 32.39% in the treated AM with respect to control. Whereas the control CA showed a $\Delta$H of 169.44 J/g and it was increased to 207.55 J/g in treated CA. The result showed 22.49% increase in $\Delta$H of treated CA as compared to control. It is assumed that biofield may altered the internal energy of the treated compounds (AM and CA) which led to significant change in $\Delta$H with respect to control samples.

## TGA studies

Thermal stability of the control and treated AM and CA compounds were evaluated using TGA. TGA thermogram of control and treated AM are presented in Figures 7 and 8. The TGA thermogram of control AM showed (Figure 7) one step thermal degradation pattern. The thermal degradation commence at around 148°C (onset) and degradation stopped at around 195°C (end set). This step showed a major weight loss and the sample lost 58.82% of its weight. However, DTA thermogram of control AM showed an endothermic peak at 86°C which may be associated with the melting temperature of the sample. The thermogram showed another endothermic peak at 174°C that may be due to thermal decomposition or breaking of intermolecular hydrogen bonding between AM. Derivative thermogravimetry (DTG) of the control AM exhibited the maximum thermal decomposition temperature ($T_{max}$) at 164°C. However, the treated AM thermogram also showed (Figure 8) single step thermal decomposition pattern. The thermal degradation started at around 135°C (onset) and terminated at around 190°C (end set). During this step the treated AM showed 68.68% of weight loss. DTA thermogram of the treated AM showed melting peak at 85°C and second endothermic was seen at 176°C (thermal decomposition). The DTG thermogram of treated AM showed no change in $T_{max}$ value (164°C) as compared to control. It may be corroborated to no alteration in thermal stability of treated AM with respect to control after biofield treatment. The TGA thermogram of control and treated CA are shown in Figures 9 and 10. The control CA showed occurrence of one step thermal degradation pattern. Thermal degradation commenced at around 120°C (onset) and completed at around 190°C (end set). During this process the sample showed (Figure 9) major weight loss (66.09%) that might be due to thermal decomposition of the CA chain. DTA thermogram of CA showed an endothermic peak at 115°C; associated with the melting of the sample. Another endothermic event was noticed at 162°C that may be due to thermal decomposition. The $T_{max}$ of control CA was observed at 150°C as shown by the DTG thermogram. TGA thermogram of treated CA showed (Figure 10) one step thermal degradation between 130-210°C. During this thermal event sample showed rapid thermal degradation and weight loss (61.16%). However, the DTA thermogram of treated

CA showed two endothermic peaks at 116°C and 176°C. The former endothermic was due to melting and later peak was attributed to thermal decomposition. The DTG thermogram of treated CA showed increase in $T_{max}$ value (167°C) as compared to control sample (150°C). This increase in thermal decomposition may be correlated to high thermal stability of the treated CA.

## Surface area analysis

Surface area of AM and CA was investigated using BET method. The surface area result of control and treated compounds (AM and CA) are presented in Table 3. The control AM showed a surface area of 0.42 m²/g, however, the treated AM showed 0.55 m²/g. The percentage increase in surface area was 31.6% in the treated AM

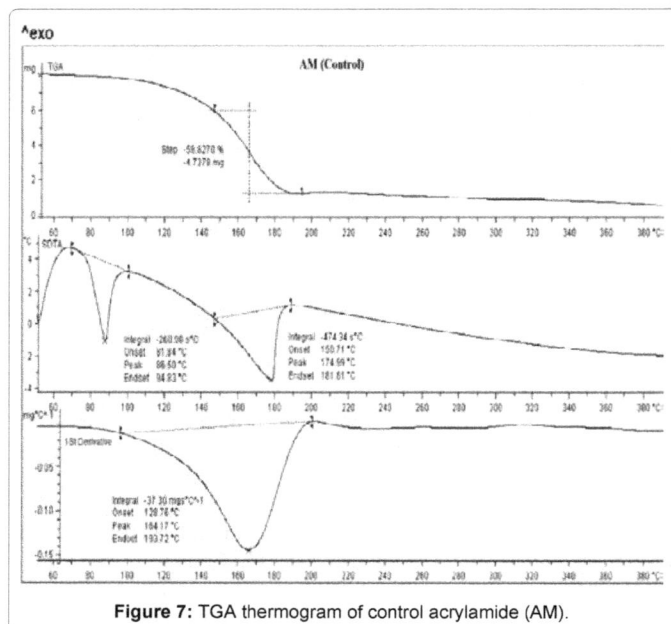

**Figure 7:** TGA thermogram of control acrylamide (AM).

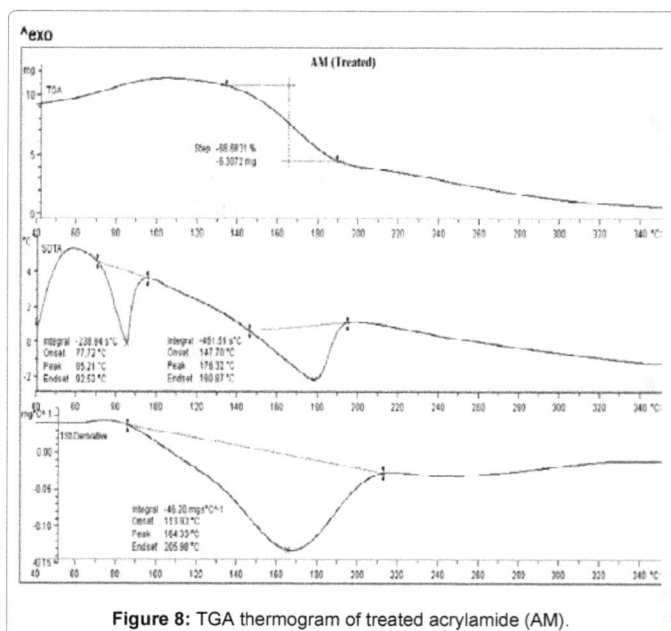

**Figure 8:** TGA thermogram of treated acrylamide (AM).

| Sample | Control (m²/g) | Treated (m²/g) | % Change in surface area |
|---|---|---|---|
| Acrylamide | 0.42 | 0.55 | 31.6 |
| 2-Chloroacetamide | 0.65 | 0.45 | -30.9 |

**Table 3:** Surface area data of acrylamide (AM) and 2-chloroacetamide (CA).

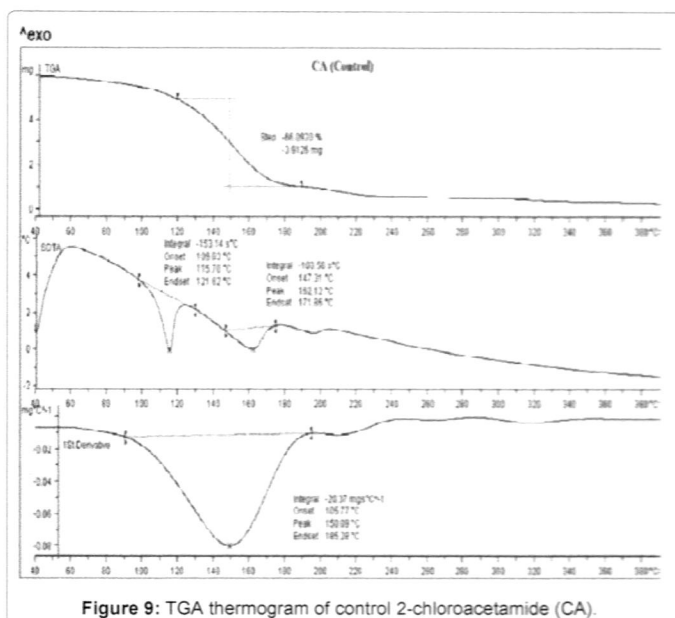

**Figure 9:** TGA thermogram of control 2-chloroacetamide (CA).

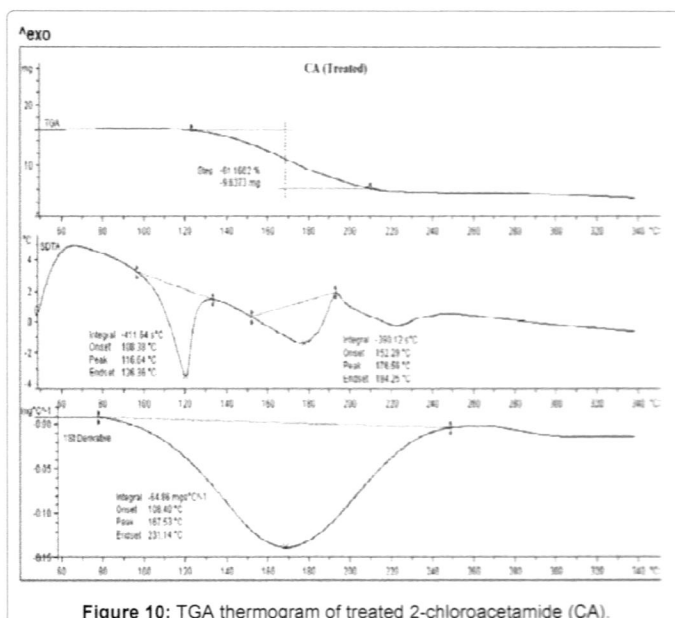

**Figure 10:** TGA thermogram of treated 2-chloroacetamide (CA).

sample as compared to control. It is hypothesized that energy milling provided by biofield treatment might cause fracturing in particles and thus reduction in particle size. Hence, reduction in particle size causes increase in surface area of the treated AM [36,37]. The control CA showed surface area 0.65 m²/g; however, it decreased to 0.45 m²/g in treated sample. The percentage decrease in surface area was 30.9% in treated CA as compared to control sample. The biofield energy may increase the particle size that lead to decrease in surface area.

## Conclusion

The present work investigated the influence of biofield treatment on physical and thermal properties of AM and CA. The XRD results showed the decrease in crystallinity of treated AM as compared to control. The crystallite size was also decreased in treated AM with respect to control. Whereas, treated CA showed significant increase in crystallite size as compared to control. DSC showed increase in melting temperature of both treated AM and CA with respect to control. Additionally, latent heat of fusion of treated compounds was substantially changed as compared to control. TGA showed enhanced thermal decomposition temperature in CA as compared to control that indicated the higher thermal stability after biofield treatment. However, no change in thermal stability was observed in treated AM with respect to control. Overall, the results indicated that biofield treatment has significant impact in alteration of the physical and thermal properties of AM and CA.

### Acknowledgement

The authors would like to thank all the laboratory staff of MGV Pharmacy College, Nashik for their assistance during the various instrument characterizations. We thank Dr. Cheng Dong of NLSC, institute of physics, and Chinese academy of sciences for permitting us to use Powder X software for analyzing XRD results.

### References

1. Raymond S, Weintraub L (1959) Acrylamide gel as a supporting medium for zone electrophoresis. Science 130: 711.

2. Yang TH (2008) Recent applications of polyacrylamide as biomaterials. Recent Pat Mater Sci 1: 29-40.

3. Bernfeld P, Wan J (1963) Antigens And Enzymes Made Insoluble By Entrapping Them Into Lattices Of Synthetic Polymers. Science 142: 678-679.

4. Abrahám M, Horváth L, Simon M, Szajáni B, Boross L (1985) Characterization and comparison of soluble and immobilized pig muscle aldolases. Appl Biochem Biotechnol 11: 91-100.

5. Ratner BD, Hoffmann AS, Schoen FJ, Lemons J (1996) Biomaterial Science. Academic Press, New York.

6. Langer R1 (1998) Drug delivery and targeting. Nature 392: 5-10.

7. Davis BK (1972) Control of diabetes with polyacrylamide implants containing insulin. Experientia 28: 348.

8. Hussain MD, Rogers JA, Mehvar R, Vudathala GK (1999) Preparation and release of ibuprofen from polyacrylamide gels. Drug Dev Ind Pharm 25: 265-271.

9. Sairam M, Babu VR, Vijaya B, Naidu K, Aminabhavi TM, et al. (2006) Encapsulation efficiency and controlled release characteristics of crosslinked polyacrylamide particles. Int J Pharm 320: 131-136.

10. Soppimath KS, Kulkarni AR, Aminabhavi TM (2001) Chemically modified polyacrylamide-g-guar gum-based crosslinked anionic microgels as pH-sensitive drug delivery systems: preparation and characterization. J Control Release 75: 331-345.

11. Murakami Y, Maeda M (2005) DNA-responsive hydrogels that can shrink or swell. Biomacromolecules 6: 2927-2929.

12. Christensen LH, Breiting VB, Aasted A, Jørgensen A, Kebuladze I, et al. (2003) Long-term effects of polyacrylamide hydrogel on human breast tissue. Plast Reconstr Surg 111: 1883-1890.

13. Zheng J, Li R, Zhu J, Zhang J, He J, et al. (2012) Degradation of the chloroacetamide herbicide butachlor by Catellibacterium caeni sp. nov DCA-1T. Int Biodeter Biodegr 73: 16-22.

14. Sonmez HB, Bicak N (2002) Quaternization of poly (4-vinyl pyridine) beads with 2-chloroacetamide for selective mercury extraction. React Funct Polym 51: 55-60.

15. Allen Jr HC (1952) The pure quadrupole spectra of solid chloroacetic acids and substituted chloroacetic acid. J Am Chem Soc 74: 6074-6076.

16. Allen Jr HC (1953) Pure quadruple spectra of molecular crystals. J Phys Chem 57: 501-504.

17. Trivedi MK, Patil S, Tallapragada RM (2013) Effect of biofield treatment on the physical and thermal characteristics of vanadium pentoxide powders. J Material Sci Eng S11: 001.

18. Trivedi MK, Patil S, Tallapragada RM (2013) Effect of biofield treatment on the physical and thermal characteristics of silicon, tin and lead powders. J Material Sci Eng 2: 125.

19. Trivedi MK, Patil S, Tallapragada RMR (2015) Effect of biofield treatment on the physical and thermal characteristics of aluminium powders. Ind Eng Manag 4: 151.

20. Trivedi MK, Patil S, Tallapragada RM (2014) Atomic, crystalline and powder characteristics of treated zirconia and silica powders. J Material Sci Eng 3: 144.

21. Korotkov K (2002) Human Energy Field: study with GDV bioelectrography. NY, Backbone publishing.

22. Myers R (2003) The basics of chemistry. Greenwood Press, Westport, Connecticut.

23. Movaffaghi Z, Farsi M (2009) Biofield therapies: biophysical basis and biological regulations? Complement Ther Clin Pract 15: 35-37.

24. Shinde V, Sances F, Patil S, Spence A (2012) Impact of biofield treatment on growth and yield of lettuce and tomato. Aust J Basic & Appl Sci 6: 100-105.

25. Sances F, Flora E, Patil S, Spence A, Shinde V, et al. (2013) Impact of biofield treatment on ginseng and organic blueberry yield. Agrivita J Agric Sci 35: 22-29.

26. Lenssen AW (2013) Biofield and fungicide seed treatment influences on soybean productivity, seed quality and weed community. Agricultural Journal 8: 138-143.

27. Altekar N, Nayak G (2015) Effect of biofield treatment on plant growth and adaptation. J Environ Health Sci 1: 1-9.

28. Trivedi MK, Patil S (2008) Impact of an external energy on Staphylococcus epidermis [ATCC –13518] in relation to antibiotic susceptibility and biochemical reactions – An experimental study. J Accord Integr Med 4: 230-235.

29. Trivedi MK, Patil S (2008) Impact of an external energy on Yersinia enterocolitica [ATCC –23715] in relation to antibiotic susceptibility and biochemical reactions: An experimental study. Internet J Alternative Med 6.

30. Trivedi MK, Bhardwaj Y, Patil S, Shettigar H, Bulbule A, et al. (2009) Impact of an external energy on Enterococcus faecalis [ATCC – 51299] in relation to antibiotic susceptibility and biochemical reactions – An experimental study. J Accord Integr Med 5: 119-130.

31. Patil SA, Nayak GB, Barve SS, Tembe RP, Khan RR (2012) Impact of biofield treatment on growth and anatomical characteristics of Pogostemon cablin (Benth.). Biotechnology 11: 154-162.

32. Rashidi AM, Amadeh A (2009) The effect of saccharin addition and bath temperature on the grain size of nanocrystalline nickel coatings. Surf Coat Technol 204: 353-358.

33. Katayama M (1956) The crystal structure of an unstable form of chloroacetamide. Acta Crystallogr 9: 986-991.

34. Gusain D, Srivastava V, Singh VK, Sharma YC (2014) Crystallite size and phase transition demeanor of ceramic steel. Mater Chem Phys 146: 320-326.

35. Szabo L, Cik G, Lensy J (1996) Thermal stability increase of doped poly (hexadecylthiophene) by ?-radiation. Synt Met 78: 149-153.

36. Mennucci B, Martinez JM (2005) How to model solvation of peptides? Insights from a quantum-mechanical and molecular dynamics study of N-methylacetamide. I. Geometries, infrared, and ultraviolet spectra in water. J Phys Chem B 109: 9818-9829.

37. Bendz D, Tüchsen PL, Christensen TH (2007) The dissolution kinetics of major elements in municipal solid waste incineration bottom ash particles. J Contam Hydrol 94: 178-194.

# 14

# Anti-Tumor and Anti-Leishmanial Evaluations of Novel Thiophene Derivatives Derived from the Reaction of Cyclopentanone with Elemental Sulphur and Cyano-Methylene Reagents

**Rafat M. Mohareb[1,2]\* and Fatma O. Al-farouk[3]**

[1]*Department of Chemistry, Faculty of Science, Cairo University, Giza, A. R. Egypt*
[2]*Department of Organic Chemistry, Faculty of Pharmacy, October University for the Modern science and Arts (MSA), Elwahaat Road, October City, Egypt*
[3]*Department of Chemistry, Faculty of Science, American University in Cairo, 5th Settlement, A.R., Egypt*

### Abstract

The reaction of cyclopentanone (**1**), elemental sulfur and either malononitrile or ethyl cyanoacetate gave the cyclopenta[*b*]thiophene derivatives **3a** and **3b**, respectively. The reaction of either **3a** or **3b** with either **2a** or **2b** afforded the cyclopenta[4,5]thieno[2,3-b]pyridine derivatives **5** and **6**, respectively. The reactivity of the latter products toward different reagents was studied to give pyrazole, pyridine, pyrimidine derivatives. The antitumor evaluation of the newly synthesized products against the three cancer cells namely breast adenocarcinoma (MCF-7), non-small cell lung cancer (NCI-H460) and VNS cancer (SF-268) showed that some of them have high inhibitory effect towards three cell lines which is higher than the standard. Moreover, the anti-leishmanial activity of the newly synthesized products was tested on Leishmania amastigotes showed that some compounds have high activity.

**Keywords:** Cyclopenta[*b*]thiophene; Coumrin; anticonvulsant; neurotoxicity; CNS depressant

## Introduction

Sulphur containing heterocycles paved the way for active research in pharmaceutical Chemistry. Nowadays benzothiophene derivatives in combination with other ring systems are extensively used in pharmaceuticals such as antiallergic [1], analgesic [2], anti-inflammatory [3] and occular hypotensive agents [4]. Raloxifene, a drug based on benzo[b]thiophene has been approved by the U.S Food and Drug Administration for the prevention and treatment of osteoporosis associated with menopause [5]. On the other hand, compounds bearing N-containing rings are very well known to exhibit powerful antimicrobial [6], anticonvulsant [7], antidepressant [8], and analgesic [9] activities. Moreover, some nitrogen and sulfur containing compounds are associated with diverse pharmacological activities [10-14]. Furthermore, various oxadiazoles [15], pyrazolin-5-ones [16], and diaryl pyrazole derivatives [17] also exhibit wide spectrum pharmacological activities.

On the way of continuing our work [18-20] on the synthesis of new heterocyclic compounds with expected biological activities, we hereby report the synthesis of some new cyclopenta[*b*]thiophene derivatives and their characterization by IR, NMR & Mass spectrometry techniques. Newly synthesized compounds were also screened for their antitumor evaluation against the three cancer cells namely breast adenocarcinoma (MCF-7), non-small cell lung cancer (NCI-H460) and VNS cancer (SF-268). Moreover, the anti-leishmanial activity of the newly synthesized products was tested on Leishmania amastigotes.

## Results and Discussion

In this research work, the reaction of cyclopentanone **1** and elemental sulfur with cyanomethylene reagents was studied. Thus, the former two reagents have been allowed to react with either malononitrile **2a** or ethyl cyanoacetate **2b** and gave the cyclopenta[*b*] thiophenes **3a,b**, respectively [21].

The reaction of **3a** with malononitrile **2a** in ethanolic triethylamine solution gave the fused pyridine derivative **5** [22]. Formation of the latter compound is suggested to have taken place via the intermediate formation of **4** followed by intramolecular cyclization, this is explained in terms of activation of the $CH_2$ between the electronegative CN group and the sp$^2$ C. And the assigned structures for the previous compounds are consistent with analytical and spectral data. However, the reaction of **3a** with ethyl cyanoacetate **2b** gave the amide derivative **6**, whose H$^1$NMR spectrum showed a multiplet at δ 2.22-2.32 corresponding to the hydrogens of the three $(CH_2)$ groups in the cyclopentene moiety, a singlet at δ 4.75 and another at δ 8.23 corresponding to the methylene group protons and the NH hydrogen, respectively (scheme 3). Compound **6** underwent ready cyclization when heated in 1,4-dioxane and triethylamine to yield the cyclopenta[4,5]thieno[2,3-b]pyridine derivative **7** (scheme 1).

The reaction of compound **6** with benzaldehyde **8** gave the benzalidine derivative **9**, while its reaction with salicaldehyde **10** produced the coumarin derivative **12**, supposedly through the formation of the arylidine derivative **11** followed by intramolecular cyclization and hydrolysis of the C=NH group to a C=O group. Formation of the latter coumarin derivative **12** through the reaction of salicaldehyde **10** with cyanomethylene reagents was previously reported in literature [23]. On the other hand, the reaction of compound **6** with benzenediazonium chloride **13** gave the phenylhydrazone derivative **14** whose analytical and spectral data are in agreement with the proposed structure. Compound **14**, upon reacting with either hydrazine hydrate

---

**\*Corresponding author:** Rafat M. Mohareb, Department of Organic Chemistry, Faculty of Pharmacy, October University for the Modern science and Arts (MSA), Elwahaat Road, October City, Egypt, E-mail: raafat_mohareb@yahoo.com

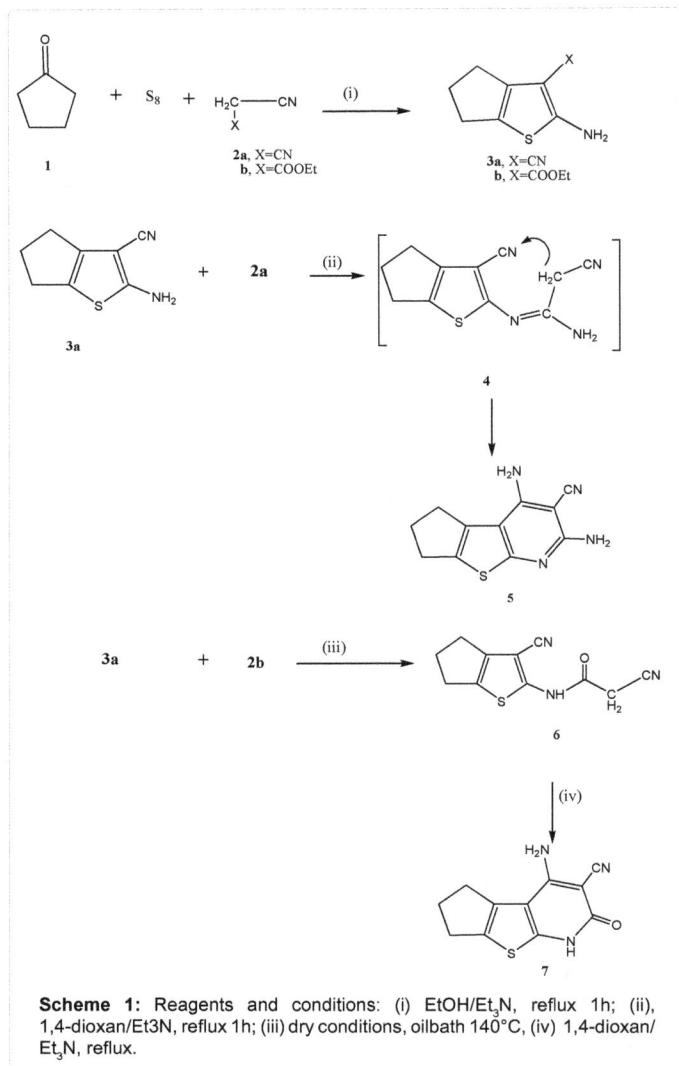

**Scheme 1:** Reagents and conditions: (i) EtOH/Et₃N, reflux 1h; (ii), 1,4-dioxan/Et3N, reflux 1h; (iii) dry conditions, oilbath 140°C, (iv) 1,4-dioxan/Et₃N, reflux.

15a or phenyl hydrazine **15b**, gave the pyrazole derivatives **17a** and **17b**, respectively. Formation of the latter compounds took place, supposedly, through the intermediates **16a,b** followed by water elimination (Scheme 2).

The reaction of compound **6** with either malononitrile **2a** or ethyl cyanoacetate **2b** gave the pyridine derivatives **19a,b**, respectively. Formation of the latter pyridines is explained in terms of the intermediate formation of **18a,b** followed by intramolecular cyclization, and structure elucidation is based on analytical and spectral data. Thus, the H¹NMR spectrum of **19a**, for instance, showed a multiplet at δ 2.24-2.39 corresponding to the hydrogens of the three (CH₂) groups in the cyclopentene moiety, two singlets (D₂O-exchangeable) for hydrogens of the two amino groups at δ 4.27 and 5.31 and another D₂O-exchangeable singlet at δ 6.51 corresponding to H-3 on the pyridine ring. Moreover, the ¹³C NMR spectrum showed δ values of 22.1, 26.2, 32.7 (3 CH₂), 116.0 (CN), 119.8, 121.0, 121.3, 122.2, 123.7, 129.0, 139.9, 143.8, 152.3, 155.0 (thiophene C, pyridine C). Compound **3a** has also reacted with phenylisothiocyanate **20** in 1,4-dioxane and triethylamine to give the phenyl thiourea derivative **21** (Scheme 3). Compound **21** has been readily cyclized by heating in 1,4-dioxane in the presence of a catalytic amount of triethylamine, to give the thioxo-pyrimidine

derivative **22**. The analytical data of the latter product was consistent with the proposed structure (see experimental section).

## Biological Evaluation

### Materials and methods

**Materials, methods & reagents:** Fetal bovine serum (FBS) and L-glutamine, were from Gibco Invitrogen Co. (Scotland, UK). RPMI-1640 medium was from Cambrex (New Jersey, USA). Dimethyl sulfoxide (DMSO), doxorubicin, penicillin, streptomycin and sulforhodamine B (SRB) were from Sigma Chemical Co. (Saint Louis, USA). Samples: Stock solutions of compounds **3a-9d** were prepared in DMSO and kept at -20 °C. Appropriate dilutions of the compounds were freshly done just prior to the assays. Final concentrations of DMSO did not interfere with the cell growth.

**Cell cultures:** Three human tumor cell lines, MCF-7 (breast adenocarcinoma), NCI-H460 (non-small cell lung cancer), and SF-268 (CNS cancer) were used. MCF-7 was obtained from the European Collection of Cell Cultures (ECACC, Salisbury, UK) and NCI-H460 and SF-268 were kindly provided by the National Cancer Institute (NCI, Cairo, Egypt). They grow as monolayer and routinely maintained in RPMI-1640 medium supplemented with 5% heat inactivated FBS, 2 mM glutamine and antibiotics (penicillin 100 U/mL, streptomycin 100 µg/mL), at 37 °C in a humidified atmosphere containing 5% CO₂. Exponentially growing cells were obtained by plating 1.5 X 10⁵ cells/mL for MCF-7 and SF-268 and 0.75 X 10⁴ cells/mL for NCI-H460, followed

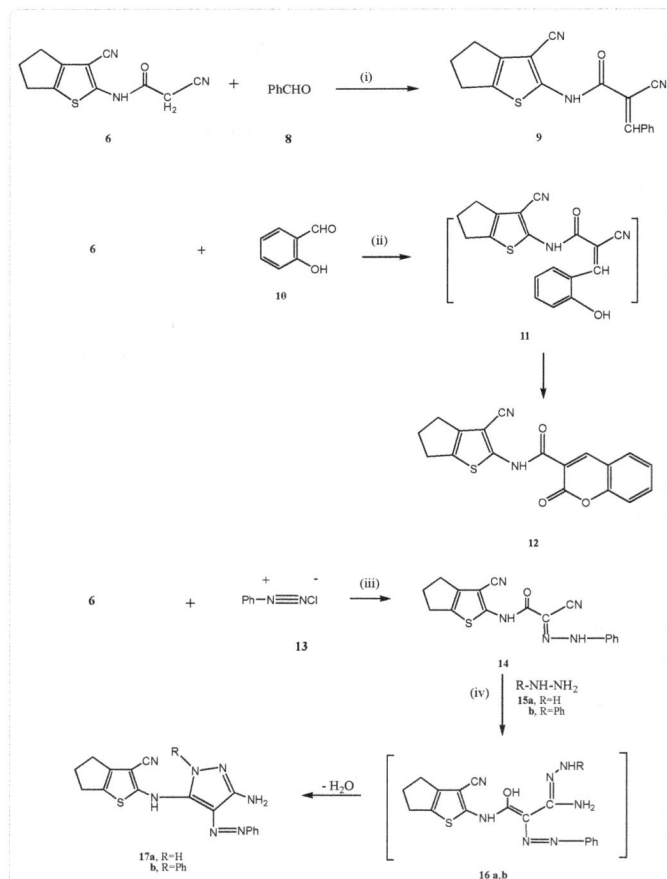

**Scheme 2:** Reagents and conditions: (i) 1,4-dioxane/Et₃N, reflux; (ii) 1,4-dioxane/Et3N; (iii) EtOH, NaOH. 0°C; (iv) 1,4-dioxane, reflux.

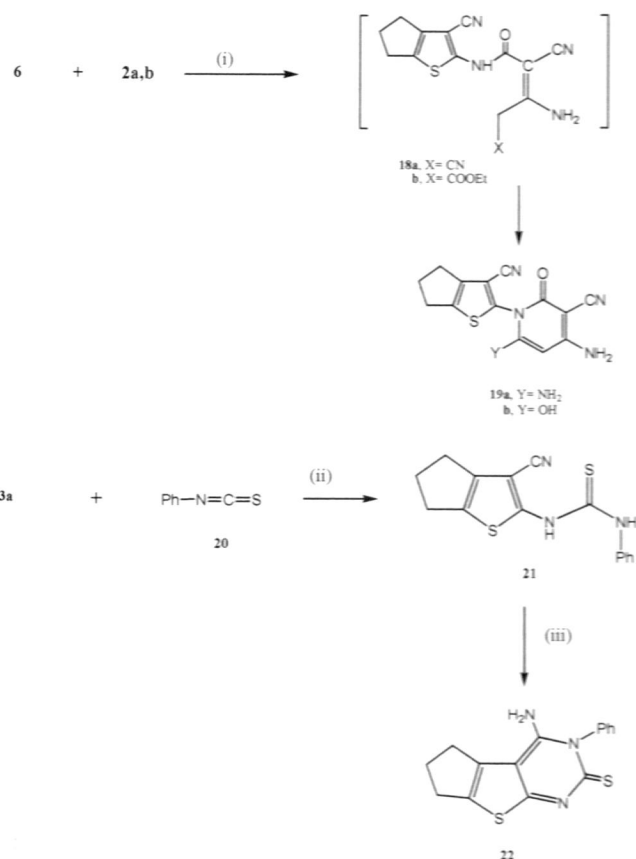

**Scheme 3:** Reagents and conditions: (i) & (ii) 1,4-dioxane, Et₃N, reflux; (iii) EtOH/EtONa, water bath 4h.

by 24 h of incubation. The effect of the vehicle solvent (DMSO) on the growth of these cell lines was evaluated in all the experiments by exposing untreated control cells to the maximum concentration (0.5%) of DMSO used in each assay.

## Anti-tumor activity

### Effect of the synthesized compounds on the growth of human tumor cell lines

All the synthesized compounds were evaluated on the *in vitro* growth of three human tumor cell lines representing different tumor types, namely, breast adenocarcinoma (MCF-7), non-small cell lung cancer (NCI-H460) and CNS cancer (SF-268), after a continuous exposure of 48 h. The results are summarized in Table 1. It is clear that compound 22 showed the best inhibition towards the three cell lines.

### Anti-leishmanial activity

Anti-leishmanial activity was tested on L. donovani amastigotes growing in macrophages at concentrations, which showed less than 40% cytotoxicity for the macrophage cell line THP-1. The compounds 7, 9, 14, 17, 19a and 21 showed high activity on L. donovani amastigotes growing in macrophages similar to that seen with axenic amastigotes at 50 µM, 95 %, 68 %, 80 %, 88 %, 86 % and 98 %, respectively. It is also obvious that compounds 7 and 21, namely 5-Amino-7-oxo-3,4-dihydro-2H-cyclopenta[4,5]thieno[2,3-b]pyridine-6-carbonitrile

and 1-(3-Cyano-5,6-dihydro-4H-cyclopenta[*b*]thiophen-2-yl)-3-phenylthiourea, showed maximum average inhibition values that are higher than the positive control (Table 2). Moreover, it could be deduced that substitution by a nitro group enhances the anti-leishmanial activity. This becomes evident by comparing the activities of 3a, 5 and 56 with average inhibitions of 22%, 20% and 24%, respectively, and those of 3b, 12, 17b and 19b with average inhibitions of 32%, 42%, 56% and 55% respectively.

## Experimental

### Chemistry

All melting points were determined in open capillaries and are uncorrected. IR spectra were measured using KBr discs on a Pye Unicam SP-1000 spectrophotometer. ¹H-NMR and ¹³CNMR

| Compound | GI₅₀ (m mol L⁻¹) | | |
|---|---|---|---|
| | MCF-7 | NCI-H460 | SF-268 |
| 3a | 30 ± 0.6 | 17.3 ± 1.4 | 22.3 ±1.5 |
| 3b | 20 ± 0.4 | 24.3 ± 0.8 | 32 ± 0.8 |
| 5 | 70.6 ± 16.9 | 38.9 ± 10.8 | 50.8 ± 8.6 |
| 6 | 40.6 ± 12.2 | 32.6 ± 8.6 | 60.4 ± 14.8 |
| 7 | 0.4 ± 0.2 | 0.1 ± 0.08 | 0.9 ± 0.08 |
| 9 | 11.8 ± 0.6 | 14.5 ± 0.8 | 16.7 ± 1.6 |
| 12 | 72.7 ± 17.5 | 40.2 ± 12.8 | 50.0 ± 9.01 |
| 14 | 50.1 ± 0.7 | 23.2 ± 4.8 | 18.4 ± 1.8 |
| 17a | 22.0 ± 0.2 | 30.6 ± 1.4 | 38.4 ± 0.6 |
| 17b | 38.0 ± 1.8 | 44.0 ± 0.8 | 20.5 ± 1.1 |
| 19a | 20.0 ± 0.6 | 22.0 ± 0.4 | 31.5 ± 8.0 |
| 19b | 0.01 ± 0.006 | 0.04 ± 0.002 | 0.03 ± 0.005 |
| 21 | 70.9 ± 0.9 | 43.6 ± 1.8 | 56.8 ± 0.8 |
| 22 | 0.02 ± 0.009 | 0.06 ± 0.008 | 0.08 ± 0.001 |
| Doxorubicin | 0.04 ± 0.008 | 0.09 ± 0.008 | 0.09 ± 0.007 |

Results are given in concentrations that were able to cause 50 % of cell growth inhibition (GI₅₀) after a continuous exposure of 48 h and show means ± SEM of three-independent experiments performed in duplicate.

**Table 1:** Effect of compounds 3a-21b on the growth of three human tumor cell lines.

| Compound | Average inhibition (%) | GI₅₀ᵃ (µ M) |
|---|---|---|
| 3a | 22 | 20 |
| 3b | 32 | 28 |
| 5 | 20 | - |
| 6 | 24 | 10.2 |
| 7 | 95 | - |
| 9 | 68 | 26 |
| 12 | 42 | - |
| 14 | 80 | 24 |
| 17a | 88 | 12.2 |
| 17b | 56 | 30 |
| 19a | 86 | - |
| 19b | 55 | 10.6 |
| 21 | 98 | 24 |
| 22 | 58 | 8.2 |
| Positive controlᵇ | 95 | |
| Negative controlᶜ | 0 | |

ᵃGI₅₀ = concentration for 50% growth inhibition
ᵇAmphotericin B (1µ M)
ᶜCulture medium and DMSO

**Table 2:** Anti-leishmanial activity of compounds 3a-9d at 50 µM against L. donovani axenic amastigotes.

spectra were measured on a Varian EM390-200 MHz instrument in CD$_3$SOCD$_3$ as solvent using TMS as internal standard, and chemical shifts are expressed as δ in units of parts per million (ppm). Analytical data were obtained from the Micro analytical data unit at Cairo University. MS spectra were determined using a Shimadzu GC–MS-2010P. Compounds **3a** and **3b** were synthesized according to the reported procedure.

5,7-Diamino-3,4-dihydro-2H-cyclopenta[4,5]thieno[2,3-b] pyridine-6-carbonitrile (**5**)

Malononitrile (0.12 g, 1.817×10$^{-3}$ mol) and 0.5 mL of triethylamine were added to a solution of 2-amino-5,6-dihydro-4H-cyclopenta[b] thiophene-3-carbonitrile **3a** (0.3 g, 1.817×10$^{-3}$ mol) in 25 mL of 1,4-dioxane and the reaction mixture was heated under reflux for 3 h. It was then poured on ice and a few HCl (18.0 mol) drops (till pH$_6$) stirred in. Filtration of the observed precipitate was then carried out.

Light brown crystals from ethanol, yield 52.38 %, 0.22 g, m.p. 138-141 °C. IR (KBr): υ/cm$^{-1}$= 3482-3346 (2 NH$_2$), 2892 (CH$_2$), 2223 (CN), 1648 (C=C), 1632 (C=N). H$^1$NMR (DMSO) δ = 2.24-2.38 (m, 6H, 3CH$_2$), 4.80, 5.31 (2s, 4H, 2NH$_2$). $^{13}$C NMR δ = 20.4, 26.9, 32.5 (3 CH$_2$), 116.8 (CN), 127.2, 134.4, 136.3, 144.4, 144.8, 154.8, 156.9 (thiophene C, pyridine C). Calcd for C$_{11}$H$_{10}$N$_4$S (230.293): C, 57.37; H, 4.38; N, 24.33, S, 13.93 %. Found: C, 57.62; H, 4.40; N, 24.43; S, 14.12 %. MS (relative intensity) m/z: 230 (M$^+$, 100 %), 231 (M+1, 13 %), 232 (4 %), 176 (65 %).

2-Cyano-N-(3-cyano-5,6-dihydro-4H-cyclopenta[b]thiophen-2-yl)acetamide (**6**)

Ethyl cyanoacetate (2.07 g, 0.018 mol) was fused at a temperature of 250 °C with 2-amino-5,6-dihydro-4H-cyclopenta[b]thiophene-3-carbonitrile **3a** (2.96 g, 0.018 mol) in a solvent-free catalyst-free reaction which yielded the target amide in 1 h. Water was then added and the precipitate filtered out. Grey crystals from ethanol yield 86.54 %, 3.60 g, m.p. 290-293 °C. IR (KBr): υ/cm$^{-1}$= 3469-3322 (NH), 2988 (CH$_2$), 2228, 2222 (2 CN), 1689 (CO), 1641 (C=C). H$^1$NMR (DMSO) δ = 2.20-2.33 (m, 6H, 3CH$_2$), 4.75 (s, 2H, CH$_2$), 8.23 (s, 1H, NH). $^{13}$C NMR δ = 19.9, 26.7, 32.8 (3 CH$_2$), 27.8 (CH$_2$), 116.2, 117.0 (2CN), 127.0, 134.6, 137.0, 143.9, 144.5 (thiophene C), 169.0 (C=O). Calcd for C$_{11}$H$_9$N$_3$OS (231.278): C, 57.13; H, 3.92; N, 18.17; S, 13.87 %. Found: C, 57.25; H, 3.98; N, 18.39; S, 13.97 %. MS (relative intensity) m/z: 231 (M$^+$, 100%), 232 (M+2, 5%), 191 (70%), 163 (36%).

5-Amino-7-oxo-3,4-dihydro-2H-cyclopenta[4,5]thieno[2,3-b] pyridine-6-carbonitrile (**7**)

Compound **6** (0.5 g, 2.162×10$^{-3}$ mol) was allowed to cyclize by heating under reflux for 2 h in a mixture of 1,4-dioxane (15 mL) and 0.5 mL of triethylamine as a catalyst. The reaction mixture was then poured on ice to which a few drops of HCl (18.0 mol) were added (till pH$_6$) and the formed precipitate was then filtered out.

Grey crystals from ethanol, yield 94 %, 0.47 g, m.p. <300 °C. IR (KBr): υ/cm$^{-1}$= 3464-3332 (NH$_2$, NH), 2893 (CH$_2$), 2225 (CN), 1692 (CO), 1633 (C=C). H$^1$NMR (DMSO) δ = 2.20-2.36 (m, 6H, 3CH$_2$), 4.46 (s, 2H, NH$_2$), 8.88 (s, 1H, NH). $^{13}$C NMR δ = 21.2, 25.8, 32.9 (3 CH$_2$), 116.9 (CN), 90.7, 126.9, 136.8, 136.0, 144.8, 145.3, 158.5, 159.7 (thiophene C, pyridine C). Calcd for C$_{11}$H$_9$N$_3$OS (231.278): C, 57.13; H, 3.92; N, 18.17; S, 13.86 %. Found: C, 57.15; H, 3.99; N, 18.25; S, 13.97 %. MS (relative intensity) m/z: 231 (M$^+$, 100 %), 233 (M+2, 6%), 214 (42%), 186 (16%).

2-Cyano-N-(3-cyano-5,6-dihydro-4H-cyclopenta[b]thiophen-2-yl)-3-phenylacrylamide (**9**).

Compound **6** (0.3 g, 1.30×10$^{-3}$ mol) was dissolved in 1,4-dioxane (10 mL) and allowed to condense with benzaldehyde (0.14 g, 1.30×10$^{-3}$ mol) by heating the two under reflux for 3 h with 0.5 mL of piperidine added as a catalyst. All contents were then poured on ice, a few HCl (18.0 mol) drops added (till pH$_6$), and the reaction mixture stirred for a few minutes after which the precipitate was filtered out.

Brown crystals from ethanol, yield 66.67 %, 0.28 g, m.p. 292-296 °C. IR (KBr): υ/cm$^{-1}$= 3422-3320 (NH), 2890 (CH$_2$), 2227, 2222 (2CN), 1688 (CO), 1643 (C=C), 3055 (CH aromatic). H$^1$NMR (DMSO) δ = 2.23-2.35 (m, 6H, 3CH$_2$), 6.21 (s, 1H, CH=C), 8.28 (s, 1H, NH), 7.25-7.34 (m, 5H, C$_6$H$_5$). $^{13}$C NMR δ = 21.7, 26.3, 32.6 (3 CH$_2$), 98.8, 153.6 (C=CH), 116.2, 117.3 (2CN), 122.8, 125.8, 126.4, 128.7, 129.4, 136.6, 136.0, 144.8 (thiophene C, pyridine C), 168.8 (C=O). Calcd for C$_{18}$H$_{13}$N$_3$OS (319.38): C, 67.69; H, 4.10; N, 13.16; S, 10.04 %. Found: C, 67.87; H, 4.13; N, 13.41; S, 10.18 %. MS (relative intensity) m/z: 319 (M$^+$, 100%), 191 (20%), 163 (32%), 92 (70 %).

N-(3-Cyano-5,6-dihydro-4H-cyclopenta[b]thiophen-2-yl)-2-oxo-2H-chromene-3-carboxamide (**12**)

Compound **6** (0.3 g, 1.30×10$^{-3}$ mol) was dissolved in 1,4-dioxane (10 mL) and allowed to condense with salicaldehyde (0.16 g, 1.30×10$^{-3}$ mol) by heating the two under reflux for 2 h with piperidine (0.5 mL) added as a catalyst. All contents were then poured on ice, a few HCl (18.0 mol) drops added (till pH$_6$), and the reaction mixture stirred for a few minutes after which the precipitate was filtered out.

Dark brown crystals from ethanol, yield 81.82 %, 0.36 g, m.p. 289-292 °C. IR (KBr): υ/cm$^{-1}$= 3467-3341(NH), 3055 (CH aromatic), 2890 (CH$_2$), 2225 (CN), 1693, 1688 (2CO), 1646 (C=C). H$^1$NMR (DMSO) δ = 2.22-2.33 (m, 6H, 3CH$_2$), 6.89 (s, 1H, coumarin H-4), 7.26-7.35 (m, 4H, C$_6$H$_4$), 8.30 (s, 1H, NH). $^{13}$C NMR δ = 21.9, 26.7, 32.2 (3 CH$_2$), 116.0 (CN), 120.4, 124.5, 125.3, 126.4, 127.9, 139.0, 140.8, 152.0 (thiophene C, coumarin C), 159.4, 166.8 (2 C=O). Calcd for C$_{18}$H$_{12}$N$_2$O$_3$S (336.36): C, 64.27; H, 3.60; N, 8.33; S, 9.53 %. Found: C, 64.42; H, 3.75; N 8.45; S, 9.79 %. MS (relative intensity) m/z: 336 (M$^+$, 68%), 338 (M+2, %), 173 (100%), 145 (70%).

2-(3-Cyano-5,6-dihydro-4H-cyclopenta[b]thiophen-2-ylamino)-2-oxo-N'-phenylacetohydrazonoyl cyanide (**14**)

To a solution of compound **6** (0.3 g, 1.30×10$^{-3}$ mol) in ethanol (15 mL), containing sodium hydroxide pellets (0.5 g), benzenediazonium chloride (0.18 g, 1.30×10$^{-3}$ mol) [prepared by adding an aqueous (20 mL of water) sodium nitrite solution (0.09 g, 1.30×10$^{-3}$ mol) to a cold solution of aniline (0.14 g, 1.30×10$^{-3}$ mol) in 30 mL of conc. HCl (18.0 mol) at 0-5 °C, with continuous stirring] was added and the coupling reaction was allowed to proceed. The formed precipitate after 1 h was collected by filtration.

Light brown crystals from ethanol, yield 29.55 %, 0.13 g, m.p. 296-299 °C. IR (KBr): υ/cm$^{-1}$= 3486-3338 (2NH), 3060 (CH aromatic), 2893 (CH$_2$), 2223-2220 (2CN), 1687 (CO), 1648 (C=C), 1634 (C=N). H$^1$NMR (DMSO) δ = 2.24-2.38 (m, 6H, 3CH$_2$), 7.28-7.38 (m, 5H, C$_6$H$_5$), 8.32, 8.89 (2s, 2H, 2NH). $^{13}$C NMR δ = 21.7, 26.7, 32.0 (3 CH$_2$), 115.8, 116.8 (2 CN), 119.6, 120.5, 121.7, 122.6, 126.4, 125.3, 126.4, 139.0, 143.4 (thiophene C, C$_6$H$_5$), 163.5 (C=O), 168.4 (C=N). Calcd for C$_{17}$H$_{13}$N$_5$OS (335.38): C, 60.88; H, 3.91; N, 20.88; S, 9.56 %. Found: C, 60.93; H, 3.99; N, 20.99; S, 9.64 %. MS (relative intensity) m/z: 335 (M$^+$, 43%), 337 (M+2, 4%), 258 (15%0, 234 (55%), 177 (80%), 92 (100%).

2-(3-Amino-4-(phenyldiazenyl)-1H-pyrazol-5-ylamino)-5,6-dihydro-4H-cyclopenta[b]thiophene-3-carbonitrile (**17a**)

To a solution of the hydrazo compound **14** (0.3 g, 8.95×10$^{-4}$ mol) in 1,4-dioxane (40 mL), hydrazine hydrate (0.06 g, 8.95×10$^{-4}$ mol) was added and the reaction mixture was subjected to heating under reflux for 3 h, after which it was poured on ice to which a few drops of HCl (18.0 mol) were added (till pH$_6$). The precipitated crystals were then filtered out.

Yellow crystals from ethanol yield 32.26 %, 0.1 g, m.p. >300 °C. IR (KBr): υ/cm$^{-1}$ = 3522-3332 (NH$_2$, 2NH), 3057 (CH aromatic), 2895 (CH$_2$), 2222 (CN), 1645 (C=C), 1637 (C=N). H$^1$NMR (DMSO) δ = 2.22-2.37 (m, 6H, 3CH$_2$), 4.51 (s, 2H, NH$_2$), 7.30-7.39 (m, 5H, C$_6$H$_5$), 8.33, 8.37 (2s, 2H, 2NH). $^{13}$C NMR δ = 22.0, 26.7, 32.4 (3 CH$_2$), 116.9 (CN), 90.6, 120.3, 120.5, 121.9, 123.8, 125.6, 125.9, 128.5, 139.7, 144.0, 150.8, 154.6 (thiophene C, pyrazole C, C$_6$H$_5$). Calcd for C$_{17}$H$_{15}$N$_5$S (349.41): C, 58.44; H, 4.33; N, 28.06; S, 9.18 %. Found: C, 58.64; H, 4.41; N, 28.30; S, 9.31 %. MS (relative intensity) m/z: 349 (M$^+$, 68%), 350 (12%), 332 (15%), 303 (25%), 187 (100%), 84 (80%).

### 2-(3-Amino-1-phenyl-4-(phenyldiazenyl)-1H-pyrazol-5-ylamino)-5,6-dihydro-4H-cyclopenta[b]thiophene-3-carbonitrile (17b)

To a solution of the hydrazo compound **14** (0.3 g, 8.95×10$^{-4}$ mol) in 1,4-dioxane (40 mL), phenyl hydrazine (0.1 g, 8.95×10$^{-4}$ mol) was added and the reaction mixture was subjected to reflux for 2.5 h, after which it was poured on ice to which a few drops of HCl (18.0 mol) were added till a solid precipitate is being formed.

Yellow crystals from ethanol yield 34.21 %, 0.13 g, m.p. 289-292 °C. IR (KBr): υ/cm$^{-1}$ = 3530-3348 (NH$_2$, NH), 3059 (CH aromatic), 2893 (CH$_2$), 2220 (CN), 1647 (C=C), 1639 (C=N). H$^1$NMR (DMSO) δ = 2.21-2.36 (m, 6H, 3CH$_2$), 4.48 (s, 2H, NH$_2$), 7.29-7.43 (m, 10H, 2C$_6$H$_5$), 8.36 (s, 1H, NH). $^{13}$C NMR δ = 22.2, 26.9, 32.2 (3 CH$_2$), 116.3 (CN), 92.8, 120.4, 120.8, 121.3, 122.9, 124.2, 125.9, 128.5, 129.0, 129.6, 139.9, 144.4, 152.3, 156.2 (thiophene C, pyrazole C, 2C$_6$H$_5$). Calcd for C$_{23}$H$_{19}$N$_5$S (425.51): C, 64.92; H, 4.50; N, 23.04; S, 7.54 %. Found: C, 65.10; H, 4.56; N, 23.19; S, 7.65 %. MS (relative intensity) m/z: 425 (M$^+$, 28%), 427 (M+2), 348 (40%), 320 (57%), 236 (100%), 163 (52%).

### 4,6-Diamino-1-(3-cyano-5,6-dihydro-4H-cyclopenta[b]thiophen-2-yl)-2-oxo-1,2-dihydropyridine-3-carbonitrile (19a)

To a solution of compound **6** (0.3 g, 1.30×10$^{-3}$ mol) in 1,4-dioxane (50 mL), malononitrile (0.09 g, 1.30×10$^{-3}$ mol) and 0.5 mL of triethylamine were added and the reactants were heated under reflux for 2 h, after which the mixture was poured on ice to which a few HCl (18.0 mol) drops were added (till pH$_6$). The precipitate was allowed to coagulate and was then filtered out.

Dark brown crystals from ethanol, yield 89.74 %, 0.35 g, m.p. 285-289 °C. IR (KBr): υ/cm$^{-1}$ = 3496-3343 (2NH$_2$), 3059 (CH aromatic), 2887 (CH$_2$), 2229, 2220 (2CN), 1649 (C=C), 1692 (CO). H$^1$NMR (DMSO) δ = 2.24-2.39 (m, 6H, 3CH$_2$), 4.27, 5.31 (2s, 4H, 2NH$_2$), 6.51 (s, 1H, pyridine H-3). $^{13}$C NMR δ = 22.1, 26.2, 32.7 (3 CH$_2$), 116.0 (CN), 119.8, 121.0, 121.3, 122.2, 123.7, 129.0, 139.9, 143.8, 152.3, 155.0 (thiophene C, pyridine C). Calcd for C$_{14}$H$_{11}$N$_5$OS (297.34): C, 56.55; H, 3.73; N, 23.55; S, 10.78 %. Found: C, 56.71; H, 3.63; N, 23.61; S, 10.66 %. MS (relative intensity) m/z: 297 (M$^+$, 100%), 299 (M+2, 6%), 281 (28%), 231 (60%).

### 4-Amino-1-(3-cyano-5,6-dihydro-4H-cyclopenta[b]thiophen-2-yl)-6-hydroxy-2-oxo-1,2-dihydropyridine-3-carbonitrile (19b)

To a solution of compound **6** (0.3 g, 1.30×10$^{-3}$ mol) in 1,4-dioxane (50 mL), ethyl cyanoacetate (0.15 g, 1.30×10$^{-3}$ mol) and 0.5 mL of

triethylamine were added and the reactants were heated under reflux for 3 h, after which the mixture was poured on ice to which a few HCl (18.0 mol) drops were added (till pH$_6$). The precipitate was allowed to coagulate and then filtered out.

Pale brown crystals from ethanol, yield 51.28 %, 0.2 g, m.p. <300 °C. IR (KBr): υ/cm$^{-1}$= 3585-3333 (OH, NH$_2$), 3049 (CH aromatic), 2893 (CH$_2$), 2227, 2222 (2CN), 1646 (C=C), 1692 (CO). H$^1$NMR (DMSO) δ = 2.22-2.36 (m, 6H, 3CH$_2$), 4.29 (s, 2H, NH$_2$), 6.69 (s, 1H, pyridine H-3), 10.41 (s, 1H, OH). $^{13}$C NMR δ = 22.1, 26.2, 32.7 (3 CH$_2$), 116.0 (CN), 119.8, 121.0, 121.3, 122.2, 123.7, 129.0, 139.9, 143.8, 158.9, 164.3 (thiophene C, pyridine C). Calcd for C$_{14}$H$_{10}$N$_4$O$_2$S (298.32): C, 56.37; H, 3.38; N, 18.78; S, 10.75 %. Found: C, 56.58; H, 3.51; N, 18.87; S, 10.94 %. MS (relative intensity) m/z: 298 (M+, 100%), 300 (6%), 280 (18%), 232 (40%).

### 1-(3-Cyano-5,6-dihydro-4H-cyclopenta[b]thiophen-2-yl)-3-phenylthiourea (21)

To a solution of 2-amino-5,6-dihydro-4H-cyclopenta[b] thiophene-3-carbonitrile **3a** (1.5 g, 9.13×10$^{-3}$ mol) in 1,4-dioxane (80 mL), phenylisothiocyanate (1.24 g, 9.13×10$^{-3}$ mol) and 0.5 mL of triethylamine were added and the reactants were subjected to heating under reflux for 2.5 h. The contents were poured on ice and a few HCl (18.0 mol) drops (till pH$_6$) stirred in to enhance precipitate formation. The latter was then filtered out.

Grey crystals from ethanol, yield 69.34 %, 1.90 g, m.p. 223-226 °C. IR (KBr): υ/cm$^{-1}$= 3491-3338 (2 NH), 3056 (CH aromatic), 2887 (CH$_2$), 2225 (CN), 1641 (C=C), 1204-1190 (C=S). H$^1$NMR (DMSO) δ = 2.20-2.35 (m, 6H, 3CH$_2$), 7.32-7.45 (m, 5H, C$_6$H$_5$), 8.26, 8.46 (2s, 2H, 2NH). $^{13}$C NMR δ = 22.0, 26.9, 32.6 (3 CH$_2$), 116.0 (CN), 120.3, 120.5, 122.9, 124.2, 125.9, 128.5, 129.0, 137.0, 139.9, (thiophene C, C$_6$H$_5$). Calcd for C$_{15}$H$_{13}$N$_3$S$_2$ (299.41): C, 60.17; H, 4.38; N, 14.03; S, 21.42 %. Found: C, 60.34; H, 4.42; N, 14.11; S, 21.68 %. MS (relative intensity) m/z: 299 (M$^+$, 20%), 207 (40%), 163 (100), 92 (76%).

### 5-Amino-6-phenyl-7-thioxo-3,4-dihydro-2H-cyclopenta[4,5]thieno[2,3-d]pyrimidine (22)

Compound **21** (0.4 g, 1.34×10$^{-3}$ mol) was allowed to undergo cyclization by being heated under reflux for 3 h in 1,4-dioxane (40 mL) to which 0.5 mL of triethylamine were added. The contents were poured on ice and a few HCl (18.0 mol) drops (till pH$_6$) stirred in to enhance precipitate formation. The latter was then filtered out.

Orange crystals from ethanol, yield 12.5 %, 0.05 g, m.p. <300 °C. IR (KBr): υ/cm$^{-1}$= 3453-3420 (NH$_2$), 3062 (CH aromatic), 2886 (CH$_2$), 1641 (C=C), 1210-1192 (C=S), 1633 (C=N). H$^1$NMR (DMSO) δ = 2.21-2.39 (m, 6H, 3CH$_2$), 3.88 (s, 2H, NH$_2$), 7.30-7.39 (m, 5H, C$_6$H$_5$). $^{13}$C NMR δ = 22.4, 26.6, 32.7 (3 CH$_2$), 122.3, 122.9, 124.0, 123.8, 126.9, 128.3, 134.8, 138.0, 144.7, 154.2, 165.2 (thiophene C, C$_6$H$_5$). Calcd for C$_{15}$H$_{13}$N$_3$S$_2$ (299.41): C, 60.17; H, 4.38; N, 14.03; S, 21.42 %. Found: C, 60.34; H, 4.51; N, 14.20; S, 21.66 %. MS (relative intensity) m/z: 299 (100%), 301 (M+2, 3%), 222 (50%).

### References

1. Connor DT, Cetenko WA, Mullican MD, Sorenson RJ, Unangst PC, et al. (1992) Novel benzothiophene-, benzofuran-, and naphthalenecarboxamidotetrazoles as potential antiallergy agents. J Med Chem 35: 958-965.

2. Wardakhan WW, Abdel-Salam OM, Elmegeed GA (2008) Screening for antidepressant, sedative and analgesic activities of novel fused thiophene derivatives. Acta Pharm 58: 1-14.

3. Mohamed AAR, Shehab MA, El-Shenawy, SM (2009) Monatsh Chem 140: 445-459.

4. Graham SL, Shepard KL, Anderson PS, Baldwin JJ, Best DB, et al. (1989) Topically active carbonic anhydrase inhibitors. 2. Benzo[b]thiophenesulfonamide derivatives with ocular hypotensive activity. J Med Chem 32: 2548-2554.

5. Jones CD, Jevnikar MG, Pike AJ, Peters MK, Black LJ, et al. (1984) Antiestrogens. 2. Structure-activity studies in a series of 3-aroyl-2-arylbenzo[b]thiophene derivatives leading to [6-hydroxy-2-(4-hydroxyphenyl)benzo[b]thien-3-yl] [4-[2-(1-piperidinyl)ethoxy]-phenyl]methanone hydrochloride (LY156758), a remarkably effective estrogen antagonist with only minimal intrinsic estrogenicity. J Med Chem 27: 1057-1066.

6. Isloor AM, Kalluraya B, Shetty P (2009) Eur J Med Chem 44: 3784–3787.

7. Hussain MI, Amir MJ (1986) Indian Chem Soc 63: 317-320.

8. Chiu SH, Huskey SW (1998) Species differences in N-glucuronidation. Drug Metab Dispos 26: 838-847.

9. Turan-Zitouni G, Kaplancikli ZA, Erol K, Kiliç FS (1999) Synthesis and analgesic activity of some triazoles and triazolothiadiazines. Farmaco 54: 218-223.

10. Holla BS, Kalluraya B (1988) Indian J Chem Sect B 27: 683-685.

11. Prasad AR, Ramalingam T, Rao AB, Diwan PW, Sattur PB Eur J Med Chem 24: 199-201.

12. El-Dawy MA, Omar AM, Ismail AM, Hazzaa AA (1983) Potential broad spectrum anthelmintics IV: design, synthesis, and antiparasitic screening of certain 3,6-disubstituted-(7H)-s-triazolo-[3,4-b][1,3,4]thiadiazine derivatives. J Pharm Sci 72: 45-50.

13. Giri S, Singh H, Yadav LDS, Khare RK (1978) J Indian Chem Soc 55: 168-171.

14. Mody MK, Prasad AR, Ramalingham T, Suttur PB (1982) J Indian Chem Soc 59: 769-770.

15. Omar F, Mahfouz N, Rahman M (1996) Design, synthesis and antiinflammatory activity of some 1,3,4-oxadiazole derivatives. Eur J Med Chem 31: 819-25.

16. El-Hawash SA, Badawey el-SA, El-Ashmawey IM (2006) Nonsteroidal antiinflammatory agents-part 2 antiinflammatory, analgesic and antipyretic activity of some substituted 3-pyrazolin-5-ones and 1,2,4,5,6,7-3H-hexahydroindazol-3-ones. Eur J Med Chem 41: 155-165.

17. Isloor AM, Kalluraya B, Rao M (2000) J Saudi Chem Soc 4: 265-270.

18. Shams HZ, Mohareb RM, Helal MH, Mahmoud Ael-S (2011) Design and synthesis of novel antimicrobial acyclic and heterocyclic dyes and their precursors for dyeing and/or textile finishing based on 2-N-acylamino-4,5,6,7-tetrahydro-benzo[b]thiophene systems. Molecules 16: 6271-6305.

19. Mohareb RM, Schatz J (2011) Anti-tumor and anti-leishmanial evaluations of 1,3,4-oxadiazine, pyran derivatives derived from cross-coupling reactions of β-bromo-6H-1,3,4-oxadiazine derivatives. Bioorg Med Chem 19: 2707-2713.

20. Mohareb RM, Ahmed HH, Elmegeed GA, Abd-Elhalim MM, Shafic RW (2011) Development of new indole-derived neuroprotective agents. Bioorg Med Chem 19: 2966-2974.

21. Wang T, Huang X, Liu J, Li B, Wu Chen JK, et al. (1010) Synlett 9: 1351-1354.

22. Junek H, Thierrichter B, Wibmer P (1979) Monatsh Chem 110: 483-485.

23. Zhou JF, Gong GX, Zhu FX, Zhi S (2009) J Chinese Chemical Letters 20: 37-45.

# 15

# Composition of Essential Oils from Five Endemic *Hypericum* Species of Turkey

**Moussa Ahmed[1]\*, Noureddine Djebli[2], Saad Aissat[1], Baghdad Khiati[1], Salima Douichene[2], Abdelmalek Meslem[1] and Abdelkader Berrani[1]**

[1]*Institute of Veterinary Sciences University Ibn-Khaldoun, Tiaret, Algeria*
[2]*Departments of Biology, Faculty of Sciences, Mostaganem University, Algeria*

## Abstract

Nearly 350 *Hypericum* species exist on the earth, widely spreading in Europe, Asia and North Africa. 32 of 80 species growing in Turkey are endemic. The chemical composition of the essential oils obtained from the aerial parts of the *H. uniglandulosum* Hausskn. ex Bornm., *H.scabroides* Robson and Poulter, *H.kotschyanum* Boiss., *H.salsugineum* Robson and Hub.-Mor. and *H.thymopsis* Boiss. By using the hydro-distillation method is identified by GC and GC/MS. Finally the results are compared with each other. The differences between the results of the *H.thymopsis* and *H.scabroides* obtained in this study and the previous studies show that the chemical compositions of the essential oils are different for the same species obtained at different locations. The essential oil compositions of these species, except for the *H. thymopsis* and *H.scabroides* are identified for the first time.

**Keywords:** *Hypericum species*; Essential oil composition; Turkey

## Introduction

The genus Hypericum (Hypericaceae) is represented by nearly 100 taxa grouped under 19 sections in Turkey. Among them, 45 taxa are endemic. In the traditional medicine of Turkey, the genus is known as "sarı kantaron, kantaron, binbirdelik otu, mayasıl otu" and most of them, especially H. perforatum, have been used for the treatment of burns, wounds, haemorroids, diarrhorea and ulcers [1-5]. Moreover, aqueous extracts prepared from the flowering aerial parts of the Hypericum species are being used in the treatment of psychological diseases such as neuralgia, anxiety, neurosis and depression [6]. The preparative forms of the Hypericum perforatum (St. John's Wort) are sold for the treatment of mild to moderate depression in the USA and Europe.

The chemical composition of the Hypericum species is composed of naphthodianthrones (especially hypericin and pseudohypericin), acylphloroglucinol derivatives (especially hyperforin and adhyperforin), flavonoids (especially quercetin, quercitrin, hyperoside and biapigenin), tannins, n-alkanes, xanthones and essential oils [7-9].

The essential oil compositions of about 50 different Hypericum species have so far been identified [10-13]. In this study, the oils of 5 endemic Hypericum species were obtained by hydro-distillation method. The oil compositions were identified by GC and GC/MS. Except for the *H. thymopsis* and *H. scabroides* the essential oil of which were obtained by hydro-distillation and the oils composition of which were reported previously [13,14], the species investigated in this study were studied for the first time to the knowledge of the authors. In this study, a comparison between the volatile oil compositions of *H. thymopsis* and *H. scabroides* species obtained from different locations were performed.

## Materials and Methods

### Plant material

Flowering aerial parts of *H. uniglandulosum* were collected from east Anatolia, namely, Erzincan: Erzincan-Eski Çayırlı road, 10 km to Eski Çayırlı, 1450 m, 15.07.2006, that of *H. scabroides* were collected from east Anatolia, namely, Erzincan: Erzincan-Kelkit, 15 km to Kelkit, 1550 m, 14.07.2006, that of *H. kotschyanum* were collected from south

Anatolia, namely, İçel: North-west of Arslanköy, 1840 m, 15.06.2006, that of *H. salsugineum* were collected from central Anatolia, namely, Konya: around The Salt Lake, on 01.07.2005, that of *H. thymopsis* were collected from central Anatolia, namely, Sivas: Sivas-Malatya road, Ziyarettepe, 1350 m, 10.07.2005. Specimens were identified and vouchers were deposited in the Herbarium of Istanbul University, Faculty of Pharmacy (İstanbul Üniversitesi Eczacılık Fakültesi Herbaryumu, İstanbul, Turkey) under code numbers of ISTE 85344, ISTE 85343, ISTE 83979, ISTE 85341 and ISTE 85342, respectively.

### Isolation of the essential oil by hydro-distillation method

Air dried and powdered plant materials were subjected to hydro-distillation in a Clevenger-type apparatus according to the method recommended in the European Pharmacopoeia [15]. The oils obtained for this study were stored at +4°C by avoiding the daylight contact before the analysis.

### Gas chromatography/mass spectrometry (GC/MS)

The GC/MS analysis was carried out with an Agilent 5975 GC-MSD system. Innowax FSC column (60 m×0.25 mm, 0.25 μm film thickness) was used with helium as carrier gas (0.8 mL/min). GC oven temperature was kept at 60°C for 10 min and programmed to 220°C at a rate of 4°C/min and kept constant at 220°C for 10 min and then programmed to 240°C at a rate of 1°C/min. Injection volume was 1 μL (10%) in hexane. Split ratio was adjusted at 40:1. The injector temperature was set at 250°C. Mass spectra were recorded at 70 eV. Mass range was from m/z 35 to 450.

\*Corresponding author: Ahmed Moussa, Institute of Veterinary Sciences, University Ibn-Khaldoun, Tiaret, Algeria, E-mail: moussa7014@yahoo.fr

## Gas chromatography (GC)

The GC analysis was carried out using an Agilent 6890N GC system. FID detector temperature was 300°C. To obtain the same elution order with GC-MS, simultaneous auto-injection was done on a duplicate of the same column applying the same operational conditions. Relative percentage amounts of the separated compounds were calculated from FID chromatograms. The analysis results are given in table 1.

## Identification of components

Identification of the essential oil components were carried out by comparison of their relative retention times with those of authentic samples or by comparison of their relative retention index (RRI) to series of n-alkanes. Computer matching against commercial (Wiley GC/MS Library, Adams Library, MassFinder 3 Library) [16,17], and in-house "Başer Library of Essential Oil Constituents" built up by genuine compounds and components of known oils, as well as MS literature data [18,19], was also used for the identification.

## Results and Discussion

By hydro-distillation, volatile oils were obtained from the aerial parts of the *H. uniglandulosum, H. scabroides, H. kotschyanum, H. salsugineum* and *H. thymopsis* with yields of 0.67% (v/w), trace (in hexane), 0.67% (v/w), trace (in hexane), 0.67% (v/w), respectively. The analyses were performed by using GC and GC/MS. The list of the essential oil components where the oils were obtained by hydro-distillation, are given in table 1. Fifty-eight constituents corresponding to the 72.7% of the oil (UN) from the *H. uniglandulosum,* thirty-two constituents corresponding to the 75.3% of the oil (SB) from *H. scabroides,* forty-five constituents corresponding to the 92.4% of the oil (KC) from the *H. kotschyanum,* fifty-four constituents corresponding to the 96.9% of the oil from the *H. salsugineum,* seventeen constituents corresponding to the 99.9% of the oil (TP) from the *H. thymopsis* were identified. 2,6-Dimethyl-3,5-heptadien-2-one (40.7%), nonacosane (3.2%), hexadecanoic acid (2.7%) and α-pinene (2.7%) were characterized as the main components of the *H. uniglandulosum* (UN).

Hexadecanoic acid (17.7%), spathulenol (5.3%), nonacosane (4.4%), dodecanoic acid (4.1%), baeckeol (4.1%) and γ-muurolene (3.9%) were characterized as the main components of the *H. scabroides* (SB). α-pinene (14.4%), nonacosane (11.1%), hexadecanoic acid (9.2%), β-pinene (8.7%), spathulenol (6.3%) and limonene (5.1%) were characterized as the main components of the *H. kotschyanum* (KC). Nonacosane (42.7%), hexadecanoic acid (23.2%) and baeckeol (6.1%) were characterized as the main components of the *H. salsugineum* (SG). α-pinene (44.0%), baeckeol (32.9%), spathulenol (8.0%), limonene (7.6%) and camphene (5.2%) were characterized as the main components of the *H. thymopsis* (TP). The chemical class distribution of the volatile oils of 5 different species obtained by hydro-distillation is given in table 2.

It has been observed that, the UN oil was rich in terms of carbonylic compounds and fatty acids. The oil of SB was rich in terms of sesquiterpene hydrocarbons and fatty acids. The KC oil, on the other hand, was dominated by monoterpene hydrocarbons, alkanes, oxygenated sesquiterpene hydrocarbons and fatty acids. The SG oil contained alkanes and fatty acids. The TP oil was found to be rich in monoterpene hydrocarbons and a phenolic-compound. As a result of this research α-pinene, 2,6 dimethyl-3,5-heptadien-2-one, baeckeol, nonacosane and hexadecanoic acid were identified as major volatile constituents (>10%) in *Hypericum* species.

| Compound | RRI | UN | SB | KC | SG | TP |
|---|---|---|---|---|---|---|
| α -Pinene | 1032 | 2.7 | 3.1 | 14.4 | - | 44.0 |
| Camphene | 1076 | - | - | 0.5 | - | 5.2 |
| Hexanal | 1093 | - | - | - | - | - |
| Undecane | 1100 | 1.9 | 0.3 | 0.1 | tr | - |
| β- pinene | 1118 | - | - | 8.7 | - | 1.7 |
| Limonene | 1203 | 0.2 | 1.0 | 5.1 | - | 7.6 |
| (Z)-3-hexenal | 1225 | - | - | - | - | - |
| γ - terpinene | 1255 | - | - | 0.7 | - | - |
| p- cymene | 1280 | - | 0.3 | 1.5 | - | - |
| Terpinolene | 1290 | - | - | - | - | - |
| 6-methyl-5-hepten-2-one | 1348 | 0.9 | - | - | - | - |
| Hexanol | 1360 | - | - | - | - | - |
| 2,6 Dimethyl-3,5 heptadien-2-one* | 1377 | 40.7 | - | - | - | - |
| Nonanal | 1400 | 0.1 | - | - | - | - |
| trans-linalool oxide (furanoid) | 1450 | - | - | - | 0.2 | - |
| α,p-dimestyrene | 1452 | - | - | - | - | - |
| cis-linalool oxide (furanoid) | 1478 | 0.9 | - | - | 0.1 | - |
| Longipinene | 1482 | - | - | - | - | - |
| Bicycloelemene | 1495 | - | - | - | 0.7 | - |
| α- copaene | 1497 | 0.1 | 0.5 | 0.6 | 0.3 | - |
| α- campholene aldehyde | 1499 | 0.2 | 0.2 | 0.6 | - | - |
| β- bourbonene | 1535 | - | - | - | 0.3 | - |
| Linalool | 1553 | 1.2 | - | - | 0.1 | - |
| Octanol | 1562 | - | - | - | - | - |
| Pinocarvone | 1586 | - | - | 0.3 | - | - |
| Fenchylalcohol | 1591 | - | - | 0.5 | - | - |
| β-ylangene | 1589 | - | - | - | 0.3 | - |
| β- copaene | 1597 | - | - | - | 0.5 | - |
| Camphene hydrate | 1598 | - | - | - | - | - |
| β-elemene | 1600 | - | - | - | 0.3 | - |
| Nopinone | 1601 | - | - | - | - | - |
| 6-methyl-3,5 heptadien-2-one | 1602 | - | - | - | - | - |
| β- caryophyllene | 1612 | - | - | - | 0.2 | - |
| Aromadendrene | 1628 | 0.1 | 1.2 | - | 0.3 | - |
| Myrtenal | 1648 | 0.2 | - | 0.7 | - | - |
| (E)-2-Decenal | 1655 | - | - | - | 0.1 | - |
| γ- Gurjunene | 1659 | - | - | - | 0.1 | - |
| Alloaroma dendrene | 1661 | - | - | - | 0.1 | - |
| trans-pinocarveol | 1670 | - | - | 1.0 | - | - |
| Acetophenone | 1671 | - | 1.5 | - | - | - |
| trans- Verbenol | 1683 | 0.2 | - | 0.4 | - | - |
| Drima-7,9(11)-diene | 1694 | tr | - | - | - | - |
| γ- Muurolene | 1704 | 1.0 | 3.9 | 1.8 | 0.4 | - |
| α -terpineol | 1706 | - | - | - | - | - |
| Borneol | 1719 | 0.2 | - | 0.3 | 0.1 | tr |
| Verbenone | 1725 | 0.3 | 0.8 | - | 0.2 | - |
| Germacrene -D | 1726 | - | - | 0.9 | 1.4 | tr |
| p-mentha-1,5-dien-8-ol | 1738 | - | - | - | - | - |
| Valencene | 1740 | 0.2 | - | - | - | - |
| α -Muurolene | 1740 | 0.2 | 0.8 | 0.3 | 0.2 | - |
| Geranial | 1740 | 0.1 | - | - | - | - |
| Carvone | 1751 | 0.1 | - | - | - | - |
| Bicyclogermacrene | 1755 | - | - | - | 1.0 | - |
| 1-decanol | 1766 | - | - | - | - | - |
| Decanol | 1766 | - | - | 0.3 | - | - |
| δ-Cadinene | 1773 | tr | 1.6 | 0.6 | 0.4 | tr |
| γ -Cadinene | 1776 | 0.4 | 2.2 | 0.6 | 0.3 | tr |
| p-methyl acetophenone | 1797 | - | - | - | - | - |
| Myrtenol | 1804 | 0.2 | - | 0.4 | - | - |
| α-cadinene | 1807 | - | - | - | - | - |

| Compound | RRI | UN | SB | KC | SG | TP |
|---|---|---|---|---|---|---|
| Methyl dodecanoate | 1815 | - | - | - | 0.2 | - |
| trans -carveol | 1845 | 0.3 | 0.9 | 0.3 | - | - |
| **Table I. Continued** | | | | | | |
| Compound | RRI | UN | SB | KC | SG | TP |
| Calamenene | 1849 | 1.0 | 1.8 | 0.9 | 0.3 | - |
| Geraniol | 1857 | 1.0 | - | - | - | - |
| p-simen-8-ol | 1864 | - | - | - | - | - |
| (E)-Geranylacetone | 1868 | 0.2 | - | 0.4 | 0.1 | - |
| α -calacorene | 1941 | 0.1 | 1.2 | 0.6 | - | - |
| 1,5- Epoxysalvial-4(14)-ene | 1945 | - | - | 1.0 | - | - |
| (E)-β-Ionone | 1958 | 0.1 | - | - | - | - |
| 1 -Dodecanol | 1973 | 0.5 | - | - | 1.5 | - |
| Eicosane | 2000 | 0.1 | - | - | - | - |
| Caryophyllene oxide | 2008 | 0.3 | 1.7 | 1.3 | 0.1 | - |
| Methyl eugenol | 2030 | - | - | - | - | - |
| Salvial -4(14)-en-1-one | 2037 | 0.6 | - | - | tr | - |
| 1-Tridecanol | 2077 | - | - | - | 0.2 | - |
| Octanoic acid | 2084 | 0.1 | 1.0 | - | - | - |
| Globulol | 2098 | 0.2 | - | - | - | - |
| Heneicosane | 2100 | - | - | - | 0.2 | - |
| Viridiflorol | 2104 | - | - | - | 0.2 | - |
| Salviadienol | 2130 | - | - | - | - | - |
| Hexahydrofarnesylacetone | 2131 | 0.6 | 1.4 | 0.8 | 1.1 | tr |
| Spathulenol | 2144 | 1.5 | 5.3 | 6.3 | - | 8.0 |
| 1-Tetradecanol | 2179 | 0.3 | - | - | 0.5 | - |
| T -cadinol | 2187 | - | tr | - | - | - |
| Nonanoic acid | 2192 | 0.3 | 2.2 | - | - | - |
| Docosane | 2200 | 0.1 | - | - | - | - |
| T- Muurolol | 2209 | 0.2 | - | - | 0.1 | - |
| Methyl hexadecanoate | 2226 | 0.1 | - | - | 0.2 | - |
| Carvacrol | 2239 | - | - | 2.4 | - | - |
| α-cadinol | 2255 | 0.3 | 1.8 | 0.8 | 0.2 | tr |
| Cadalene | 2256 | 0.4 | 3.0 | 0.8 | 0.2 | 0.5 |
| Ethyl hexadecanoate | 2262 | 0.2 | - | - | - | - |
| Decanoic acid | 2298 | 0.6 | - | - | - | - |
| Tricosane | 2300 | 0.9 | 0.6 | 0.8 | 0.6 | - |
| Eudesma-4(15),7-dien-1β-ol | 2369 | - | - | 0.9 | 0.7 | tr |
| 1-Hexadecanol | 2384 | - | - | - | 0.1 | - |
| Undecanoic acid | 2400 | 0.2 | - | - | - | - |
| Tetracosane | 2400 | - | - | - | 0.4 | - |
| Pentacosane | 2500 | 0.2 | 2.6 | 3.9 | 1.9 | - |
| Dodecanoic acid | 2503 | 2.3 | 4.1 | 3.0 | 0.1 | tr |
| Hexacosane | 2600 | - | - | - | 0.2 | - |
| Phytol | 2622 | - | - | tr | 0.7 | - |
| Benzyl benzoate | 2655 | 0.1 | - | 1.8 | - | - |
| Baeckeol | 2668 | 0.9 | 4.1 | 2.4 | 6.1 | 32.9 |
| Tetradecanoic acid | 2670 | 0.8 | 2.9 | 2.4 | 2.5 | - |
| Heptacosane | 2700 | 0.1 | 1.2 | 1.0 | 2.2 | - |
| Octacosane | 2800 | - | - | - | 1.3 | - |
| Pentadecanoic acid | 2822 | - | - | - | 1.2 | - |
| Nonacosane | 2900 | 3.2 | 4.4 | 11.1 | 42.7 | tr |
| Hexadecanoic acid | 2931 | 2.7 | 17.7 | 9.2 | 23.2 | tr |
| **TOTAL** | | 72.7 | 75.3 | 92.4 | 96.9 | 99.9 |

**UN:** *H. uniglandulosum*, **SB:** *H. scabroides*, **KC:** *H. kotschyanum*, **SG:** *H. salsugineum*
**TP:** *H. thymopsis*
RRI Relative retention indices calculated against *n*-alkanes
% calculated from FID data
tr Trace (<0.1%)
*Tentative identification

**Table 1:** Percentage of volatiles of 5 endemic *Hypericum* species.

| Chemical Class | UN (#/%) | SB (#/%) | KC (#/%) | SG (#/%) | TP (#/%) |
|---|---|---|---|---|---|
| Monoterpene Hydrocarbons | 2 / 2.9 | 3 / 4.4 | 6 / 30.9 | - | 4 / 58.5 |
| Oxygenated Monoterpenes | 12 / 4.3 | 4 / 5.0 | 11 / 7.5 | 6 / 0.7 | 2 / Tr |
| Sesquiterpene Hydrocarbons | 11 / 3.5 | 9 / 16.2 | 9 / 7.1 | 17 / 7.3 | 4 / 0.5 |
| Oxygenated Sesquiterpenes | 6 / 3.0 | 4 / 7.2 | 5 / 10.1 | 5 / 1.1 | 2 / 8.0 |
| Alkanes + Alkenes | 7 / 6.5 | 5 / 9.1 | 5 / 16.9 | 9 / 49.5 | 1 / Tr |
| Fatty acids | 8 / 7.9 | 5 / 27.9 | 3 / 14.6 | 6 / 27.3 | 2 / Tr |
| Others | 12 / 44.6 | 2 / 5.5 | 5 / 5.3 | 11 / 11.0 | 2 / 32.9 |
| TOTAL | 58 / 72.7 | 32 / 75.3 | 45 / 92.4 | 54 / 96.9 | 17 / 99.9 |

#: number of compound, % relative percentage

**Table 2:** The chemical class distribution of the oil components of 5 endemic *Hypericum* species.

Comparing the main constituent of the *H. uniglandulosum* oil to the other studies, the followings are observed; 2,6 Dimethyl-3,5-heptadien-2-one was found to be the major component only in *H. tetrapterum* from Serbia [20].

Hexadecanoic acid and spathulenol were detected in high amount in *H. scabroides* in this study, however, the δ-3-carene and sabinene were reported as the main constituents of the *H. scabroides* collected from different locations [21]. Nonacosane was found to be the major component only in two species, namely *H. salsugineum* and *H. davisii* [14].

It was observed that α-pinene was found to be the main component in *H. kotschyanum* and *H. thymopsis*. Among the previous studies about Hypericum essential oils from Turkey, many taxa were characterized by the high amount of α-pinene [11,18,22-29], namely *H. calycinum*, *H. cerastoides*, *H. montbretii*, *H. scabrum*, *H. perforatum* [30,31], *H. hyssophyfolium subsp. elongatum var. elongatum* [32], *H. capitatum var. capitatum*, *H. aviculariifolium subsp. depilatum var. depilatum* [32], *H. apricum* [19], α-pinene, baeckeol, limonene and spathulenol were identified as major components in *H. thymopsis*, although baeckeol and limonene were not determined in a previous study. α-pinene, germacrene D, δ-cadinene, γ-cadinene, spathulenol, α-cadinol, eudesma-4(15),7-dien-1β-ol, nonacosane and hexadecanoic acid were identified both in this study and the previous one [13]. Essential oil compositions of species show difference in the sense of collecting regions and dates. Chemical profiling using volatiles may be useful in taxonomical classifications.

### Acknowledgements

The authors would like to thank Tuba KIYAN for her assistance.

### References

1. Baytop T (1999) *Türkiye'de Bitkilerle Tedavi* (2ndedn). Nobel Tıp Kitabevi, İstanbul, Turkey.

2. Honda G, Yeşilada E, Tabata M, Sezik E, Fujita T, et al. (1996) Traditional medicine in Turkey. VI. Folk medicine in west Anatolia: Afyon, Kütahya, Denizli, Muğla, Aydin provinces. J Ethnopharmacol 53: 75-87.

3. Tuzlaci E, Aymaz PE (2001) Turkish folk medicinal plants, Part IV: Gönen (Balikesir). Fitoterapia 72: 323-343.

4. Sezik E, Yesilada E, Honda G, Takaishi Y, Takeda Y, et al. (2001) Traditional medicine in Turkey X. Folk medicine in Central Anatolia. J Ethnopharmacol 75: 95-115.

5. Kültür S (2007) Medicinal plants used in Kirklareli Province (Turkey). J Ethnopharmacol 111: 341-364.

6. Blumenthal M, Goldberg A, Brinckmann J (2000) Integrative Medicine Communications. Herbal Medicine, Newton 359-366.

7. Bombardelli E, Morazzoni P (1995) Hypericum perforatum. Fitoterapia 66: 43-68.

8. Bruneton J (1995) Pharmacognosy, Phytochemistry, Medicinal Plants. Lavoisier Publishing, Paris 367-370.

9. Di Carlo G, Borrelli F, Ernst E, Izzo AA (2001) St John's wort: Prozac from the plant kingdom. Trends Pharmacol Sci 22: 292-297.

10. Erken S, Malyer H, Demirci F (2001) Chemical Investigations on Some Hypericum Species Growing In Turkey-I. Chem Nat Comp 37: 434-438.

11. Demirci B, Başer KHC, Crockett SL, Khan IA (2005) Analyses of the Volatile Constituents of Asian Hypericum L. (Clucsiaceae, Hypericoideae) Species. J Essent Oil Res 17: 659-663.

12. Crockett S, Demirci B, Baser KHC, Khan IA (2008) Volatile Constituents of Hypericum L. Section Myriandra (Clusiaceae): Species of the H. fasciculatum Lam. Alliance. J Essent Oil Res 20: 244-249.

13. Gençler Özkan AM, Demirci B, Baser KHC (2009) Essential Oil Composition of Hypericum thymopsis Boiss. J Essent Oil Res 21: 149-153.

14. Bağcı E, Yüce E (2010) Essential Oils of the Aerial Parts of Hypericum apricum Kar. and Kir. and Hypericum davisii Robson (Guttiferae) Species from Turkey. Asian J Chem 22: 7405-7409.

15. European Pharmacopoeia (2005) Council of Europe. (5thedn), Strasbourg, France 1: 217.

16. McLafferty FW, Stauffer DB (1989) The Wiley/NBS Registry of Mass Spectral Data. J Wiley and Sons, New York, NY.

17. Koenig WA, Joulain D, Hochmuth DH (2004) Terpenoids and Related Constituents of Essential Oils. MassFinder 3, In: D.H. Hochmuth (ed). Convenient and Rapid Analysis of GC/MS, incorporating W.A. Hamburg, Germany.

18. Joulain D, Koenig WA (1998) The Atlas of Spectra Data of Sesquiterpene Hydrocarbons. E.B. Verlag, Hamburg, Germany.

19. ESO 2000 (1999) The Complete Database of Essential Oils, Boelens Aroma Chemical Information Service. The Netherlands.

20. Smelcerovic A, Spiteller M, Ligon AP, Smelcerovic Z, Raabe N (2007) Essential Oil Composition of Hypericum L. Species from Southeastern Serbia and their Chemotaxonomy. Biochem Syst Ecol 35: 99-113.

21. Bağcı E, Bekçi F (2010) Variation in essential oil composition of Hypericum scabrum L. and H. scabroides N. Robson & Poulter (Hypericaceae) aerial parts during its phenological cycle. Acta Bot Gallica 157: 247–254.

22. Başer KHC, Özek T, Nuriddinov HR, Demirci B (2002) Chem Nat Compd 38: 54-57.

23. Coudalis M, Baziou P, Petrakis PV, Harvala C (2001) Essential oil Composition of Hypericum perforatum L. Growing Different Locations in Greece. Flav Fragr J 16: 204-206.

24. Couladis M, Chinou IB, Tzakou O, Petrakis PV (2003) Composition and antimicrobial activity of the essential oil of Hypericum rumeliacum subsp. apollinis (Boiss. & Heldr.). Phytother Res 17: 152-154.

25. Sajjadi SE, Rahiminezhad MR, Mehregan I, Poorassar A (2001) Constituents of Essential Oil of Hypericum dogonbadanicum Assadi. J Essent Oil Res 13: 43-44.

26. Petrakis PV, Couladis M, Roussis V (2005) A method for detecting the biosystematic significance of the essential oil composition: The case of five Hellenic Hypericum L. Species. Biochem Syst Ecol 33: 873-898.

27. Pavlović M, Tzakou O, Petrakis PV, Coudalis M (2006) The essential oil of Hypericum perforatum L., Hypericum tetrapterum Fries and Hypericum olympicum L. growing in Greece. Flav Fragr J 21: 84-87.

28. Saroglou V, Marin PD, Rancic A, Veljic M, Skaltsa H (2007) Composition and Antimicrobial Activity of the Essential Oil Six Hypericum Species from Serbia. Biochem Syst Ecol 35: 146-152.

29. Crockett S, Demirci B, Baser KHC, Khan IA (2007) Analysis of the Volatile Constituents of Five African and Mediterranean Hypericum L. (Clucsiaceae, Hypericoideae) Species. J Essent Oil Res 19: 302-306.

30. Çakır A, Duru ME, Harmandar M, Criminna R, Passannati S, et al. (1997) Comparison of the Volatile Oils of Hypericum scabrum L. and Hypericum perforatum L. from Turkey. Flav Fragr J 12: 285-287.

31. Erken S, Malyer H, Demirci F (2001) Chemical Investigations on Some Hypericum Species Growing In Turkey-I. Chem Nat Comp 37: 434-438.

32. Çakır A, Kordali S, Zengin H, Izumi S, Hirata T (2004) Composition and Antifungal Activity of Essential Oils Isolated from Hypericum hyssopifolium and Hypericum heterophyllum. Flav Fragr J 19: 62-68.

# 16

# Base Catalyzed Glycerolysis of Ethyl Acetate

**Pranita P. Kore\*, Snehal D. Kachare, Sandip S. Kshirsagar and Rajesh J. Oswal**

*Department of Pharmaceutical Chemistry, JSPM's Charak College of Pharmacy and Research, Gat No. 720/1&2, Wagholi, Pune-Nagar Road, Pune-412207, Maharashtra, India*

### Abstract

Glycerol is a simple polyol compound widely used as a green solvent, having backbone to all lipids known as triglycerides. In our present study, we used glycerol as an acyl acceptor for transesterification of aliphatic acetate by conventional and using microwave assisted synthesis. It was found that increase in the concentration of substrate and catalysts with reaction time, increases yield of aliphatic alcohols. The reaction was followed by separation of product from glycerol by simple extraction with diethyl ether and studied for their HPLC and GC analysis.

**Keywords:** Aliphatic alcohol; Acetate; Polyol; Transesterification

## Introduction

The hydrolysis of esters is a basic organic transformation [1]. Traditionally, the reaction is performed in an acidic or basic aqueous solution under mild conditions yielding the corresponding carboxylic acid and alcohol. Alternatively, due to low miscibility of most esters in water and since acidic or basic conditions lead to equipment corrosion, catalytic transesterification of ester in the presence of an alcohol that removes the carboxylic group from the ester and releases the corresponding alcohol (alcoholysis) was also extensively studied. The alcoholysis of ester is an equilibrium reaction, which usually requires excess amounts of alcohol to yield high conversion, and the presence of catalyst that can be either homogenous or heterogeneous [2-5]. Solid acids and bases as well as immobilized lipase are often employed for this purpose as they have the advantage of being easily separated from the reaction mixture, recycled and reused. Reactions of alcohols with ethyl acetate (transesterification) and propylene oxide were investigated by use of a variety of solid base catalysts to elucidate the activity determining factors of the catalyst in relation with type of alcohol. Solid base catalysts examined were alkaline earth oxides, hydroxides, and carbonates, alumina supported KF and KOH, rare earth oxide, zirconium oxide, etc. Reaction rate of the alcoholysis of ethyl acetate varies with the combination of type of alcohol and basic strength of solid base catalyst [6,7]. Over strongly basic catalysts such as CaO, SrO, BaO, 2-propanol reacted much faster than methanol. On the other hand, over weakly basic catalysts such as alkaline earth hydroxides, methanol reacted faster than 2-propanol. 2-Methyl-2-propanol reacted only over strongly basic catalysts, and much slower than methanol and 2-propanol. Alcoholysis with propylene oxide was catalyzed only by strongly basic catalysts such as alkaline earth oxides, and KF/alumina, alkaline earth hydroxides scarcely showed activity [8]. γ-Alumina, however, showed a high activity, though the selectivity of products was different from those for alkaline earth oxides. Reactives of alcohols with propylene oxide were in the order, methanol>ethanol>2-propanol>2-methyl-2-propanol, regardless of the type of catalyst. One of the characteristic features observed for both alcoholyses is that the catalysts are tolerant to air exposure, which is caused by strong absorptivity of alcohol competitive to that of carbon dioxide and water [9-11]. In this paper, we report on our study about the glycerolysis of aliphatic acetate, using sodium hydroxide and magnesia as representative soluble and solid base catalysts, using glycerol as a green solvent and as an acyl acceptor (Figure 1). The effects of reaction conditions and catalyst type and loading on aliphatic alcohol yield were studied. In addition, product separation procedure and catalyst recycling were also examined.

## Experimental

In a typical procedure, 0.1 g of ethyl acetate was added together with 0.01 g of catalyst to a vial with 5 g of alcohol (all purchased from Merck Pvt. Ltd). The mixture was placed in a preheated oil bath and heated to the required temperature (45-100°C) after which it was magnetically stirred for 1-5 h. At the end of the reaction, the reaction mixture was cooled and extracted with 3×10 mL diethyl ether. The organic phase was concentrated under reduced pressure, and the resulting crude product was analyzed by HPLC and GC analysis using a C-18 and HP-5 column (30 m×0.25 mm, 0.25 μm thick).

### By conventional method

Two catalyst recycling methods were tested. The first reaction cycle in both methods was run as follows: 1 g of ethyl acetate and 0.1 g of solid catalyst were added to a vial with 10 g of glycerol. The vial was then heated at 75°C for 1 h. For the first catalyst recycling method, the catalyst was filtrated at the end of the reaction and the product was extracted by 5×10 mL diethyl ether and analyzed by GC. The filtrated catalyst was then added to a fresh mixture of ethyl acetate in glycerol, and the reaction was repeated. For the second method, the product was extracted by 5×10 mL diethyl ether from the glycerol-catalyst mixture without the filtration of the catalyst, to which was added fresh ethyl acetate.

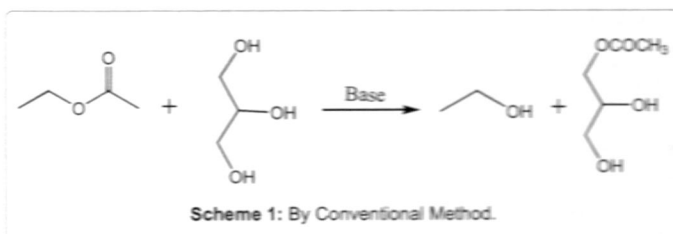

**Scheme 1:** By Conventional Method.

**\*Corresponding author:** Pranita P. Kore, Department of Pharmaceutical Chemistry, JSPM's Charak College of Pharmacy and Research, Gat No. 720/1&2, Wagholi, Pune-Nagar Road, Pune-412207, Maharashtra, India
E-mail: sanpals24@rediffmail.com

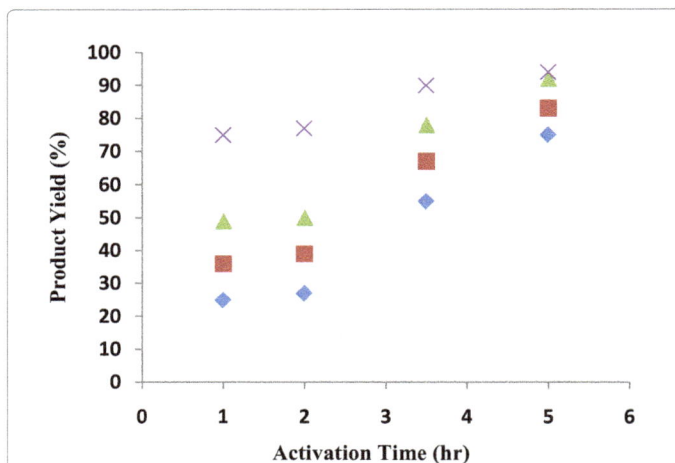

**Figure (1a):** Effect of reaction temperature on the reaction progress with time of ethyl acetate in glycerol using NaOH as catalyst. Reaction conditions: 0.1 g ethyl ester, 0.01 g NaOH, 5 g glycerol. (♦) 45°C; (■) 55°C; (▲) 75°C; (x) 100°C.

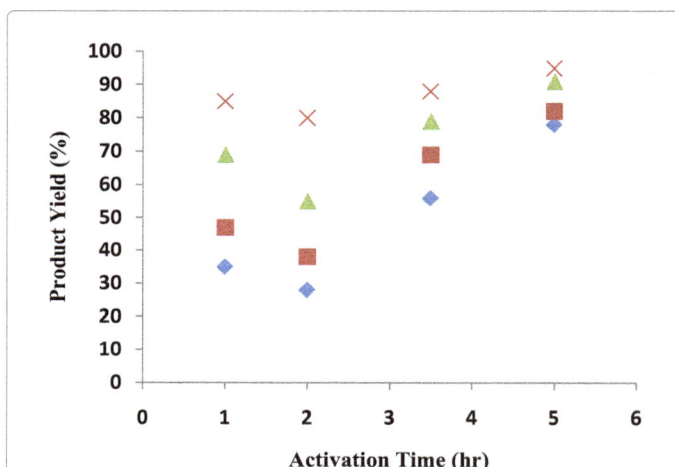

**Figure (1b):** Effect of reaction temperature on the reaction progress with time of propyl acetate in glycerol using NaOH as catalyst. Reaction conditions: 0.1 g propyl ester, 0.01 g NaOH, 5 g glycerol. (♦) 45°C; (■) 55°C; (▲) 75°C; (x) 100°C.

## By microwave assisted method

Microwave assisted reactions were conducted at atmospheric pressure in a domestic microwave (800 W) in a vial, which was covered with a watch glass. The substrate was dissolved in 5 g glycerol followed by addition of the catalyst. After the vial was covered with the watch glass, the reaction mixture was heated in the microwave oven at low intensity from 26°C to 56°C for duration of 40 sec and at full intensity from 26°C to 61°C for 5 sec. At the end of the reaction, the vial was cooled to room temperature in ice, and the reaction mixture was extracted with petroleum ether for HPLC and GC analysis.

## Results and Discussion

The synthesis of aliphatic alcohol is usually produced *via* synthesis of aliphatic chloride by chlorination of toluene that is then hydrolyzed under alkali conditions [12]. This multistage process includes many steps of separation and purification after the respective reactions. Thus,

it is not advantageous economically and environmentally. Moreover, the hydrolysis step requires an excess amount of alkali base resulting in a large amount of a salt solution contaminated with organic compounds that has to be treated. However, aliphatic alcohol can also be produced by oxyacetoxylation from toluene, acetic acid, and oxygen to aliphatic acetate followed by hydrolysis of the ester. As ester alcoholysis can be accomplished using heterogeneous acids and bases and alcohol as an acyl acceptor it is more favored to avoid byproduct salts formation as in aqueous hydrolysis [13]. As previously mentioned, esters can be alcoholized through transesterification using alcohol, which can simultaneously be used as a solvent and as an acyl acceptor, using various homogeneous and heterogeneous catalysts [3-5]. Hence, the investigation began by testing the progress of aliphatic alcohol yield with time in glycerol in several temperatures (Figures 1a and 1b). In general, as expected, increasing the reaction time or temperature in the range of 1-5 h and 45-100°C respectively, increased the product yield. Yet, it can be seen from the results in figures 1a and 1b that increasing the temperature behind 75°C did not significantly change the product yield. At the end of the reaction, aliphatic acetate was easily separated from the reaction mixture by simple extraction with diethyl ether. As glycerol was used as both a solvent and an acyl acceptor it may yield glycerol monoacetate, glycerol diacetate, and/or triacetin as by-products. However, the high boiling points and high solubilities of these potential by-products in glycerol allow the product to be easily separated from the reaction mixture. Magnesia is commonly used solid base catalyst. It is well known that the basic sites on magnesia surface adsorb oxygen and moister from air that affect catalytic performance. Hence, the effect of pretreatment temperature and time on magnesia performance in the transesterfication of aliphatic acetate was first examined (Figures 2a and 2b). As illustrated in figures 2a and 2b, both pretreatment time and temperature affect the product yield, and increase in oven temperature decreases the required time to release the adsorbed molecules and activate the basic sites.

On the surface, Thereby, increasing the product yield. Therefore, based on energy efficiency measurements, the catalyst was preheated at 300°C for 2 h for all subsequent experiments of the study. The progress of aliphatic alcohol yield with time in glycerol in several temperatures was also studied over magnesia (Figures 3a and 3b). As previously detected with NaOH, increasing either the reaction temperature or

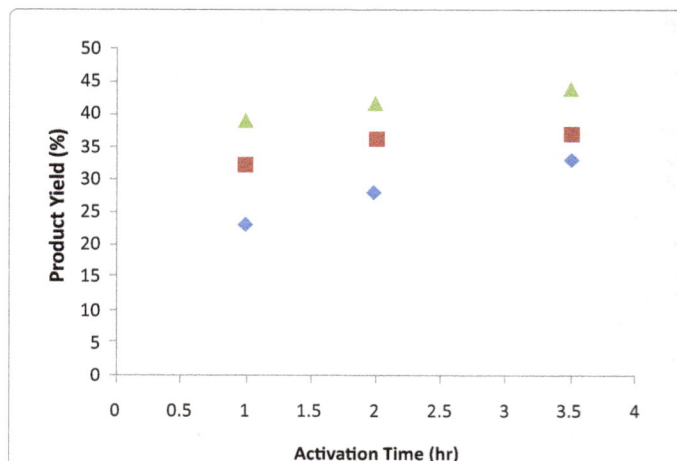

**Figure (2a):** Effect of MgO activation conditions on the yield of ethyl acetate in glycerol. Reaction conditions: 0.1 g ethyl ester, 0.01 g MgO, 5 g glycerol. (♦) 80°C; (■) 300°C; (▲) 450°C.

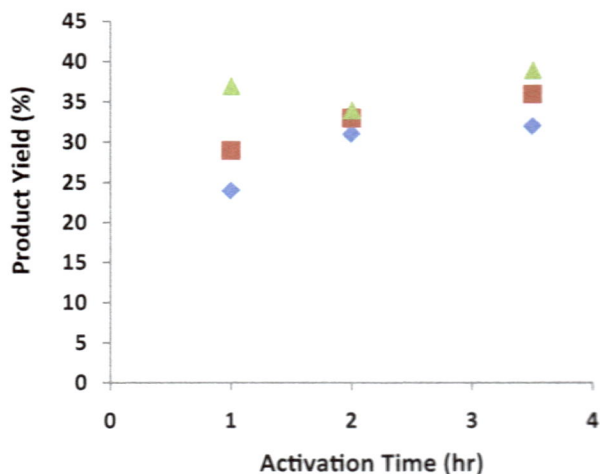

**Figure (2b):** Effect of MgO activation conditions on the yield of propyl acetate in glycerol. Reaction conditions: 0.1 g propyl ester, 0.01 g MgO, 5 g glycerol. (♦) 80°C; (■) 300°C; (▲) 450°C.

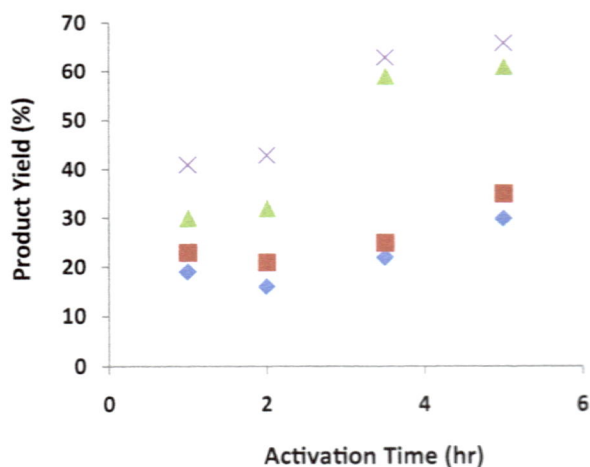

**Figure (3a):** Effect of reaction temperature on the reaction progress with time of ethyl acetate in glycerol using MgO as catalyst. Reaction conditions: 0.1 g ethyl esters, 0.01 g MgO, 5g glycerol. (♦) 45°C; (■) 55°C; (▲) 75°C; (x) 100°C.

tested using two different methods: running the first reaction cycle and filtration of the catalyst that was subsequently employed in second cycle with fresh.

Aliphatic acetate and glycerol, or running the second catalytic cycle by extraction of the product from the glycerol-catalyst mixture without the filtration of the catalyst, to which was added fresh aliphatic acetate. Thus, to test the viability of MgO reuse, the first reaction cycle was run for 1 h at 75°C after which the catalyst was filtrated and the product and the residual substrate were fully extracted by diethyl ether. After the first reaction cycle, the catalyst was filtrated and added to a fresh mixture of aliphatic acetate in glycerol and the reaction was run again under similar conditions. Two cycles of catalyst reuse were done by this method resulting in a small decrease of yield in each cycle, probably due to the catalyst lost during recycling procedure. Since, glycerol was used both as a solvent and an acyl acceptor, recycling of the glycerol together with the heterogeneous catalyst was also tested by extraction

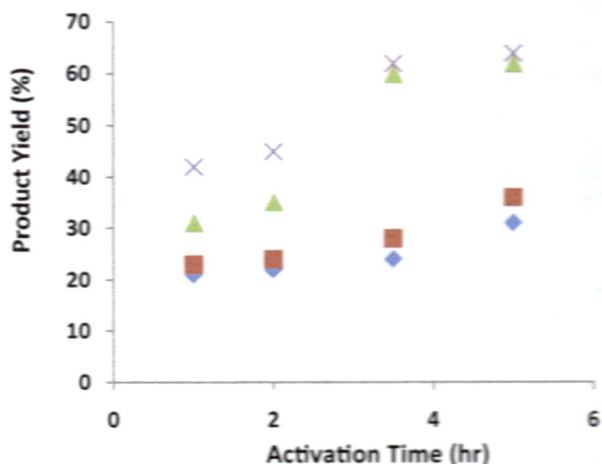

**Figure (3b):** Effect of reaction temperature on the reaction progress with time of propyl acetate in glycerol using MgO as catalyst. Reaction conditions: 0.1 g propyl esters, 0.01 g MgO, 5 g glycerol. (♦) 45°C; (■) 55°C; (▲) 75°C; (x) 100°C.

the reaction time increased the yield of aliphatic alcohol, and again, at temperature above 75°C the effect of temperature was negligible after 2 h. As the active sites of the two representatives' homogeneous and heterogeneous catalysts are different by strength and amount the comparison of the performance of the two catalysts was done on equal mass basis. It was found that the soluble base is more active than the solid base and using NaOH yielded 37% yield after 1 h at 55°C while employing MgO required.

75°C to reach 33% yield after an hour. Yet, though the performance of the soluble base was higher than the solid one, solid catalyst is advantageous from industrial point of view, as it can be easily separated from the reaction mixture by filtration and recycled. Increasing the substrate to catalyst ratio (S/C), by either increasing aliphatic acetate amount or decreasing MgO loading, increased the initial reaction rate, this was calculated by the amount of reacted substrate per g of catalyst divided by reaction time (Figure 4). The recycling of magnesia was also

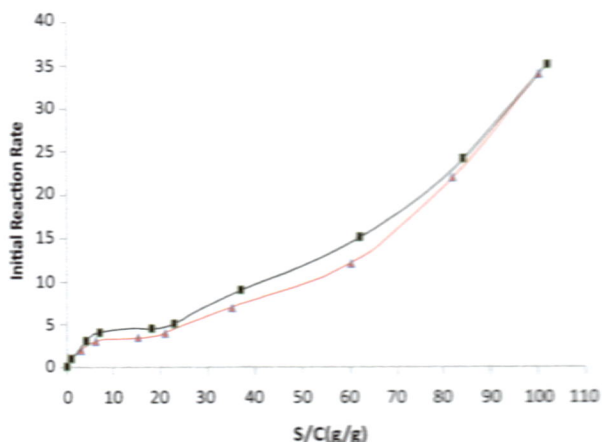

**Figure (4):** Effect of S/C on initial reaction rate in the transesterification of ethyl and propyl acetate in glycerol using MgO as catalyst. Reaction conditions: 0.1 g aliphatic esters, 0.01 g MgO, 5 g glycerol, 75°C, 1h.

of the product and the residual substrate from the glycerol-MgO mixture with diethyl ether followed by the addition of fresh aliphatic acetate. Supporting our previous assumption was the loss of catalyst during filtration was the reason for the decrease in the product yield in the former recycling procedure.

Finally, microwave-promoted heating was recently reported to enhance organic reactions, including the hydrolysis of esters, relative to conventional heating [14]. Thus, using an unmodified home microwave, the transesterification of ethyl acetate in glycerol under microwave irradiation was also tested in an open reaction vessel with MgO as the catalyst. Heating the reaction mixture for up to 40 s at low intensity (from 26°C to 56°C) or for 5 s at high intensity (from 26°C to 61°C) resulted in 23% yield, while under conventional heating at 55°C for 1 h the yield was 27%. It seems that glycerol that has high dielectric constant and three hydroxyl groups adsorbs microwave irradiation efficiently. Furthermore, the markedly low vapor pressure and high boiling temperature of glycerol make it an attractive solvent for microwave-promoted organic synthesis under atmospheric pressure [10].

## Conclusions

Glycerol can be successfully used as a green solvent and with this concept we successfully developed an acyl acceptor in the transesterification of ethyl ester to ethyl alcohol by using soluble and solid base. It allowed the conventional heating method to be replaced with the more efficient microwave-promoted heating. The yield of aliphatic alcohol was increased by increasing the reaction time and the temperature while maintaining the loading of the catalyst. Heating the reaction in an unmodified home microwave instead of using conventional heating resulted in higher reaction rate. Transesterification of glycerol mechanism helps to develop efficient separation techniques for product from catalyst and recycling of catalyst.

## References

1. Smith MB, March J (2007) Advanced organic chemistry: reactions, mechanisms, and structure. (6thedn), Wiley & Sons 397.

2. Francois N, Lionel L, Lionel D, Albert D (2008) Microwave-Assisted Synthesis of Vinyl Esters through Ruthenium-Catalyzed Addition of Carboxylic Acids to Alkynes. Aust J Chem 62: 227-231.

3. Helwani Z, Othman MR, Aziz N, Kim J, Fernando WJN (2009) Solid heterogeneous catalysts for transesterification of triglycerides with methanol: A review. Appl Catal A Gen 363: 1-10.

4. Weiyang Z, Konar SA, David GB (2003) Ethyl esters from the single-phase base-catalyzed ethanolysis of vegetable oils. Journal of the American Oil Chemists Society 80: 367.

5. Dlugy C, Wolfson A (2007) Lipase catalyse glycerolysis for kinetic resolution of racemates. Bioprocess Biosyst Eng 30: 327-330.

6. Wolfson A, Saidkarimov D, Dlugy C, Tavor D (2009) Glycerol as a sustainable solvent for green chemistry. Green Chemistry Letters and Reviews 2: 107.

7. Wolfson A, Atyya A, Dlugy C, Tavor D (2010) Glycerol triacetate as solvent and acyl donor in the production of isoamyl acetate with Candida antarctica lipase B. Bioprocess Biosyst Eng 33: 363-366.

8. Hideshi H, Masaomi S, Hujime K (2000) Alcoholysis of ester and epoxide catalyzed by solid bases. Stud Surf Sci Catal 130: 3507-3512.

9. Nichele TZ, Faverio C, Monterio AL (2009) Pd-catalyzed Suzuki cross-coupling reactions of glycerol arylboronic esters. Catal Commun 1015: 693-696.

10. Wibowo TY, Zakuria R, Abdullah AZ (2010) Selective Glycerol Esterification over Organomontorillonite Catalysts. Sains Malaysiana 39: 811.

11. Wang TT, Huang TC, Yeh MY (1990) Benzyl Alcohol from Transfer Catalyzed Weakly Alkaline Hydrolysis of Benzyl Chloride. Chem Eng Commun 92: 139-151.

12. Miyake T, Hattori A, Hanaya M, Tokumaru S, Hamaji H, et al. (2000) Benzylacetate synthesis by oxyacetoxylation of toluene on Pd–Bi binary catalyst. Topics in Catalysis 13: 243-248.

13. Miyake T, Asakawa T (2005) Recently developed catalytic processes with bimetallic catalysts. Appl Catal A Gen 280: 47-53.

14. Escalante J, Diaz-Coutino FD (2009) Synthesis of gamma-nitro aliphatic methyl esters via Michael additions promoted by microwave irradiation. Molecules 14: 1595-1604.

# 17

# Molecular Docking of HIV-1 env gp120 Using Diterpene Lactones from *Andrographis paniculata*

**Kabir OO\*, Abdulfatai TA and Akeem AJ**

*Chemistry Unit, Department of Chemical, Geological and Physical Sciences, Kwara State University, Malete, Ilorin, Nigeria*

## Abstract

The search for effective drugs to treat HIV/AIDS has been the major task of most researchers since several years of its discovery. Most synthetic drugs such as Efavirenz, Tenofovir, Emtricibatine among others are employed in the antiretroviral treatment which have dangerous effects on patients. Thus, herbal medicine can be used as an alternative source of treatment for HIV positive patients as they exhibit little or no side effects when compared to synthetic drugs. This research work sought to examine whether plant diterpene lactones isolated from *Andrographis paniculata* exhibit anti-HIV activity using molecular docking studies. The HIV-1 env gp120 was docked by two diterpene lactones namely; andrographolide and neoandrographolide using docking tool (igemdock v2.1) after retrieving protein structure from Protein Data Bank (PDB). The result indicates that neoandrographolide is a more promising drug against HIV-1 than andrographolide due to its low interaction energy for the formation of ligand-receptor complex.

**Keywords:** Andrographolide; Env gp120; HIV-1; Docking; PDB

## Introduction

Plants serve as a source of new drugs for treating different kinds of ailments including HIV/AIDS. Reported that 25% of the drugs prescribed worldwide originate from plant, one hundred and twenty one such active compounds being in current use. Of the two hundred and fifty two drugs considered as basic and essential by World Health Organization (WHO), 11% are exclusively of plant origin and a significant number are synthetic drugs obtained from natural precursors. Plant based drugs are formulated to produce varieties of effective pharmaceutical formulations to enhance anti-HIV activities. Human Immunodeficiency Virus (HIV) is an endemic virus characterized by attacking the human immune system which makes the immune cells weak and unable to fight against it, thus leading to illness. The final stage in the life cycle of HIV is known as Acquired Immunodeficiency Syndrome (AIDS), a set of symptoms and infections resulting from the damage of the human immune system caused by the human immunodeficiency virus (HIV) [1]. The actual cause of virus is through sexual intercourse, exposure to infected body fluids or tissues, and from vertical transmission during pregnancy, delivery or breast feeding among others. According to WHO, 70 million people have been infected with HIV and 35 million people have died of AIDS since the inception of the epidemic. At the end of 2011, approximately 36 million people were HIV positive and 1.7 million people died of AIDS-related diseases. Recent research works show that HIV/AIDS is the leading cause of death in Sub-Saharan Africa and the fourth leading cause of death worldwide. To prevent more death and widespread of the virus, clinical trial by researchers is ongoing worldwide for novel anti-HIV drugs. The "King of bitters" (*Andrographis paniculata*), family Acanthaceae, is a small endangered medicinal plant native to India, China and Sri Lanka (Figure 1). It is widely cultivated in southern Asia, where the roots and leaves are used in traditional medicine and pharmaceutical industries [2]. Andrographolide, a bicyclic diterpenoid lactone, the main constituent of *Andrographis paniculata* is known for its pharmacological activities. Andrographolide is known to possess antihepatotoxic [3], antibiotic [4], anti-inflammatory [5], anti-snake venom [6], anticancerous [7] and anti-HIV [8] properties. Besides all it is generally used as immunostimulant [9] agent. Andrographolide content varies with plant parts and with the geographical distribution. Leaves of *A. paniculata* are reported to contain maximum andrographolide [10]. The "env" gene in HIV encodes a single protein, gp160. Cellular

enzymes attack it when it travels to the cell surface, then protein gp 160 is chopping into two pieces gp120 and gp41. If and when new virus particles bud off from the host cell, these two pieces lie on opposite sides of the virus membrane. gp 120 sits on the outside of the virus particle, forming the virus's spikes, while gp41 sits just on the inside of the membrane – each protein is being anchored to each other through the membrane [11,12]. There is no report to the best of our knowledge as regards the availability of any plant-derived drug in clinical use in the treatment of HIV/AIDS and in fact, screening these large numbers of secondary metabolites of plant origin using computational tools will help in drug design and in a very short time compared to the conventional methods. In the present study, andrographolide and neoandrographolide was checked for inhibition against env gp120 using iGemdock.

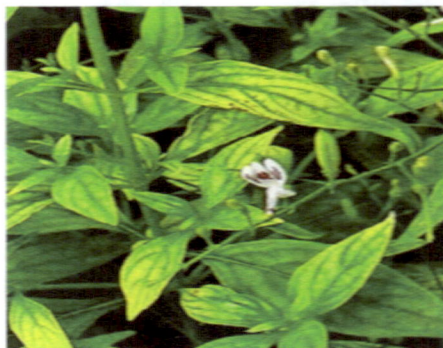

**Figure 1:** *Andrographis paniculata.*

---

**\*Corresponding author:** Kabir OO, Chemistry Unit, Department of Chemical, Geological and Physical Sciences, Kwara State University, Malete, PMB 1530, Ilorin, Nigeria, E-mail: kabir.otun@kwasu.edu.ng

## Materials and Methods

### iGemDock

The tool was developed by Jinn-Moon Yang, a Professor of the Institute of Bioinformatics, National Chiao Tung University. It is a Generic Evolutionary Method for molecular docking. iGEMDOCK is a program for computing a ligand conformation and orientation relative to the active site of target protein. It is a graphical-automatic drug discovery system used for integrating docking, screening, post-analysis, and visualization of various ligands. The main features of iGemdock are scoring function and evolutionary algorithm. In scoring function, the complicated AMBER-based energy function is replaced by local minima. The new rotamer based mutations operator is used to reduce search space in the ligand structure which is an advantage over the Gaussian and Cauchy mutation effects [13]. The core idea in it is Generic algorithm, scoring function.

### PUBCHEM and PDB SUM

PUBCHEM is used to obtain the database of the protein like primary, 2D; 3D and other literature survive of researchers. The ligand and analogues andrographolide of *Andrographis paniculata* is derived from PUBCHEM. PDB SUM is used for finding the active site, number of chains, the number of amino acids present in the each chain and motif details. Figure 2 shows the structure of HIV env gp120 obtained from PUBCHEM.

### Preparation of target protein

Protein Data Bank (PDB) is a repository of 3-D structural data of bio macromolecules [14]. In the present study, the structure of env gp120 was procured from PDB as given below.

**1G9M:** It is viral envelope glycoprotein (gp) 120 structure of strain HIV-1 hxbc2 complicated with the CD4 and induced neutralizing antibody 17b.

### Preparation of Ligand

With the numerous use of natural products in clinical research due to their medicinal and therapeutic values without any side effects as compared to the drugs. We have taken plant diterpene lactones from the *Andrograhis paniculata* into consideration since some of the phytochemicals extracted from the plant showed anti-HIV activity [8]. The ligands which were taken into consideration are andrographolide and neoandrographolide. The ligands for the study were analyzed for their hydrophobicity. The hydrophobic activity is usually calculated by Lipinski filter tool by DruLito tool. The distribution of the Log P shows the hydrophobic activity of the drug. The structures of the ligands were taken from ChemSketch and were docked by iGEMDOCK tool. Finally all the results were compared and discussed.

## Results and Discussion

### Lipinski rule of five analysis

The drug-likeness is necessary to be evaluated at the primary stage as this reduces the chances of selecting the false positive results. Various basic physicochemical properties such as log P, H-bond acceptor, H-bond donor, molecular weight ad molar refractivity were calculated to evaluate a molecule to act as drug. The value of log P should be ≤ 5; this is the distribution coefficient important for finding the solubility of the drug that is lipophilicity. Molecular weight of the compound

should not exceed 500Da as most of the drugs are small molecules [15,16] (Figures 3 and 4).

**1. Andrographolide**

Molecular Weight: 350.4492

Hydrogen bond donor: 3

Hydrogen bond acceptor: 5

Log P=2.9

Molar refractivity: 94.1

**2. Neoandrographolide**

Molecular Weight: 480.598

Hydrogen bond donor: 4

Hydrogen bond acceptor: 8

Log P=1.17

Molar refractivity: 124.83

It is evident from Table 1 that neoandrographolide has a lower Log P value than andrographolide which indicates that the former has higher hydrophobic activity than the latter. The best ligand for docking studies is determined by evaluating the interaction energy for the specific ligand-receptor complex under study (Table 2). It is evident from the above result that neoandrographolide is a more promising drug against HIV-1 than andrographolide because of its lower interaction energy (-108.61 kcal), However, none of the two ligands has violated Lipinski rule (Figures 5-7).

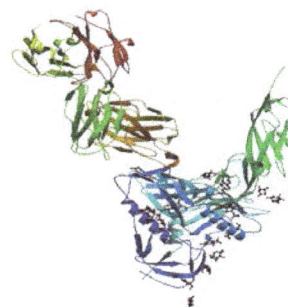

**Figure 2:** HIV env gp120, PDB id: 1G9M.

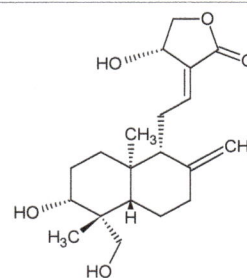

**Figure 3:** Andrographolide.

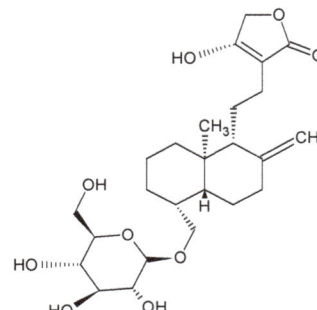

**Figure 4:** Neoandrographolide.

| S No | Ligand Name | MW (g/mol) | Log P | H-bond donor | H-bond acceptor | Molar refractivity |
|------|-------------|------------|-------|--------------|-----------------|--------------------|
| 1 | Andrographolide | 350.4492 | 2.9 | 3 | 5 | 94.1 |
| 2 | Neoandrographolide | 480.598 | 1.17 | 4 | 8 | 124.83 |

**Table 1:** Characteristics of Ligands. *MW: Molecular weight.

| S No | Ligand-receptor complex | Energy (kcal) |
|------|-------------------------|---------------|
| 1 | cav4HHB-HEM-zinc_andrographolide | -91.42 |
| 2 | cav4HHB-HEM-zinc_neoandrographolide | -108.61 |

**Table 2:** Bond Energy Interaction for specific ligand-receptor complex.

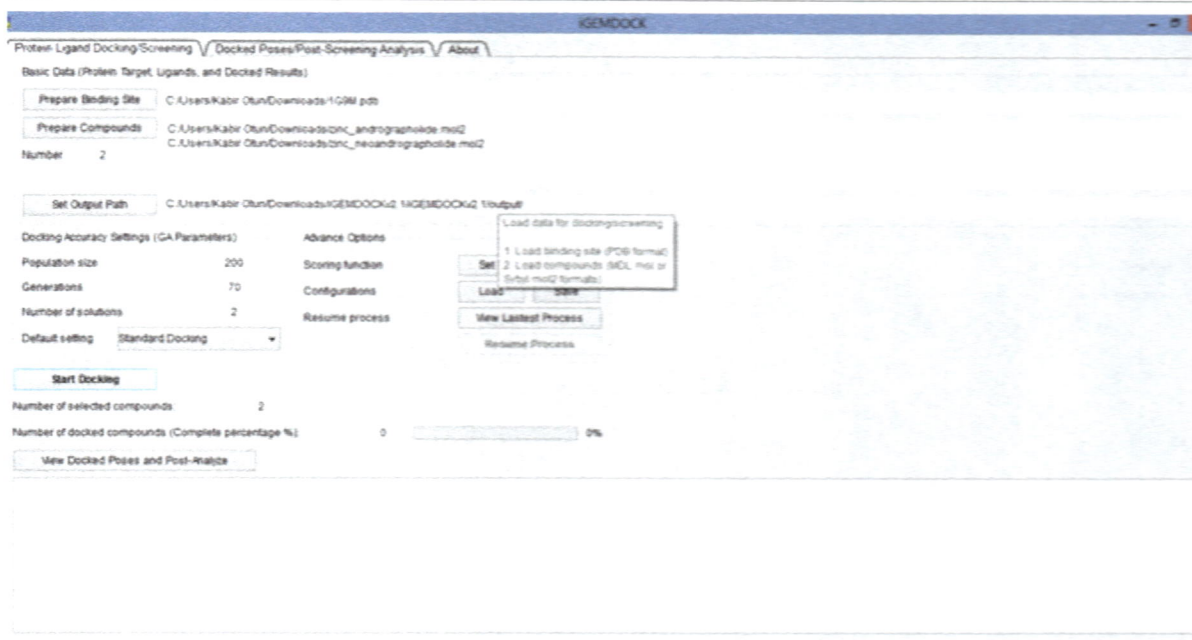

**Figure 5:** Screenshot of Preparing Binding site and Ligand in iGEMDOCK.

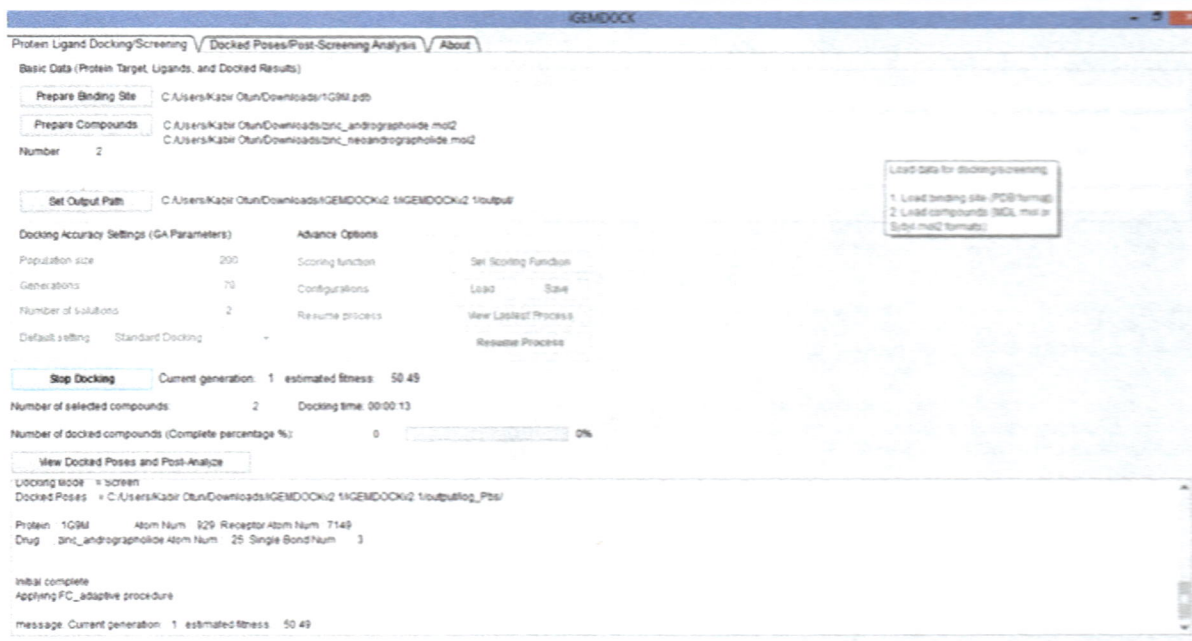

**Figure 6:** Screenshot of docking process in iGEMDOCK.

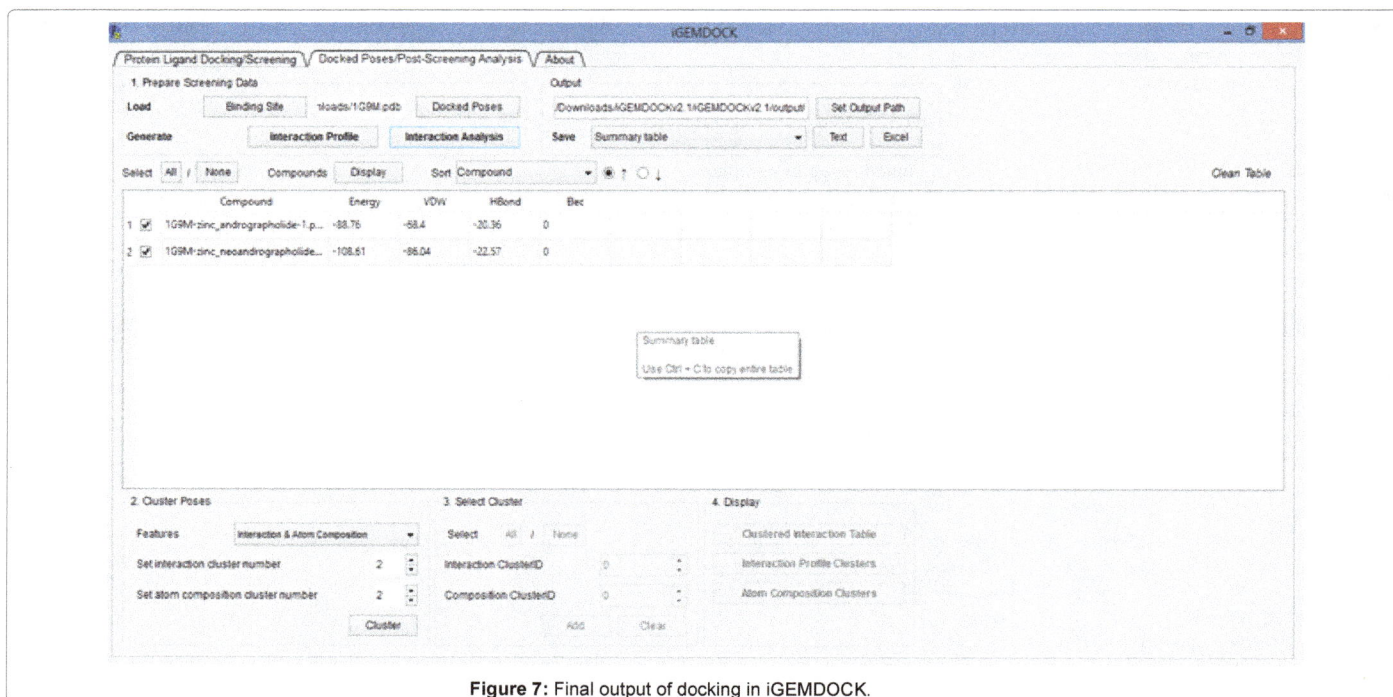

**Figure 7:** Final output of docking in iGEMDOCK.

## Conclusion

The molecular docking helps in drug design and provide a good understanding of the mechanism of interaction of the drug and target protein. From this study, we conclude that the ligand neoandrographolide, the active compound of *Andrographis paniculata* is more effective in inhibiting HIV env gp120 than andrographolide. Hence, neoandrgrapholide can be subjected to further analysis and preceding preclinical trials.

### References

1. Sepkowitz KA (2001) AIDS--the first 20 years. N Engl J Med 344: 1764-1772.

2. Kumar A, Dora J Singh A, Tripathi R (2012) A review of king of bitter (Kalmegh). Int J Res Pharmacy Chem 2: 116-124.

3. Maiti K, Mukherjee K, Murugan V, Saha BP, Mukherjee PK (2010) Enhancing bioavailability and hepatoprotective activity of andrographolide from Andrographis paniculata, a well-known medicinal food, through its herbosome. J Sci Food Agric 90: 43-51.

4. Sule A, Ahmed QU, Samah OA, Omar MN (2010) Screening for antibacterial activity of Andrographis paniculata used in Malaysia folkloric medicine: a possible alternative for the treatment of skin infection. Ethnobot Leaflets. 14: 445-456.

5. Lee KC, Chang HH, Chung YH, Lee TY (2011) Andrographolide acts as an anti-inflammatory agent in LPS-stimulated RAW264.7 macrophages by inhibiting STAT3-mediated suppression of the NF-κB pathway. J Ethnopharmacol 135: 678-684.

6. Premendran SJ, Salwe KJ, Pathak S, Brahmane R, Manimekalai K (2011) Anti-cobra venom activity of plant Andrographis paniculata and its comparison with polyvalent anti-snake venom. J Nat Sci Biol Med 2: 198-204.

7. Lim JC, Chan TK, Ng DS, Sagineedu SR, Stanslas J, et al. (2012) Andrographolide and its analogues: versatile bioactive molecules for combating inflammation and cancer. Clin Exp Pharmacol Physiol 39: 300-310.

8. Tang C, Liu Y, Wang B, Gu G, Yang L, et al. (2012) Synthesis and biological evaluation of andrographolide derivatives as potent anti-HIV agents. Arch Pharm (Weinheim) 345: 647-656.

9. Radhika P, Annapurna A, Rao SN (2012) Immunostimulant, cerebroprotective & nootropic activities of Andrographis paniculata leaves extract in normal & type 2 diabetic rats. Indian J Med Res 135: 636-641.

10. Mishra S, Tiwari SK, Kakkar A, Pandey AK (2010) Chemoprofiling of Andrographis paniculata (Kalmegh) for its andrographolide content in Madhya Pradesh, India. Int J Pharm Biosci 1:1-5.

11. Chan DC, Fass D, Berger JM, Kim PS (1997) Core structure of gp41 from the HIV envelope glycoprotein. Cell 89: 263-273.

12. Coon JT, Ernst E (2004) Andrographis paniculata in the treatment of upper respiratory tract infections: a systematic review of safety and efficacy. Planta Med 70: 293-298.

13. Kimura T, Nishikawa M, Ohyama A (1994) Intracellular membrane traffic of human immunodeficiency virus type 1 envelope glycoproteins: vpu liberates Golgi-targeted gp160 from CD4-dependent retention in the endoplasmic reticulum. J Biochem 115: 1010-1020.

14. Robertson TA, Varani G (2007) An all-atom, distance-dependent scoring function for the prediction of protein-DNA interactions from structure. Proteins 66: 359-374.

15. Vaibhav M, Nidhi M (2013) Molecular Docking Studies of anti-HIV drug BMS-488043 derivatives using HEX and GP120 Interaction Analysis using Pymol. International Journal of Scientific and Research Publications 3: 1-7.

16. Rates SM (2001) Plants as source of drugs. Toxicon 39: 603-613.

# 18

# The Behavior of 3-Anilinoenone and N-Phenyl Cinnamamide toward Carbon Nucleophiles: Spectroscopy and X-Ray Studies Reveal Interesting New Synthesis Routes to Nicotinonitriles and Tetrahydropyridine-3-Carbonitrile

**Al-Omran F\* and El-Khair A**

*Department of Chemistry, Faculty of Science, Kuwait University, P.O. Box 12613, Safat 13060, Kuwait*

### Abstract

The reactions of 3-anilinoenones with active methylene nitrile either in acid or base media were investigated. Reasonable mechanisms to account for the formation of the nicotinonitrile, ethyl nicotinate, nicotinic acid and dienamide derivatives were suggested. A one-pot multicomponent reactions (MCRs) of enaminone, aniline and either malononitrile or ethyl cyanoacetate in acid or base media afforded 1,3,5-triacycl benzene derivative. Treatment of N-phenyl cinnamamide with malononitrile in refluxing sodium ethoxide lead to tetrahydropyridine derivative. The structures of the synthesized compounds were elucidated by elemental analyses, X-ray and a variety of spectroscopic methods, including proton and carbon nuclear magnetic resonance spectroscopy ('H-NMR and ¹³C-NMR), correlation spectroscopy (COSY), heteronuclear single quantum coherence spectroscopy (HSQC), and mass spectrometry (MS).

**Keywords:** Enaminones; 3-Anilinoenones; 2-Anilinonicotinonitriles; Ethoxy nicotinate; *N*-Phenyl cinnamamides; Nicotinic acids; Tetrahydropyridines 1,3,5-Triacyl benzene

## Introduction

The chemistry of enaminones is recognized as a significant field of study because enaminones serve as both starting materials for the synthesis of various heterocyclic compounds that have potential agrochemical use [1-10] and as intermediates in dye and pharmaceutical industries [11-15]. Due to their biological activities and pharmacological properties, pyridines and pyridones represent an important class of compounds that have been developed using functionally substituted enaminones [16-22]. In continuation of our interest in the synthesis of functionally substituted heteroaromatic compounds such as pyridines and pyridones, utilizing enaminones as starting materials [22-33], we report here the behavior of 3-anilinoenone and *N*-phenyl cinnamamide derivative toward malononitrile in the synthesis of pyridine derivatives. The 3-anilinoenone derivative **2a** have been used previously, by Al-Saleh et al. [19]. They have reported that 3-anilinoenone **2a** reacted with malononitrile in acetic acid afforded non isolated *N*-phenyl-1,2-dihydropyridine **3**. In this article, it was assumed that malononitrile initially dimerizes and then condenses to a carbonyl group in **2a,** followed by cyclization. The authors failed to isolate **3,** until it was reacted with piperidine to afford **4** (Scheme 1). To the best of our knowledge, this was the only paper reported that discuss the reactions of 3-anilinoenone **2a** with active methylene nitriles (Scheme 1).

On the other hand, Abdelrazek and Elsayed [20] have reported that the reaction of enaminones **1a-b** with active methylene nitriles afforded pyridone **8** through the isolated intermediate **7.** The authors proposed a synthetic pathway for the formation of 2-pyridone **8** includes 1,4-addition of active methylene in malononitrile to dimethylamino group of enaminone **1** rather than 1,2-addition to the carbonyl group (Scheme 2). These assumptions have attracted our attention and have raised doubts about whether enaminone **1** have the same behavior as 3-anilinoenone **2** toward carbon nucleophile or have different behavior. Also we would investigate whether anilino group is a good leaving group as *N,N*-dimethyl amino group. Therefore, we

decided to investigate the reactions of a compound **2a-b** with active methylene nitriles [such as i.e. malononitrile and ethyl cyanoacetate] in acetic acid and extend the studies in sodium ethoxide and ethanolic piperidine. We reported here new synthesis routes to nicotinonitriles, ethyl nicotinate, nicotinic acid, and dienamide with comprehensive spectroscopic analysis of these compounds and supplies with X-ray crystallography.

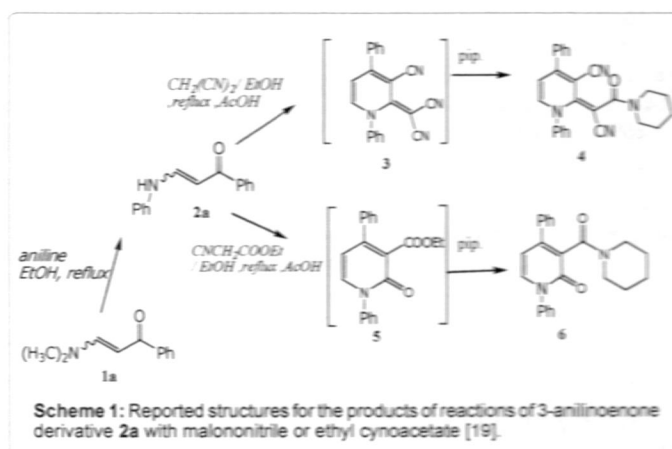

**Scheme 1:** Reported structures for the products of reactions of 3-anilinoenone derivative **2a** with malononitrile or ethyl cyanoacetate [19].

**\*Corresponding author:** Al-Omran F, Department of Chemistry, Faculty of Science, Kuwait University, P.O. Box 12613, Safat 13060, Kuwait E-mail: fatima.alomran@ku.edu.kw

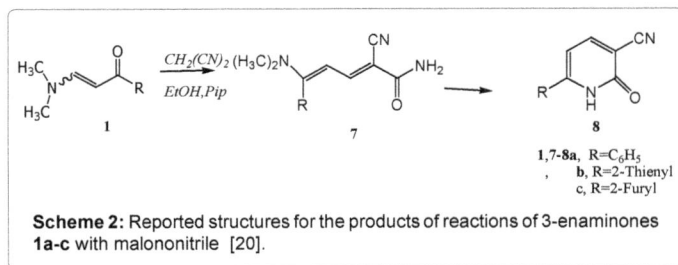

**Scheme 2:** Reported structures for the products of reactions of 3-enaminones **1a-c** with malononitrile [20].

## Experimental Section

### Reagents and apparatus

All m.p values are reported uncorrected and were determined on a Gallenkamp apparatus. The Fourier Transform-Infrared (FT-IR) spectra were recorded on a FT-IR (Jasco FT/IR-6300) using a KBr disc. The $^1$H and $^{13}$C-NMR spectra were recorded on a Bruker DPX 400 MHz spectrometer, with DMSO-d$_6$ or CDCl$_3$ as solvent and tetramethylsilane (TMS) as internal standard. Chemical shifts are reported as d unit (ppm). Mass spectra were measured on GC/MS DFS, THERMO instrument. Microanalyses were performed on a CHNS-Vario Micro Cube Analyzer (Germany), a single-crystal X-ray crystallography instrument (Rigaku, Rapid II, Japan), and a Bruker X$_8$ Prospector (Bruker, Germany) in the Chemistry Department of Kuwait University. Compound **2** was prepared by a method reported in the literature [19].

### Synthesis

**6-Aryl-2-anilino-3-nicotinonitrile derivatives (9a-b): Method A:** A mixture of **2b** (2.29 gm, 10.0 mmol) and malononitrile (0.66 gm, 10.0 mmol) in acetic acid (20 mL) was refluxed for 6 hours. The reaction mixture was allowed to cool to room temperature for 24 hours. The solid product so formed was collected by filtration and crystallized from ethanol.

**Method B:** A mixture of **1a-b** (10.0 mmol) and aniline (0.92 gm, 10.0 mmol) in ethanol (20mL) was refluxed 6 hours. To a stirred reaction solution, malononitrile (0.66 gm, 10.0 mmol) in acetic acid (10 mL) was added and refluxed for 6 hours. The solid product, so formed, was collected by filtration and crystallized from ethanol.

**2-(phenylamino)-6-phenyl-2-yl-3- nicotinonitrile (9a):** Yellow crystals: Yield: 2.03 (75%), m.p 252-253ºC. FTIR: $\nu_{max/cm}^{-1}$ 3454(NH) and 2189(CN). $^1$H-NMR (DMSO-d$_6$): $\delta_H$ 7.03-8.16 (12H, m, phenyl-H) and 9.18 ppm (s, 1H, NH, D$_2$O exchangeable). $^{13}$C- NMR (DMSO-d$_6$): $\delta_c$ 160.6 and 153.1(C-2 and C-6), 135.1, 130.2, 129.7, 129.4, 129.2, 129.1, 128.9, 127.8, 125.4, 121.2 (phenyl carbons), 118.6(CN) and 92.1(C-3) ppm. MS $m/z$ (%): 271 [M$^+$ 65%]. *Anal.* Calcd. for C$_{18}$H$_{13}$N$_3$ (271.31): C, 79.68, H, 4.83, N, 15.48%. Found: C, 79.57, H, 4.72, N, 15.31%.

**2-(Phenylamino) -6-thiophen-2-yl) nicotinonitrile (9b):** Yellow crystals: Yield: 1.96 (71%), m.p 159–160ºC. FTIR: $\nu_{max/cm}^{-1}$: 3308 (NH) and 2228 cm$^{-1}$ (CN). $^1$H-NMR (DMSO-d$_6$): $\delta_H$ 7.06 (t, 1H, J=8.0 Hz, H-4"), 7.20 (t, 1H, J=4.0 Hz, H-4'), 7.35 (t, 2H, J=8.0 Hz, H-3" and H-5"), 7.46 (d, 1H, J=8.0 Hz, H-5), 7.73 (d, 2H, J=8.0 Hz, H-2" and H-6"), 7.75 (d, 1H, J=4.0 Hz, H-3'), 7.92 (d, 1H, J=4.0 Hz, H-5'), 8.01 (d, 1H, J=8.0 Hz, H-4) and 9.13 ppm (s,1H, NH, D$_2$O exchangeable). $^{13}$C-NMR (DMSO-d$_6$): $\delta_c$ 155.0 and 153.4 (C-2 and C-6), 144.0 (C-4), 143.6 (C-1"), 139.9 (C-2'), 130.8 (C-3'), 128.8 (C-4'), 128.3 (C-3" and C-5"), 127.7 (C-5), 122.7 (C-4"), 120.8 (C-2" and C-6"), 116.8 (CN), 109.2 (C-5), and 90.6 (C-3) ppm. MS $m/z$ (%): 277 [M$^+$, 100%]. *Anal.*

Calced. for C$_{16}$H$_{11}$N$_3$S (277.34): C, 69.29, H, 3.99, N, 15.17%. Found: C, 69.23, H, 3.82, N, 15.18%.

**Ethyl-2-(phenylamino)-6-thiophen-2-yl-3-nicotinate (9c):** A mixture of **1b** (1.81 gm, 10.0 mmol) and aniline (0.92 gm, 10.0 mmol) in ethanol (20 mL) was refluxed for 6 hours. To a stirred reaction solution, ethyl cyanoacetate (1.13 gm, 10.0 mmol) in acetic acid (10 mL) was added and refluxed for 6 hours. The solid product so formed was collected by filtration and crystallized from the ethanol as yellow crystals. Yield: 2.36 (73%), mp 101-103ºC. FTIR: $\nu_{max/cm}$ 3297(NH) and1679 cm$^{-1}$ (CO). $^1$H-NMR (DMSO-d$_6$): $\delta_H$ 1.32(t, 3H, J=7.2Hz, CH$_3$), 4.31(q, 2H, J=7.2 Hz, OCH$_2$-), 7.03-8.18 (m, 10 H, phenyl and thiophene-H), 10.26 (s, 1H, NH, D$_2$O exchangeable). $^{13}$C- NMR (DMSO-d$_6$): $\delta_c$ 167.1(CO), 155.1, 154.7, 144.4, 141.4, 140.0, 130.1, 129.2, 127.9, 125.4, 122.8, 120.3, 109.1, 105.5 (phenyl and thiophene carbons), 61.6(-OCH$_2$), 14.5(CH$_3$) ppm. MS $m/z$ (%): 324 [M$^+$ 100%]. *Anal.* Calcd. for C$_{18}$H$_{16}$N$_2$O$_2$S (324.40): C, 66.64, H, 4.97, N,8.63%. Found: C, 66.52, H, 4.88, N, 8.52%.

**1,3,5,-Tri-[(thiophen-2-yl)methanoyl]benzene (10):** A mixture of **1b** (1.81 gm, 10.0 mmol), aniline (0.92 gm, 10 mmol) and molonitrile (0.66 gm, 10 mmol) or ethyl cyanoacetate (1.13 gm, 10.0 mmol) in acetic acid (20 mL) or ethanol piperidine (20 mL) was refluxed for 5 hours. The reaction mixture was allowed to cool to room temperature for 24 hours. The solid product so formed was collected by filtration and crystallized from ethanol as yellow crystals. Yield: 3.06 gm (75%), m.p: 203-205ºC. FTIR: $\nu_{max/cm}^{-1}$: 1642 cm$^{-1}$ (CO). $^1$H-NMR (DMSO-d$_6$): $\delta_H$ 7.32 (t, 3H, J=4.0 Hz, H-4'), 7.89 (d, 3H, J=4.0 Hz, H-3'), 8.16 (d, 3H, J=4.0 Hz, H-5'), and 8.40 ppm (s, 3H, phenyl-H). $^{13}$C-NMR (DMSO-d$_6$): $\delta_c$ 185.8 (3CO), 142.1 (C-2'), 138.1, 136.6, 136.4, 132.2, 129.1 (phenyl and thiophene carbons). MS $m/z$ (%) 408 [M$^+$, 15%]. *Anal.* Calcd. for C$_{21}$H$_{12}$O$_3$S$_3$ (408.51): C, 61.74, H, 2.96%. Found: C, 61.62, H, 3.01%.

**6-Aryl -2-Ethoxy-3-nicotinonitrile (15a-b):** A mixture of **1b-c** (10.0 mmol) and malononitrile (0.66 gm, 10.0 mmol) in sodium ethoxide (30 mL) was refluxed for 5 hours. The reaction mixture was allowed to cool to room temperature, poured into ice-cold water, and neutralized with HCl (10%). The solid product so formed was collected by filtration and crystallized from the ethanol.

**2-Ethoxy-6-(thiophen-2yl)-3-nicotinonitrile (15a):** Yellow crystals: Yield: 1.74 (76%), m.p 81–83ºC. Lit [24] m.p 80-82ºC. FTIR: $\nu_{max/cm}^{-1}$: 2221 cm$^{-1}$ (CN). $^1$H-NMR (DMSO-d$_6$): $\delta_H$ 1.37 (t, 3H, J=8Hz, CH$_3$), 4.44 (q, 2H, J=8Hz, OCH$_2$), 7.18 (t, 1H, J=4.0 Hz, H-4'), 7.54 (d, 1H, J=8.0 Hz, H-5), 7.75 (d, 1H, J=4.0 Hz, H-3'), 7.89 (d, 1H, J=4.0 Hz, H-5'), and 8.12 ppm (d, 1H, J=8.0 Hz, H-4). $^{13}$C-NMR (DMSO-d$_6$): $\delta_c$ 162.6 (C-2), 153.4 (C-6), 144.0 (C-4), 142.5 (C-2'), 130.9 (C-3'), 128.8 (C-4'), 128.1 (C-5'), 115.6 (CN), 111.3 (C-5), and 92.7 (C-3), 62.4 (CH$_2$) and 14.1 (CH$_3$) ppm. MS $m/z$ (%) 230[M$^+$, 100%]. *Anal.* Calcd. for C$_{12}$H$_{10}$N$_2$OS (230.28): C, 62.58, H, 4.37, N, 12.16%. Found: C, 62.43, H, 4.26, N, 12.00%.

**2-Ethoxy-6-(2-furyl)-3-nicotinonitrile (15b):** Dark brown crystals: Yield: 1.56 (73%), m.p 111-112ºC. Lit [24] m.p 110-112ºC. FTIR: $\nu_{max/cm}^{-1}$ 2216 cm$^{-1}$ (CN). $^1$H-NMR (DMSO-d$_6$): $\delta_H$ 1.50 (t, 3H, J=8 Hz, CH$_3$), 4.50 (q, 2H, J=8 Hz, OCH$_2$), 6.40 (dd, 1H, J=3.6 Hz, H-4'), 7.32 (t, 1H, J=3.6 Hz, H-3'), 7.43 (d, 1H, J=8 Hz, H-5). 7.96 (d, 1H, J=1.2 Hz, H-5') and 8.25 (d, 1H, J=8 Hz, d H-4). $^{13}$C-NMR (DMSO-d$_6$): $\delta_c$ 163.4, 152.0, 150.0, 146.6, 145.2, 116.1 (CN), 113.4, 113.2, 111.3, 93.6 (furyl and pyridine carbons), 63.4 (OCH$_2$) and 14.2 (CH$_3$) ppm. MS $m/z$ (%): 214 [M$^+$ 100%]. *Anal.* Calcd. for C$_{12}$H$_{10}$N$_2$O$_2$ (214.22) C, 67.28, H, 4.70, N, 13.07%. Found: C, 67.38, H, 4.63, N, 13.20%.

(2E,4E)-2-Cyano-5-phenyl-5-(phenylamino) pent-2,4-dienamide (16): A mixture of 1a (1.75 gm, 10.0 mmol) and aniline (0.92 gm, 10.0 mmol) in ethanol (20 ml) was refluxed 6 hours. To a stirred reaction solution, malononitrile (0.66 gm, 10.0 mmol) in sodium ethoxide (20 mL) was added and refluxed for 6 hours. The reaction mixture was allowed to cool to room temperature, poured into ice-cold water and neutralized with HCl (10%). The solid product so formed was collected by filtration and crystallized from the ethanol as yellow crystals. Yield: 2.22 (77%), m.p 264-266°C. FTIR: $v_{max/cm}$ $^{-1}$ 3467, 3362 ($NH_2$ and NH), 1642(CO), 2189 cm$^{-1}$(CN). $^1$H-NMR (DMSO-d$_6$): $\delta_H$ 6.20 (d, J=13.0 Hz, 1H, CH), 7.15 (s, 2H, $NH_2$ D$_2$O exchangeable), 7.21-7.60 (m, 11H, phenyl-H and CH), 9.79 ppm (s, 1H, NH, D$_2$O exchangeable). $^{13}$C- NMR (DMSO-d$_6$): $\delta_c$ 164.7 (CO $_{amide}$), 160.6, 153.1, 139.9, 135.1, 130.2, 129.8, 129.2, 125.4, 121.2, 123.6, 118.7, 98.5, 92.0 ppm (Phenyl and alkenes carbons). MS m/z (%): 289 [M$^+$ 100%]. Anal. Calcd. for C$_{18}$H$_{15}$N$_3$O (289.33) C, 74.23, H, 5.22, N,14.52%. Found: C, 74.44, H, 5.33, N, 14.43.

2-oxo-6-phenyl-1,2-dihydropyridine-3-carboxylic acid (17): Compound 16 (2.89 gm, 10.0 mmol) in ethanol (20 ml) and hydrochloric acid (5 mL) was refluxed 30 minutes. The reaction mixture was allowed to cool to room temperature. The solid product, so formed, was collected by filtration and crystallized from the ethanol as beige crystals. Yield: 1.45 (68%), m.p 259-261°C. Lit [34-37] m.p 260-262°C. FTIR: $v_{max/cm}$ $^{-1}$ 3450 (NH), 1710 (CO), 1625 cm$^{-1}$ (CO). $^1$H-NMR (DMSO-d$_6$): $\delta_H$ 7.36 (d, 1H, J=8Hz, H-4), 7.55-7.99 (m, 5H, phenyl-H), 8.00 (s, 1H, NH, D$_2$O exchangeable), 8.45 (d, 1H, J=8 Hz, H-5), 9.10 ppm (br. 1H, OH, D$_2$O exchangeable). $^{13}$C-NMR (DMSO-d$_6$): $\delta_c$ 163.3 (CO), 161.7 (CO $_{amide}$), 149.3, 132.4, 130.7, 129.8, 129.4, 127.9, 126.6, 116.6 ppm (aromatic carbons). MS m/z (%): 215 [M$^+$ 100%]. Anal. Calcd. for C$_{12}$H$_9$NO$_3$ (215.20) C, 66.97, H, 4.22, N,6.51%. Found: C, 67.09, H, 4.33, N, 6.66.

2-cyano-5-dimethylamino-5-phenylpenta-2,4-dienoic acid amide (18a): A mixture of 1a (1.75 gm, 10.0 mmol) and aniline (0.92 gm, 10.0 mmol) in ethanol (20 ml) was refluxed for 6 hours. To a stirred reaction solution, malononitrile (0.66 gm, 10.0 mmol) in ethanolic piperidine (20 mL) was added and refluxed for 6 hours. The reaction mixture was allowed to cool to room temperature. The solid product so formed was collected by filtration and crystallized from the ethanol as yellow crystals. Yield: 1.9 (76%), m.p 261.262°C. Lit [20] m.p 260-262°C. FTIR: $v_{max/cm}$ 3437 and 3292 ($NH_2$) and 2195 (CN) and 1661 cm$^{-1}$ (CO). $^1$H-NMR (DMSO-d$_6$): $\delta_H$ 2.93 (S, 6H, 2CH$_3$), 5.66 (d, 1H, J=13 Hz, CH), 6.78 (s, 2H, $NH_2$ D$_2$O exchangeable), 7.13-7.56 ppm (m, 6H, phenyl -H and CH).$^{13}$C-NMR (DMSO-d$_6$): $\delta_c$ 165.2, 165.1 153.8, 134.4, 130.1,129.3, 129.2,129.1 (phenyl carbon), 118.91(CN), 96.81 (C-2), 87.58 (C-4), 41.92 (2CH$_3$) ppm. MS m/z (%): 241 [M$^+$ 100%]. Anal. Calcd. for C$_{14}$H$_{15}$N$_3$O (241.12): C, 69.69, H, 6.27, N,17.41%. Found: C, 69.62, H, 6.14, N, 17.40%.

2-cyano-5-dimethylamino-5-thiophen-2-yl-pent 2, 4- dienoic acid amide (18b): A mixture of 1b (1.81 gm, 10.0 mmol) and aniline (0.92 gm, 10.0 mmol) in ethanol (20 ml) was refluxed 6 hours. To a stirred reaction solution, malononitrile (0.66 gm, 10.0 mmol) in ethanolic piperidine (20 mL) was added and refluxed for 6 hours. The reaction mixture was allowed to cool to room temperature. The solid product so formed was collected by filtration and crystallized from the ethanol as yellow crystals. Yield: 2.12 (86%), m.p 254-255°C. Lit [20] m.p 253-254°C. FTIR: $v_{max/cm}$ $^{-1}$ 3402 and 3326 ($NH_2$) and 1644 (C=O) and 2193 cm$^{-1}$ (CN). $^1$H-NMR (DMSO-d$_6$): $\delta_H$ 3.00 (s, 6H, 2CH$_3$), 5.69 (d, 1H, J=13 Hz, CH), 6.78(s, 2H, $NH_2$ D$_2$O exchangeable), 7.18-

7.25 (m, 2H, thiophene-H), 7.40 (d, 1H, J=13 Hz, CH), 7.89 (d, 1H, J=4.8 Hz, thiophene-H). $^{13}$C-NMR (DMSO-d$_6$): $\delta_c$ 164.4 (CO), 157.4, 152.6, 133.2, 130.7, 129.7, 127.7 (thiophene carbons, C-3 and C-5), 118.6 (CN), 98.8(C-2), 89.5(C-4), 41.0 ppm (2CH$_3$). MS m/z (%): 247 [M$^+$ 100%]. Anal. Calcd. for C$_{12}$H$_{13}$N$_3$OS (247.32): C, 58.28, H, 5.30, N,14.99%. Found: C, 58.32, H, 5.41, N, 16.88%.

N-Phenyl cinnamamide (26): A mixture of aniline (0.92 gm, 10.0 mmol) and cinnamoyl chloride (1.66 gm, 10.0mmol) in ethanol (20 mL) was stirred at room temperature for 2 hours. The solid product, so formed was collected by filtration and crystallized from the ethanol as white crystals. Yield: 1.73 (78%), m.p 134-136°C. FTIR: $v_{max/cm}$ $^{-1}$3270 (2NH) and 1661 cm$^{-1}$ (CO). $^1$H-NMR (DMSO-d$_6$): $\delta_H$ 6.54 (1H, d, J=16 Hz, H-3), 6.86 (1H, d, J=12 Hz, H-2), 7.06-7.73 (m, 10H, Ar-H), 10.23 ppm (s, 1H, NH, D$_2$O exchangeable). $^{13}$C-NMR (DMSO-d$_6$): $\delta_c$ 167.5 (CO), 143.9, 140.1, 139.3, 134.7, 129.8, 129.0, 128.8, 123.3, 122.3, 119.2 (phenyl and alkenes carbons). MS m/z (%): 223 [M$^+$ 100%]. Anal. Calcd. for C$_{15}$H$_{13}$NO (223.27): C, 80.69, H, 5.86, N,6.27. Found: C, 80.55, H, 5.82, N, 6.16%.

6-oxo-4-phenyl-2(phenyl amino) 1,4,5,6-tetrahydropyridine-3-carbonitrile (32): A mixture of compound 26 (2.23 gm, 10.0 mmol) and malononitrile (0.66 gm, 10.0 mmol) in sodium ethoxide (30 mL) was refluxed for 6 hours. The reaction mixture was allowed to cool to room temperature, poured into ice- water and neutralized with HCl (10%). The solid product so formed was collected by filtration and crystallized from the ethanol as white crystals. Yield: 2.19 (76%), m.p 204-206°C. FTIR: $v_{max/cm}$ 3214 (NH) and 1642 (CO) and 2189 cm$^{-1}$ (CN). $^1$H-NMR (DMSO-d$_6$): $\delta_H$ 2.68 (dd, 1H, J$_{5a,5b}$=16 Hz, H-5$_a$), 3.10 (dd, 1H, J$_{5b,4}$=6.8 Hz, H-5$_b$), 3.94 (dd, 1H, J$_{5a,4}$=5.2 Hz, H-4), 6.96-7.43 (m, 10H, Ar-H), 8.86 (s, 1H, NH, D$_2$O-exchangeable), 10.22 (s, 1H, NH, D$_2$O-exchangeable). $^{13}$C-NMR (DMSO-d$_6$): $\delta_c$ 170.1 (CO), 149.5 (C-2), 142.1 and 141.0 (C-1' and C-1''), 129.4, 129.2, 127.6, 127.3, 122.2, 120.2 (aromatic carbons), 119.5 (CN), 70.1 (C-3), 39.3 (C-5), 37.6 ppm (C-4). MS m/z (%): 289 [M$^+$ 100%]. Anal. Calcd. for C$_{18}$H$_{15}$N$_3$O (289.33) C, 74.23, H, 5.22, N,14.52%. Found: C, 74.33, H, 5.23, N, 14.44%.

## Results and Discussion

The 3-anilinoenones 2a-b used in our experiments have been prepared, by reactions of 3-enaminones 1a-b with aniline in ethanol. The reactivity of the 3-anilinoenones 2a-b toward active methylene nitrile was investigated. Therefore, treatment of 2a-b with malononitrile in acetic acid according to the method and under the same reaction conditions reported by Elnagdi et al. [19] afforded 2-anilinonicotinonitrile derivatives 9 a-b (Scheme 3) in good yield and not compound 4 (Scheme 1). The structures of 9a and 9b were established via their analytical and spectroscopic data. The mass spectra (MS) of the compounds 9a-b revealed a molecular ion peak (M$^+$) with a m/z value of 271 and 277 which correspond to molecular formulae C$_{18}$H$_{13}$N$_3$ and C$_{16}$H$_{11}$N$_3$S, respectively. The infrared (IR) spectrum of the isolated product 9b showed an absorption band at $v_{max}$ 3308 cm$^{-1}$ for the amino group in addition to a strong absorption band at $v_{max}$ 2228 cm$^{-1}$ for the cyano group. The chemical shifts of protons of 9b were assigned using correlation spectroscopy (COSY) measurements, provided proton-proton coupling data (Figure 1). The $^1$H-NMR spectrum of 9b showed in addition to phenyl and thiophene protons, two doublets at $\delta_H$ 8.01 and 7.46 ppm with J=8 Hz, as required for the pyridine protons H-4 and H-5, respectively. Also there is a singlet signal integrated for one proton that appears at $\delta_H$ 9.13 ppm assigned to NH group. The latter signal underwent facile hydrogen-deuterium exchange upon the addition of deuterium oxide. Moreover, the chemical shifts of carbon of compound **9b** were assigned using heteronuclear single quantum coherence (HSQC) (Figure 2). The $^{13}$C-NMR spectrum for **9b** is characterized by two signals at d 155.0 and 153.4 ppm assigned to the C-2 and C-6, respectively. The assignments of $^1$H and $^{13}$C chemical shifts of **9b** are presented in Figure 3 (cf. Experimental Section). The X-ray crystallography pictures [The CCDC files 972454 and 934259 contain the supplementary crystallographic data for compounds **9a-b** in this paper. These data can be obtained free of charge from the Cambridge Crystallographic Data Center *via* www.ccdc.cam.ac.uk] afforded an unambiguous evidence of structures **9a-b** (Figures 4 and 5). It shows that the anilino group is attached to the 2- position of the pyridine ring that affords a conclusive evidence to 2-anilinopyridine structures **9**. The formation of nicotinonotriles **9** from the reactions of 3-anilinoenone **2a-b** with malononitrile is assumed to occur *via* the sequence depicted in Scheme 4. The nicotinonotrile **9** formed most likely *via* initial 1,4- addition of malononitrile to the activated double bond in **2**, followed by the elimination of the aniline molecule under acidic conditions to give the intermediate malononitrile derivative **12**. The latter then undergoes ring closure *via* its enolized form to yield the imidoester **13**. The aniline molecules that is still present in the reaction medium is attacked to C-2 of imidoester **13** to yield the intermediate **14**, which then undergoes cyclization *via* elimination of water to give the final nicotinonotrile derivative **9**. To the best of our knowledge a rearrangement reaction of aniline group has not been reported. Elnagdi and co-workers [19] claim that enaminone **1** and 3-anilinoenone **2** have the same behaviour with regard to their reaction with active methylene reagent. If this conclusion is true, we would expect the formation of the 2-*H*-pyridone derived **8** from the reaction of **1** with malononitrile, as

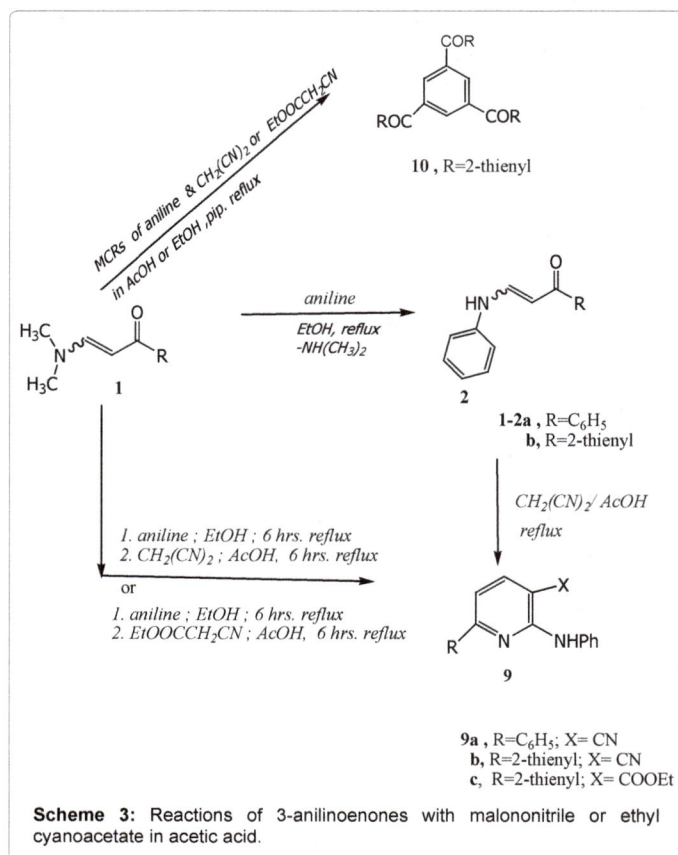

**Figure 1:** COSY spectrum of 2-anilino nicotinonitrile derivative **9b**.

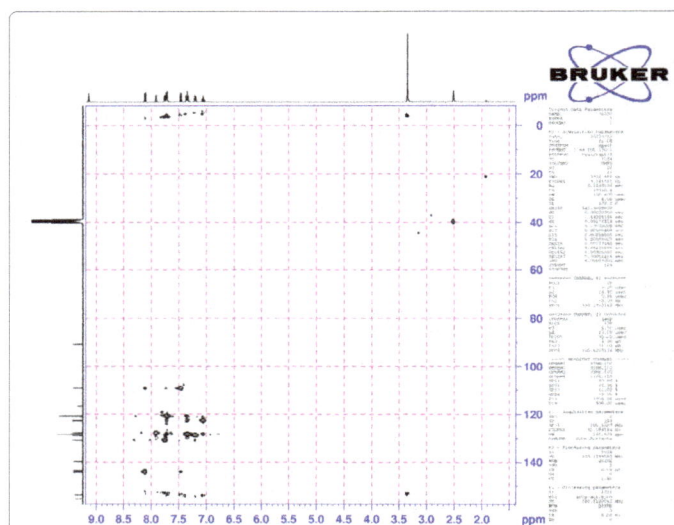

**Figure 2:** HSQC spectrum of 2-anilino nicotinonitrile derivative **9b**.

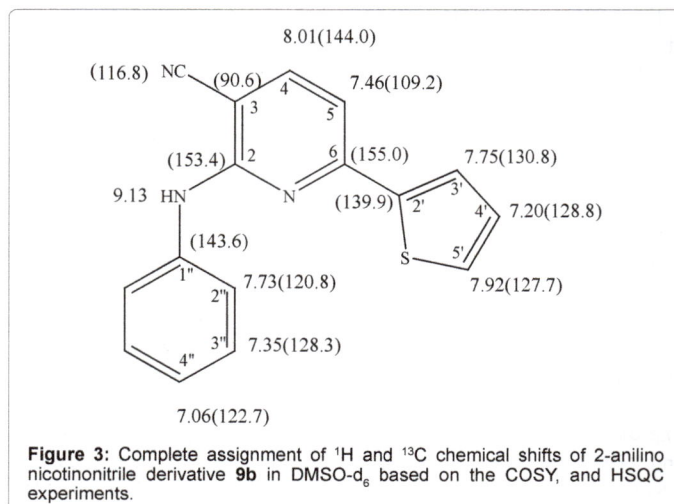

**Scheme 3:** Reactions of 3-anilinoenones with malononitrile or ethyl cyanoacetate in acetic acid.

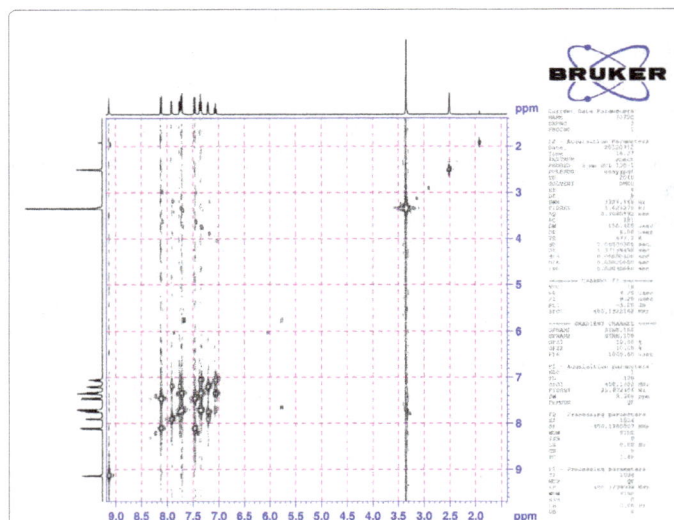

**Figure 3:** Complete assignment of $^1$H and $^{13}$C chemical shifts of 2-anilino nicotinonitrile derivative **9b** in DMSO-d$_6$ based on the COSY, and HSQC experiments.

reported recently by Abdelrazek et al. [20]. The alternative pathway is assumed for the formation of nicotinonotrile **9** take place *via* the attack of aniline molecule to one of the nitrile groups in **11** to give intermediate **14**, which then undergoes cyclization *via* elimination of water under the reaction condition to afford the final nicotinonotrile derivative **9**. On the other hand, a different route has been also observed such as a one-pot two step reactions of the enaminones **1a-b** with aniline in refluxing ethanol followed by treating the reaction mixture with malononitrile or with ethyl cyanoacetate in acetic acid to give **9a-b** and **9c** respectively in good yield (Scheme 3). The analytical and spectral data of the latter reaction products are all consistent with the proposed structure. The X-ray crystallography afforded an unambiguous evidence of structure **9c** (Figures 6). [The CCDC file 972453 contains the supplementary crystallographic data for compounds **9c** in this report. These data can be obtained free of charge from the Cambridge Crystallographic Data Center *via* www.ccdc.cam.ac.uk]. The X-ray of the compound **9c** (Scheme 3) shows that anilino group attached to carbon-2 of pyridine. This result is in contrast to what Elnagdi et al. [19] has been reported for the formation of **6** (Scheme 1).

The 2-anilinonicotinate **9c** has represented further evidence for the suggested mechanism aforementioned. It is apparent that ethyl cyanoacetate followed the same sequence of reaction mechanism as reaction 3-anilinoenone with malononitrile in acetic acid to afford 2-anilinonicotinate through non isolated intermediates 11-14 (Scheme 4).

On the other hand, one-pot method involving multicomponent reactions (MCRS) of enaminone **1b**, aniline with either malononitrile or ethyl cyanoacetate in refluxing acetic acid or ethanolic piperidine, gave the 1,3,5-triacyl benzene derivatives **19** [36] in good yield (Scheme 3). The mass spectrum revealed a molecular ion peak [M+], with $m/z=408$, which corresponds to a molecular weight consistent with a formula of $C_{21}H_{12}O_4S_3$. The structure of tri-thiophen-2-yl benzene derivative **10** was unambiguously confirmed by X-ray crystallography as 1,3,5,-tri-[(thiophen-2-yl)methanoyl]benzene **10** (Figure 7). The CCDC file 959796 contains the supplementary crystallographic data for compound **10** in this paper. These data can be obtained free of

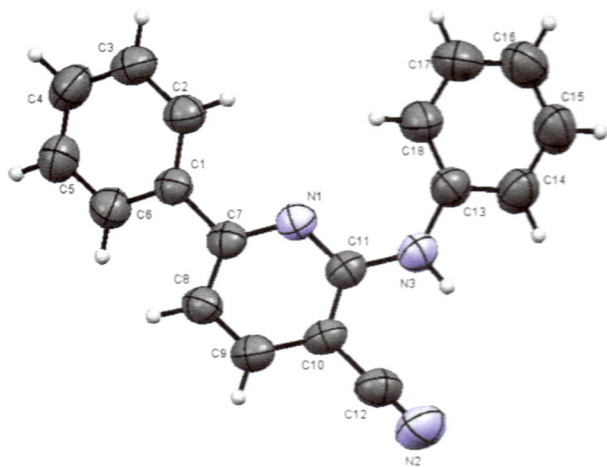

**Figure 5:** Perspective view and atom labeling of X-ray structure of 2-anilino nicotinonitrile derivative **9b**.

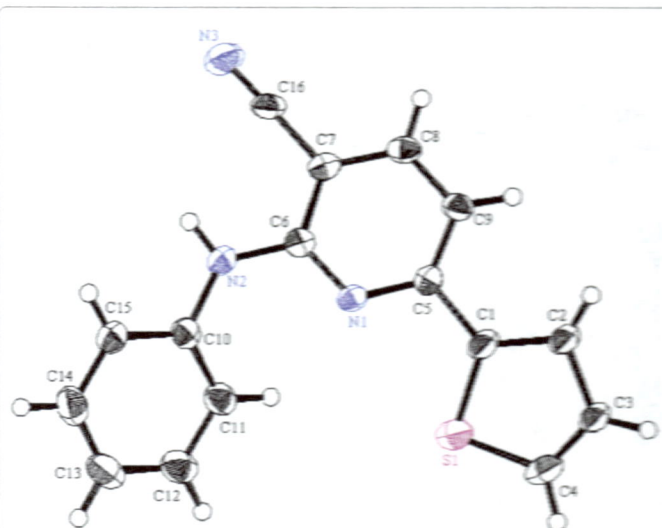

**Figure 6:** Perspective view and atom labeling of X-ray structure of 2-anilinonicotinate derivative **9c**.

charge from the Cambridge Crystallographic Data Center *via* www. ccdc.cam.ac.uk.

When enaminones **1b-c** was allowed to react with malononitrile in refluxing sodium ethoxide for 6 hours, 2-ethoxy nicotinonitrile derivatives **15 a-b** was afforded (Scheme 5). The mass spectra of these compounds revealed a molecular ion peak (M+) with a $m/z$ value of 230 and 214 respectively. These masses are compatible with the molecular ion peak (M+) for $C_{12}H_{10}N_2SO$ and $C_{12}H_{10}N_2O_2$ respectively as described previously [24]. The complete assignment of $^1H$ and $^{13}C$ chemical shift for **15a** are presented in Figure 8. The chemical shifts of protons for **15a** were assigned using COSY measurements, which provided the proton-proton coupling (Figure 9). The $^1H$-NMR spectrum of **15a** revealed two doublets at $\delta_H$ 8.12 and 7.54, respectively, with a *J* value of 8 Hz that could be attributed to the pyridine protons H-4 and H-5, respectively. In addition, thiophene protons and a signal for ethoxy protons at d 1.37

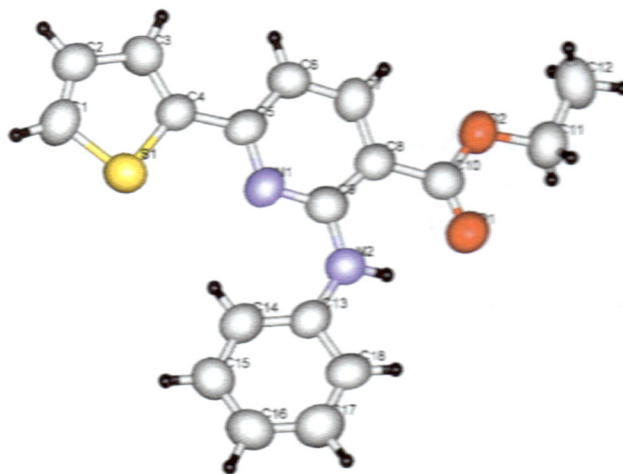

**Figure 4:** Perspective view and atom labeling of X-ray structure of 2-anilino nicotinonitrile derivative **9a**.

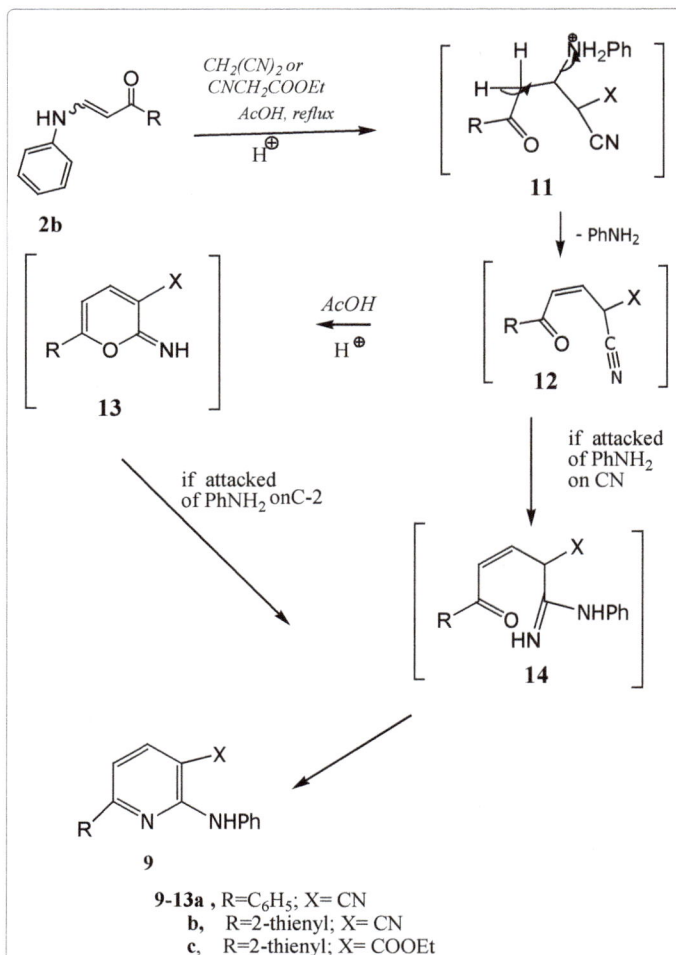

**Scheme 4:** Proposed pathway for formation of 2-anilino nicotinonitriles **9a-b** and ethyl nicotinate derivative **9c**.

**9-13a** , R=C$_6$H$_5$; X= CN
**b**,    R=2-thienyl; X= CN
**c**,    R=2-thienyl; X= COOEt

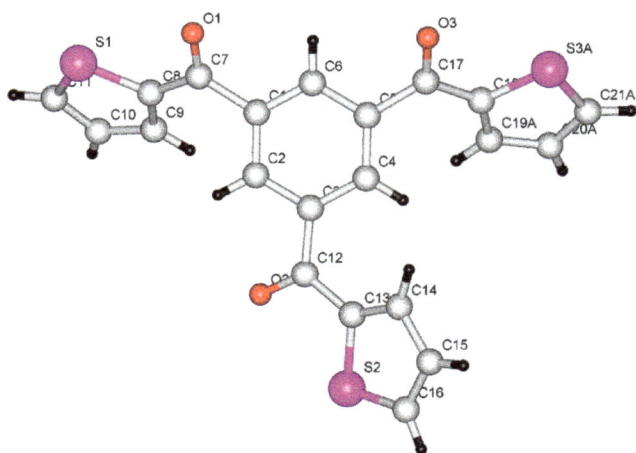

**Scheme 5:** Synthesis 2-ethoxy nicotinonitrile, 2,4-diamonds and 2-nicotinic acid derivatives in base media.

**15a** , R=2-thienyl
**b** ,R=2-furyl

**18a** , R=Ph
**b** R=2-thienyl

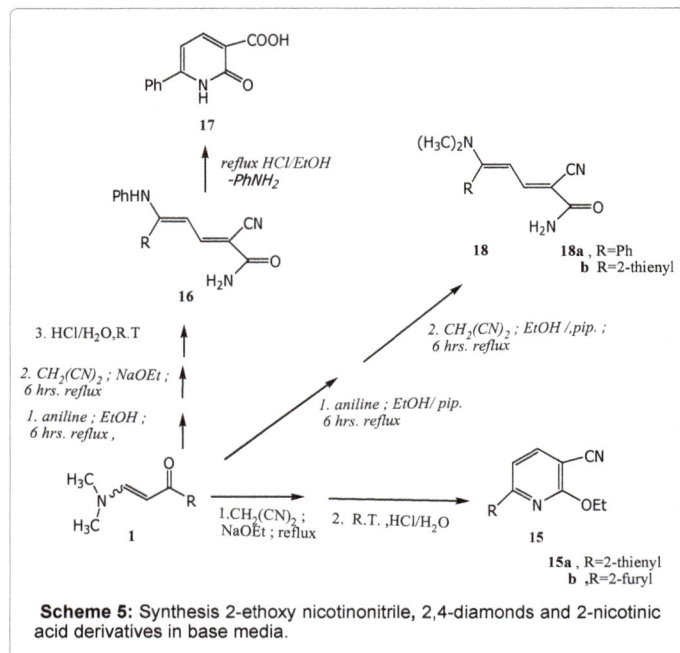

**Figure 8:** Complete assignment of $^1$H and $^{13}$C chemical shifts of 2-ethoxy **nicotinonitrile** derivative **6** based on COSY and HSQC experiments.

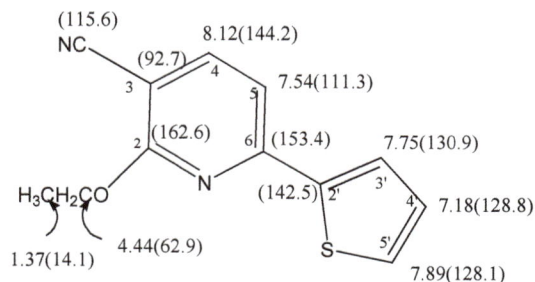

**Figure 7:** Perspective view and atom labeling of X-ray structure of 1,3,5-triacylbenzene derivative **10**.

and 4.44 ppm ($J$=8 Hz) were characterized as a triplet and a quartet for methyl and methylene protons, respectively. The $^{13}$C-NMR chemical shift assignments for **15a** were assigned using HSQC measurement (Figure 10). The $^{13}$C-NMR spectrum for **15a** showed two downfield signals at $\delta_c$ 162.6 and 153.4 ppm assignable to C-2 and C-6 respectively. The structure of compound **15a** was unambiguously confirmed by X-ray crystallography as 2-ethoxy-6-(thiophen-2-yl) nicotinonitrile **15a** (Figure 11). (The CCDC file 934259 contains the supplementary crystallographic data for compound **15** in this paper. These data can be obtained free of charge from the Cambridge Crystallographic Data Center *via* www.ccdc.cam.ac.uk). The structure of the isolated product **15** from the reaction of enaminone **1b** with malononitrile in basic medium is assumed to take place *via* the sequence described in Scheme 6. The active methylene group of malononitrile undergoes addition to the double bond of **1b** to generate the intermediate **19**, under this condition we expected that ethoxide ion will attack one of the nitrile groups in **20** to give the imidoester **21**, which then undergoes cyclization *via* elimination of water under acidic media to afford the final nicotinonitrile derivative **15**.

Furthermore, a one -pot, three-step synthesis of 2,4-dienamide derivative **16** has been achieved by allowing enaminones **1a-b** to

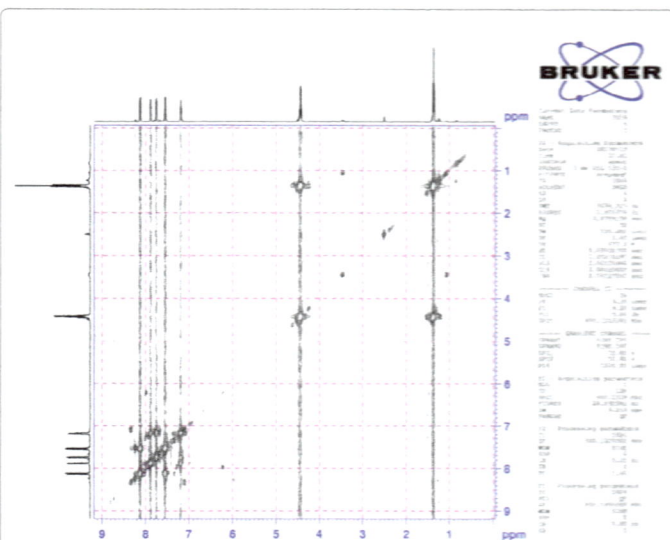

Figure 9: COSY spectrum of 2-ethoxy nicotinonitrile derivative 6.

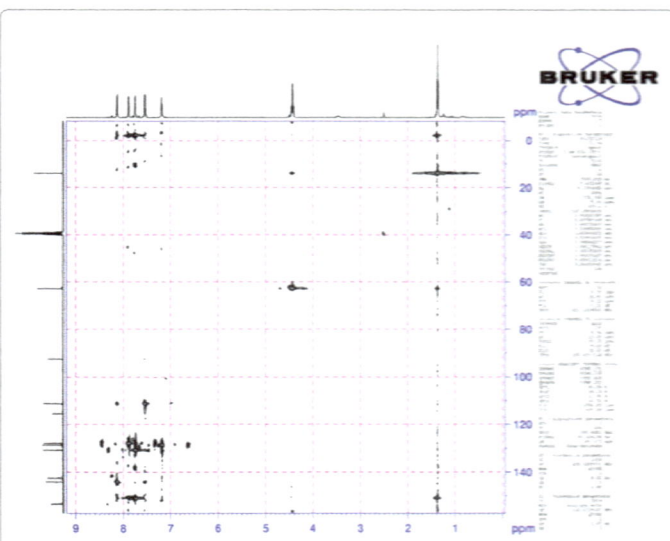

Figure 10: HSQC spectrum of 2-ethoxy nicotinonitrile derivative 6.

group of malononitrile undergoes *via* initial 1,4- addition activated double bond in **2b** to give intermediate **22**. The base removes a proton from α-carbon followed by the elimination of the aniline molecule to afford the intermediate malononitrile derivative **22**. The latter then undergoes ring closure *via* its enolized form to yield the imidoester **23**. The aniline molecule that is still present in the reaction medium is attacked to C-6 of iminopyran **24** and undergoes ring opening to afford the isolated 2,4-dienamide derivative **16**. Conversions of **16** into nicotinic acid derivatives **17** were achieved by boiling in EtOH/ HCl 17, *via* elimination of anilino group.

Interesting, when the enaminones **1a-b** allowed to react with aniline in refluxing ethanol, followed by treatment the reaction mixture with malononitrile in ethanolic piperidine instead of sodium ethoxide, in one pot reactions, unexpected product with formula of $C_{14}H_{15}N_3O$ and $C_{12}H_{13}N_3OS$ was produced (Scheme 5). The X-ray crystallographic

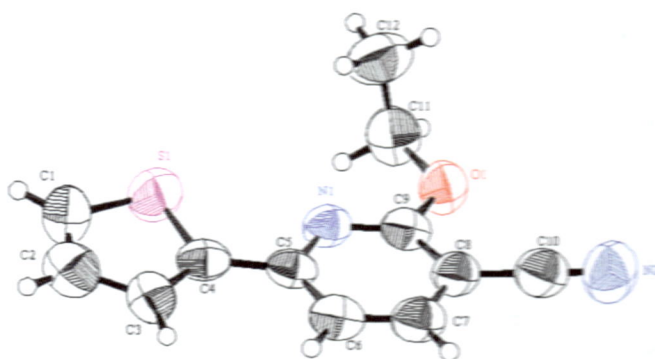

Figure 11: Perspective view and atom labeling of X-ray structure of 2-ethoxy nicotinonitrile derivative 6.

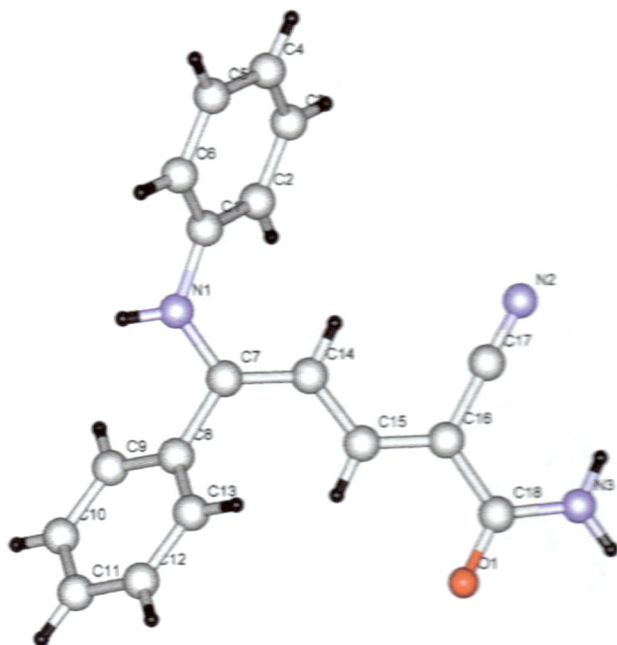

Figure 12: Perspective view and atom labeling of X-ray structure of compound 16.

reacted with aniline in refluxing ethanol, followed by treatment the reaction mixture with malononitrile in sodium ethoxide and then neutralized with HCl at room temperature. However, the mass spectra of obtained product revealed a molecular ion peak (M+) with *m/z* 289 and was compatible with the molecular formula $C_{18}H_{15}N_3O$. The X-rays crystallographic picture (the CCDC file 972452 contains the supplementary crystallographic data for compound **16** in this paper. These data can be obtained free of charge from the Cambridge Crystallographic Data Center *via* www.ccdc.cam.ac.uk) afforded an unambiguous evidence of structure **16**. It shows that anilino group attached to the same carbon atom (C-5) which carrying the phenol group and the terminus carrying the cyano and the amide group on (C-2) as shown in Figure 12 and Scheme 5.

The formation of 2,4-dienamide derivative **16** from the reaction of **2a** with malononitrile in sodium ethoxide as catalyst is assumed to occur *via* the sequence depicted in Scheme 6. The active methylene

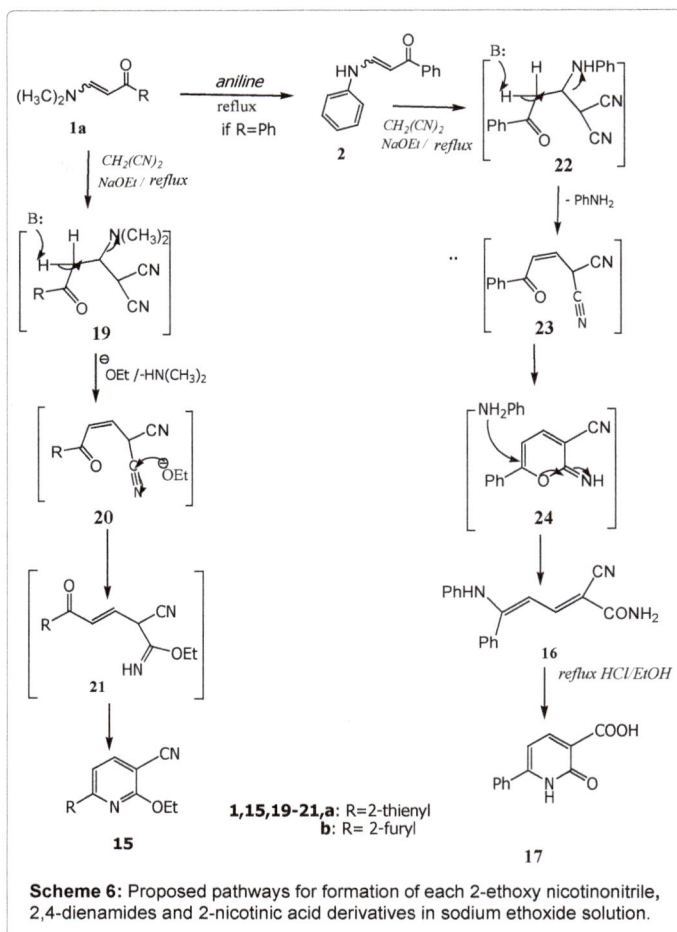

**Scheme 6:** Proposed pathways for formation of each 2-ethoxy nicotinonitrile, 2,4-dienamides and 2-nicotinic acid derivatives in sodium ethoxide solution.

protons confirmation of methylene protons at position 5 will not in the same environment. These germinal pairs, H-5$_a$, H-5$_b$ and H-4 are resonating at δ$_H$ 2.68, 3.10 and 3.94 ppm. Each signal appears as a doublet of doublets with coupling constant $^2J_{5a,5b}$=16Hz, $^3J_{5b,4}$,=6.8Hz, $^3J_{5a,4}$=5.2 Hz, the germinal protons has largest coupling constant. Moreover, the chemicals of the carbons for compound **32** were assigned using HSQC measurement. The $^{13}$C NMR spectrum for 32 is revealed a low field signal at dc170.1ppm corresponding to carbonyl carbon. While the high field signals at δ$_C$ 39.3 and 37.6 ppm corresponding to methylene and methine carbons respectively. The X-rays crystallographic picture (the CCDC file 9724451 contains the supplementary crystallographic data for compound **32** in this paper. These data can be obtained free of charge from the Cambridge Crystallographic Data Center *via* www.

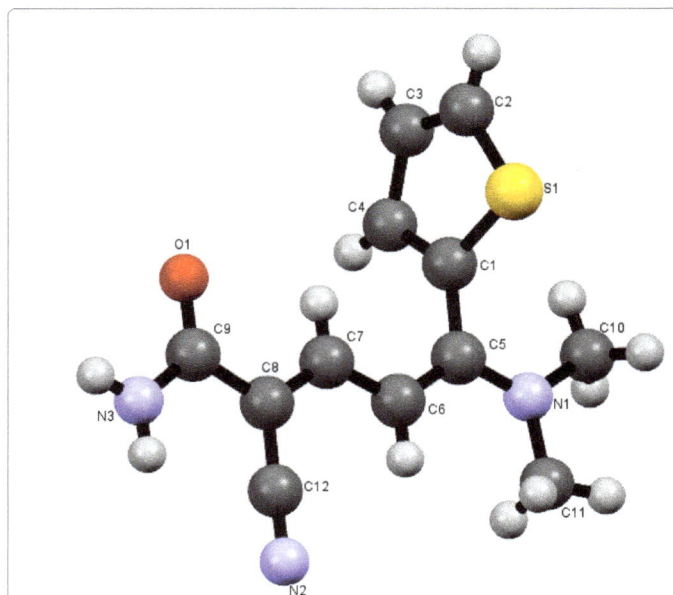

**Figure 13:** Perspective view and atom labeling of X-ray structure of compound **18b.**

analysis (Figure 13) was used to unambiguously assign the structure of this substance as 2,4-dienamide derivative **18b** (the CCDC file 974230 contains the supplementary crystallographic data for compound **18b** in this paper. These data can be obtained free of charge from the Cambridge Crystallographic Data Center *via* www.ccdc.cam.ac.uk)

It is believed that **18** is produced *via* mechanistic route displayed in Scheme 6, involving intermediates 22-23, by changing the solvent in the second step from sodium ethoxide to ethanolic piperidine solvent change encourage the *N,N*-dimethyl amine molecule to be much faster than an aniline molecule (which they are still present in the reaction medium) to attack to C-2 of imidoester **24** and undergoes ring opening to afford the isolable compound **18**.

On the other hand, treatment at room temperature of cinnamoyl chloride with aniline in ethanol afforded the corresponding the *N*-phenyl cinnamide **26**. The structure of the product **26** was confirmed on the basis of elemental analysis and spectral data. The compound **26** reacts with malononitrile in refluxing ethanolic sodium ethoxide afforded the tetrahydropyridine-3-carbonitrile derivative **32** in good yield (Scheme 7). The mass spectrum of the reaction product revealed a molecular ion peak [M$^+$], with *m/z* 289 which corresponds to a molecular weight consistent with a formula of C$_{18}$H$_{15}$N$_3$O. The IR spectrum of **32** shows the two NH, nitrile and carbonyl amide absorption bands in the region of ν$_{max}$ 3214, 2189 and 1642 cm$^{-1}$ respectively. The chemical shift of protons for **32** were assigned using the COSY (correlation spectroscopy) measurement which provided the

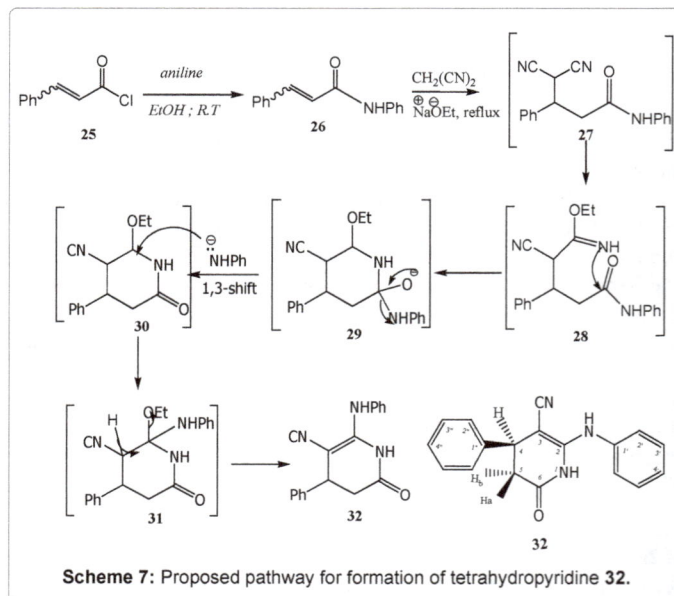

**Scheme 7:** Proposed pathway for formation of tetrahydropyridine **32.**

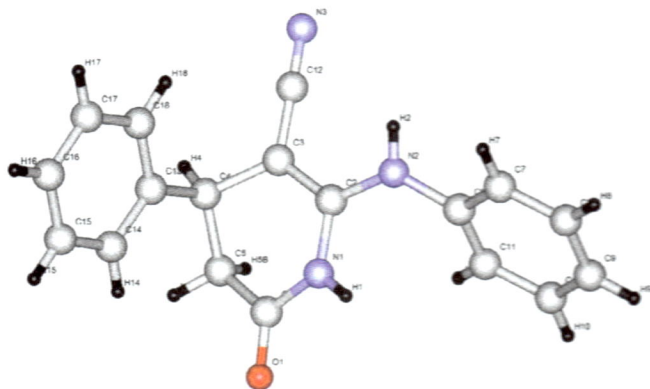

**Figure 14:** Perspective view and atom labeling of X-ray structure of compound **32.**

ccdc.cam.ac.uk) afforded an unambiguous evidence of structure **32**. The formation of tetrahedropyridine **32** from the reaction of compound **26** with malononitrile in sodium ethoxide solution is assumed to take place *via* the sequence depicted in Scheme 6 through non-isolated intermediates from **27-31** involving migration of an aniline molecule *via* 1,3-shift (Scheme 6).

An attempt to react compound **26** with malononitrile in refluxing either acetic acid, or ethanolic piperidine solution for 2-5 hours failed.

## Conclusion

We have suggested a reasonable mechanism that explains the behavior of 3-anilinoenone with active methylene nitriles in acid or base media. We have verified that enaminones have different behavior than 3-anilinoenone with regard to their reactions with methylene nitriles in acid and base solution and corrected some literature wrong structures. Also we have suggested a reasonable mechanism that explains the behavior *N*-phenyl cinnamide with malononitrile in sodium ethoxide. The structures of the synthesized compounds were elucidated by elemental analyses, ¹H-NMR; ¹³C-NMR spectra; COSY; HSQC; MS and X-ray investigations.

### Acknowledgment

This work was financed by University of Kuwait research grant SC02/08. We are grateful to the Faculty of Science, Chemistry Department, SAF Facility for the spectral and analytical data (Project GS01/01, GS03/08, GS01/03).

## References

1. Yu Y-Y, Bi L, Georg GI (2013) Palladium-catalyzed direct C–H arylation of cyclic enaminones with aryl iodides. J Org Chem 7: 6163–6169.

2. Zohreh N, Alizadeh A, Babaki M (2013) One-pot solvent-free three- component synthesis of conjugated enaminones containing three alkyl carboxylate groups. J Chem 2013: 1-6.

3. Alexander MS, Scott KR, Harkless J, Butcher RJ, Jackson-Ayotunde PL (2013) Enaminones 11. An examination of some ethyl ester enaminone derivatives as anticonvulsant agents. Bioorg Med Chem 21: 3272-3279.

4. Josefík F, Svobodová M, Bertolasi V, Simunek P (2013) A simple, enaminone-based approach to some bicyclic pyridazinium tetrafluoroborates. Beilstein J Org Chem 9: 1463-1471.

5. Yunyun L, Rihui Z, Jie-Ping W (2013) Water-promoted synthesis of enaminones: mechanism investigation and application in multicomponent reactions. Synthetic Comm 43: 2475-2483.

6. Al-Awadhi H, Al-Omran F, Elnagdi MH, Infantes L, Foces-Foces C, et al. (1995) New synthetic approaches to condensed pyridazinones: alkyl pyrid- azinyl carbonitriles as building blocks for the synthesis of condensed pyridazinones. Tetrahedron 51: 12745-12762.

7. Abou-Shanb F, Elnagdi MH, Ali FA, Wakefield BJ (1994) a,a–Dioxoketene dithioacatals as starting materials for synthesis of polysubstituted pyridines. J Chem Soc perkin Trans I: 1445-1452

8. Dawood KM, Farag AM, Kandeel ZE ( 1999) Heterocyclic synthesis via enaminonitriles: One-pot synthesis of some new pyrazole, isoxazole, pyrimidine, pyrazolo[1,5-a]pyrimidine, pyrimido[1,2-a]benzimidazole and pyrido[1,2-a]benzimidazole derivatives. J Chem Res: 88-89.

9. Al-Mousawi SM, Moustafa MS, Abdelkhalik MM, Elnagdi MH (2009) Enaminones as building blocks in organic syntheses: on the reaction of 3-dimethylamino-2-propenones with malononitrile. Arkivoc 2009: 1-10.

10. Al-Mousawi SM, El-Apasery MA, Elnagdi MH (2010) Enaminones in heterocyclic synthesis: A novel route to tetrahydropyrimidines, dihydropyridines, triacylbenzenes and naphthofurans under microwave irradiation. Molecules 15: 58-67.

11. Bartoli G, Cupone G, Dalpozzo R, Nino AD, Maiuolo L, et al. (2002) Stereoselective reduction of enaminones to syn γ-aminoalcohols.Terahedron Lett 43 :7441-7444.

12. Macho S, Miguel D, Neo A G, Rodriguez T, Torroba T (2005) Cyclopentathiadiazines, new heterocyclic materials from cyclic enaminonitriles. Chem Comm 3: 334-336.

13. Sheikhshoaie I, Fabian WMF (2009) Theoretical insights into material properties of schiff bases and related azo compounds. Curr Org Chem 13:149-171.

14. El-Apasery MA, SAI-Mousawi1 SM, Mahmoud H, Elnagdi MH (2011) Novel routes to biologically active enaminones, dienoic acid amides, arylazonicotinates and dihydropyridazines under microwave irradiation. Inter Res J Pure and App Chem 1: 69-83.

15. Al-Mousawi S M, El-Apasery M A, Mahmoud H, Elnagdi MH (2012) Studies with biologically active enaminones: an Easy method for structural elucidation of products produced from enaminone starting materials through pathways employing microwave irradiation. Inter Res J Pure and App Chem 2: 77-90.

16. Shawali A S (2012) Bis-enaminones as versatile precursors for terheterocycles: synthesis and reactions. Arkivoc J (i): 383-431.

17. Abu-Shanab FA, Sherif SM, Sayed, Mousa AS (2009) Dimethylformamide dimethyl acetal as a building block in heterocyclic synthesis. J Heterocyclic Chem 46: 801-827.

18. Kidwai M, Bhardwaj S, Mishra NK, Bansal V, Kumar A, Mozumder S (2009) A novel method for the synthesis of β-enaminones using Cu-nanoparticles as catalyst. Catalysis Comm 10: 1514-1517.

19. Al-Saleh B, El-Apasery MA, Abdel-Aziz RS, Elnagdi MH (2005) Enaminones in heterocyclic synthesis: Synthesis and chemical reactivity of 3-anilino-1-substituted-2-propene-1-one. J Heterocyclic Chem 42: 563-566.

20. Abdelrazek FM, Elsayed AN (2009) About the reaction of β-dimethyl- amino-a,β-enones with active methylene nitriles. J Heterocyclic Chem 46: 949-953.

21. Al-Saleh B, Abdelkhalik M M M, Eltoukhy A M, Elnagdi M H (2002) Enaminones in heterocyclic synthesis: A new regioselective synthesis of 2,3,6-trisubstituted pyridines, 6-substituted-3-aroylpyridines and 1,3,5-triaroylbenzenes. J Heterocyclic Chem 39: 1035-1038.

22. Abu-Shanab FA, Wakefield, B, Al-Omran F, Abdel-khalek M M, Elnagdi MH (1995) Alkyl(oxo)pyridazine carbonitriles as building block in heterocyclic synthesis: Novel synthesis of pyrido[3,4-c] pyridazine, pyrido[3,4-d] pyridazine and thiopyrano[3,4-c]pyridazine and 1,3,4-thiadiazacacena- phthylenes a,a-dioxoketene dithioacetals as starting materials for synthesis of polysubstituted pyridines. J Chem Res: 2924-24946.

23. AL-Omran FA, Al-Awadi NA (1995) Studies of polyfunctionally substituted heteroaromatic: synthesis of new poly functionally substituted azabiaryls. J Chem Res 329-393.

24. Al-Omran FA, Al-Awadi N, El-khair AA, Elnagdi MH (1997) Synthesis of new aryl and heteroaromatic substituted pyridines,pyrazoles,pyrimidines, and pyrazolo[3,4-d]pyridazines. OPPI 29: 285-292.

25. Al-Etaibi A, Awadi N, Al-Omran F, Abdel Khalik MM, Elnagdi MH (1999) Novel C-alkylation reaction of condensed thiophenes with enaminones, enaminonitriles, ethoxymethylenemalononitrile, aryl vinyl ketones and ω-nitrostyrenes. J Chem Res 4-5.

26. Al-Omran F (2000) Studies with 1-functionally substituted alkylbenzotriazoles: An efficient route for the synthesis of 1-azolyl- benzotriazoles, benzotriazolylazines and benzotriazolylazoloazines. J heterocyclic Chem 37: 1219-1223.

27. Al-Omran FA, Abd El-Hay OY, El-khair AA (2000) New routes to synthesis of pyridazinone, ethoxypyridine, pyrazole and pyrazolo[1,5-a]pyrimidine derivatives incorporating a benzotriazole moiety. J Heterocyclic Chem 37 (6): 1617- 1622. DOI: 10.1002/jhet. 5570370635.

28. Al-Omran FA, Elkair AA (2001) Synthesis of 2,6-dimethyl -4-substituted pyridine-3,5- dicarbonitriles from β-aminoacryloniriles. Indian J Chem 40 B: 608-611.

29. Al-Omran F, Elassar A-Z, El-Khair AA (2003) Novel synthesis and biological effects of new derivatives of azines incorporating coumarin. J Heterocyclic Chem 40: 249-254.

30. Al-Omran FA, El-khair AA (2005) Heterocyclic synthesis via enaminones: Novel synthesis of (1H)-pyridin-2-one, pyrazolo[1,5-a]pyrimidine and isoxazole derivatives incorporating a N-methylphthalimide and their biological evaluation. J Heterocyclic Chem 42: 307-312.

31. Al-Omran F, El-Khair AA (2007) Synthesis of new derivatives of thieno[2,3-b] pyridine fused with octyl ring. Heterocyclic Chem 44: 561-568.

32. Al-Omran FA, El-Khair AA (2008) Studies with S-alkylpyrimidine: New route for the synthesis derivatives of isoxazol, pyrazolo[1,5-a]pyrimidine, pyridazine,,pyridine, thieno[2,3-b]pyridine and thiophene of potential anti-HIV. J Heterocyclic Chem 45: 1057-1063.

33. Al-Omran F, El-Khair AA (2009) Studies with 2-(acetonylthio)- benzothiazole. New routes to isoxazoles, isoxazolo[3,4-b]pyridines, pyrazolo[1,5-a] pyrimidines, pyridines and quinolizines. J Chem Res 2009(7): 433-436.

34. Helmy NM, El-Baih FEM, Al-Alshaikh M, Moustafa MS (2011) A route to dicyanomethylene pyridines and substituted benzonitriles utilizing malononitrile dimer as a precursor. Molecules 16: 298-306.

35. Moustafa MS, Al-Mousawi SM, Hilmy NM, Ibrahim YA, Liermann JC, et al. ( 2013) Unexpected behavior of enaminones: Interesting new routes to 1,6-naphthyridines, 2-oxopyrrolidines and pyrano [4,3,2-de][1,6] naphthyridines. Molecules 18(1): 276-286.

36. Elghamry I (2003) Cyclotrimerization of enaminones: An efficient method for the synthesis of 1,3,5-triaroylbenzenes. Synthesis 15: 2301-2303.

37. Alnajjar A-Z, Abdelkhalik MM, Al-Enezi A, Elnagdi MH (2008) Enaminones as Building Blocks in Heterocyclic Syntheses:Reinvestigating the Product Structures of Enaminones with Malononitrile. A Novel route to 6-substituted-3-oxo-2,3-dihydropyridazine-4-carboxylic acids. Molecules 14: 68-77.

# 19

# Ranking and Screening Hazardous Chemicals for Human Health in Southeast China

**Jining Liu[1], Chen Tang[2], Deling Fan[1], Lei Wang[1], Linjun Zhou[1] and Lili Shi[1]***

*[1]Nanjing Institute of Environmental Sciences, Ministry of Environmental Protection, Nanjing, China*
*[2]Nanjing Entry-Exit Inspection and Quarantine Bureau, Nanjing, China*

### Abstract

Copeland and comprehensive multi-index comparison methods were used to rank and screen hazardous chemicals using original and pre-treatment data sets. The results show that the Copeland method can yield similar results for the two data sets. The results of a comprehensive multi-index comparison with the pretreatment dataset also show some similarities to those obtained using Copeland method. The results of the two methods show 18 common chemicals that belong in the top 20 chemicals. Of these chemicals, six are different types of dichlorodiphenyltrichloroethane, seven are POPs, three are polycyclic aromatic hydrocarbons, and two are pesticides. These substances should be regarded as chemicals of concern, and appropriate handling should be followed. Overall, the Copeland method with the original dataset can rapidly, reasonably, and effectively rank and screen hazardous chemicals.

**Keywords:** Chemical risk ranking; Priority setting; Copeland method; Comprehensive multi-index comparison; Human health

## Introduction

With the recent rapid development and change in technologies, the chemical industry frequently produces new materials that can pose a threat to public health and the environment if improperly handled. Approximately 46,000 chemicals have been registered in China. However, a significantly higher number of chemicals have been produced and used, with some produced in significant amounts. The chemicals have brought considerable convenience but can also have drastic effects to the environment and human health. Thus, the management of chemical hazards has become increasingly important. Screening, ranking, and scoring systems are key technologies to the determination of hazardous chemicals.

Ranking, decision-support, and scoring systems can be used to determine potential risks. In environment chemistry, ranking and scoring systems are always used to identify the hazardous chemicals as well as the project to consider. Although numerous ranking and scoring systems have been developed, a consensus on the effective ranking methods has been made. In recent years, chemical risk-ranking programs have been implemented in China, for example, to identify which chemical should be placed in the priority pollutant lists.

To date, chemical risk ranking and scoring methods have been developed in countries such as the US, Canada, EU members, Japan, and Germany. In general, ranking methods can be classified as a "scalar approach method" and "vector performance" method by Halfon and Reggiani [1]. The "scalar approach method" means that an overall rank or score is determined by its own characters. Each object can obtain a score according to the indicators used in the ranking model and the score will not change. The objects are then ranked according to these scores. "Vector performance" is based on the elements of the vector and uses mathematical analysis to obtain the scores. The vector is created by using the indicators of objects. An increase or decrease in the objects will affect the scores. The variability of the score is the main difference between the two methods. Some examples of the "scalar approach method" are CHEMS-1 by Swanson et al. [2], CHEMS-2 by Dunn [3], EURAM by Hansen et al. [4], and SCRAM by Snyder et al. [5]. Some examples of "vector performance" are the Hasse diagram by Halfon et al. [1], Copeland score method (Al-Sharrah [6]), and the comprehensive multi-index comparison (Ren and Xiong [7]).

The aim of this paper is to identify the chemicals that are hazardous to human health in Southeast China and to determine the hazard ranking of these chemicals. The results may provide realistic information that can be used in developing hazard control policies and management. In this study, human health effects, environmental effects, octanol–water partitioning, bioaccumulation, human exposure concentration, and frequency of detection were used as indicators to determine the chemical hazards. The human exposure concentration data of these chemicals were obtained by the project team that completed the previous measurement. The other data for the study can be found in the "Case study" section. The Copeland score method and comprehensive multi-index comparison method were used in this study.

## Principle of the Method

### Copeland score method

The Copeland score method was proposed by A. H. Copeland (1951) in a seminar on applications of mathematics to the social sciences at the University of Michigan [8]. It is a simple nonparametric method that has been used to evaluate the election results after voting. To use the method outside the voting field, candidates are replaced by objects, and votes are replaced by indicators. Two papers reported on the investigation of the properties and flaws of the Copeland score method [9,10].

The Copeland score method is based on a comparison of one

*****Corresponding author:** Lili Shi, Nanjing Institute of Environmental Sciences, Ministry of Environmental Protection, Nanjing, China. E-mail: sll@nies.org

indicator with another for each pair of objects. For example, assume that the number of chemicals is $m$, and each chemical has n indicators. From there, we can create a matrix X, as follows:

$$X = \begin{bmatrix} x_{11} & x_{12} & \cdots & x_{1n} \\ x_{21} & \cdots & \cdots & \vdots \\ \vdots & \cdots & \cdots & \vdots \\ x_{m1} & \cdots & \cdots & x_{mn} \end{bmatrix}$$

The next step is to compare each indicator for chemical i and j. $S_{k,ij}$ is the result of the $k$ indicator comparison. The original Copeland has elements of the comparison matrix of (1,0,-1), so we quote this matrix in our method. Equivalent to the matrix are the (1, 1/2, 0) and (1/3, 1/6, 0) [9].

$$s_{k,ij} = \begin{cases} 1 & x_{ik} > x_{ik} \\ 0 & x_{ik} = x_{ik} \\ -1 & x_{ik} < x_{ik} \end{cases} \quad k = 1, 2 \cdots n$$

The sum of the comparisons is set as a comparison matrix A. $a_{ij}$ is the sum of comparisons for chemicals $i$ and $j$. The comparison matrix is a special matrix; it has all zeros as its diagonal elements because a comparison of a variable to itself always results in zero. The element in any row i and j is the negative of the element in rows j and i, respectively [6].

$$A = \begin{bmatrix} a_{11} & a_{12} & \ldots & a_{1m} \\ a_{12} & \cdots & \cdots & \vdots \\ \vdots & \cdots & \cdots & \vdots \\ a_{ml} & \cdots & \cdots & a_{mm} \end{bmatrix}$$

$$a_{ij} = \sum_{k=1}^{n} s_{k,ij}$$

After evaluating all the matrix elements, the sum of the row forms the Copeland score for each chemical:

$$CS_i = \sum_{j=1}^{m} a_{ij}$$

Consequently, the chemicals are ranked according to the Copeland score values. However, the method does not guarantee that the chemicals would have different scores; some objects may have the same rank. The Copeland scores can be easily calculated using a computer.

### Comprehensive multi-index comparison method

Again, assume that the number of chemicals is m, and each chemical has n indicators. From there, we can create a matrix X:

$$X = \begin{bmatrix} x_{11} & x_{12} & \cdots & x_{1n} \\ x_{21} & \cdots & \cdots & \vdots \\ \vdots & \cdots & \cdots & \vdots \\ x_{m1} & \cdots & \cdots & x_{mn} \end{bmatrix}$$

The minimum and maximum values of each column of matrix X is normalized to [0 1], as follows:

$$Z = \begin{bmatrix} z_{11} & z_{12} & \cdots & z_{1n} \\ z_{21} & \cdots & \cdots & \vdots \\ \vdots & \cdots & \cdots & \vdots \\ z_{m1} & \cdots & \cdots & z_{mn} \end{bmatrix}$$

$$z_{ij} = \frac{x_{ij} - \min x_j}{\max x_j - \min x_j}$$

We then create the transposed matrix $Z^T$:

$$Z^T = \begin{bmatrix} z_{11} & z_{21} & \cdots & z_{m1} \\ z_{12} & \cdots & \cdots & \vdots \\ \vdots & \cdots & \cdots & \vdots \\ z_{1n} & \cdots & \cdots & z_{mn} \end{bmatrix}$$

We then determine the weights of each indicator. In our case study, the indicators were given equal weight by assigning a value of 1.

$$W = \begin{pmatrix} w_1 & w_2 & \cdots & w_n \end{pmatrix}^T$$

We then create the comparison matrix A.

$$A = \begin{bmatrix} a_{11} & a_{21} & \cdots & a_{m1} \\ a_{12} & \cdots & \cdots & \vdots \\ \vdots & \cdots & \cdots & \vdots \\ a_{1n} & \cdots & \cdots & a_{mn} \end{bmatrix}$$

$$= \begin{bmatrix} w_1 a_{11} & w_1 a_{21} & \cdots & w_1 a_{m1} \\ w_2 a_{12} & \cdots & \cdots & \vdots \\ \vdots & \cdots & \cdots & \vdots \\ w_n a_{1n} & \cdots & \cdots & w_n a_{mn} \end{bmatrix}$$

Ideal point:

$$P^* = \max_j \{ a_{ij} \quad i = 1, 2, \cdots, m \}$$

$$= \begin{pmatrix} p_1^* & p_2^* & \cdots & p_n^* \end{pmatrix}^T$$

We can then calculate the di and Ti values, as follows:

$$d_i = \sqrt{\sum_{j=1}^{n} \left( a_{ij} - p_j^* \right)^2} \quad i = 1, 2, \cdots, m$$

$$T_i = 1 - \frac{\sum_{j=1}^{n} a_{ij} p_j^*}{\sum_{j=1}^{n} \left( p_j^* \right)^2} \quad i = 1, 2, \cdots, m$$

Consequently, the chemicals are ranked according to Ti. The lower the Ti value, the more accurate the ranking. When the $T_i$ are equal in size, then we can use $d_i$ to rank, the lower the more accurate the ranking [7].

### Case Study

A total of 79 chemicals were selected for screening; all these chemicals were requested by the 2008 Commonwealth and

Environmental Protection Project of the Ministry of Environmental Protection of the People's Republic of China (MEP): "Bioconcentration of Toxic Hazardous Substances in body adipose tissues and risk analysis on human health.. Sample collection and detection were performed in previous. The human exposure concentration and detection frequency data were reported in two articles [11,12]. Human exposure concentration experiments show that eight chemicals were not detected in human adipose tissue samples. Therefore, these eight chemicals are not potential hazards. Thus, at first screening, these eight chemicals can be ignored. The remaining chemicals were then used in the subsequent experiments.

Table 1 presents the toxicological and exposure endpoints used in this article. Four human health effects, two environmental effects, octanol–water partition, bioaccumulation, human exposure concentration, and frequency of detection are included. Human health effects and environmental effects are important indicators that reflect the hazards or toxicities of chemicals; these effects are indispensable in risk assessment. Octanol–water partition is also an important parameter. Octanol is a long-chain alcohol that can reflect the transmission and distribution capacity of organisms.

The experimental data were obtained from the Hazardous Substances Data Bank, Pesticide Properties Database, U.S EPA Aggregated Computational Toxicology Resource, and U.S EPA ECOTOX Database whenever possible. Structure-activity relationships (SARs) and quantitative structure-activity relationships (QSARs) such as EPI Suite™ v3.20 and ECOSAR™ v 1.00 were used to estimate missing data. This estimation depends on the availability of reliable SARs or QSARs. If an SAR or QSARs was not available, the missing data were decided through expert judgment. The human exposure concentration and detection frequency data were quoted from previously published two articles.

Toxicological values such as rodent oral LD50, fish LC50, and bird LD50 have negative correlations with the hazards of human health. To simplify the comparison calculations, we used the negative number of the values. Moreover, the octanol–water partition (Kow) and bioconcentration factor (BCF) values were sometimes too large to calculate. Thus, we used log Kow and log BCF. The properties of each chronic effect were divided into three classes: recognized, suspected, and not likely. These parameters were qualitative and could not be calculated using formulas. For this reason, the parameters were

assigned the quantitative values of 5, 3, and 0, respectively. At this step, the original dataset was established.

To apply the Copeland method and the multi-index comparison method, programs were written using the MATLAB mathematical software.

After the original data set was constructed, the Copeland and multi-index comparison methods were used to determine the ranking order. The results are listed in Table 2.

We calculated the correlation coefficient of these two methods from the results. The correlation coefficient can explain the similarities between the ranking orders. If the correlation coefficient is close to 1, the two ranking orders are highly similar. However, the correlation coefficients of the Copeland and multi-index comparison methods are 0.8424, the two ranking orders are not much similar. Why is there a big difference between the Copeland method and the multi-index comparison method? The principles of these two methods can provide some explanations. The Copeland method only focuses on the numerical magnitude between two values. Therefore, the results of 1<100 and 1<1000 are equivalent. On the other hand, in the multi-index comparison method, the numerical magnitude and numerical distribution significantly affect the result. Calculation of the ideal point is a key step in this method. The ideal point is directly related to the maximum values of each indicator. When a value is considerably bigger than the others, a lower T value may be obtained but may have a smaller effect on the Copeland method.

Value pretreatment was performed to improve the results of these two methods. Each indicator to the oral LD50, fish LC50, and bird LD50 toxicity terms can range from zero to five. A cutoff value set for each indicator so that the hazard value for very high or low toxicities would not exceed five or be below zero.

The hazard value for the acute oral toxicity (HVOR) was based on the oral LD50 and was calculated using a continuous, linear function, with the cutoff values at 5,000 and 5 mg/kg:

$$HV_{OR} = 6.165 - 1.667\log(oral\ LD50) \text{ for } 5 \text{ mg/kg} < \text{oral LD50} \leq 5,000 \text{ mg/kg}$$

$$HV_{OR} = 0 \text{ for oral LD50} > 5,000 \text{ mg/kg}$$

$$HV_{OR} = 5 \text{ for oral LD50} \leq 5 \text{ mg/kg}$$

| Type | Endpoint | Definition |
|---|---|---|
| Human health effects | | |
| Acute | Rodent oral LD50 | The mass of the substance administered per unit mass of the test subject that will kill half of the test subjects within 14 days when orally administered as a single dose. |
| Chronic | Carcinogenicity | |
| Chronic | Reproductive and Developmental Toxicity | Based on supporting evidence. |
| Chronic | Endocrine Toxicity | |
| Environmental effects | | |
| Aquatic, acute | Fish LC50 | The concentration of a substance in water that will cause 50% of fish deaths in the 96 h test. |
| Terrestrial, acute | Bird LD50 | The mass of the substance administered per unit mass of the test subject that will kill half of the test subjects within 14 days when orally administered as a single dose. |
| Exposure potential | | |
| Partition | Octanol–water Partition (Kow) | The ratio of the distribution of a substance between octanol and water. |
| Bioaccumulation | Bioconcentration factor (BCF) | The ratio of the concentration of a chemical in a biological tissue to that in the water surrounding that tissue. |
| | Human exposure concentration | The concentration of a chemical in human adipose tissues. |
| | Frequency of detection | Frequency of detection in human adipose tissue samples. |

Table 1: Toxicological and exposure endpoints.

| NO. | Chemical | Copeland method | | Multi-index comparison | |
|---|---|---|---|---|---|
| | | Original data | Pretreatment data | Original data | Pretreatment data |
| 1 | 2,4'- DDD | 19 | 19 | 25 | 15 |
| 2 | 2,4'- DDE | 13 | 13 | 12 | 12 |
| 3 | 2,4'- DDT | 1 | 1 | 1 | 1 |
| 4 | 4,4'- DDD | 10 | 10 | 9 | 11 |
| 5 | 4,4'- DDE | 3 | 3 | 6 | 9 |
| 6 | 4,4'- DDT | 2 | 2 | 3 | 2 |
| 7 | alpha-Hexachlorocyclohexane | 35 | 35 | 26 | 28 |
| 8 | beta-Hexachlorocyclohexane | 8 | 9 | 2 | 6 |
| 9 | gamma-Hexachlorocyclohexane | 24 | 23 | 32 | 21 |
| 10 | Lindane | 44 | 44 | 33 | 45 |
| 11 | Hexachlorobenzene | 7 | 6 | 4 | 3 |
| 12 | Mirex | 18 | 18 | 14 | 18 |
| 13 | Aldrin | 9 | 8 | 10 | 8 |
| 14 | Endrin | 5 | 5 | 15 | 5 |
| 15 | Heptachlor | 12 | 11 | 8 | 10 |
| 16 | Chlordane | 36 | 37 | 35 | 47 |
| 17 | Chlordane | 26 | 26 | 31 | 41 |
| 18 | Acenaphthylene | 63 | 63 | 67 | 64 |
| 19 | Acenaphthene | 58 | 58 | 49 | 58 |
| 20 | Anthracene | 49 | 46 | 46 | 48 |
| 21 | 1,2-Benzanthracene | 25 | 22 | 29 | 23 |
| 22 | Benzo[a]pyrene | 6 | 7 | 7 | 7 |
| 23 | Benzo[b]fluorathene | 38 | 38 | 30 | 31 |
| 24 | Benzo(ghi)perylene | 29 | 27 | 39 | 40 |
| 25 | Benzo[k]fluorathene | 11 | 12 | 17 | 13 |
| 26 | Dibenz[a,h]anthracene | 23 | 25 | 21 | 25 |
| 27 | Fluoranthene | 37 | 36 | 59 | 37 |
| 28 | Fluorene | 53 | 53 | 41 | 57 |
| 29 | Indeno[1,2,3-cd]Pyrene | 16 | 15 | 18 | 14 |
| 30 | naphthalene | 32 | 32 | 11 | 24 |
| 31 | Phenanthrene | 22 | 24 | 23 | 32 |
| 32 | Pyrene | 40 | 41 | 42 | 46 |
| 33 | Butyl benzyl phthalate | 52 | 52 | 28 | 43 |
| 34 | | 59 | 59 | 38 | 59 |
| 35 | Dibutyl phthalate | 30 | 30 | 27 | 30 |
| 36 | Dicofol | 14 | 14 | 13 | 17 |
| 37 | Methamidophos | 51 | 51 | 65 | 56 |
| 38 | Chlordimeform | 56 | 56 | 34 | 50 |
| 39 | Acetamiprid | 67 | 67 | 70 | 67 |
| 40 | Alachlor | 43 | 42 | 22 | 35 |
| 41 | Amitraz | 45 | 45 | 20 | 39 |
| 42 | Buprofezin | 60 | 60 | 58 | 60 |
| 43 | Machette | 61 | 61 | 61 | 62 |
| 44 | Carbofuran | 57 | 57 | 63 | 55 |
| 45 | Chlorothalonil | 42 | 43 | 24 | 29 |
| 46 | Clorpyrifos | 17 | 16 | 36 | 16 |
| 47 | Clomazone | 71 | 71 | 68 | 71 |
| 48 | Cyfluthrin | 46 | 48 | 54 | 44 |
| 49 | Cypermethrin | 28 | 28 | 45 | 22 |
| 50 | Deltamethrin | 48 | 49 | 56 | 51 |
| 51 | Diazinon | 55 | 54 | 50 | 52 |
| 52 | Thiosulfan I | 34 | 34 | 48 | 38 |
| 53 | Thiosulfan II | 31 | 31 | 47 | 36 |
| 54 | Mocap | 21 | 20 | 37 | 20 |
| 55 | Phenvalerate | 41 | 39 | 57 | 34 |
| 56 | Esfenvalerate | 15 | 17 | 40 | 26 |
| 57 | Hexythiazox | 64 | 65 | 44 | 63 |
| 58 | Isoproturon | 68 | 68 | 64 | 68 |
| 59 | Cyhalothrin | 39 | 40 | 53 | 42 |

| 60 | Metolachlor | 62 | 62 | 52 | 61 |
| 61 | Nitrfen | 33 | 33 | 16 | 27 |
| 62 | O,O-Dimethyl-S-methylcarbamoylmethyl phosphorothioate | 65 | 64 | 66 | 65 |
| 63 | Oxyfluorfen | 47 | 47 | 43 | 49 |
| 64 | Parathion-methyl | 50 | 50 | 51 | 53 |
| 65 | Pirimicarb | 66 | 66 | 71 | 66 |
| 66 | Prometryn | 70 | 70 | 62 | 69 |
| 67 | Pyridaben | 54 | 55 | 60 | 54 |
| 68 | Triazophos | 27 | 29 | 55 | 33 |
| 69 | Tricyclazole | 69 | 69 | 69 | 70 |
| 70 | Trifluralin | 20 | 21 | 19 | 19 |
| 71 | Chlorobiphenyl | 4 | 4 | 5 | 4 |

**Table 2:** Ranking order of the Copeland method and the multi-index comparison method after data pretreatment.

| | R |
| --- | --- |
| Copeland method on original data vs. pretreatment data | 0.9989 |
| multi-index comparison of original data vs. multi-index comparison of pretreatment data | 0.8784 |
| Copeland method on original data vs. multi-index comparison of original data | 0.8424 |
| Copeland method on original data vs. multi-index comparison of pretreatment data | 0.9718 |
| Copeland method on pretreatment data vs. multi-index comparison of original data | 0.8419 |
| Copeland method on pretreatment data vs. multi-index comparison of pretreatment data | 0.9736 |

**Table 3:** Correlation coefficients of the Copeland method and the multi-index comparison method.

The hazard value for the acute aquatic toxicity ($HV_{FA}$) was based on the acute fish LC50 and was calculated using a continuous, linear function, with the cutoff values at 100 and 0.01 mg/l:

$HV_{FA} = 2.5 - 1.25\log(LC50)$ for 0.01 mg/l < LC50 ≤ 100 mg/l

$HV_{FA} = 0$ for LC50 > 100 mg/l

$HV_{FA} = 5$ for LC50 ≤ 0.01 mg/l

The hazard value for the acute bird toxicity ($HV_{BA}$) was based on the acute bird LD50 and was calculated using a continuous, linear function, with the cutoff values at 5,000 and 5 mg/kg:

$HV_{BA} = 6.165 - 1.667\log(bird\ LD50)$ for 5 mg/kg < bird LD50 ≤ 5,000 mg/kg

$HV_{BA} = 0$ for bird LD50 > 5,000 mg/kg

$HV_{BA} = 5$ for bird LD50 ≤ 5 mg/kg

Kow and BCF can range from one to five. Cutoff values were also set for the indicators so that the hazard value would not exceed five or be less than one. The Kow hazard value ($HV_{Kow}$) was calculated using

$HV_{Kow} = 0.6667\log Kow + 0.3333$ for 1 < log Kow ≤ 7

$HV_{Kow} = 1$ for log Kow ≤ 1 and for LD50 > 7

The BCF hazard value ($HV_{BCF}$) was calculated using

$HV_{BCF} = 1.3333\log BCF - 0.3333$ for 1 < log Kow ≤ 4

$HV_{BCF} = 1$ for log Kow ≤ 1 and for LD50 > 4

The hazard value of the frequency of detection ($HV_{FD}$) was also calculated using continuous, linear functions, with the cutoff values at 0.001 and 1:

$HV_{FD} = \log(FD) + 3$ for 0.001 < FD < 1 and for FD ≤ 0.001

On the other hand, the hazard value of human exposure concentration ($HV_{HEC}$) was calculated using continuous, linear functions without cutoff values.

$HV_{HEC} = \log(HEC)$

The carcinogenicity, reproductive and developmental toxicity, and endocrine toxicity data did not need pretreatment; the values used in this step are the same as those in the original data.

After data pretreatment, a new dataset was created. The Copeland method and the multi-index comparison method were then used.

Table 2 shows the ranking order of the Copeland method and the multi-index comparison method after data pretreatment, as well as the ranking order of the Copeland method using original data. The correlation coefficients were calculated and are listed in Table 3. The correlation coefficient between the original data and pretreatment data was much higher when the Copeland method was used, indicating that the two results have a high degree of similarity (Figure 1). The correlation coefficient between the original data and pretreatment data in multi-index comparison method is 0.8784, it seem that the pretreatment of the data have much influence on the ranking result. Besides, the correlation coefficient between the multi-index comparison with original data and Copeland method are all about 0.84. While the correlation coefficient between the multi-index comparison with pretreatment data and Copeland method are all about 0.97, which also indicated that the pretreatment of the data have significant influence on the ranking result.

Although the correlation coefficients between the Copeland method and the multi-index comparison method reached 0.97, the ranking orders were not very similar, particularly the middle part of the rank orders (Figure 2). However, the top and last parts of the rank order are highly similar. This similarity is sufficient in identifying the hazardous chemicals.

As a result, the Copeland method seems much more convenient, it can give a good ranking result without data pretreatment. The multi-index comparison method is not very good at deal with original data, but it also can give a good ranking result with data pretreatment.

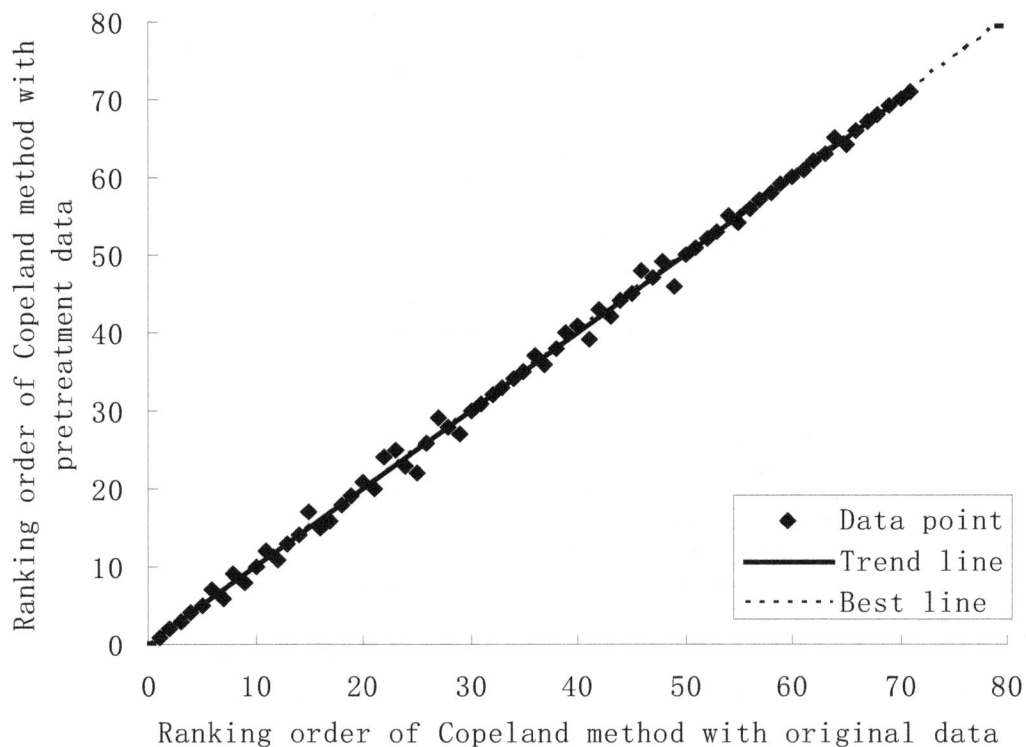

**Figure 1:** Relationship between the Copeland method results using original data and pretreatment data.

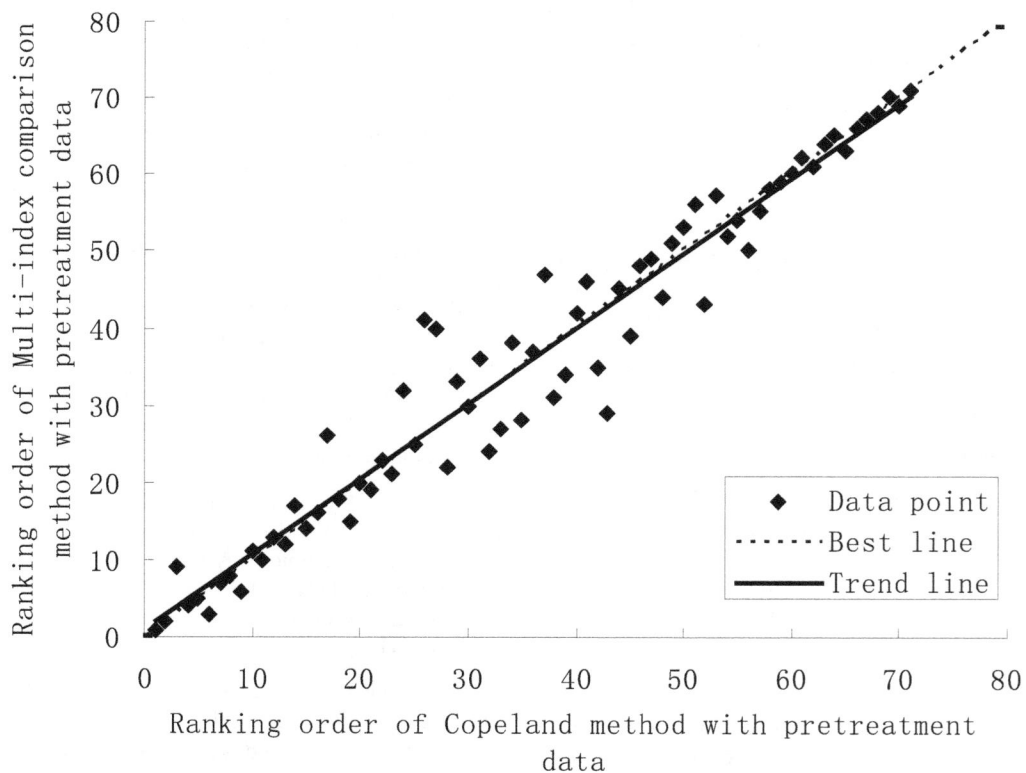

**Figure 2:** Relationship between the Copeland method and the multi-index comparison method results using pretreatment data.

| Rank | Copeland method | | Multi-index comparison | |
|---|---|---|---|---|
| | Original data | Pretreatment data | Original data | Pretreatment data |
| 1 | 2,4'- DDT | 2,4'- DDT | 2,4'- DDT | 2,4'- DDT |
| 2 | 4,4'- DDT | 4,4'- DDT | beta-Hexachlorocyclohexane | 4,4'- DDT |
| 3 | 4,4'- DDE | 4,4'- DDE | 4,4'- DDT | Hexachlorobenzene |
| 4 | Chlorobiphenyl | Chlorobiphenyl | Hexachlorobenzene | Chlorobiphenyl |
| 5 | Endrin | Endrin | Chlorobiphenyl | Endrin |
| 6 | Benzo[a]pyrene | Hexachlorobenzene | 4,4'- DDE | beta-Hexachlorocyclohexane |
| 7 | Hexachlorobenzene | Benzo[a]pyrene | Benzo[a]pyrene | Benzo[a]pyrene |
| 8 | beta-Hexachlorocyclohexane | Aldrin | Heptachlor | Aldrin |
| 9 | Aldrin | beta-Hexachlorocyclohexane | 4,4'- DDD | 4,4'- DDE |
| 10 | 4,4'- DDD | 4,4'- DDD | Aldrin | Heptachlor |
| 11 | Benzo[k]fluorathene | Heptachlor | naphthalene | 4,4'- DDD |
| 12 | Heptachlor | Benzo[k]fluorathene | 2,4'- DDE | 2,4'- DDE |
| 13 | 2,4'- DDE | 2,4'- DDE | Dicofol | Benzo[k]fluorathene |
| 14 | Dicofol | Dicofol | Mirex | Indeno[1,2,3-cd]Pyrene |
| 15 | Esfenvalerate | Indeno[1,2,3-cd]Pyrene | Endrin | 2,4'- DDD |
| 16 | Indeno[1,2,3-cd]Pyrene | Clorpyrifos | Nitrfen | Clorpyrifos |
| 17 | Clorpyrifos | Esfenvalerate | Benzo[k]fluorathene | Dicofol |
| 18 | Mirex | Mirex | Indeno[1,2,3-cd]Pyrene | Mirex |
| 19 | 2,4'- DDD | 2,4'- DDD | Trifluralin | Trifluralin |
| 20 | Trifluralin | Mocap | Amitraz | Mocap |

Table 4: Top 20 Hazardous Chemicals.

## Discussion

The top 20 ranked chemicals are presented in Table 4. In the top 20, the three ranking results share 18 common chemicals. These chemicals were then classified into four general groups: six kinds of dichlorodiphenyltrichloroethane (DDTs), seven kinds of other POPs, three kinds of polycyclic aromatic hydrocarbons, and two kinds of pesticides. These substances should be regarded as chemicals of concern for human health, and appropriate management should be taken. From the result, the ranking of 2,4'-DDT, 4,4'-DDT, 4,4'-DDE, chlorobiphenyl, endrin, benzo[a]pyrene, hexachloro cyclohexane, and aldrin were relatively high because of their chronic effects and high human exposure concentrations. Some of these pesticides are still being used in some regions of China. Thus, the use of these products may increase the concern on the potential health hazards to humans.

The results of this study show that the Copeland method is a simple and effective ranking and screening method. The ranking order of the Copeland method using original data can rationally explain the hazard relationships between 71 chemicals. In our study, the indicators of human exposure concentration and detection frequency are more significant for ranking results. The Copeland method can be easily performed using software, thus making the ranking, screening, and assessing of hazardous chemicals more convenient. If necessary, expert judgment can be used to add weight to the indicators.

### Acknowledgments

Financial support from 2013 Commonweal and Environmental Protection Project of Ministry of Environmental Protection of the People's Republic of China (No.: 2013467028) and the National High Technology Research and Development Program of China (863 Program, No.: 2013AA06A308).

## References

1. Halfon E, Reggiani M (1986) On Ranking Chemicals for Environmental Hazard. Environ SciTechnol 20: 1173–1179.

2. Swanson MB, Davis GA, Kincaid LE, Terry W. Schultz, John E. Bartmess et al. (1997) A screening method for ranking and scoring chemicals by potential human health and environmental impacts. Environ Toxicol Chem 16: 372–383.

3. Dunn AM (2009) A Relative Risk Ranking of Selected Substances on Canada's National Pollutant Release Inventory. Hum Ecol Risk Assess 15: 579–603.

4. Hansen BG, van Haelst AG, van Leeuwen K, Peter van der Zandt (1999) Priority setting for existing chemicals: European Union risk ranking method. Environ ToxicolChem 18: 772–779.

5. Snyder EM, Snyder SA, Giesy JP, Blonde SA, Hurlburt GK, et al. (2000) SCRAM: A scoring and ranking system for persistent, bioaccumulative, and toxic substances for the North American Great Lakes. Environ SciPollut R 7: 52–61.

6. Al-Sharrah G (2010) Ranking using the Copeland score: a comparison with the Hasse diagram. J ChemInf Model 50: 785–791.

7. Jing Ren, YijieXiong (2010) An optimised method of weighting combination in multi-index comprehensive evaluation. International Journal of Applied Decision Sciences 3: 34–52.

8. Copeland A (1951) A "reasonable" social welfare function. Mimeographed notes from a Seminar on Applications of Mathematics to the Social Sciences; University of Michigan: Ann Arbor, MI.

9. Saari DG, Merlin VR (1996) The Copeland method: I. relationships and the dictionary. Journal of Economic Theory 8: 51–76.

10. Merlin VR, Saari DG (1997) Copeland method: II. manipulation, monotonicity, and paradoxes. Journal of Economic Theory 72: 148–172.

11. Wang N, Kong D, Cai D, Shi L, Cao Y, et al. (2010) Levels of polychlorinated biphenyls in human adipose tissue samples from southeast China. Environ SciTechnol 44: 4334-4340.

12. Wang N, Shi L, Kong D, Cai D, Cao Y, et al. (2011) Accumulation levels and characteristics of some pesticides in human adipose tissue samples from Southeast China. Chemosphere 84: 964-971.

# 20

# Synthesis and Antimicrobial Study of Pyrimidinone Substituted 4(3H)-Quinazolinone Derivatives

**Natvar A Sojitra[1], Rajesh K Patel[3], Ritu B Dixit[2] and Bharat C Dixit[1]\***

[1]*Chemistry Department, V. P. & R. P. T. P. Science College, Affiliated to Sardar Patel University, Gujarat, India*
[2]*Ashok & Rita Patel Institute of Integrated Studies and Research in Biotechnology and Allied Sciences, Gujarat, India*
[3]*Department of Life Sciences, Hemchandracharya North Gujarat University, Gujarat, India*

## Abstract

A series of structurally diverse and newly designed pyrimidinone substituted 4(3H)-Quinazolinone derivatives **6a–6j** were synthesized in a simple and facile manner under both conventional and microwave heating conditions. Entitled compounds **6a–6j** were synthesized using N-acylanthranilic acid derivatives **1a–1j** and hydrazinylbenzenesulfonamide**2** as a key starting materials, which under goes hierarchy of the reactions *via* different intermediate stapes; quinazolinone derivatives **3a–3j** and hydrazonoquinazolinone derivatives **5a–5j**. All the synthesized compounds are in good amount of yield. The structure of entitle compounds have been evaluated on the basis of various spectroscopic techniques and analytical methods as well as, all the synthesized compounds were subjected to *in vitro* antibacterial activities. Some of the compounds displayed moderate to good *in vitro* antimicrobial activity by broth micro dilution method against pathogenic bacteria (*S. aureus*,*B. subtilis*, *B. megaterium*, *E. coli*, *P. vulgaris*, *P. aeruginosa*) species.

**Keywords:** Quinazolinone; Pyrimidinone; Antimicrobial activity; Microwave heating

## Introduction

Heterocyclic scaffold has been under discovery due to their considerable therapeutic and medicinal importance. One of such heterocycles 2-pyrimidinone, while not naturally occurring, are receiving increased attention because of their wide-ranging biological activity like, an efficient inhibitor of DNA synthesis in *E. coli,* inhibitors of the enzyme *cytidinedeaminase*, and as metaphase inhibitors. Such pyrimidinone derivatives and the pharmaceutical acceptable salts thereof having remarkable antagonistic action against angiotensin II receptor, thereby, being useful in treating cardiovascular disease caused by binding angiotensin II to its receptor. Antitumor activity of pyrimidinones derivatives have shown statistically significant synergism with cyclophosphamide (CY) against antitumor / P388 leukemia etc. [1-7]. They have also shown anticancer [8,9], potent TF-FVIIa inhibitors [10] and inhibition of C5-cytosine DNA methyltransferases [11] activities.

In addition to that, quinazolinone and its derivatives comprise an important class of heterocyclic molecules due to their broad spectrum application in medicinal and pharmaceutical chemistry such as analgesic [12], antimicrobial and antitubercular [13,14], antitumor [15], anticancer [16], anti-inflammatory [17], anticonvulsant [18], antimalarial and antihistamine [19].

Moreover, there are many reports regarding synthesis of pyrimidinone derivatives [20,21] but no reports are available in which pyrimidinone is substituted at 3rd position of quinazolinone through hydrazono benzenesulfonamide linkage. Looking to the aforementioned literature and our ongoing research in the field of quinazolinone derivatization [22,23], we reports here synthesis of various pyrimidinone substituted quinazolinone derivatives linked through hydrazono benzenesulfonamide, of that some of the compounds showed good to moderate antimicrobial activity.

## Experimental Section

### Material and measurements

All the chemicals were purchased from Spectro chem. Ltd. (Mumbai, India) and were used without further purification. Solvents employed were distilled, purified and dried by standard procedures prior to use [24]. Melting points of the synthesized compounds were determined in open capillary tube method and are uncorrected. Reactions were monitored by thin-layer chromatography (TLC on alluminium plates coated with silica gel 60 $F_{254}$, 0.25 mm thickness purchased from E. Merck Ltd., Mumbai-India). The mobile phase was chloroform and methanol (9:1), and detection of the components was done under UV light or explore in Iodine chamber. Infrared (IR) spectra were recorded as potassium bromide pellets using a Shimadzu 8501 Fourier transform infrared (FTIR) Spectrophotometer. [1]H and [13]C NMR spectra were recorded on a Bruker AVANCE II 400-MHz NMR spectrometer (Bruker Corporation, Billerica, MA, USA), with chemical shift in ppm downfield from TMS as an internal reference and DMSO-$d_6$ used as solvent. Carbon, hydrogen and nitrogen elemental analysis were estimated by PerkinElmer 2400-II CHN elemental analyzer, USA. The electro-spray ionization mass spectra in positive mode were recorded on a Shimadzu LC-MS 2010 eV mass spectrophotometer using acetonitrile. All the microwave assisted reactions were carried out at atmospheric pressure using a multimode microwave reactor (Scientific Microwave Synthesis System, Model: Cata-R, Catalyst™ Systems, Pune-India) with attachment of glass vessel prolonged by a reflux condenser with constant stirring, whereby microwaves were generated by magnetron at a frequency of 2450 MHz having an output energy range of 140 to 700 Watts, and the temperature was monitored with an external flexible probe.

**\*Corresponding author:** Bharat C Dixit, Chemistry Department, V. P. & R. P. T. P. Science College, Affiliated to Sardar Patel University, VallabhVidyangar– 388 120, Anand, Gujarat, India, E-mail: dixits20002003@yahoo.co.in

## General procedure for the synthesis of 4-amino-N-(2-methyl/aryl-4-oxoquinazolin-3(4H)-yl)benzenesulfonamide derivatives(3a-3j)

Compounds 3a–3j were prepared according to the reported method [22,23]. The possible synthetic route is given in Scheme 1.

-3-yl)benzenesulfonamide (5b) m.p. 145–147°C, ESI MS (m/z): 520.1,

522.3 [M]⁺.

4-[N'-(1-Acetyl-2-oxo-propylidene)hydrazino]-N-(6,8-dibromo-2-methyl-4-oxo-4H-quinazolin-3-yl)-benzenesulfonamide (5c) m.p. 172–174°C, ESI MS (m/z): 597.2, 599.1, 601.3 [M]⁺.

4-[N'-(1-Acetyl-2-oxopropylidene)hydrazino]-N-(2-methyl-6-nitro-4-oxo-4H-quinazolin-3-yl)benzenesulfonamide (5d) m.p. 189–190°C, ESI MS (m/z): 549.4 [M+H]⁺.

4-Amino-N-(2-methyl-4-oxoquinazolin-3(4H)-yl)benzenesulfonamide (3a)mp 217-218°C. ESI MS (m/z): 330.8 [M+H]⁺.

4-Amino-N-(6-bromo-2-methyl-4-oxoquinazolin-3(4H)-yl)

benzenesulfonamide (3b)mp 187-189°C. ESI MS (m/z): 409.2, 411.3 [M]⁺.

4-Amino-N-(6,8-dibromo-2-methyl-4-oxoquinazolin-3(4H)-yl)benzenesulfonamide (3c) mp 212-214°C. ESI MS (m/z): 488.4, 490.4, 492.4 [M]⁺.

4-Amino-N-(6-nitro-4-oxo-2-phenylquinazolin-3(4H)-yl)benzenesulfonamide (3d)mp 206-209°C. ESI MS (m/z): 438.2 [M+H]⁺.

4-Amino-N-(4-oxo-2-phenylquinazolin-3(4H)-yl)benzenesulfonamide (3e)mp 166-168°C. ESI MS (m/z): 393.4 [M+H]⁺.

4-Amino-N-(6-bromo-4-oxo-2-phenylquinazolin-3(4H)-yl)benzenesulfonamide (3f)mp 214-217°C. ESI MS (m/z): 471.3, 473.3 [M]⁺.

4-Amino-N-(6,8-dibromo-4-oxo-2-phenylquinazolin-3(4H)-yl)benzenesulfonamide (3g)mp 219-221°C. ESI MS (m/z): 548.3, 550.3, 552.3 [M]⁺.

4-Amino-N-[2-(4-chlorophenyl)-4-oxoquinazolin-3(4H)-yl]benzenesulfonamide (3h)mp 175-179°C. ESI MS (m/z): 426.9, 428.9 [M]⁺.

4-Amino-N-[6-bromo-2-(4-chlorophenyl)-4-oxoquinazolin-

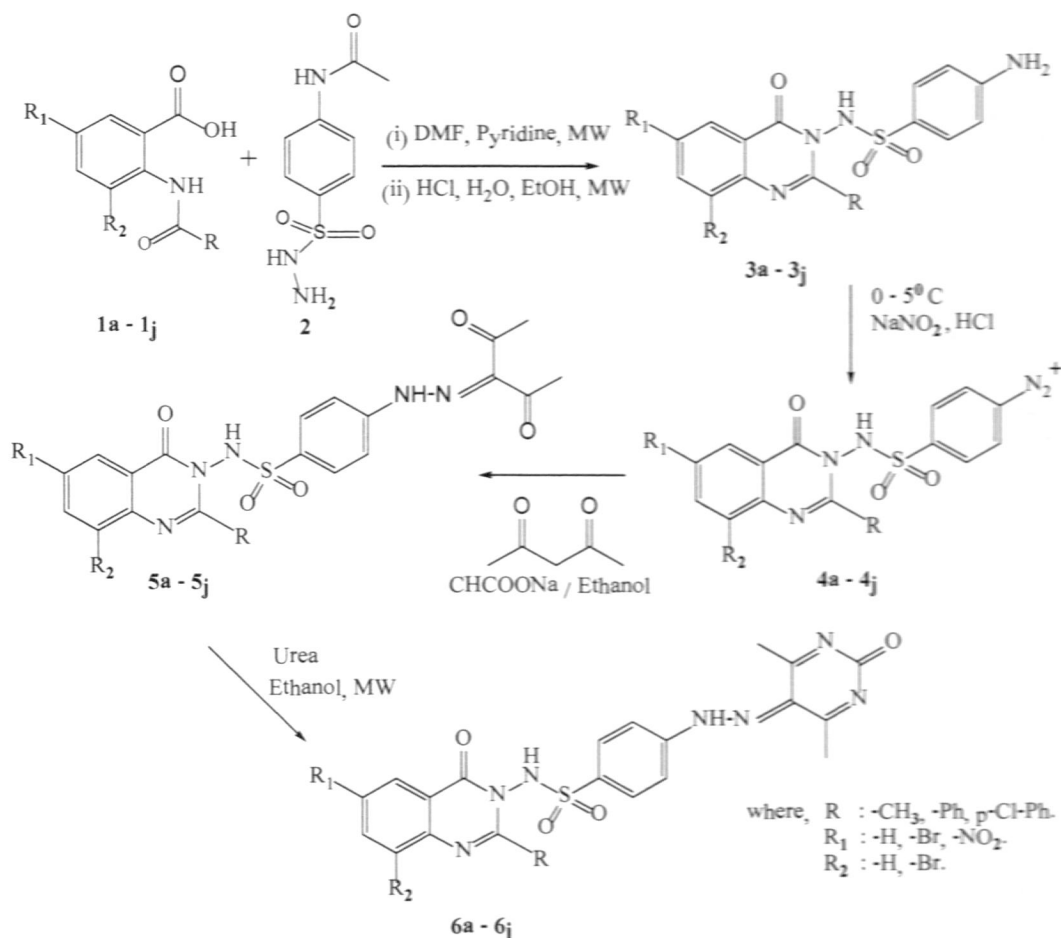

**Scheme 1:** Proposed synthetic route for the preparation of pyrimidinone substituted 4(3H)-quinazolinone derivatives compound (6a-6j).

3(4*H*)-yl]benzene-sulfonamide (3i)mp 186-187°C. ESI MS (m/z): 503.8, 505.8, 507.8 [M]⁺.

### 4-Amino-*N*-[6,8-dibromo-2-(4-chlorophenyl)-4-oxoquinazolin-3(4*H*)-yl]benzene

sulfonamide (3j)mp 183-184°C. ESI MS (m/z): 582.5, 584.5, 586.5, 588.5 [M]⁺.

## General procedure for the diazotization of 4-amino-*N*-(2-methyl/aryl-4-oxoquinazolin-3(4H)-yl)benzenesulfonamide derivatives(4a-4j)

The diazotized derivatives **4a-4j** were prepared by following the reported method [23,25]. Schematic diagram to produce diazotized compounds **4a-4j** are shown in Scheme 1. The reaction mass was cooled at 0-5°C for 1 hours and was directly used for the next step.

## General procedure for the synthesis of hydrazono derivatives of compounds 4a-4j using acetyl acetone as an active methylene compound (5a-5j)

Hydrazonoquinazolinone compounds **5a-5j** were prepared by following the reported method [23,26]. The synthetic route is shown in Scheme 1.

**4-[N'-(1-Acetyl-2-oxo-propylidene)hydrazino]-N-(2-methyl-4-oxo-4H-quinazolin-3-yl)benzenesulfonamide (5a)** m.p. 167–168°C, ESI MS (m/z): 442.2 [M+H]⁺.

**4-[N'-(1-Acetyl-2-oxo-propylidene)hydrazino]-N-(6-bromo-2-methyl-4-oxo-4H-quinazolin-[N'-(1-Acetyl-2-oxopropylidene)hydrazino]-N-(4-oxo-2-phenyl-4H-quinazolin-3-yl)benzenesulfonamide (5e)** m.p. 164–166°C, ESI MS (m/z): 504.2 [M+H]⁺.

**4-[N'-(1-Acetyl-2-oxopropylidene)hydrazino]-N-(6-bromo-4-oxo-2-phenyl-4H-quinazolin-3-yl)benzenesulfonamide (5f)** m.p. 183–184°C, ESI MS (m/z): 582.3, 584.4 [M]⁺.

**4-[N'-(1-Acetyl-2-oxo-propylidene)hydrazino]-N-(6,8-dibromo-4-oxo-2-phenyl-4H-quinazolin-3-yl)benzenesulfonamide (5g)** m.p. 193–195°C, ESI MS (m/z): 660.0, 662.2, 663.9 [M]⁺.

**4-[N'-(1-Acetyl-2-oxopropylidene)hydrazino]-N-[6-bromo-2-(4-chloro-phenyl)-4-oxo-4H-quinazolin-3-yl]benzenesulfonamide (5h)** m.p. 178–180°C, ESI MS (m/z): 538.0, 539.8 [M]⁺.

**4-[N'-(1-Acetyl-2-oxopropylidene)hydrazino]-N-[6-bromo-2-(4-chlorophenyl)-4-oxo-4H-quinazolin-3-yl]benzenesulfonamide (5i)** m.p. 203–204°C, ESI MS (m/z): 616.2, 617.9, 620.0 [M]⁺.

**4-[N'-(1-Acetyl-2-oxopropylidene)hydrazino]-N-[6,8-dibromo-2-(4-chlorophenyl)-4-oxo-4H-quinazolin-3-yl]benzenesulfonamide (5j)** m.p. 214–215°C, ESI MS (m/z): 693.8, 695.7, 697.7, 699.9 [M]⁺.

## Synthesis of pyrimidinone substituted quinazolinone derivatives 6a-6j

It was prepared using the appropriate hydazono compound **5a-5j**(0.01 mol) and urea (0.015) in 40 ml absolute ethanol using anhydrous sodium acetate as a catalyst (0.001 mol). The reaction mixture was heated under reflux for 5 hours. Completion of reaction was checked by TLC using 9:1 CHCl₃:CH₃OH solvent system. After cooling to room temperature the reaction mixture was poured into crushed ice and the mixtures was stirred for 1 hours. Separated product was collected by filtration, washed with water, dried and recrystallized from aqueous

ethanol. The colors of the compounds were yellowish to dark orange and are soluble in polar organic solvents like methanol, ethanol and pyridine. The synthetic route is shown in Scheme 1. Yield was in the range of 55-68%.

## General microwave assisted procedure for the synthesis of pyrimidinone substituted quinazolinone derivatives 6a–6j

Appropriate hydazono compound **5a–5j** (0.01 mol), urea (0.015 mol), anhydrous sodium acetate (0.001 mol) and 2 ml ethanol were taken in a two-neck round bottomed flask fitted with a device condenser. The mixture was heated under microwave irradiation at 350 W for 3-5 minutes. After cooling to room temperature the reaction mixture was poured into crushed ice and the mixtures was stirred for 30 minutes. Separated product was collected by filtration, washed with water, dried and recrystallized from aqueous ethanol. The colors of the compounds were yellowish to dark orange and are soluble in polar organic solvents like methanol, ethanol and pyridine. Yield was in the range of 71-83%.

### 4-[N'-(4,6-Dimethyl-2-oxo-2H-pyrimidin-5-ylidene)hydrazino]-N-(2-methyl-4-oxo-4H-quinazolin-3-yl)benzenesulfonamide (6a)

Off white solid (Yield 76%).Mp: 145-147°C. IR (KBr): 3108 (C-H$_{Ar}$), 1694 (C=O), 1600 (C=N), 1339, 1156 (S=O) cm⁻¹. ¹H NMR (400 MHz, DMSO-*d₆*) δ 10.54 (s, 1H, SO₂NH), 8.70 (s, 1H, -NH-), 7.79 (dd, J = 7.71, 1.56 Hz, 1H, Ar-H), 7.71 (td, J = 13.16, 1.52 Hz, 1H, Ar-H), 7.49 (dd, J = 7.88, 1.41 Hz, 1H, Ar-H), 7.37 (td, J = 7.17, 1.40 Hz, 1H, Ar-H), 7.25 (dd, J = 8.64, 2.63 Hz, 2H, Ar-H), 6.44 (dd, J = 10.68, 2.14 Hz, 2H, Ar-H), 2.43 (s, 3H, CH₃), 2.18 (s, 6H, CH₃). ¹³C NMR (400 MHz, DMSO-*d₆*) δ 172.36, 158.76, 158.10, 157.37, 155.10, 146.51, 142.63, 134.45, 130.25, 130.10, 127.17, 126.81, 123.70, 121.03, 113.06, 24.8, 23.12.MS *m/z*: 466.40 (M⁺). Anal.Calcd for C₂₁H₁₉N₇O₄S: C-54.19, H-4.11, N-21.06, S-6.89. Found: C-54.35, H-4.22, N-20.98, S-5.70.

### N-(6-bromo-2-methyl-4-oxoquinazolin-3(4H)-yl)-4-(2-(4,6-dimethyl-2-oxopyrimidin-5(2H)-ylidene)hydrazinyl)benzenesulfonamide (6b)

Pale yellow (Yield 83%).Mp: 237-238 °C. IR (KBr): 3104 (C-H$_{Ar}$), 1697 (C=O), 1602 (C=N), 1336, 1154 (S=O) cm⁻¹. ¹H NMR (400 MHz, DMSO-*d₆*) δ 10.86 (s, 1H, SO₂NH), 8.77 (s, 1H, -NH-), 7.79 (dd, J = 7.71, 1.56 Hz, 1H, Ar-H), 7.71 (td, J = 13.16, 1.52 Hz, 1H, Ar-H), 7.49 (dd, J = 7.88, 1.41 Hz, 2H, Ar-H), 7.64 (td, J = 7.17, 1.40 Hz, 1H, Ar-H), 6.67 (dd, J = 10.68, 2.14 Hz, 2H, Ar-H), 2.20 (s, 3H, CH₃), 2.54 (s, 6H, CH₃). ¹³C NMR (400 MHz, DMSO-*d₆*) δ 171.46, 159.15, 158.62, 158.11, 152.56, 148.82, 142.34, 137.18, 131.34, 127.56, 126.21, 125.78, 123.83, 122.29, 115.11, 24.7, 23.08. MS *m/z*: 545.1 (M⁺). Anal.Calcd for C₂₁H₁₈BrN₇O₄S: C-46.33, H-3.33, N-18.01, S-5.89. Found: C-46.10, H-3.47, N-17.88, S-5.62.

### N-(6,8-Dibromo-2-methyl-4-oxo-4H-quinazolin-3-yl)-4-[N'-(4,6-dimethyl-2-oxo-2H-pyrimidin-5-ylidene)hydrazino]benzenesulfonamide (6c)

Dark yellow (Yield 79%).Mp: 180-182°C. IR (KBr): 3102 (C-H$_{Ar}$), 1691 (C=O), 1598 (C=N), 1340, 1156 (S=O) cm⁻¹. ¹H NMR (400 MHz, DMSO-*d₆*) δ 10.74 (s, 1H, SO₂NH), 8.83 (s, 1H, -NH-), 8.32 (dd, J = 7.71, 1.56 Hz, 1H, Ar-H), 8.01 (td, J = 13.16, 1.52 Hz, 1H, Ar-H), 7.29 (dd, J = 7.88, 1.41 Hz, 1H, Ar-H), 7.37 (td, J = 7.17, 1.40 Hz, 1H, Ar-H), 6.78 (dd, J = 8.64, 2.63 Hz, 2H, Ar-H), 2.48 (s, 3H, CH₃), 2.19 (s, 6H, CH₃). ¹³C NMR (400 MHz, DMSO-*d₆*) δ 172.55, 162.17, 159.45, 157.76, 152.57, 151.62, 143.67, 138.32, 133.67, 131.29, 127.42, 125.10,

119.21, 113.87, 112.13, 24.04, 22.88. MS $m/z$: 623 (M⁺). Anal.Calcd for $C_{21}H_{17}Br_2N_7O_4S$: C-40.47, H-2.75, N-15.73, S-5.14. Found: C-40.29, H-2.54, N-15.88, S-5.25.

#### 4-(2-(4,6-dimethyl-2-oxopyrimidin-5(2H)-ylidene)hydrazinyl)-N-(2-methyl-6-nitro-4-oxoquinazolin-3(4H)-yl) benzenesulfonamide (6d)

Bright yellow (Yield 75%).Mp: 194-195°C. IR (KBr): 3110 (C-H$_{Ar}$), 1690 (C=O), 1594 (C=N), 1335, 1154 (S=O) cm⁻¹. ¹H NMR (400 MHz, DMSO-$d_6$) δ 10.79 (s, 1H, SO₂NH), 8.78 (s, 1H, -NH-), 8.67 (dd, J = 7.71, 1.56 Hz, 1H, Ar-H), 7.83 (td, J = 13.16, 1.52 Hz, 1H, Ar-H), 7.49 (dd, J = 7.88, 1.41 Hz, 1H, Ar-H), 7.37 (td, J = 7.17, 1.40 Hz, 1H, Ar-H), 7.42 (dd, J = 8.64, 2.63 Hz, 1H, Ar-H), 6.87 (dd, J = 10.68, 2.14 Hz, 2H, Ar-H), 2.48 (s, 3H, CH₃), 2.23 (s, 6H, CH₃). ¹³C NMR (400 MHz, DMSO-$d_6$) δ 171.32, 159.11, 159.11, 158.93, 156.43, 154.88, 152.21, 144.67, 141.21, 132.65, 130.21, 127.43, 125.76, 123.47, 121.51, 113.11, 24.10, 22.83. MS $m/z$: 511 (M⁺). Anal.Calcd for $C_{21}H_{18}N_8O_5S$: C-49.41, H-3.55, N-21.95, S-6.28. Found: C-49.22, H-3.31, N-22.09, S-6.47.

#### 4-[N'-(4,6-Dimethyl-2-oxo-2H-pyrimidin-5-ylidene) hydrazino]-N-(4-oxo-2-phenyl-4H-quinazolin-3-yl) benzenesulfonamide (6e)

Cream solid (Yield 72%).Mp: 177-179°C. IR (KBr): 3103 (C-H$_{Ar}$), 1689 (C=O), 1601 (C=N), 1337, 1153 (S=O) cm⁻¹. ¹H NMR (400 MHz, DMSO-$d_6$) δ 10.32 (s, 1H, SO₂NH), 8.34 (s, 1H, -NH-), 7.75 (dd, J = 7.71, 1.56 Hz, 1H, Ar-H), 7.45 (m, 4H, Ar-H), 7.38 (dd, J = 7.88, 1.41 Hz, 1H, Ar-H), 7.37 (m, 5H, Ar-H), 7.17 (dd, J = 8.64, 2.63 Hz, 1H, Ar-H), 6.31 (dd, J = 10.68, 2.14 Hz, 1H, Ar-H), 2.09 (s, 6H, CH₃). ¹³C NMR (400 MHz, DMSO-$d_6$) δ 169.12, 161.57, 157.87, 153.38, 152.39, 149.24, 139.22, 137.22, 132.45, 131.45, 130.27, 129.71, 127.21, 126.77, 126.02, 125.66, 123.42, 122.21, 115.18, 23.12. MS $m/z$: 528 (M⁺). Anal.Calcd for $C_{26}H_{21}N_7O_4S$: C-59.19, H-4.01, N-18.59, S-6.08. Found: C-58.87, H-4.26, N-18.43, S-6.17.

#### N-(6-bromo-4-oxo-2-phenylquinazolin-3(4H)-yl)-4-(2-(4,6-dimethyl-2-oxopyrimidin-5(2H)-ylidene)hydrazinyl) benzenesulfonamide (6f)

Light yellow (Yield 80%). Mp: 192-193°C. IR (KBr): 3112 (C-H$_{Ar}$), 1698 (C=O), 1605 (C=N), 1332, 1152 (S=O) cm⁻¹. ¹H NMR (400 MHz, DMSO-$d_6$) δ 10.41 (s, 1H, SO₂NH), 8.37 (s, 1H, -NH-), 7.71 (dd, J = 7.71, 1.56 Hz, 1H, Ar-H), 7.39 (m, 4H, Ar-H), 7.31 (dd, J = 7.88, 1.41 Hz, 1H, Ar-H), 7.15 (m, 4H, Ar-H), 6.69 (dd, J = 8.64, 2.63 Hz, 1H, Ar-H), 6.44 (dd, J = 10.68, 2.14 Hz, 1H, Ar-H), 2.11 (s, 6H, CH₃). ¹³C NMR (400 MHz, DMSO-$d_6$) δ 171.45, 163.54, 157.76, 155.53, 153.83, 148.33, 143.71, 139.46, 135.17, 132.56, 131.65, 130.45, 129.67, 129.14, 125.76, 124.32, 123.88, 117.78, 113.95, 23.07. MS $m/z$: 607 (M⁺). Anal.Calcd for $C_{26}H_{20}BrN_7O_4S$: C-51.49, H-3.32, N-16.17, S-5.29. Found: C-51.31, H-3.07, N-16.41, S-5.35.

#### N-(6,8-dibromo-4-oxo-2-phenylquinazolin-3(4H)-yl)-4-(2-(4,6-dimethyl-2-oxopyrimidin-5(2H)-ylidene)hydrazinyl) benzenesulfonamide (6g)

Brownish yellow (Yield 77%).Mp: 209-211°C. IR (KBr): 3105 (C-H$_{Ar}$), 1694 (C=O), 1595 (C=N), 1334, 1151 (S=O) cm⁻¹. ¹H NMR (400 MHz, DMSO-$d_6$) δ 10.37 (s, 1H, SO₂NH), 8.30 (s, 1H, -NH-), 7.63 (dd, J = 7.71, 1.56 Hz, 1H, Ar-H), 7.57 (m, 3H, Ar-H), 7.34 (dd, J = 7.88, 1.41 Hz, 1H, Ar-H), 6.99 (m, 3H, Ar-H), 6.54 (dd, J = 8.64, 2.63 Hz, 1H, Ar-H), 6.36 (dd, J = 10.68, 2.14 Hz, 2H, Ar-H), 2.21 (s, 6H, CH₃). ¹³C NMR (400 MHz, DMSO-$d_6$) δ 173.42, 162.74, 160.47, 158.87, 153.78, 152.32, 143.12, 137.32, 134.01, 132.65, 131.48, 130.37, 127.47, 129.54, 125.51, 124.91, 118.56, 115.11, 110.23, 23.05. MS $m/z$: 686 (M⁺). Anal.

Calcd for $C_{26}H_{19}Br_2N_7O_4S$: C-45.57, H-2.79, N-14.31, S-4.68. Found: C-45.35, H-2.96, N-14.18, S-4.77.

#### N-(2-(4-chlorophenyl)-4-oxoquinazolin-3(4H)-yl)-4-(2-(4,6-dimethyl-2-oxopyrimidin-5(2H)-ylidene)hydrazinyl) benzenesulfonamide (6h)

Creamy white solid (Yield 71%).Mp: 215-217°C. IR (KBr): 3113 (C-H$_{Ar}$), 1688 (C=O), 1603 (C=N), 1329, 1158 (S=O) cm⁻¹. ¹H NMR (400 MHz, DMSO-$d_6$) δ 10.41 (s, 1H, SO₂NH), 8.25 (s, 1H, -NH-), 7.81 (dd, J = 7.71, 1.56 Hz, 1H, Ar-H), 7.45 (m, 3H, Ar-H), 7.37 (dd, J = 7.88, 1.41 Hz, 1H, Ar-H), 7.22 (d, J = 7.88 Hz, 2H, Ar-H), 6.82 (d, J = 7.88 Hz, 2H, Ar-H), 6.70 (dd, J = 8.64, 2.63 Hz, 1H, Ar-H), 6.38 (dd, J = 10.68, 2.14 Hz, 2H, Ar-H), 2.17 (s, 6H, CH₃). ¹³C NMR (400 MHz, DMSO-$d_6$) δ 173.32, 164.58, 155.11, 153.65, 152.85, 149.75, 144.65, 137.83, 136.27, 132.17, 130.55, 129.41, 126.81, 126.61, 126.17, 126.02, 125.24, 122.66, 113.76, 23.19. MS $m/z$: 563 (M⁺). Anal.Calcd for $C_{26}H_{20}ClN_7O_4S$: C-55.57, H-3.59, N-17.45, S-5.71. Found C-55.29, H-3.68, N-17.27, S-5.53.

#### N-(6-bromo-2-(4-chlorophenyl)-4-oxoquinazolin-3(4H)-yl)-4-(2-(4,6-dimethyl-2-oxopyrimidin-5(2H)-ylidene)hydrazinyl) benzenesulfonamide (6i)

Pale yellow (Yield 79%).Mp: 188-190°C. IR (KBr): 3104 (C-H$_{Ar}$), 1698 (C=O), 1606 (C=N), 1337, 1155 (S=O) cm⁻¹. ¹H NMR (400 MHz, DMSO-$d_6$) δ 10.47 (s, 1H, SO₂NH), 8.31 (s, 1H, -NH-), 7.83 (dd, J = 7.71, 1.56 Hz, 1H, Ar-H), 7.49 (m, 3H, Ar-H), 7.38 (d, J = 7.88 Hz, 1H, Ar-H), 7.26 (d, J = 7.88 Hz, 2H, Ar-H), 6.85 (d, J = 7.88 Hz, 1H, Ar-H), 6.72 (dd, J = 8.64, 2.63 Hz, 1H, Ar-H), 6.41 (dd, J = 10.68, 2.14 Hz, 2H, Ar-H), 2.18 (s, 6H, CH₃). ¹³C NMR (400 MHz, DMSO-$d_6$) δ 172.67, 161.32, 159.11, 155.41, 152.35, 147.63, 142.14, 137.51, 136.71, 135.57, 134.17, 132.43, 130.64, 129.56, 126.33, 125.81, 123.09, 114.12, 113.59, 23.22. MS $m/z$: 641 (M⁺). Anal.Calcd for $C_{26}H_{19}BrClN_7O_4S$: C-48.73, H-2.99, N-15.30, S-5.00. Found: C-48.56, H-2.73, N-15.42, S-5.17.

#### N-(6,8-dibromo-2-(4-chlorophenyl)-4-oxoquinazolin-3(4H)-yl)-4-(2-(4,6-dimethyl-2-oxopyrimidin-5(2H)-ylidene)hydrazinyl) benzenesulfonamide (6j)

Yellowish orange (Yield 81%).Mp: 224-225°C. IR (KBr): 3110 (C-H$_{Ar}$), 1699 (C=O), 1603 (C=N), 1338, 1154 (S=O) cm⁻¹. ¹H NMR (400 MHz, DMSO-$d_6$) δ 10.45 (s, 1H, SO₂NH), 8.27 (s, 1H, -NH-), 7.82 (dd, J = 7.71, 1.56 Hz, 1H, Ar-H), 7.52 (m, 3H, Ar-H), 7.39 (d, J = 7.88 Hz, 1H, Ar-H), 7.24 (d, J = 7.88 Hz, 2H, Ar-H), 6.86 (d, J = 7.88 Hz, 1H, Ar-H), 6.36 (dd, J = 10.68, 2.14 Hz, 2H, Ar-H), 2.21 (s, 6H, CH₃). ¹³C NMR (400 MHz, DMSO-$d_6$) δ 171.78, 162.84, 160.78, 157.72, 154.35, 152.51, 143.56, 137.78, 136.32, 132.21, 133.17, 132.45, 131.42, 130.77, 128.21, 124.43, 117.69, 116.21, 110.18, 23.12. MS $m/z$: 720 (M⁺). Anal.Calcd for $C_{26}H_{18}Br_2ClN_7O_4S$: C-43.38, H-2.52, N-13.62, S-4.45. Found: C-43.22, H-2.71, N-13.48, S-4.60.

### In vitro antimicrobial studies

All the synthesized compounds were screened for their in vitro antimicrobial activities against selected pathogenic bacterial strains to determine minimum inhibitory concentrations (MIC) by broth micro dilution method [27] using Sulfanilamide as reference drugs of parent moieties, while Streptomycin was employed as a standard antibacterial drug.

The in vitro antimicrobial activities of all the synthesized compounds were screened for their antibacterial against Staphylococcus aureus, Bacillus subtilis, Bacillus megaterium (Gram Positive) and Escherichia coli, Proteus vulgaris, Pseudomonas aeruginosa (Gram

| Compound | Minimum Inhibition Concentration in μg/mL | | | | | |
| --- | --- | --- | --- | --- | --- | --- |
| | Gram positive bacteria | | | Gram negative bacteria | | |
| | B. megaterium | S. aureus | B. subtilis | E. coli | P. vulgaris | P. aeruginosa |
| 6a | 300 | 200 | 300 | 225 | 200 | 250 |
| 6b | 375 | 350 | 350 | 300 | 225 | 350 |
| 6c | 400 | 395 | 400 | 250 | 300 | 420 |
| 6d | 375 | 400 | 350 | 300 | 275 | 350 |
| 6e | 225 | 200 | 225 | 175 | 200 | 225 |
| 6f | 300 | 300 | 275 | 225 | 225 | 225 |
| 6g | 250 | 200 | 225 | 225 | 250 | 200 |
| 6h | 175 | 125 | 150 | 50 | 50 | 75 |
| 6i | 200 | 175 | 200 | 75 | 50 | 150 |
| 6j | 200 | 150 | 100 | 100 | 75 | 100 |
| Sf* | 425 | 375 | 400 | 375 | 400 | 425 |
| Sm* | 20 | 20 | 20 | 20 | 20 | 20 |

*Sf= Sulfanilamide, Sm = Streptomycin

**Table 1:** Antimicrobial activities of the synthesized compounds 6a-6j.

Negative), using the broth dilution method [27]. All the ATCC culture was collected from institute of microbial technology, Bangalore. 2% Luria broth solution was prepared in distilled water while, pH of the solution was adjusted to 7.4 ± 0.2 at room temperature and sterilized by autoclaving at 15 lb pressure for 25 min. The tested bacterial and fungal strains were prepared in the luria broth and incubated at 37°C and 200 rpm in an orbital incubator for overnight. Sample solutions were prepared in DMSO for various concentration. The standard drug solution of Streptomycin (antibacterial drug) was prepared in DMSO. Serial broth micro dilution was adopted as a reference method. 10 μL solution of test compound was inoculated in 5 mL luria broth for each concentration respectively and additionally one test tubes was kept as control. Each of the test tubes was inoculated with a suspension of standard microorganism to be tested and incubated at 35°C for 24 h. At the end of the incubation period, the tubes were examined for the turbidity. Turbidity in the test tubes indicated that microorganism growth has not inhibited by the antibiotic contained in the medium at the test concentration. The antimicrobial activity tests were performed in triplicate and the deviation for any triplicate results was not more than ± 1% to 5% while average MIC values of the compounds are represented in Table 1.

## Results and Discussion

The intention of the present study was to synthesize and investigate the antimicrobial activities of pyrimidinone ring substituted at 3rd position of quinazolinone derivatives (6a-6j) through hydrazono benzenesulfonamide linkage in a single molecular frame work. Synthetic rout of the target molecules is outlined in the Scheme 1 under both conventional and microwave heating condition. When cyclization for the formation of pyrimidinone derivatives was performed under microwave irradiation, the reaction was taken place in 3 minutes in an improved yield.

All the synthesized compounds were characterized by their physical, analytical and spectral properties. The IR, ESI-MS and NMR (¹H and¹³C) spectral data of all the synthesized compounds were in good agreement with the structure assigned. Further, in the MS-ESI with positive mode spectra exhibited molecular ion peak ([M+H]+), appeared at different intensities, confirmed the exact mass or molecular weights of the examined compounds 6a-6j, while appearance of a characteristic two isotope peak ([(M+H)+2]+) along with molecular ion peak ([M+H]+) in an intense ratio almost 3:1 or 1:1 to the molecular ion peak confirmed the presence of halogen (Cl or Br) atoms of high

abundance nature. The IR spectrum of 6a-6j showed the presence of C=O group at ~1690 cm⁻¹due to —CONH and —COCH₃ groups. Two sharp bands at ~1340 and ~1160 cm⁻¹ were due to asymmetric and symmetric stretching vibrations of SO₂ group respectively. The ¹H NMR of entitled compounds 6a-6j showed singlet around δ ~10.50 ppm due to highly deshielded proton of —NHSO₂ group. One singlet was observed about δ ~8.77 ppm due to secondary amine group. The two more singlet appeared in the aliphatic region at δ 2.43 and 2.18 ppm due to —CH₃(second position of the quinazolinone ring) and —CH₃ (two methyl group of pyrimidinone ring) protons. All the eight aromatic protons resonated in the region of δ7.79 to 6.44 ppm. The ¹³C NMR of final compounds 6a-6j showed two signals of carbonyls carbon around δ ~170 (CO of pyrimidinone ring), ~160 (CO of quinazolinone ring). The aromatic carbons appeared between δ 113-158 ppm. The signals due to aliphatic carbons appeared about ~24 and ~22.5 due to presence of two different methyl groups in quinazolinone ring and pyrimidinone ring.

### *In vitro* antimicrobial activities

Inspection of the data of Table 1, on the preliminary *in vitro* antimicrobial evaluations of the entitled compounds 6a-6j revealed that all the screened compounds have a varied degree of antibacterial activity, evident from their MIC values in μg/mL. Among the screened compounds, most of the compounds have shown more or equal antimicrobial activities compare to the reference drug Sulfanilamide (Sf), while a very few of the screened compounds found to be equipotent to the standard drugs, Streptomycin (Sm). From the results of *in vitro* antibacterial activity data, it was observed that compounds 6e, 6h, 6i and 6j demonstrated excellent activity against *E. coli* and *P. vulgaris* bacterial species. In general, compounds showed more selectivity against Gram negative over Gram positive species amongst all the bacterial strains. All the compounds were screened for their *in vitro* antimicrobial activities and results revealed that the group 4-chlorophenyl substitution at the 2nd position in quinazolin-4(3*H*)-one nucleus led to increase their biological activities.

## Conclusion

In summary, a series of new pyrimidinone substituted Quinazolin-4(3*H*)-one derivatives have been synthesized by adapting a simple and facile manner using both conventional and microwave heating condition in a good amount of yields. All the compounds were screened for their *in vitro* antimicrobial activity and results revealed that compounds 6e, 6h, 6i and 6j demonstrated excellent activity against *E. coli* and *P. vulgaris* bacterial species.

## Acknowledgements

We are thankful to the Principal, V. P. & R. P. T. P. Science College, VallabhVidyanagar for providing necessary research facilities. We are also grateful to SICART, VallabhVidyanagar for FT-IR and Mass analysis; CDRI, Lucknow for NMR analyses.

## References

1. Li LH, Wallace TL, Wierenga W, Skulnick HI, DeKoning TF (1987) Antitumor activity of pyrimidinones, a class of small-molecule biological response modifiers. J Biol Response Mod 6: 44-55.

2. Li LH, Wallace TL, Hamilton RD, DeKoning TF (1987) Pharmacological evaluation of combination therapy of P388 leukemia with cyclophosphamide and pyrimidinones. Int J Immunopharmacol 9: 31-39.

3. Eggermont AM, Marquet RL, De Bruin RW, Weimar W, Jeekel J (1986) Site-specific antitumour effects of 2-pyrimidinone compounds in rats. Br J Cancer 54(2): 337-339.

4. Li LH, Johnson MA, Moeller RB, Wallace TL (1984) Chemoimmunotherapy of B 16 melanoma and P388 leukemia with cyclophosphamide and pyrimidinones. Cancer Res 44: 2841-2847.

5. Li LH, Wallace TL, Richard KA, Tracey DE (1985) Mechanism of antitumor action of pyrimidinones in the treatment of B16 melanoma and P388 leukemia. Cancer Res 45: 532-538.

6. Driscoll JS, Marquez VE, Plowman J, Liu PS, Kelley JA, et al. (1991) Antitumor properties of 2(1H)-pyrimidinoneriboside (zebularine) and its fluorinated analogues. J Med Chem 34: 3280-3284.

7. Li LH, DeKoning TF, Nicholas JA, Kramer GD, Wilson D, et al. (1987) Effect of mouse hepatitis virus infection on combination therapy of P388 leukemia with cyclophosphamide and pyrimidinones. Lab AnimSci 37: 41-44.

8. Rashad AE, Shamroukh AH, Yousif NM, Salama MA, Ali HS, et al. (2012) New pyrimidinone and fused pyrimidinone derivatives as potential anticancer chemotherapeutics. Archiv der Pharmazie 345: 729-738.

9. Holla BS, Rao BS, Sarojini BK, Akberali PM (2004) One pot synthesis of thiazolodihydropyrimidinones and evaluation of their anticancer activity. Eur J Med Chem 9: 777-783.

10. Zhang X, Glunz PW, Jiang W, Schmitt A, Newman M, et al. (2013) Design and synthesis of bicyclic pyrazinone and pyrimidinone amides as potent TF-FVIIa inhibitors. Bioorg Med Chem Lett 23: 1604-1607.

11. Paul JH, Alan JW, Geoffrey SB, Sharon MK, Jonathan PW, et al. (1999) Mechanism-based inhibition of C5-cytosine DNA methyltransferases by 2-H pyrimidinone. J Mol Bio 286: 389-401.

12. Aly MM, Mohamed YA, El-Bayouki KA, Basyouni WM, Abbas SY (2010) Synthesis of some new 4(3H)- quinazolinone-2-carboxaldehyde thiosemicarbazones and their metal complexes and a study on their anticonvulsant, analgesic, cytotoxic and antimicrobial activities - part-1. Eur J Med Chem 45: 3365-3373.

13. Kidwai M, Saxena S, Khan MKR, Thukral SS (2005) Synthesis of 4-aryl-7,7-dimethyl-1,2,3,4,5,6,7,8-octahydroquinazoline-2-one/thione-5-one derivatives and evaluation as antibacterials. Eur J Med Chem40: 816-819.

14. Dahiya R, Kumar A (2008) Synthesis and biological activity of peptide derivatives of iodoquinazolinones/nitroimidazoles. Molecules 13: 958-976.

15. El-Azab AS, Al-Omar MA, Sayed-Ahmed MM, Abdel-Hamide SG (2010) Design, synthesis and biological evaluation of novel quinazoline derivatives as potential antitumor agents: molecular docking study. Eur J Med Chem 45:4188-4198.

16. Giri RS, Thaker HM, Giordano T, Chen B, Vasu KK (2010) Synthesis and evaluation of quinazolinone derivatives as inhibitors of NF-kappaB, AP-1 mediated transcription and eIF-4E mediated translational activation: inhibitors of multi-pathways involve in cancer. Eur J Med Chem 45: 3558-3563.

17. Kumar A, Rajput CS (2009) Synthesis and anti-inflammatory activity of newer quinazolin-4-one derivatives. Eur J Med Chem44: 83-90.

18. Kashaw SK, Kashaw V, Mishra P, Jain NK, Stables JP (2009) Synthesis, anticonvulsant and CNS depressant activity of some new bioactive 1-(4-substituted-phenyl)-3-(4-oxo-2-phenyl/ethyl-4H-quinazolin-3-yl)-urea. Eur J Med Chem 44: 4335-4343.

19. Ghorab MM, Gawad SMA, El-Gaby MS (2000) Synthesis and evaluation of some new fluorinated hydroquinazoline derivatives as antifungal agents. Farmaco 55: 249-255.

20. Mei H, Chun C (2009) Three-component one-pot synthesis of pyrimidinone derivatives in fluorous media: Ytterbium bis(perfluorooctanesulfonyl)imide complex catalyzed Biginelli-type reaction. J Het Chem 46: 1430-1432.

21. Hossan ASM, Abu-Melha HMA, Al-Omar MA, Amr AEGE (2012) Synthesis and antimicrobial activity of some new pyrimidinone and oxazinone derivatives fused with thiophene rings using 2-chloro-6-ethoxy-4-acetylpyridine as starting material. Molecules 17: 13642-13655.

22. Jagani CL, Sojitra NA, Vanparia SF, Patel TS, Dixit RB, Dixit BC (2011) A comparative study of solution phase as well as solvent free microwave assisted synthesis of 3-benzothiazole/isoxazole substituted 2-styryl-4(3H)-quinazolinones.ARKIVOC ix: 281-294.

23. Sojitra NA, Dixit RB, Patel RK, Patel JP, Dixit BC (2012) Classical and microwave assisted synthesis of new-(3,5-dimethyl-1-phenyl-1H-pyrazol-4-ylazo)-N-(2-substituted-4-oxo-4H-quinazolin-3-yl)benzenesulfonamide derivatives and their antimicrobial activities. J Saudi Chem Soc.

24. Vogel AI (2004) A Textbook of Practical Organic Chemistry.(5thedn) Singapore.

25. Dixit BC, Patel HM, Dixit RB, Desai DJ (2010) Synthesis, characterization and dyeing assessment of novel acid azo dyes and mordent acid azo dyes based on 2- hydroxy-4- methoxybenzophenone on wool and silk fabrics. J Serb ChemSoc 75: 605-614.

26. Singh SP, Parmar SS, Raman K, Stenberg VI (1981) Chemistry and biological activity of thiazolidinones. Chem Rev 81: 175-203.

27. Jani DH, Patel HS, Keharia H, Modi CH (2010) Novel drug-based Fe(III) heterochelates: synthetic, spectroscopic, thermal and in-vitro antibacterial significance Applorganometalchem 24: 99-111.

# 21

# The Chemistry of Alkenyl Nitriles and its Utility in Heterocyclic Synthesis

**Abdel-Sattar S Hamad Elgazwy\* and Mahmoud R Mahmoud Refaee**

*Department of Chemistry Faculty of Science, Ain Shams University, Abbassia 11566, Cairo, Egypt*

### Abstract

This review deals with the synthesis involving alkenyl nitriles of heterocyclic systems arranged by increasing ring size and the heteroatoms. Reagents containing alkenyl nitriles and aryl nitriles centers are very important in organic synthesis since they can be versatile and effective species for the efficient construction of rather complex structures from relatively simple starting materials. These reagents have proven to be valuable tools in the synthesis of a wide variety of molecular heterocyclic systems. Their importance stems from the facile bond formation at cyanide centers which can react selectively under suitable conditions. The aim of this review is the analysis and comparison of the various models having evolved on the basis of alkenyl nitriles and their application toward stereoselective synthesis.

**Keywords:** Alkenyl Nitriles; Annelated heterocycles; Oxygen nucleophile; Nitrogen nucleophile; Sulfur nucleophile; Carbon nucleophile

## Introduction

The first compound of the homolog row of nitriles, the nitrile of formic acid, hydrogen cyanide was first synthesized by C.W. Scheele in 1782 [1]. In 1811 J. L. Gay-Lussac was able to prepare the very toxic and volatile pure acid. The nitrile of benzoic acids was first prepared by Friedrich Wöhler and Justus von Liebig, but due to minimal yield of the synthesis neither physical nor chemical properties were determined or a structure suggested. Théophile-Jules Pelouze synthesized propionitrile in 1834 suggesting it to be ether of propionic alcohol and hydrocyanic acid [2]. The synthesis of benzonitrile by Hermann Fehling in 1844, by heating ammonium benzoate, was the first method yielding enough of the substance for chemical research. He determined the structure by comparing it to the already known synthesis of hydrogen cyanide by heating ammonium formate to his results. He coined the name nitrile for the newfound substance, which became the name for the compound group [3].

Nitriles occur naturally in a diverse set of plant and animal sources with over 120 naturally occurring nitriles being isolated from terrestrial and marine sources. Nitriles are most commonly encountered in fruit pits, especially almonds, and during cooking of Brassica crops (such as cabbage, brussel sprouts, and cauliflower) which lead to nitriles being released through hydrolysis. Mandelonitrile, a cyanohydrin produced by ingesting almonds or some fruit pits, releases cyanide as the main degradation pathway and is responsible for the toxicity of cyanogenic glycosides [4].

Historically over 30 nitrile-containing pharmaceuticals are currently marketed for a diverse variety of medicinal indications with more than 20 additional nitrile-containing leads in clinical development. The nitrile group is quite robust and, in most cases, is not readily metabolized but passes through the body unchanged. The types of pharmaceuticals containing nitriles are diverse, from Vildagliptin a recently released antidiabetic drug to Anastrazole which is the gold standard in treating breast cancer. In many instances the nitrile mimics functionality present in the natural enzyme substrate while in other cases the nitrile increases water solubility or decreases susceptibility to oxidative metabolism in the liver [5].

Alkenyl nitrile is one of the most versatile reagents in Organic Chemistry. It has been used as a precursor for producing nucleotides and for synthesising a wide variety of heterocyclic compounds [6]

including purines [7,8], pyrimidines [9], pyrazines [10] (some which are widely employed in the fluorescent dye industry [11]), imidazoles [12], biphenylenes [13], porphyrazines (which have great potential in optical sensor technology) [14] and diimines that are used as catalysts [15]. This review highlights the alkenyl nitrile chemistry with the focus on the utility of heterocyclic compounds. The synthesis and chemistry of the highly strained aryl nitrile is also briefly reviewed [16].

Heterocycles are ubiquitous in all kind of compounds of interest, and among all the possible synthetic methods of achieving their introduction into an structure, probably the use of a alkenyl nitrile analogue is the most direct one. The present review deals with the generation and synthetic uses of alkenyl nitriles and aryl nitriles formed in heterocyclic synthesis, and can be considered as an update of our revision published in 2007 on this topical [17]. Therefore, only references published from the second quarter of 2003 until the third quarter of 2010 are included, and the same restrictions to the literature coverage applied. Thus, only heterocycles compounds which are found applicability are considered. As previously, the present review is organized by the type of ring members and subdivided by the type of heterocycles fused compounds, including methods for their preparation and their synthetic uses. New developments in the utilities of some alkenyl nitriles in heterocyclic synthesis are reviewed. General synthetic routes based on the utilization of alkenyl nitriles of active imines are discussed. The major methods and modifications are analyzed [18-21].

In this review, which covers the literature up to date, we describe the new and improved methods for the construction of the skeletons, with a particular emphasis on the four, five and six membered ring of heterocyclic compounds. Some of these procedures have clear technical advantages over older methods in terms of yield and versatility, but do not employ new chemistry in the construction of the ring systems. The

**\*Corresponding author:** Abdel-Sattar S. Hamad Elgazwy, Department of Chemistry Faculty of Science, Ain Shams University, Abbassia 11566, Cairo, Egypt, E-mail: elgazwy@sci.asu.edu.eg

use of combinatorial synthesis, microwave enhanced processes and new catalytic methodologies in the preparation of these heterocycles is a clear indication that significant advancement has been made in recent years. The syntheses of both on the four, five and six membered ring of fused and polyheterocyclic compounds will be classified into the following five categories, based on the substitution patterns of the ring system: New approaches for synthesis of different mono and polyheterocyclic derivatives arranged by increasing ring size and the heteroatoms utilizing activated nitriles are surveyed. Activated nitriles are very important in organic synthesis since they can be used as effective species for efficient construction of rather complex structures from relatively simple starting materials. The scope and limitation of the most important of these approaches are demonstrated.

## Preparation of Alkenyl Nitrile and Aryl Nitrile Derivatives

Industrially, the main methods for producing nitriles 2 are ammoxidation and hydrocyanation. Both routes are green in the sense that they do not generate stoichiometric amounts of salts. In ammonoxidation, a hydrocarbon is partially oxidized in the presence of ammonia. This conversion is practiced on a large scale for acrylonitrile, as shown below [22].

$$\text{1} \quad + 3/2O_2 \quad + NH_3 \longrightarrow \quad NC\text{-}\!\!\!=\!\!\!\text{ } \text{2} \quad + 3H_2O$$

An example of hydrocyanation is the production of adiponitrile 4 from 1,3-butadiene 3, as outlined below.

$$\text{3} \quad + 2HCN \longrightarrow NC\text{-}\!\!\!\text{-}\!\!\!\text{-}CN \quad \text{4}$$

Usually for more specialty applications in organic synthesis, nitriles can be prepared by a wide variety of other methods: Dehydration of primary amides. Many reagents are available, the combination of ethyl dichlorophosphate and DBU just one of them in this conversion of benzamide to benzonitrile [23]. Two intermediates in this reaction are amide tautomer **A** and their phosphate adducts **B**, as summarized diagrammatically in Scheme 1.

Scheme 1

In one study an aromatic or aliphatic aldehyde is reacted with hydroxylamine and anhydrous sodium sulfate in a dry media reaction for a very small amount of time under microwave irradiation through an intermediate aldoxime [24], as shwon in Scheme 2. A commercial source for the cyanide group is diethylaluminum cyanide $Et_2AlCN$ [25], which can be prepared from triethylaluminium and HCN, it has been used as nucleophilic addition into ketones [26].

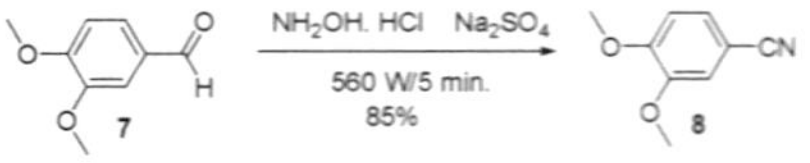

Scheme 2

For an example of its use Kuwajima Taxol total synthesis of cyanide ions facilitate the coupling of dibromides. Reaction of $\alpha,\alpha\beta$-dibromo adipic acid with sodium cyanide in ethanol yields the cyano cyclobutane [27], as shown in Scheme 3.

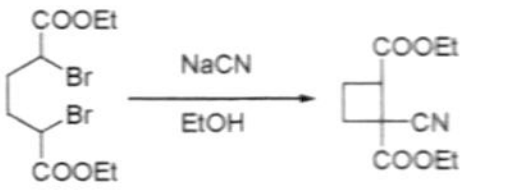

Scheme 3

In the so-called Franchimont Reaction (A. P. N. Franchimont, 1872) $\alpha$-bromocarboxylic acid is dimerized after hydrolysis of the cyan group and decarboxylation. Aromatic nitriles can be prepared from base hydrolysis of trichloromethyl aryl ketimines ($RC(CCl_3)=NH$) in the Houben-Fischer synthesis [28-31]. In reductive decyanation the nitrile group is replaced by a proton [32]. An effective decyanation is by a dissolving metal reduction with HMPA and potassium metal in tert-butanol. $\alpha$-Amino nitriles can be decyanated with lithium aluminium hydride. Nitriles self-react in presence of base in the Thorpe reaction in a nucleophilic addition. In organometallic chemistry nitriles are known to add to alkynes in carbocyanation, as summarized diagrammatically in Scheme 4 [33].

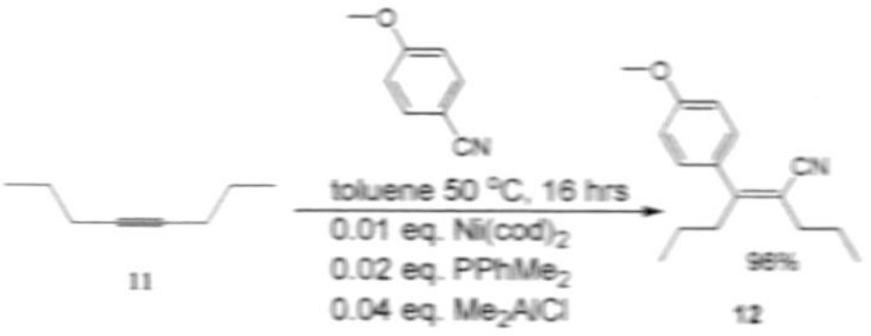

Scheme 4

## Synthesis

### Four membered rings

Organic cyano compounds are versatile reagents, which have been extensively utilized in heterocyclic synthesis. Alkenyl nitriles behaves as a typical stable organic molecule, the stability of alkenyl nitriles and aryl nitriles arises from the fact that it has an aromatic delocalized $\pi$-electron system. Enormous number of reports [34-43], on the utility of these compounds in synthesis of heterocycles has been reported. It is our intention in this review, therefore, to fill the gaps and report on the utilities of $\alpha-\beta$-unsaturated nitriles. Such as arylidene malononitrile 13 which successfully used to prepare 4-Aryl-2-iminothietane-3-carbonitrile 14 in a moderate yield via the reaction [44] of with ammonium benzyl dithio-carbamate 15, as outlined in Scheme 5.

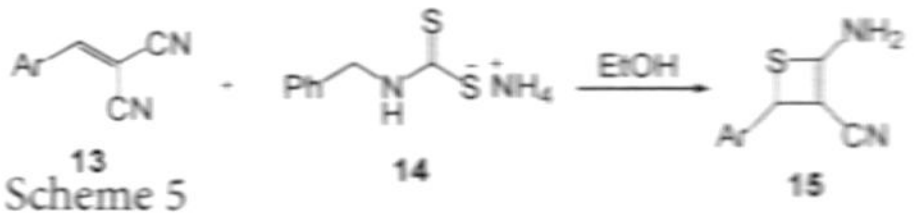

Scheme 5

### Five membered rings

#### Five membered ring with one heteroatom:

**Synthesis of thiophene and fused thiophene derivatives:** The $\alpha-\beta$-unsaturated nitriles with active methylene group at $\beta$-carbon 16 could react with elemental sulphur to yield an intermediate mercapto derivative 17, which cyclizes into the most isolable stable aminothiophene derivative 18, as outlined in sheme 6 [45-49].

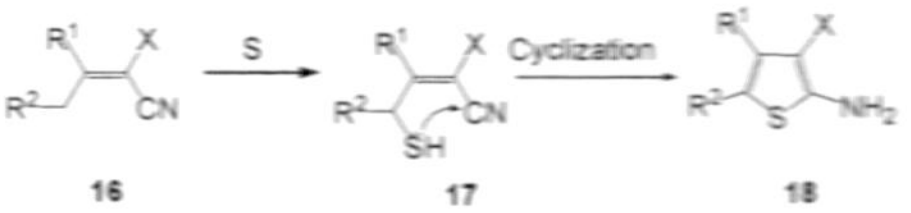

Scheme 6

$\alpha-\beta$-unsaturated nitriles were thiolated into thiophene derivatives. For example, the arylidene derivative of cyclohexanone 19 was converted into the enaminothiophene derivative 20 on treatment with elemental sulphur [45]. The enamines can also be formed on heating mixtures of the ketone, the activated nitrile and elemental sulphur in the presence of a basic catalyst.

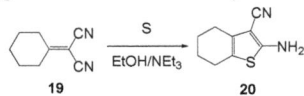

Scheme 7

Formation of thiophenes **22** from the reaction of $\alpha-\beta$-unsaturated nitriles with thioglycollic acid has been reported [50-53].

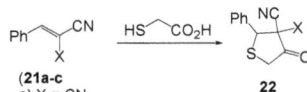

Scheme 8

Tetracyanoethylene **23** has been reported to react with hydrogen sulphide [54,55] to produce the thiophene derivative **24** in moderate yields 68%.

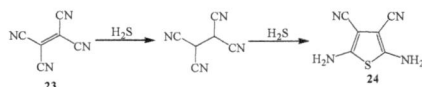

Scheme 9

Another similar synthesis that affords thiophene derivatives **26** utilizing thioanilides of the type as starting component is the reaction of **25** with active methylene reagents [56].

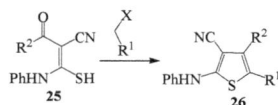

Scheme 10

Formation of thiophenes via the reaction of arylidene derivatives of 3-oxoalkanenitriles has been reported by El-Nagdy et al. [51,52,56-58]. For example, the thiophene derivatives **28** were formed from the reaction of **27** with ethyl thioglycollate. On the other hand, the thiophene derivative **29** was isolated on using thioglycollic acid together with Michael adduct **30**, as outlined in Scheme 11.

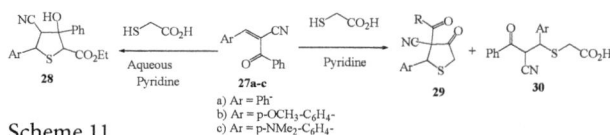

Scheme 11

**Synthesis of furan derivatives:** To be considered as an update of our revision published in 1998 on this topic such as photochemical transformations of 2(5$H$) furanones [59]. In the last decade, it was reported by Aran and Soto [60] for the formation of furan derivatives **31** by heating 2-benzoyl-3-phenyl-acrylonitrile **27** with cyanide ion.

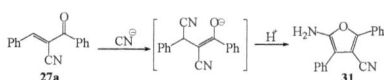

Scheme 12

**Synthesis of pyrrole and condensed pyrrole derivatives:** Several synthesis of pyrrole derivatives utilizing organic cyano compounds as starting components were reported [60-68]. The most interesting results of that are demonstrated in Scheme 13 [68-73].

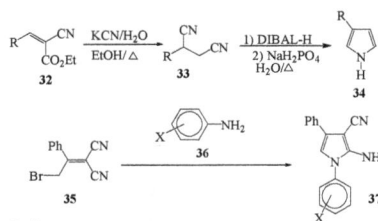

Scheme 13

Hydrazinolysis of $Z$ or $E$-2-methyl-3-cyano-4-pentenoate **38** afforded the pyrrole derivatives shown below **39, 40** [74-76].

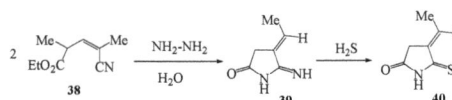

Scheme 14

Nucleosides pyrrole derivatives have also been synthesized utilizing $\alpha-\beta$-unsaturated nitriles [77], an interesting example is depicted in Scheme 15.

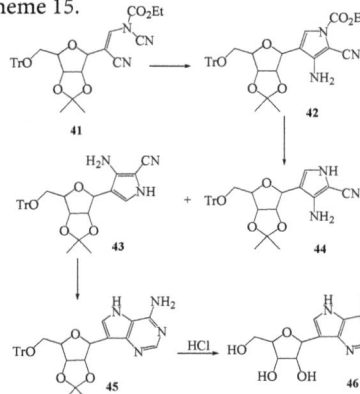

Scheme 15

Several indole syntheses have been reported [78-82]. For example, heating o-azidocinnamonitrile **47** in DMSO at 140°C afforded 2-cyanoindole **49** [83] in good yield. The latter could also be obtained on heating o-nitrocinnamonitrile **48** in triethylorthophosphate at 160°C [84].

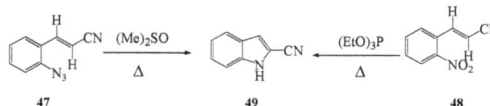

Scheme 16

The 5-Amino-26-diethylcarboxy-3-substituted-4-(4 chlorophenyl)-6-iminothieno[3,2-5,6], thiopyrano[2,3-b]pyrrols **51** were synthesized in a one-pot procedure via the reaction of $\beta$-substituted arylidene malononitrile **50** with CS$_2$ and ethylchloroacetate in 1:1:2 molar ratio under PTC conditions (dioxane/K$_2$CO$_3$/tetrabutylammonium bromide TBAB) in good yield [85].

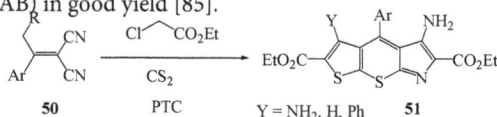

Scheme 17

Heating pyrrole-2-carbodithionates **52** with anions of C-H acids generated from malononitrile or cyanoacetamide in KOH/DMSO (room temperature, 0.5 h). Interaction of the resulting enethiolates **53** with haloacetylenes **54** afforded the pyrrolothiazolidines **55**, as outlined in Scheme 18 [86-91].

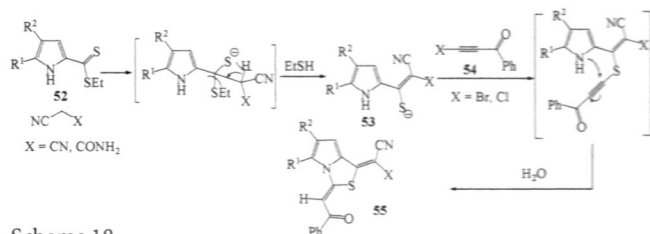

Scheme 18

**Five membered rings with two heteroatoms:**

**Synthesis of 1,2-oxazole derivatives:** The $\alpha-\beta$-Unsaturated nitriles are extensively utilized for the syntheses of 1,2-oxazoles [92-95]. For example, the dimer of ethyl cyanoacetate **56** reacted with hydroxylamine hydrochloride to yield **57**.

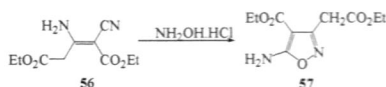

Scheme 19

Other examples for the syntheses of amino-1,2-oxazole are shown in Scheme 20 [96-98].

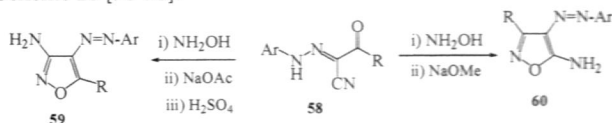

Scheme 20

Compound **27a** was examined against hydroxylamine hydrochloride to yield a mixture of the aldoxime **61** and the aminoisoxazole derivatives **62** [99].

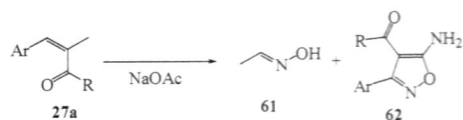

Scheme 21

In contrast to the previously reported behaviour of 2-pyrazolin-5-one [100], 2-thiazoline-4-one [101], and 2-thiohydantion derivatives [102], towards the action of arylidenemalononitrile, 3-phenyl-2-isoxazolin-5-one **64** reacted with the cinnamonitrile derivative **63** to yield only the arylidene derivative **65** [103], as shown in Scheme 22.

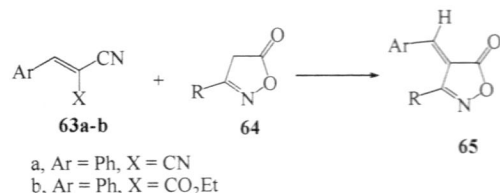

a, Ar = Ph, X = CN
b, Ar = Ph, X = CO$_2$Et

Scheme 22

3-Oxime-4-phenyl-1(H)-1,5-benzodiazepin-2-one **66** was allowed to react with different arylidenenitriles in the presence of triethylamine and yielded spirobenzodia-zepine isoxazole derivatives **67** and **68**, as outlined in Scheme 23 [104].

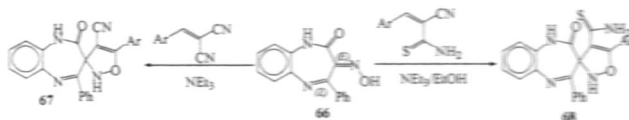

Scheme 23

**Synthesis of isothiazole derivatives:** The chemistry of isothiazoles has been reviewed by one of us [105] and the isothiazole derivatives **72** was produced by treatment of the dimercaptomethylenemalononitrile salt **69** with elemental sulfur in refluxing methanol, in a good yield. The existence of intermediates **70** and **71** has been envisioned. The former arises from nucleophilic attack by mercaptide anion on sulfur, whereas the latter involves a second nucleophilic attack on the nitrile with expulsion of the sulfur moiety by the nitrogen. Another example of this reaction involving the mononitrile derivative **73** has been described, which presumably proceeds through the same path, leading to the isothiazole derivative **74** [106], as outlined in Scheme 24.

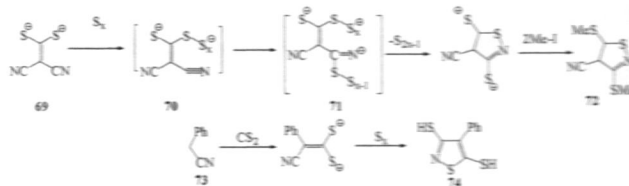

Scheme 24

**Synthesis of thiazole derivatives:** An investigation was undertaken to explore the potential utility of the reaction of some activated nitriles with mercaptoacetic acid as a route for the synthesis of thiazoles, thus, cinnamonitriles **35** react with mercaptoacetic acid to give the thiazole derivatives **75** [107,108], as outlined in Scheme 25.

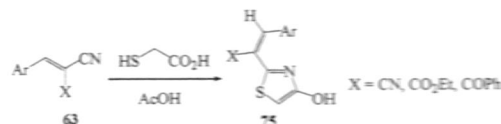

Scheme 25

**Synthesis of pyrazole and fused pyrazole derivatives:** Scission of the double bond in the arylidene derivatives of 3-oxoalkanenitriles **27** was reported to take place by the action of hydrazines in basic media, whereas the formation of 3,5-diaryl-3-pyrazolines **76** was reported to take place in acidic media [109-111]. The intermediate phenylhydrazone derivative **77** was isolated together with **78** on reaction of **27** (Ar = p-O$_2$N-C$_6$H$_4$-) with phenylhydrazine. El-Nagdy et al. [112-114] have reported that **27** (Ar = p-Me$_2$N-C$_6$H$_4$-) reacts with $\beta$-cyano-ethylhydrazine to yield the hydrazone **79**, which was cyclized to yield either **80** or **81** depending on the applied reaction conditions as shown in Scheme 26.

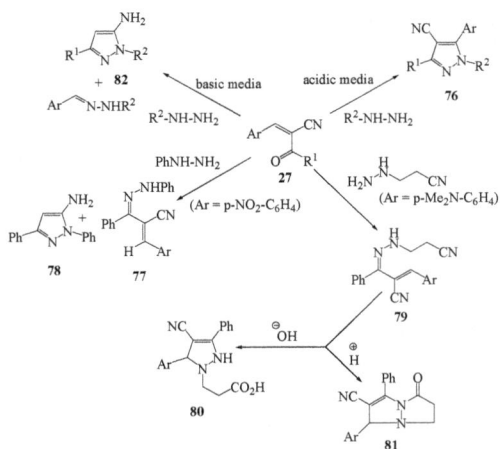

Scheme 26

Cusmano and Sprio [109,110,115,116] have shown that the double bond in compound 27 functions as a ylidenic bond even toward the action of semicarbazide hydrochloride, thus heating benzylidene-ω-cyanoacetophenone 27 with semicarbazide hydrochloride in an ethanolic solution of sodium carbonate results in the formation of benzaldehyde semicarbazone and ω-cyanoacetophenone. However, when the reaction mixture was left for several days, compound 82 (formulated by Cusmano and Sprio as 83 ($R^1$ = H, $R^2$ = Ph, $R^3$ = $CONH_2$) was formed in addition to benzaldehyde phenylhydrazono, as described in Scheme 27.

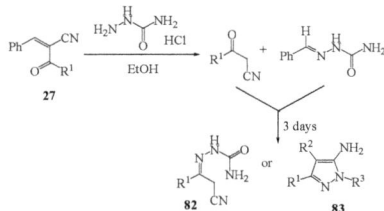

Scheme 27

It has been shown that ethyl β-trichloromethyl-β-aminomethylenecyanoacetate (84, X = $CCl_3$) reacts with hydrazine hydrate to afford the aminopyrazole derivative 86 via intermediate formation of the amidrazone 85 which could be isolated [116,117-119]. This is in contrast to the reported formation of 3-amino-4-cyano-5-trifluoromethylpyrazole 87 on treatment of β-trifluoromethyl-β-aminomethylene-malononitrile (84, X = $CF_3$) with hydrazine hydrate [120]. Synthesis of pyrazoles via similar routes has been reported [107,121], as outlined in Scheme 28.

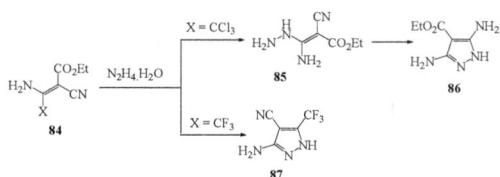

Scheme 28

Ethoxymethylenemalononitrile 88 reacted with hydrazine hydrate to yield the pyrazole derivatives 89 and 90 [122], as outlined in Scheme 29.

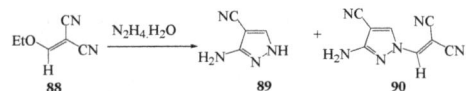

Scheme 29

In an attempt to synthesize 3-amino-4-ethoxycarbonyl-pyrazole 92 via reacting 91 with hydrazine hydrate in a manner similar to that reported for its reaction with phenyl hydrazine which is established to afford pyrazole derivatives, Midorikawa et al. [123,124] have obtained, instead of the expected pyrazole derivative 92, the pyrazolo[1,5-a]pyrimidine derivative 94. The formation of this product is expected to proceed via intermediate formation of 93, as outlined in Scheme 30.

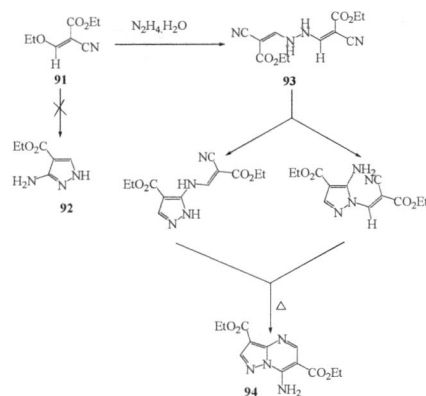

Scheme 30

4-(4-Phenyl-3-pyrazolyl)-4H-1,2,4-triazole 97 was recently prepared by the action of formylhydrazine 96 on α-phenyl-α-cyanoacetaldehyde 95 [125], as depicted in Scheme 31.

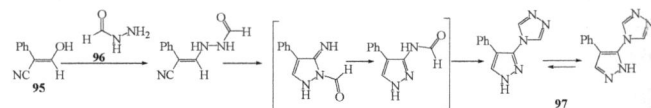

Scheme 31

β–Dimethylamino- α-(2-ribosyl) acrylonitrile 98 reacted with hydrazine hydrate to yield the aminopyrazole derivative 99. This reaction opened a new route for the synthesis of formycone and formycine analogues [126], as shown in Scheme 32.

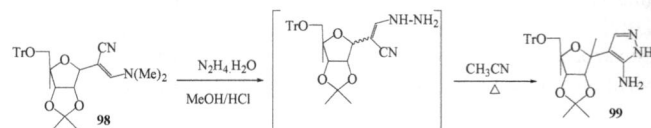

Scheme 32

A variety of new pyrazole derivatives 101-104 have been synthesized utilizing the same idea of reacting α–β-unsaturated nitriles 100a-c with hydrazine or acylated hydrazines [99,127-149]. Examples for the most interesting of these syntheses are shown in Scheme 33.

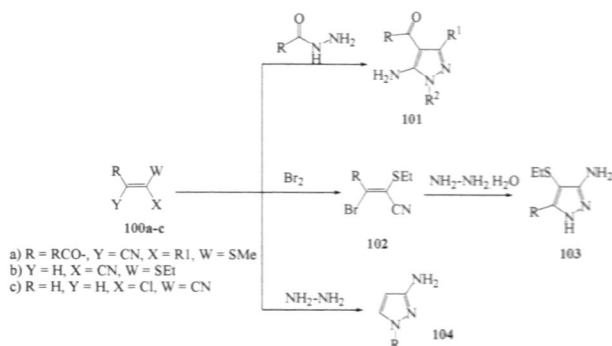

Scheme 33

3,4-Dicyano-5-aminopyrazoles have been synthesized by taking the advantage of the tetracyanoethylene 23 for Michael addition. Thus, aryl and alkyl hydrazones as well as hydrazides, semicarbazides and thiosemicarbazides have been reported to react with tetracyanoethylene to afford 1-substituted-4,5-dicyano-3-aminopyrazoles [145]. The structure assigned for the reaction product of 23 with methylhydrazine was reinvestigated by Hecht et al. [145] and Earl et al. [146] in two separate contributions. It has been shown by Hecht et al. [145] that consideration of the mechanistic routes suggested in literature for this reaction illustrates the source of structural ambiguity in the formation of these products from methylhydrazine and 23. Thus, one might for example, envision formation of the 1-methyl-4,5-dicyano-3-amino-pyrazole 105 by conjugate addition of the more nucleophilic substituted hydrazine nitrogen of the hydrazine to the cyano group, affording the observed product is depicted in scheme 34.

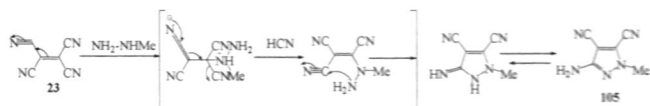

Scheme 34

Alternatively, as has been previously suggested, addition of the substituted nitrogen of methylhydrazine to the cyano group might occur first to give 106 and the reaction then proceeds are depected in Scheme 35.

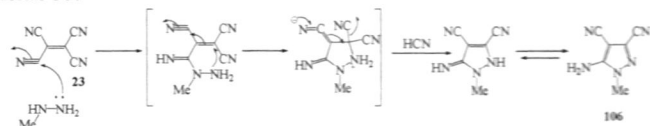

Scheme 35

Both authors on reconducting the above reaction have shown that it affords a mixture of two isomeric pyrazoles (53% and 27%) [146], (47% and 8%) [147]. These author have shown on the basis of chemical evidences as well as spectroscopic data that the major product for which the 3-amino-4,5-dicyano-1-methylpyrazole structure was formally assigned is really 1-methyl-3,4-dicyano-5-aminopyrazole. El-Nagdy et al. [113] reported that the reaction of arylhydrazono derivatives of 2,3-dioxo-3-phenylpropionitrile 107 reacted with ethyl hydrazinoacetate 108 to yield the imidazo[1,2-b]pyrazole derivatives which can be formulated as 109 or 110. Structure 110 was considered most likely for these products based on spectroscopic data. The formation of 110 in this reaction may be assumed to proceed via the sequence demonstrated in Scheme 36 and attempted to isolate

intermediates for this reaction were unsuccessful.

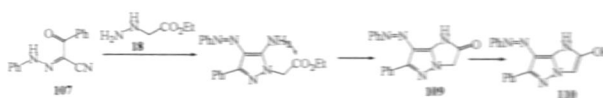

Scheme 36

Furthermore, compound 108 reacted with the dimer of malononitrile 111 to afford 112 in excellent yield. Attempted cyclization of 111 by action of 3% NaOH has afforded the carboxylic acid derivatives 113. On the other hand the hydrochloride 114 was obtained on attempted cyclization of 112 by the action of conc. HCl [113], as shown in Scheme 37.

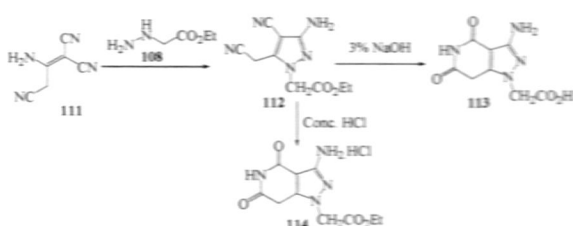

Scheme 37

Similar to the behaviour of 77, phenyl hydrazonomalononitrile 115 reacted with 108 to yield the imidazo[1,2-b]pyrazole derivative 116 is depicted in Scheme 38.

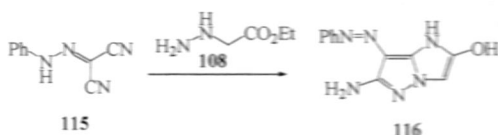

Scheme 38

The behaviour of the ethoxymethylene derivatives of cyanoacetic acid 117 has also been investigated [113]. It has been found that 117 react with 108 to yield the aminopyrazole derivatives 118 are depicted in Scheme 39.

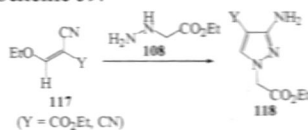

(Y = CO₂Et, CN)

Scheme 39

The nitrile 119 reacted with 4-bromo-3-methylpyrazol-5-one 120 in ethanol in the presence of catalytic amount of triethylamine to give the corresponding pyrano[2,3-c]pyrazole derivatives 121 [150] are depicted in Scheme 40.

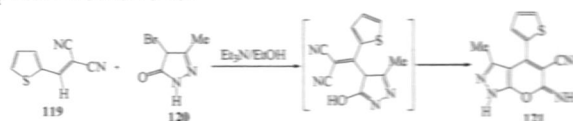

Scheme 40

Treatment of activated nitrile 122 under the above conditions gave the acyclic pyrazolone derivative 123 which could not be cyclized under the applied conditions in contrast to the previous case [150], the product is depicted in Scheme 41.

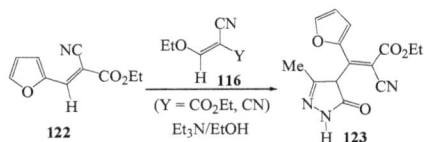

Scheme 41

The arylidene malononitrile **124a-c** has been reacted with 3-methyl-2-pyrazolin-5-one **125a** to yield the pyranopyrazoles **126a-c**, which were also obtained from the reaction of arylidene pyrazolones **127a-c** with malononitrile [100] and this reaction proved to be a general one. Thus, pyranopyrazoles **126d-i** were formed from **125a** and **124d-i** in yield (66-99%) [151-156] and the products are depicted in Scheme 42.

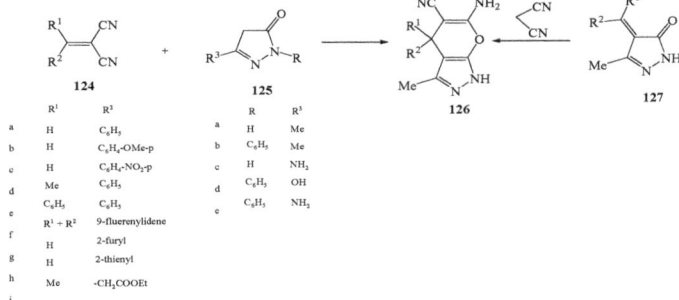

| | $R^1$ | $R^3$ | | $R$ | $R^3$ |
|---|---|---|---|---|---|
| a | H | $C_6H_5$ | a | H | Me |
| b | H | $C_6H_4$-OMe-p | b | $C_6H_5$ | Me |
| c | H | $C_6H_4$-NO$_2$-p | c | H | NH$_2$ |
| d | Me | $C_6H_5$ | d | H | OH |
| e | $R^1 + R^2$ | 9-fluerenylidene | e | $C_6H_5$ | NH$_2$ |
| f | H | 2-furyl | | | |
| g | H | 2-thienyl | | | |
| h | Me | -CH$_2$COOEt | | | |
| i | | | | | |

Scheme 42

However, attempts to extend this approach in order to enable synthesis of **128** failed. Abdo et al. [100] reinvestigated reaction of **125** with **124a,b** and obtained a product the structure of which was assigned as **129** since they proved that **128a,b** were obtained via addition of malononitrile to **100a,b** [100,140] as depiected in Scheme 43.

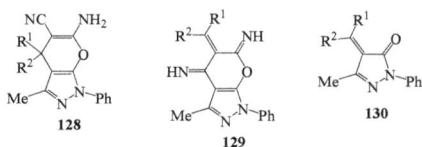

Scheme 43

The structure of the products of the reaction of **124** with **125b** has been recently shown [157] to be **133** formed most likely via decomposition of the initially formed Michael adduct **131** into **132** and addition of one molecule of **125b** to this decomposition product affording arylidene-bis-pyrazolones that react with piperidine present in the reaction mixture to yield **133**, as depicted in Scheme 44 [158].

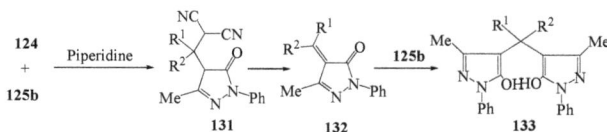

Scheme 44

Girgis et al. [150] have reported that compound **129g,h** were formed via reacting **124g,h** with **125b**. However, Abdelrazik et al. [151] have later reported that the product of the reaction of **125b** with **124g** is **129**. Similar to the behaviour of **125a**, compound **125c** reacted with

**124a** to yield **134** [159]. Similar results were obtained with **125d**, as depicted in Scheme 45 [160-162].

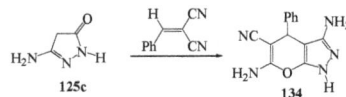

Scheme 45

El-Torgoman et al. [162] reported the formation of **137** from **125** and p-anisylidene thiocyanoacetamide **135** via elimination of hydrogen sulphide from the intermediate Michael adduct **136**, as shown Scheme 46.

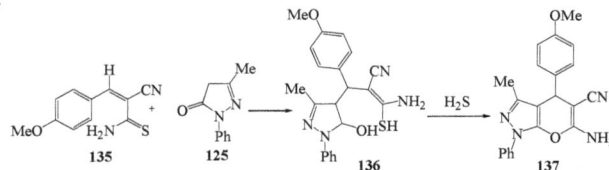

Scheme 46

Mahmoud et al. [163] reported that equimolar amounts of 1-phenyl-3-methylpyrazolin-5-one **125** and α-cyano-3,4,5-trimethoxycinnamonitrile **138** were refluxed in absolute ethanol in the presence of piperidine as a catalytic base. After 15 minutes an insoluble fraction was isolated as colorless crystals (13%) and detected to be the oxinobispyrazole **139** and the reaction was completed for 3h. Removal of most of the solvent and acidification with dilute acetic acid afforded the 1:1 adduct **140a** or **140b** as pale yellow crystals (44% and 46% yield, respectively), as outlined in Scheme 47.

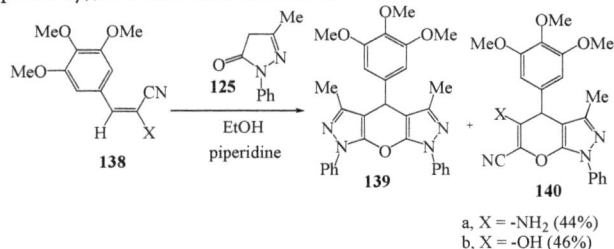

a, X = -NH$_2$ (44%)
b, X = -OH (46%)

Scheme 47

Spiropyranopyrazoles **142** have been obtained through reacting substituted cyanomethylideneindolidinones **141** with **125a,b**. It is of value to report that these products were earlier believed to be the quinoline derivatives **143**. $^{13}$C-NMR spectra have been utilized to discriminate between the two structures (Scheme 48) [159,164].

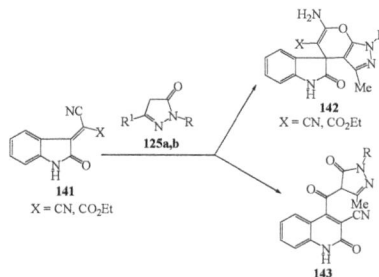

Scheme 48

Pyranopyrazoles **145** were formed via reacting **144a,b** with **125a** [100]. However, the reaction of **144c** with **125a** led to the formation of **126** [151]. Similar results have been reported on treatment of **125** derivatives with **144**, as depiected in Scheme 49 [160].

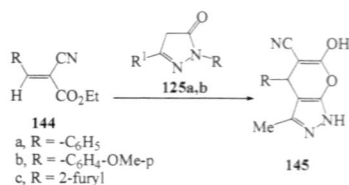

**Scheme 49**

The reaction of chalcones **146a** with **125a** yields the corresponding Michael adducts **147** [165-168]. These could be cyclized by the action of polyphosphoric acid into **148**. The reaction of α-cyanochalcone **146b** with **125a** resulted in the direct formation of **148b** (X = CN), as outlined in Scheme 50.

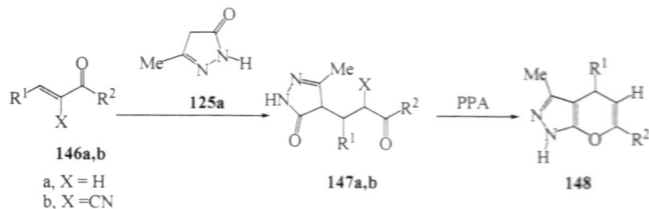

**Scheme 50**

Excellent yield of pyranopyrazole derivatives **149** were obtained upon treatment of nitrile **27** with **125** [169], and is depicted in Scheme 51.

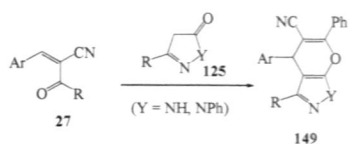

**Scheme 51**

1-Phenyl-pyrazolin-3,5-dione **150** reacts with activated nitrile derivatives to yield several pyrano[2,3-c]pyrazoles [161] **151**, **152**. Similarly 1,3-diphenylthio-hydantoin, thiazolidinethiones and isorhodanine reacts with cinnamonitriles to yield the corresponding pyranoazole derivatives, however in some cases, ylidene group exchange took place and the compound is depicted in Scheme 52 [161].

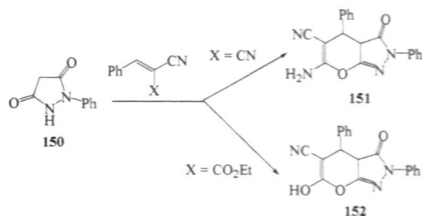

**Scheme 52**

During the course of our investigations on the use of DAMN in heterocyclic synthesis, we designed new approaches to 4-cyano-1,3-dihydro-2oxo-2H-imidazole-5-(N1-tosyl)carboxamide as a reactive precursor thiopurine [170]. In some of these cases, new DAMN derivatives, N-({[(Z)-2-amino-1,2-dicyanovinyl]amino}carbonyl)-4-methylbenzenesulfonamide, were used as the key intermediates. Since until now the preparation and characterization of the above stated sulfonamides have been mentioned only briefly, we give herein a report

on these compounds in more detail [170]. Tetracyanoethylene **23** reacted with **123** to yield product of condensation by the elimination of hydrogen cyanide, which is formulated as **124** depicted in Scheme 53 [171].

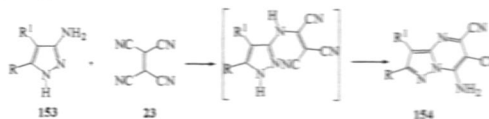

**Scheme 53**

Madkour et al. [171] has reported that hydrazinolysis of 3-(4-methoxyphenyl) and 3-(2'-thienyl)-2-cyano-2-propenoyl chlorides **155a** [172-176] and **155b** at 0°C afforded the pyrazolone derivatives **156**, bishydrazine **157** and anisylideneazine **158**, while, treatment of **155a** with benzoylhydrazine afforded the pyrazolone **159**, as outlined in Scheme 54.

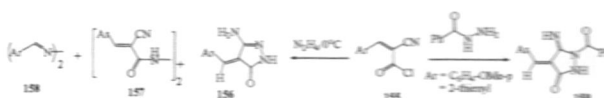

**Scheme 54**

Furthermore, the electrophilicity of the lactonic carbonyl functionality of benzoxazinone **160** has been investigated [172] via its reaction with some nitrogen and oxygen nucleophiles. Thus, stirring **160** with hydrazine hydrate at 0°C in dioxane gave the pyrazolo[5,1-b]-(1H) quinazolinone **161** in moderate yield. On the other hand, addition of hydrazine hydrate to **160** in n-butanol followed by stirring at room temperature or at reflux afforded a mixture of 2-aminopyrazolo [5,1-b] quinazolinone **162** besides the Schiff's base **163** and the azine **158**, as outlined in Scheme 55.

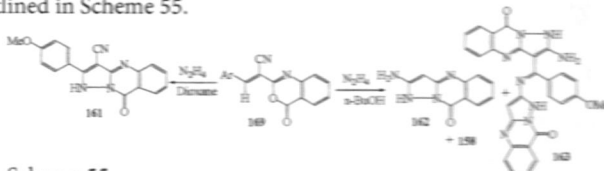

**Scheme 55**

4-Arylidene-1-phenyl-3,5-pyrazolinediones **164** [177] reacted with activated nitriles **165** (N, S-acetals) [178] to give pyrazolino-1,3-oxazine derivatives **166**, as outlined in Scheme 56.

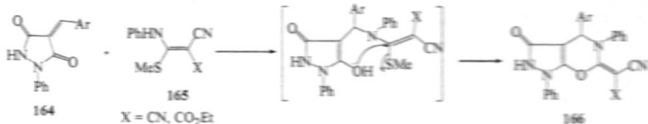

**Scheme 56**

On refluxing compound **167** with cycloalkylidenecyanoacetamide **168** in dioxane in the presence of triethylamine, the corresponding pyrazolopyridinethione derivatives **169** were obtained [179], as outlined in Scheme 57.

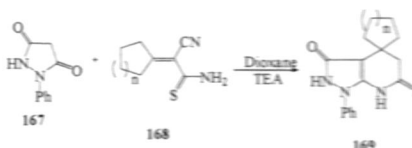

**Scheme 57**

Treatment of **170** [179] with benzylidenemalononitrile furnished the corresponding spiropyrazolopyridine derivatives **171** and are depicted in Scheme 58 [180].

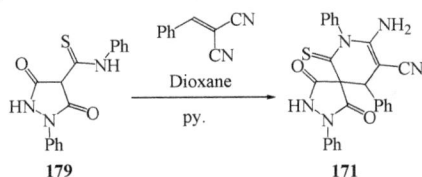

Scheme 58

The synthesis of various pyrazolo[1,5-a]pyrimidines as unique phophodiesterase inhibitors from easily available starting materials has been the subject of several publication [181-183]. In spite of enormous literature reported for the synthesis of pyrazolo[1,5-a]pyrimidines using 5-aminopyrazoles as educts, very few reports have appeared describing the utility of diaminopyrazoles as starting components for the synthesis of condensed pyrazoles. In conjuction to previous work, compound **172** was reacted with cinnamonitrile derivative to yield the pyrazolo[1,5-a]pyrimidine derivatives **173** is depicted in Scheme 59 [184].

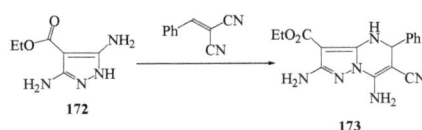

Scheme 59

New spiro heterocyclic systems attached by coumarin nucleus were synthesized by the reaction of 2-coumarylidene malononitrile **174** with some active methylene or bidentates such as hydrazine hydrate to afford **175** and **176**, respectively [179], as outlined in Scheme 60.

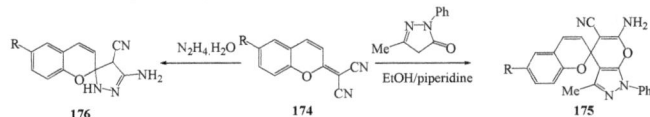

Scheme 60

**Synthesis of midazole and condensed imidazole derivatives:** A New Synthesis of 4-Cyano-1,3-dihydro-2-oxo-2H-imidazole-5-(N1-tosyl) carboxamide: A Reactive Precursor for ThioPurine Analogs Hamad et al. [169]. 2,3-Diaminodinitrile **177** has been recently utilized for the synthesis of imidazole derivatives. Thus, **177** reacted with formamidine to yield 2,3-diaminofumaronitrile **178** which could be cyclized under different conditions to yield different imidazole derivatives **179-182**, and the product is depicted in Scheme 61 [149,185-188].

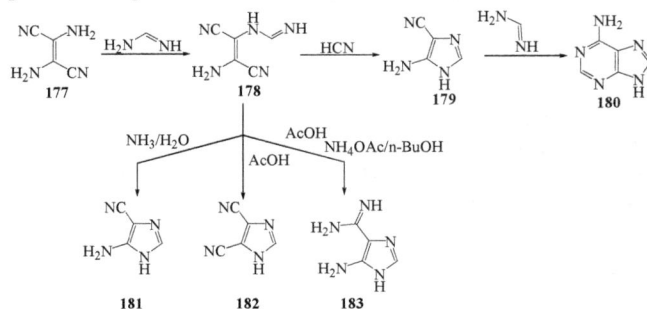

Scheme 61

Aziz et al. [102] found that the activated nitriles **184** react with 3-phenyl-2-thiohydantoin **185** to yield 1:1 adducts **186** together with the 5-benzylidene-2-thiohydantoin derivatives **187**. The same products were obtained when **187** were treated with malononitrile, as shown in Scheme 62.

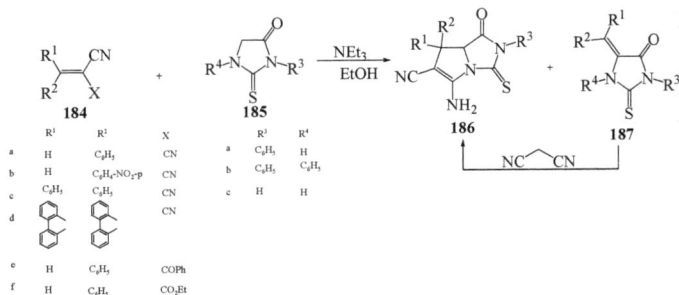

Scheme 62

In contrast to the behaviour of compound **185a** towards **184a-d**, the 2-thiohydantoin derivatives **185b,c** reacted with **184a,b** to yield 5-arylidene derivative **187ab**, **187bb** and **187ac**, respectively, as the sole isolable products and were recovered almost unaffected after treatment with **184c** under the same experimental conditions, as outlined in Scheme 63.

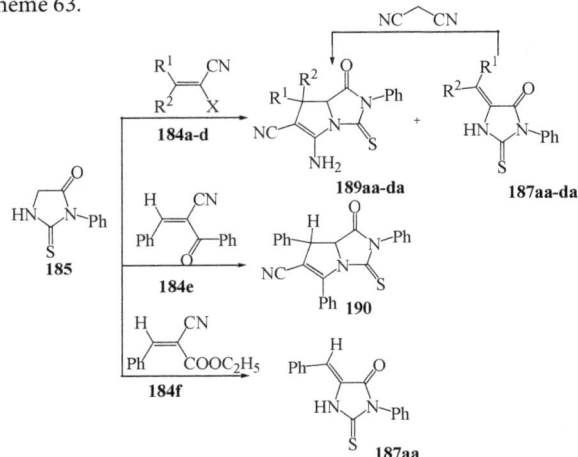

Scheme 63

Treatment of compound **187ab** and **187bb** with malononitrile afforded the pyrano[2,3-d]imidazole derivatives **188ab** and **188bb**, respectively, whereas treatment of **187ac** with malononitrile afforded the pyrrolo[1,2-c]imidazole derivative **189ac**, as outlined in Scheme 52. Compound **184e** reacted with **185a** to yield the pyrrolo[1,2-c] imidazole derivative **190**. On the other hand, the pyrano[2,3-d] imidazole derivative **188eb** was formed from the reaction of **184e** and **185b**, as outlined in Scheme 64.

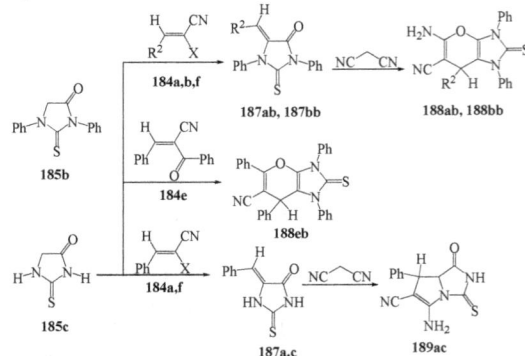

Scheme 64

In contrast to the behaviour of compound **184a-c** toward **185**, ethylbenzilidene cyanoacetate **152f** reacted with imidazolidines **185a-c** to give the benzylidene derivatives **187aa**, **187ab** and **155ac**, respectively. The reaction of thiohydantoin **185b** with **27** has been, however, shown to yield either pyrano[2,3-d]imidazoles **191** or pyrrolo[1,2-c]imidazoles **192** depending on the nature of substituents on the thiohydantoin and is depicted in Scheme 65 [102].

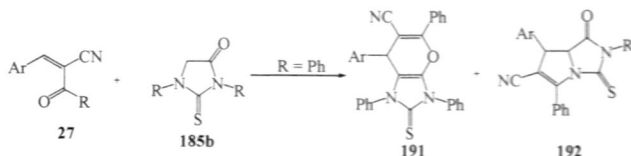

Scheme 65

**Synthesis of five membered rings with three heteroatoms**

The synthetic potentialities of 2-arylhydrazinonitriles have recently been reviewed [189]. El-Mousawi et al. have reported that 2-phenyl-hydrazono-3-oxo-butanenitrile **193** reacted with hydroxylamine hydrochloride in ethanolic sodium acetate to yield amidoxime **194** that cyclized readily into 4-acetyl-2-phenyl-1,2,3-triazol-5-amine **195** upon reflux in DMF in presence of piperidine [190], as outlined in Scheme 66.

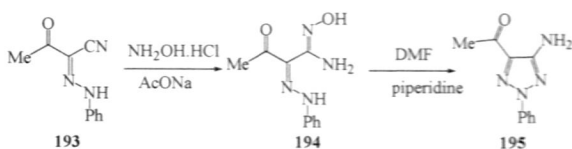

Scheme 66

$^{13}$C-NMR of the reaction product indicated that this is not the case, as it indicated the absence of the carbonyl carbon in the range $\delta$ = 180-200 ppm. Therefore, the formation of isomeric oxazole **II** was considered as the correct structure which can take place only via intermediacy of the initial product of condensation of the ketocarbonyl of **193** with hydroxylamine yielding inetrmediate **I** that could cyclize to isomeric oxazole **II**. Intermediate isomeric oxazole **II** when heated in DMF in the presence of piperidine it rearranged readily to **195** [191], as outlined in Scheme 67.

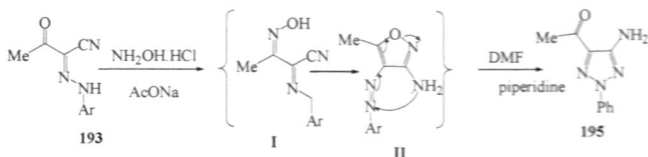

Scheme 67

Tetracyanoethylene **23** reacts with ethyldiazoacetate to yield 1,2,3-triazoles **196** is depicted in Scheme 68 [186,192]

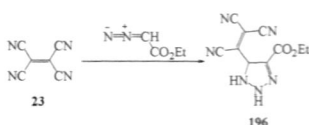

Scheme 68

Cyanamide derivatives **197** have been extensively utilized for the synthesis of 1,2,4-triazoles [193-195]. The one interesting example for

the utility of these reactions in synthesis of triazole derivatives **198** and **199** is shown below in the following Scheme 69.

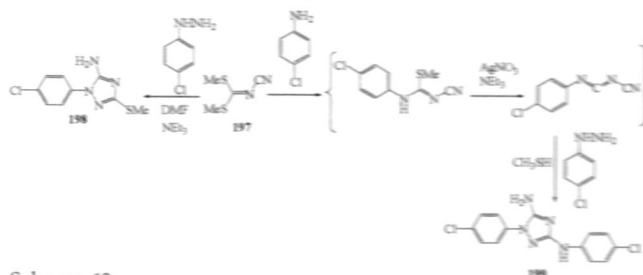

Scheme 69

Cyanamide derivatives have been utilized for the synthesis of oxadiazoles [196]. For example, benzoyldicyandiamide **200** afforded a mixture of the urido-1,2,4-oxadiazole derivatives **201** and **202** on treatment with hydroxylamine, the first was predominating, as the product depicted in Scheme 70 [194,196].

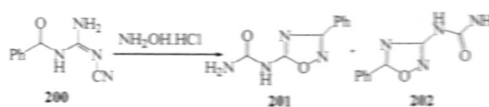

Scheme 70

Similarly, the iminoether **203** afforded the amino-oxadiazole derivative **204** on reaction with hydroxylamine and the product depicted in Scheme 71 [197].

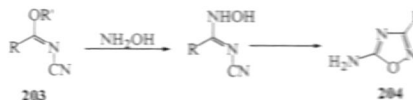

Scheme 71

Also, the cyanamide **205** reacted with hydroxylamine to yield 1,2,4-oxadiazole derivative **206** is depicted in Scheme 72 [198].

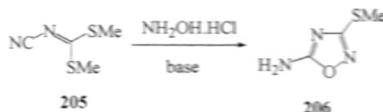

Scheme 72

1-Substituted-3-cyano-isothioureas **207**, gave mixture of the 5-amino-3-substituted-amino-1,2,4-oxadiazoles **208** and the isomeric 3-amino-5-substituted-amino-1,2,4-oxadiazoles **209** on reaction with hydroxylamine, the compound **208** usually predominated and is depicted in Scheme 73 [199].

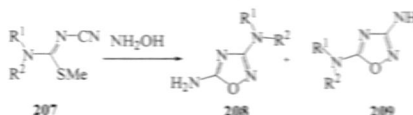

Scheme 73

**Six-membered heterocycles**

**Six membered heterocycles with one heteroatom**

**Synthesis of pyridine and condensed pyridine derivatives:** Several pyridine syntheses, utilizing nitriles as starting components have been reported [37-41]. Although a number of papers have been published concerning the synthesis of 2-oxopyridine derivatives [200-212] no preparations using 2-cyano acrylates, cycloalkanones and ammonium acetate as starting materials have been reported. Some 2-oxopyridine derivatives such as 4-aryl-3-cyano-2-oxo-7-(substituted benzylidene)-2,5,6,7-tetrahydro-1H-pyridines (38-57% yield) **212** were synthesized from ethyl 2-cyanoacrylates **210**, cycloalkanones **211** and ammonium acetate in refluxing ethanol [213], as shown in Scheme 74.

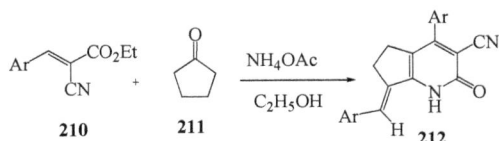

Scheme 74

An analoguous reaction between ethyl 2-cyanoacrylates (R = Me, Et) **213**, cyclopentanone and ammonium acetate gave, as products, ethyl-2-cyclopentylidene-2-cyanoacetate (major product) and the 4-alkyl-3-cyano-2-oxo-2,5,6,7-tetrahydro-1H-1-pyridines **214** instead of the expected 7-alkylidene derivatives and the product depicted in Scheme 75.

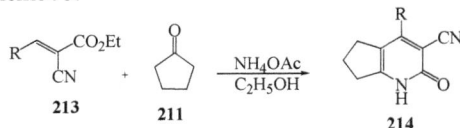

Scheme 75

The reaction of **210a-d** with **215b** gave the 4-aryl-3-cyano-2-oxo-1,2,3,4,5,6,7,8-(or 1,2,3,4,4a,5,6,7-)octahydroquinolines **216** and/or the 4-aryl-3-cyano-2-oxo-1,2,5,6,7,8-hexahydroquinoline **217**, the product depicted in Scheme 76.

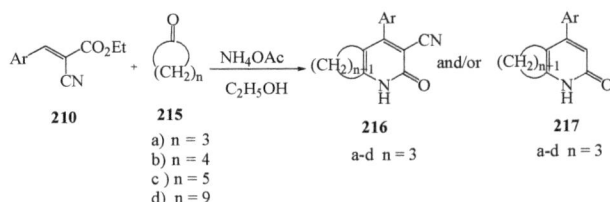

Scheme 76

The 2-iminopyridine derivatives **220** obtained in fairly good yield from the condensation of arylidene malononitrile **218** with alkyl ketones **219** in the presence of excess ammonium acetate with boiling benzene and the product depicted in Scheme 77 [214].

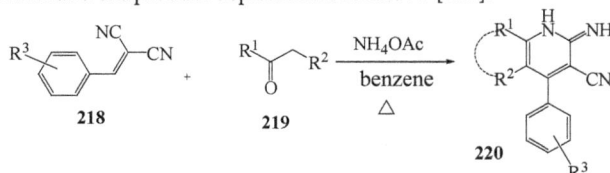

Scheme 77

The reaction of β-alkylarylidenemalononitrile **50** with phenylisocyanate or phenylisothiocyanate under PTC conditions (MeCN/K$_2$CO$_3$/TBAB) yielded pyridinone and pyridine-2-thione derivatives **221** [105,215], as shown in Scheme 78.

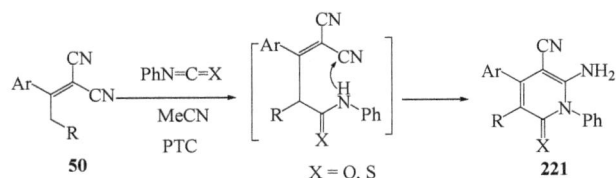

Scheme 78

However, treatment of arylidene malononitrile with some reactive halo compounds under PTC afforded the N-aminopyridine derivative **222**, [105,215]. Where with hydrazine hydrate yielded the pyridine derivatives **223** and **224** in moderate to good yields, as shown in Scheme 79.

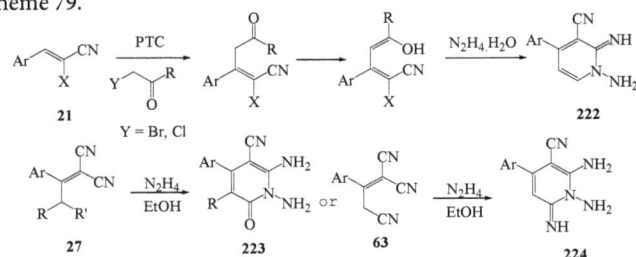

Scheme 79

The cinnamonitriles **63** react with cyanoacetic acid hydrazide **226** to afford N-aminopyridones. Soto et al. [215] reported the direct formation of **226** as sole reaction product on heating **63** with **225** for 5 min. However, El-Moghayar et al. [216] have reported that the product previously identified as **226** is really **227** which rearranged on refluxing in aqueous ethanolic triethylamine solutions into **228**, as shown in Scheme 80. Evidence afforded on this problem are not conclusive and further investigation seem to be mandatory.

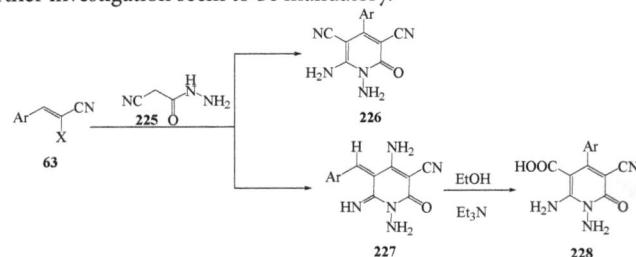

Scheme 80

The reaction of the dimers **111** and **56** with cinnamonitriles in ethanolic triethylamine solutions afforded the pyridine derivatives **229**, **230** are depicted in Scheme 81 [217-219].

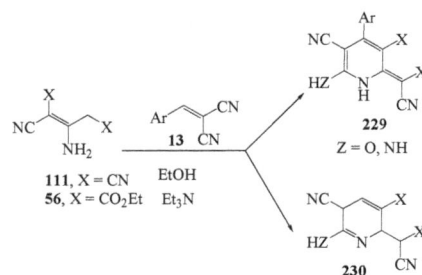

Scheme 81

Recently, this approach has been explored for the synthesis of several pyridine derivatives **231-236**, as outlined in Scheme 82 [220-223].

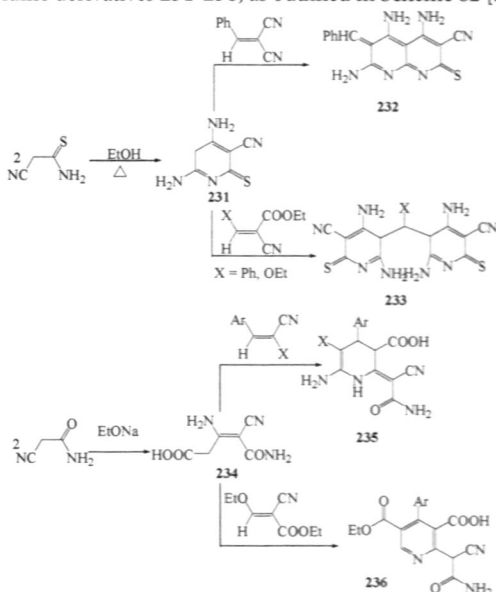

Scheme 82

Daboun et al. [223] have found that a solution of equimolar amounts of 2-amino-1,1,3-tricyanoprop-1-ene **111** and acetylacetone in ethanol was refluxed for 2 hr in the presence of piperidine as a catalyst to yield a product of molecular formula $C_{11}H_8N_4$. Two possible structures, 3-cyano-2-dicyanomethylene-4,6-dimethyl-1,2-dihydropyridine **237** and the isomeric **238**, were considered. Structure **237** was established by the results of IR and $^1$H-NMR spectra. The obtained products beer several functional substituents and appear promising for further chemical transformations, as outlined in Scheme 83.

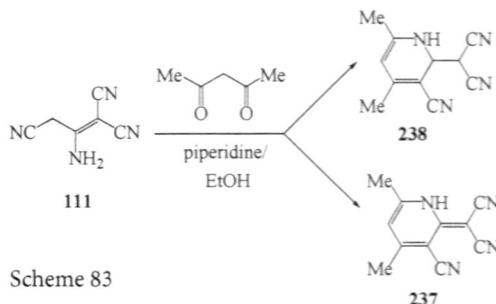

Scheme 83

The activated nitrile **240** and **241** was synthesized by El-Nagdy et al. [224] through the reaction of phenacylthiocyanate **239** with malononitrile. as outlined in Scheme 84

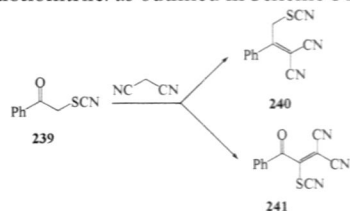

Scheme 84

The structure **241** was ruled out on the basis of IR and $^1$H-NMR spectra. Reaction of trichloroacetonitrile with **240** in refluxing toluene in the presence of a catalytic amount of piperidine yield a 1:1 gave

adduct **241** wich cyclized into produce **243** which was suggested based on spectroscopies. The compound **240** also might be gave **242** and then cyclized to give **244**, as outlined in Scheme 85.

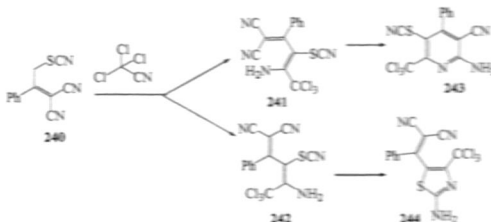

Scheme 85

Conflicting results have been reported for the reaction of cinnamonitrile **63** with cyanothioacetamide **215**. Thus, Daboun and Riad [225] reported that the dihydropyridines **246a,b** were isolated from the reaction of **63** with **245**. On the other hand, Sato et al. [226] reported that the pyridines **246a,b** were the isolable products from the reaction of **63** and **245** [227,228]. Recently, it has been shown that the thiopyrans **247** are the products of the kinetically controlled reactions of **63** with **245** (via chemical routes and inspection of the high resolution $^1$H-NMR and $^{13}$C-NMR). These products rearrange on heating in aqueous ethanol into the thermodynamically stable dihydropyridines **248**. Observed [100,101,151] dependency of the products of reactions of active methylene reagents with cinnamonitrile derivatives on the nature of reactants and reaction conditions has been reported [134,226-236], as outlined in Scheme 86.

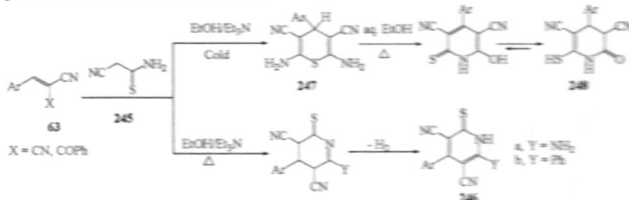

Scheme 86

Pyridine derivatives **250**, **251**, **252**, **253** and **255** were successfully synthesized via condensation of cyanothioacetamide **245** with the cinnamonitrile derivatives **249** or the acrylonitrile derivatives **253**, as outlined in Scheme 87 [237].

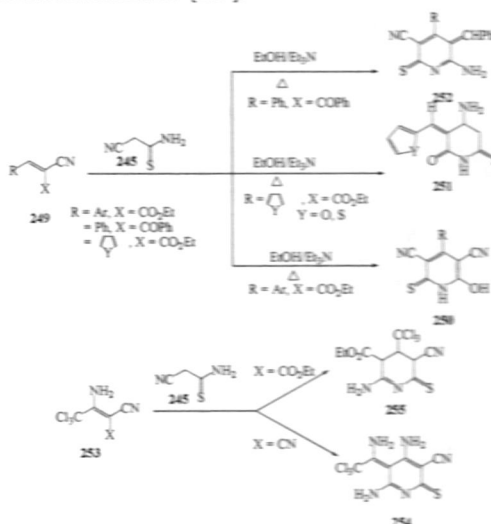

Scheme 87

Gewald et al. [237] have shown that the product of reaction of **56** and **111** with trichloroacetonitrile are really the pyridine derivatives **256** and **257**, respectively. Convincing evidence from 13C-NMR for the proposed structures were reported, as outlined in Scheme 88.

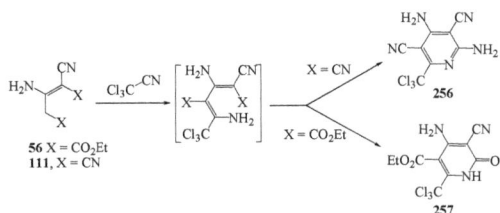

Scheme 88

A route for the synthesis of 6-[5-amino-pyrrol-4-yl]pyridines **253** and their conversion into 3-[pyridine-6-yl]pyrazolo[1,5-a]pyrimidine **263** has been reported [238]. Thus, 1-phenylethylidene malononitrile **258** was refluxed in pyridine solution with enaminonitriles **253a-c** to yield products via chloroform elimination. Structure **230** or its cyclized product **231** seemed possible as **253a-c** were assumed to add **258** affording the intermediate Michael adducts **229** which then lost chloroform to **230**. **230** may undergo cyclization into **261** under the reaction conditions. Compound **261** reacted with hydrazine hydrate affording **261**, which condensed readily with acetylacetone affording the required pyrazolo[1,5-a]pyrimidines **263**, as outlined in Scheme 89.

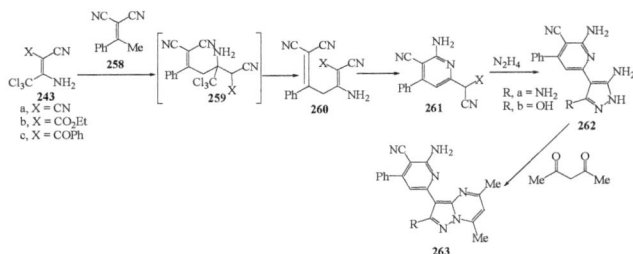

Scheme 89

El-Nagdy et al. [239] reported that 3-amino-2-cyano-4-ethoxycarbonyl crotononitrile **234** reacted with trichloroacetonitrile in refluxing ethanol in presence of triethylamine to give **235** which resemble the formation of pyridine derivative from the reaction of 2-amino-1,1,3-tricyanopropene with trichloroacetonitrile. Compound **235** reacted with hydrazine hydrate to yield hydrazine derivatives **236** which successfully cyclized into **237**, as outlined in Scheme 90.

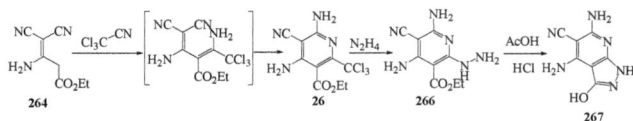

Scheme 90

It was reported that when indan-1,3-dione or 1-phenylimino-indan-3-one **268a,b** were heated with arylidenecyanothioacetamide in refluxed ethanol in the presence of a catalytic amount of piperidine, 4-aryl-3-cyano-5-oxo- or (phenylimino)-indeno[1,2-b]pyridine-2[1H] thiones [240] **269a-c** were obtaine, as outlined in Scheme 91.

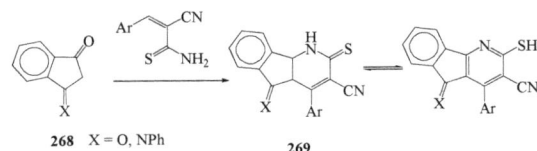

Scheme 91

When **269** was subjected to react with NaN$_3$ in DMF in the presence of NH$_4$Cl to synthesize mercaptotetrazolylindeno pyridine **271** as reported [241,242] the aminoindenopyridine derivative **270** was obtained instead of **271**, as depicted in Scheme 92.

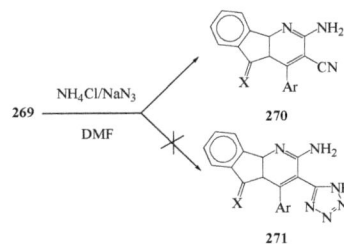

Scheme 92

1-Dicyanomethylene-3-indanone **272** was prepared by the reaction of indandione and malononitrile in ethanolic sodium ethoxide solution [243,244]. Compound **272** reacted with carbon disulphide to give the dithiocarboxylic acid derivative **273**, which in turn was alkylated with methyl iodide and ethyl chloroacetate to give **274** and **275**. Indeno[2,1-c] pyridines **276** were prepared through the reaction of **272** with phenyl (benzoyl) isothiocyanate. Compound **272** with malononitrile and aromatic aldehyde afforded **277**, which reacted with acetic acid to give **278**. Bromination of **272** with N-bromosuccinimide afforded 2-bromo derivative **279** which reacted with aniline, methylthioglycolate, and ethylglycinate to give indeno[2,1-b]pyrroles **280**, **281** [245], indeno[2,1-b]thiopyrane **282**, **283** and **284**, as outlined in Scheme 93.

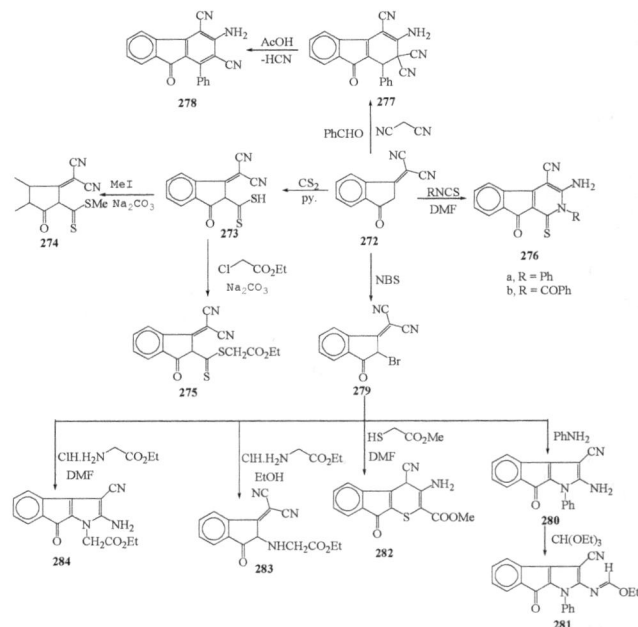

Scheme 93

Piperidylidene malononitrile **285** reacts with benzaldehyde to give **286**, which underwent an addition reaction with malononitrile to give **287**, which was also obtained by reacting **288** with one mole of benzaldehyde and two moles of malononitrile [246], as outlined in Scheme 94.

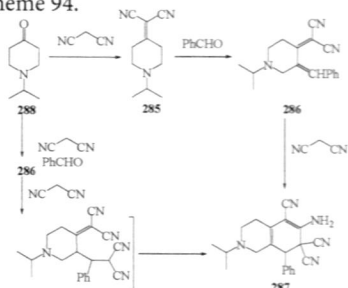

Scheme 94 & 95

Synthesis of novel 1,4,5,6,7,8-hexahydroquinoline **291** bearing amino and cyano groups on $C_2$ and $C_3$ has been carried out by refluxing equimolar amounts of the corresponding arylidene malononitrile **289**, dimedone **290** and excess of ammonium acetate in acetic acid as solvent in a similar way to that reported for other related structures [247,248]. Compounds **291** are obtained as crystalline solids in 55-75% yields, as outlined in Scheme 96.

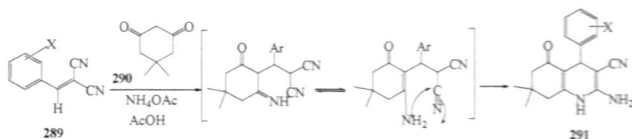

Scheme 96

A route for the synthesis of thiazolo[2,3-a]pyridine **293** from the reaction of 2-functionally substituted 2-thiazoline-4-one **292** with cinnamonitrile has been reported simultaneously and independently by El-Nagdy et al. [101] and Kambe et al. [249]. Better yields were obtained using a 2:1 molar ratio of cinnamonitrile derivative and **292**, as outlined in Scheme 97.

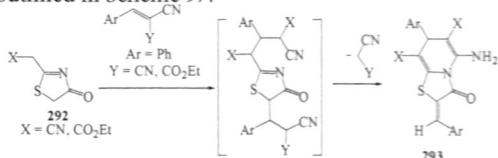

Scheme 97

In our laboratories [250] it has been found that the reaction of **292** with cinnamonitrile **294a,b** in absolute ethanol in presence of piperidine afforded a semisolid product from which 6-substituted-7H-2,3-dihydro-5-amino-8-cyano-3-oxo-7-(3,4,5-trimethoxyphenyl)-thiazolo[3,2-a] pyridines **295a,b** (ethanol soluble fraction, major yield) and 6-substituted-7H-2,3-dihydro-5-amino-8-cyano-3-oxo-2-(3,4,6-trimethoxybenzylidene)-7-(3,4,5-trimethoxyphenyl)-thiazolo [3,2-a] pyridines **296a,b** (ethanol-insoluble fraction, minor yield) can be isolated, as depicted in Scheme 98.

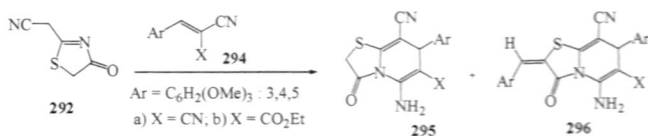

Scheme 98

E-Z-assignement of compounds containing exocyclic C=C double bonds throughout the present work were elucidated and proven by $^1$H-NMR calculations [251]. Unexpectedly, it has been found that the product **299** from the reaction of **292** with 2-oxo-3-dicyanomethylidene-2,3-dihydroindole **297a**, 2-oxo-3-cyanoethoxy carbonylmethylidene-2,3-dihydroindole **297b** and with isatin under the same experimental conditions was one and the same product **298** [252], as outlined in Scheme 99.

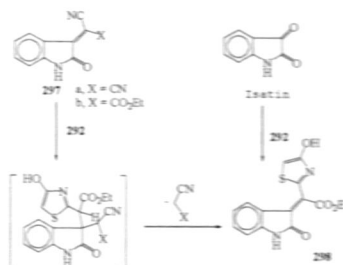

Scheme 99

It was believed that the reaction product was found via additions of **292** to the activated double bond in **297a,b** to form the corresponding intermediate Michael adducts, which then loses either malononitrile or ethyl cyanoacetate to yield the final isolable product **298**. Junek [253] has reported that salicylaldehyde **299** reacts with 2-amino-1,1,3-tricyano prop-1-ene **111** to yield the benzopyrano[3,4-c]pyridine **300** is depicted in Scheme 100.

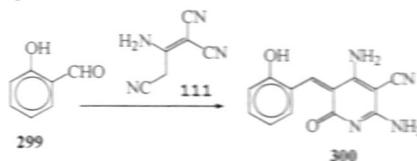

Scheme 100

Midorikawa et al. [249] have shown that the reaction of substituted amines with ylidenemalononitriles affords pyridine derivatives **301** and **302**, as outlined in Scheme 101.

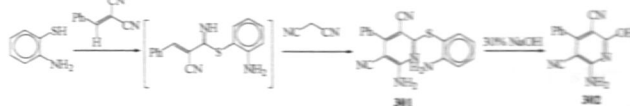

Scheme 101

Conversion of 4-acetyloxazoles **303** into pyridine derivatives **306** via reaction with malononitrile has been reported. The reaction proceeds via formation of the ylidenemalononitrile derivative **304** and then cyclized into **305** [254], as outlined in Scheme 102.

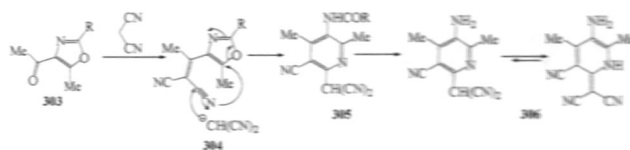

Scheme 102

Several other pyridine syntheses from activated $\alpha,\beta$-unsaturated nitriles are already available in literature; a very old example is the reaction of two fold of 2-amino crotononitrile with aromatic aldehyde **307** to yield a pyridine derivative **308** is depicted in Scheme 103 [255],

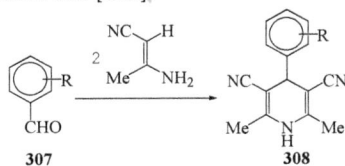

Scheme 103

Other interesting examples, 2-amino-1,1,3-tricyano-prop-1-ene **111** has been reported to react with benzalmalononitrile to yield pyridine derivative [256] **309**. Similarly, diethyl-3-amino-2-cyanopent-2-ene-1,5-dicarboxylate **56** has been reported to yield pyridines **310** utilizing almost the same idea [218], as outlined in Scheme 104.

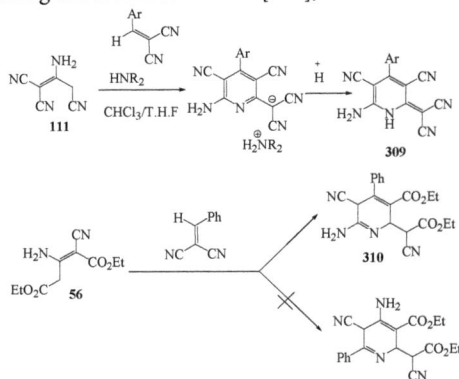

Scheme 104

2-Amino-1,1,3-tricyano-prop-1-ene **111** has been reported to condense with β-diketones **311** and β-aminoenones **312** to yield pyridine derivatives **313**, **314** respectively [40,257], as outlined in Scheme 105.

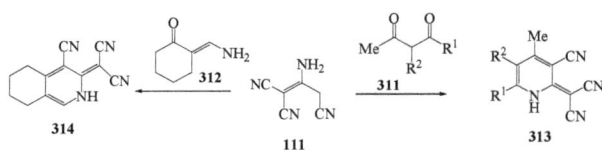

Scheme 105

The reaction of 3-aminoacrylonitriles derivative **315** with ethoxymethylene malononitrile in chloroform or dichloromethane at temperature below 0°C and 5°C for 24 hours, leads to dienaminonitriles **316** in good yields [257]. These adducts **316** are transformed into the pyridine derivatives **317** in almost quantitative yields. The reaction of compound **317** with formamide lead to pyrido[2,3-d]pyrimidine derivatives **318** in 65-98% yields as outlined in Scheme 106.

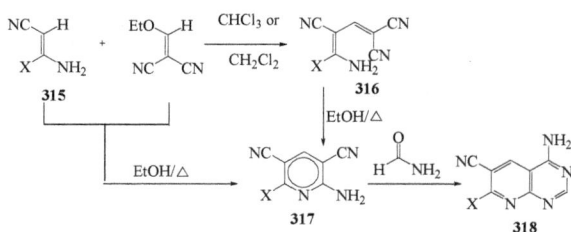

Scheme 106

The synthesis of pyridine derivatives **320** is best accomplished by cyclization of the new dienaminoester **319** in refluxing DMSO and as depicted in Scheme 107 [258].

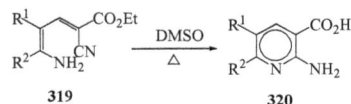

Scheme 107

**5.3.1.2. Synthesis of quinoline and fused quinolone derivatives:** The ylidene **321** could be cyclized by heating with ethylphosphite at 160-170°C into the corresponding quinoline derivatives **322** is depicted in Scheme 108 [259].

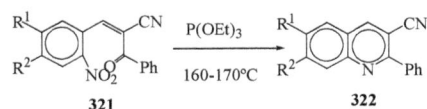

Scheme 108

Several pyrano[3,2-c]quinolines **324** were prepared [231,260] from 4-hydroxy-2(1H)quinolinones **323** and ylidene nitriles is depicted in Scheme 109.

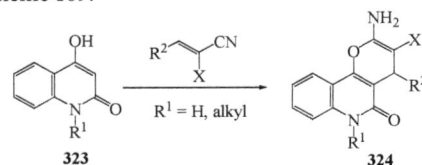

Scheme 109

Khallh et al. [261] reported that 8-quinolinol **325** reacts with cinnamonitrile derivatives in an ethanolic solution in the presence of piperidine to afford 2-amino-4-aryl-4H-pyrano[3,2-h]quinoline-3-carbonitriles **326** or ethyl 2-amino-4-aryl-4H-pyrano[3,2-h]quinoline-3-carboxylates **326** in moderate yield, as depicted in Scheme 110.

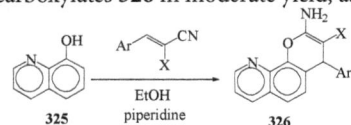

Scheme 110

Mahmoud et al. [262] reported that the same results obtained when α-cyano-3,4,5-trimethoxyphenyl cinnamonitrile **327a** and/or ethyl α-cyano-3,4,5-trimethoxy phenylcinnamate **327b** been reacted with **325**. Thus, 2-amino-3-cyano-4-(3,4,5-trimethoxyphenyl)-4H-pyrano[3,2-h]quinoline **328a** and ethyl 2-amino-3-cyano-4-(3,4,5-trimethoxyphenyl)-4H-pyrano[3,2-h]quinoline-3-carboxylate **328b** were synthesized from **325** and **327a,b**, as depicted in Scheme 111.

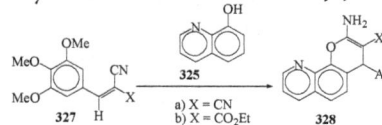

Scheme 111

The reaction of acetyl and benzoylmethyl pyridine **329** with ethoxymethylene malononitrile in DMSO in the presence of potassium carbonate gave acylcyanoimino quinolines **330** with yields 70-96%, respectively, as depicted in Scheme 112 [263].

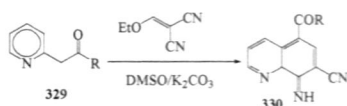

Scheme 112

Ortho-itrobenzaldehyde **307** reacted with 2-amino-1,1,3-tricyano-prop-1-ene **111** to yield the condensation product **331** which could be readily cyclized into **332** is depicted in Scheme 113 [254].

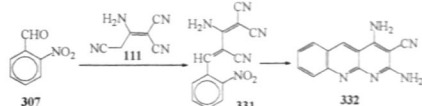

Scheme 113

High yielding syntheses of polyfunctional benzo[a]quinolizines are well-documented [264]. Abdallah et al. [265] reported a new and general one-step route affording polyfunctional substituted benzo[a]quinolizines in good yield from readily available inexpensive starting materials using isoquinoline derivatives. Thus, treatment of 1-methyl isoquinoline **334** with arylidenesulfonylacetonitriles **333** in boiling acetonitrile in the presence of an equimolar amount of piperidine leads, in each case, to the formation of only one product **337 and 338**, as indicated by TLC and 1H-NMR analysis. The formation of **337** may be explained by cyclization of the initially formed Michael adduct **335** to the unisolated product **336**, subsequent autoxidation of the latter leads to the final product **337**. When the reaction of **334** with **333** was carried out in the presence of excess piperidine (2 moles) then the product **338** were formed directly, as outlined in Scheme 114.

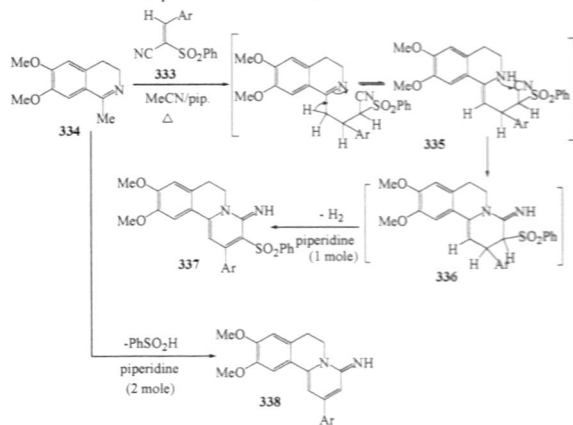

Scheme 114

Similarly, isoquinoline-1-yl acetonitrile **339** reacted with **333** to give **341** and **342** through the cyclization and dehydrogenation of **340**, as outlined in Scheme 115.

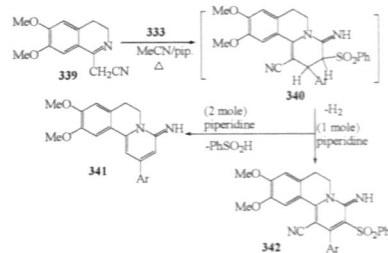

Scheme 115

**Synthesis of pyran, coumarin and condensed pyran derivatives:** Pyrans were readily obtained in good yields on treatment of ylidene derivatives of α-cyanochalcones with active methylene nitriles and active methylene ketones. Thus, benzylidene derivatives, aminomethylene and mercaptomethylene derivatives has been reported [170,235,236] to react with active methylene reagents to yield pyran derivatives [266-272].

The formation of pyrans in these reactions is assumed to proceed via additions of the reagent to the activated double bond and subsequent cyclization to the pyrane derivative, as demonstrated by the formation of **345** from the reaction of cinnamonitrile derivative **343** with active methylene reagents and as depicted in Scheme 116.

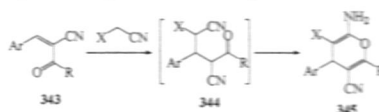

Scheme 116

The development of new method [273] for asymmetric synthesis of highly functionalized 2-amino-4-aryl-4H-pyrans **347** was achieved via Michael addition reaction of β-ketoesters **346** to arylidene malononitriles **218**, as depicted in Scheme 117 in the presence of piperidine as a base.

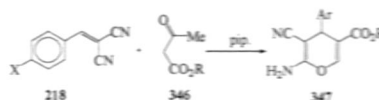

Scheme 117

4-Alkyl-2-amino-4H-pyran **349** was synthesized via Michael addition reaction of benzoyl acetonitrile to α-cyanoacrylates **348** is depicted in (Scheme 118) using piperidine as a catalyst [274].

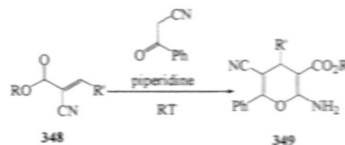

Scheme 118

Asymmetric Michael addition of cyanoacetates **45** to α-benzoylcinnamonitrile **27** in the presence of piperidine as catalyst has been studied, the resulting 3-alkoxy carbonyl-2-amino-5-cyano-4,6-diphenyl-4H-pyrans **350** have been obtained in good yield, , as depicted in Scheme 119 [275].

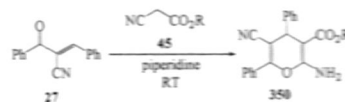

Scheme 119

The reaction of arylidene malononitrile **63** with some reactive halo compounds under phase transfer conditions (PTC) afforded the pyran derivative **351**, as depicted in Scheme 120.

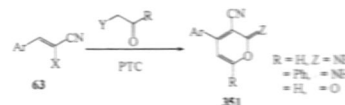

Scheme 120

Several new benzo[b]pyrans **352**, naphtho[1,2-b : 6,5-b]dipyrans **353**, naphtho [1,2-b]pyrans **354** and naphtho[2,1-b]pyrans **355** have been prepared by the reaction of cinnamonitriles **63** with resorcinol, 1,5-naphthalenediol, 1-naphthol and 2-naphthol, respectively [276,277], as outlined in Scheme 121.

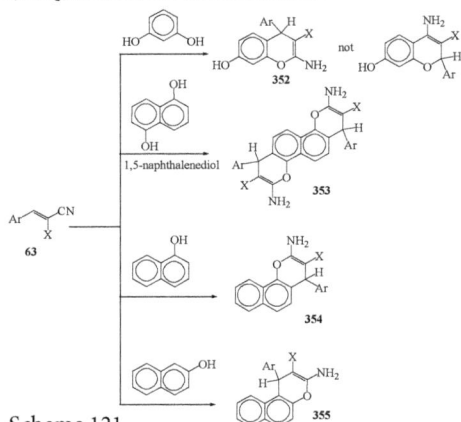

Scheme 121

It was reported [254] that salicylaldehyde reacted with diethyl 3-amino-2-cyano-pent-2-ene-1,5-dicarboxylate **111** to yield the coumarin derivative **356**. The same compound has been claimed to be obtained directly from the reaction of salicylaldehyde with ethylcyanoacetate [278], as shown in Scheme 122.

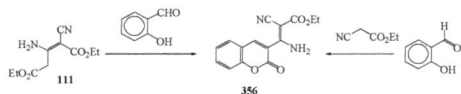

Scheme 122

Similarly, substituted salicylaldehyde have been reported [254] to afford the iminocoumarin derivative **357** when reacted with 2-amino-1,1,3-tricyanoprop-1-ene, as depicted in Scheme 123.

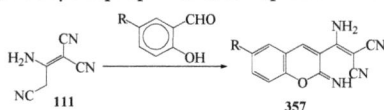

Scheme 123

3-Phenyl-7-hydroxy-3,4-dihydrocoumarin **358** [279,280] has been reported from the reaction of resorcinol with activated nitrile in catalytic amount of zinc chloride, as depicted in Scheme 124.

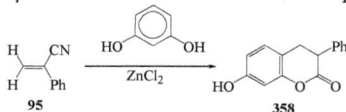

Scheme 124

Cycloaddition of substituted phenols with the nitriles derivative gave the 3-cyanocoumarin derivatives **359** is depicted in Scheme 125 [281,282].

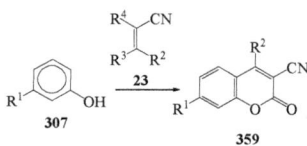

Scheme 125

Condensation of 4-hydroxycoumarin **358** has also been successful with unsaturated nitriles **210** using pyridine [283] and yielded the pyrano[3,2-c]coumarin derivatives **360** is depicted in Scheme 126.

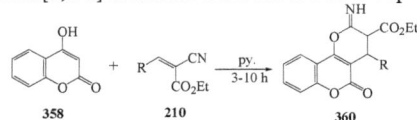

Scheme 126

4-Phenylcoumarin-3,4-dihydro-α-pyrone **362** (m.p. 183-4°C) was obtained by condensation of 4-hydroxy coumarin **358** with benzylidene malononitriles **63b** in pyridine and the resulting intermediate **361** was subsequently hydrolyzed with HCl/AcOH and finally cyclized with Ac$_2$O [284], as outlined in Scheme 127.

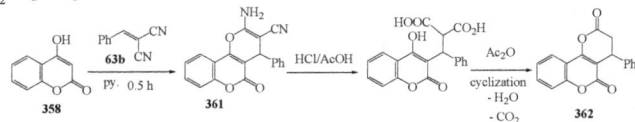

Scheme 127

Recently [285], it was reported that annulations reactions of 4-hydroxycoumarin **358** with p-anisylidene ethylcyanoacetate or p-anisylidene malononitrile **210** yielded the corresponding 2-amino-3-carbomethoxy(cyano)-4-(4'-methoxyphenyl)-5H-1-benzo pyrano-[4,3-b]-pyran-5-ones **363a,b**, as depicted in Scheme 128.

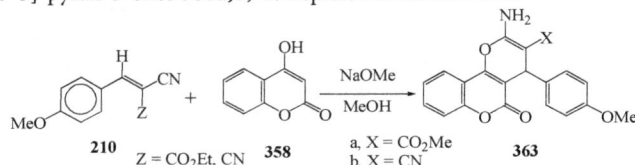

Scheme 128

It was reported that [286] thermal Michael addition reaction takes place when 6,7-dimethoxy isochromanone **334** was treated with benzylidene malononitrile **21** at 190°C to afford **264** which underwent elimination of malononitrile producing **365**, as shown in Scheme 129.

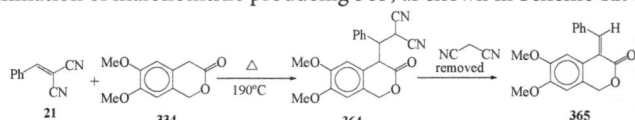

Scheme 129

New spiro pyran systems attached to coumarin nucleus **366** and **367** were synthesized by the reaction of 2-coumarylidenemalononitriles with some active methylene compounds in the presence of triethylamine [189], as shown in Scheme 130.

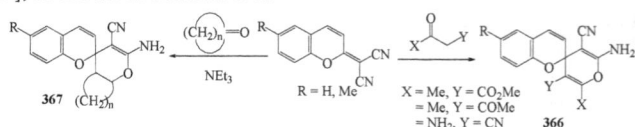

Scheme 130

**Six membered rings with two heteroatoms**

**Synthesis of pyridazine and condensed pyridazine derivatives:** An interesting approach for synthesis of pyridazines **368** has been achieved by cyclization of the intermediated of the reaction of cinnamonitrile derivatives **63** with aryldiazonium chloride [287]. This synthetic approach is summarized below in the following Scheme 131.

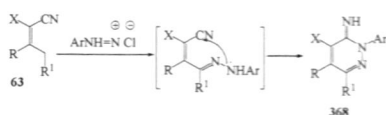

Scheme 131

El-Nagdy et al. [239] reported that 3-amino-2-cyano-4-ethoxycarbonyl but-2-enonitrile coupled with aromatic diazonium chlorides to yield **369** which converted to **370** on refluxing in acetic acid / HCl mixture, as outlined in Scheme 132.

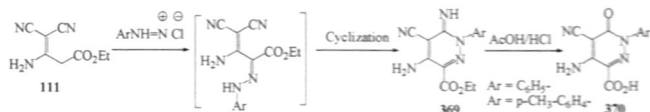

Scheme 132

Coupling of 2-amino-1,1,3-tricyanoprop-1-ene with aryldiazonium salts and subsequent cyclization of the coupling products yielded the pyridazine derivative **349**. The same pyridazine derivatives **373** could be alternatively synthesized via treatment of arylhydrazonomethylenemalononitrile derivatives **371** with malononitrile, a reaction that proceeds almost certainly via the intermediacy of the hydrazone **372** [288], as outlined in Scheme 133.

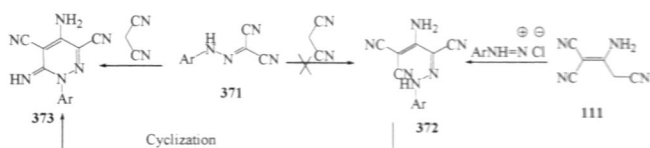

Scheme 133

Similar synthesis of pyridazine derivatives utilizing diethyl 3-amino-2-cyano-pent-2-ene-1,5-dicarboxylate has been reported [287]. Acenaphthenoquinones readily condense with malononitrile to yield the corresponding yildenemalononitrile **374**, which reacts readily with hydrazine hydrate to yield aminopyridazine derivatives **375** [41], as depicted in Scheme 134

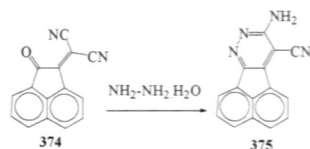

Scheme 134

Cinnoline derivatives were also reported utilizing α-hydrazononitrile as starting components. Thus, heating phenylhydrazonomalononitrile **371** with anhydrous aluminum chloride affords 4-amino-3-cyanocinnoline **376** is depicted in Scheme 135 [289].

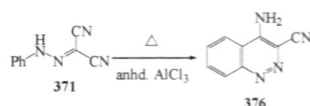

Scheme 135

**Synthesis of pyrimidine and fused pyrimidine derivatives:** α,β-

Unsaturated nitriles have been extensively utilized for the synthesis of pyrimidines. Tylor et al. [290,291] have summarized all literature in this area in more than one reference. One of the interesting examples of the utility of α,β-unsaturated nitriles for pyrimidine synthesis is the reported reaction of 2-aminomethylene malononitrile **377** with acetamidine **378** to yield pyrimidines **379** and **380** and cyclized into fused pyrimidine derivative **381**, as outlined in Scheme 136 [291].

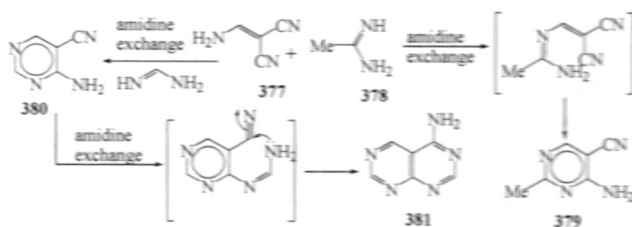

Scheme 136

Cyanoethylation [292] of cyanoguanidine in presence of lithium hydride has been reported to yield pyrimidines **382-385** in good to moderate yields, as outlined in Scheme 137.

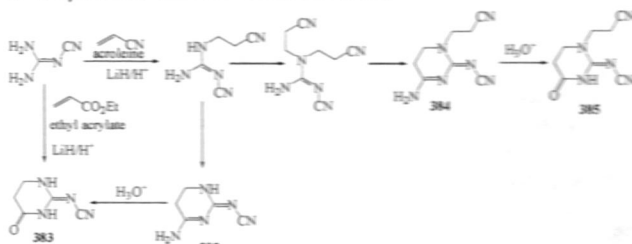

Scheme 137

4-Oxo-2-thioxopyrimidine derivative **386** [293] was obtained by the reaction of ethyl α-cyano-β-methoxy-cinnamate with thiourea in the presence of potassium carbonate, as depicted in Scheme 138.

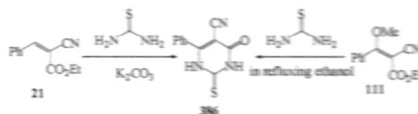

Scheme 138

The reaction of ethoxymethylene derivatives **388** with urea derivatives **387** were shown to yield carbethoxy pyrimidines [294-296]. The behaviour of **363** with thiourea [297] was demonstrated and gave the pyrimidine derivatives **389-393**, as outlined in Scheme 139.

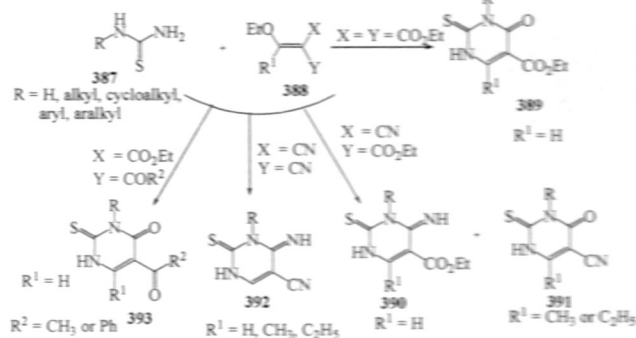

Scheme 139

A direct one-step synthesis of pyrimidines has been reported utilizing 3-oxoalkanenitriles as starting materials, thus, 2-aryl-3-oxoalkanenitriles **394** reacts with formamide and phosphoryl chloride to yield pyrimidines **395**, as depicted in Scheme 140 [297].

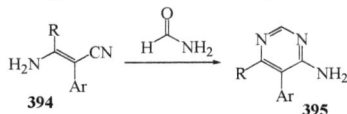

as depicted in Scheme 140

The isolation of several other side products, depending on the nature of the oxoalkanenitrile, has been reported [298-301]. Several other synthetic approaches for pyrimidines utilizing 3-oxoalkanenitriles as starting material have been reported and surveyed [301,302]. Compound **396** reacted with either phenylisothiocyanate or benzoylisothio-cyanate in refluxing dioxane to yield the pyrimidine derivative **397** is depicted in Scheme 141 [240].

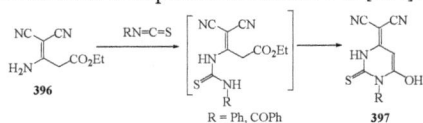

Scheme 141

The reaction of barbituric acid, thiobarbituric acid and 4-bromo-3-methyl pyrazolin-5-one with acrylonitriles **398** was reported by Abdel-Latif [303], thus, the compound **399a** reacted with **398a-c** to give the pyrano[3,2-d]pyrimidines **400a-c**. The alternative structure **401** was excluded on the basis of spectral data. Similarly, the acid **399b** reacted with **398a,d** to give the corresponding pyrano[3,2-d] pyrimidines **400de**. On the contrary, attempts to bring about addition of **399b** to **398b,c** (Ar = 2-furyl, 2-thienyl) failed and the reactions were recovered unchanged after being refluxed in ethanolic triethylamine. Thus, it can be concluded that the introductions of a π-deficient heterocycles at β-position of the acrylonitrile increases the reactivity of the double bond towards Michael type addition reaction and the introduction of a π-excessive heterocycle decrease its reactivity. In contrast to the behaviour of **398a-d** towards **399a,b** attempts to bring about addition of **373f-h** to **399a,b** resulted in the formation of ylidene derivatives **402a-d**, which assumed to be formed via elimination of a malononitrile molecule from the Michael adduct intermediate. Similar ylidene formation by the addition of α,β-unsaturated nitriles to active methylene reagents has been observed earlier in several reactions [101,250], as shown in Scheme 142.

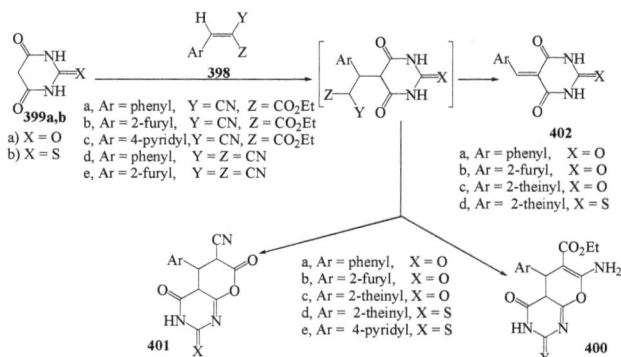

Scheme 142

Mahmoud et al. [163] found that when compound **374a,b** was

submitted to react with the cinnamonitrile derivatives **63** in refluxing pyridine afforded the arylidene derivative **378** as the sole product, as depicted in Scheme 143.

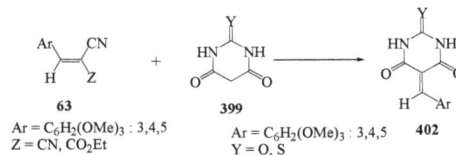

Scheme 143

A variety of pyrimidine synthesis, utilizing nitriles as starting components has been reported [291,292,304-308]. Enaminonitriles **403** and **404** react with trichloro-acetonitrile to yield the corresponding pyrimidine derivatives **405** and **406**, respectively [106,229,309], as shown in Scheme 144.

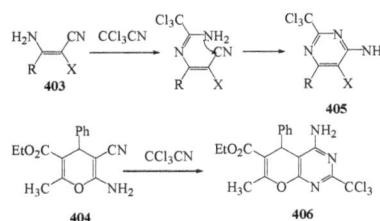

Scheme 144

Another reported pyrimidine synthesis is summarized below as depicted in Scheme 1145 [310].

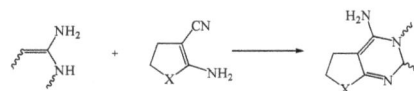

Scheme 145

Fawzy et al. [310] reported that hydrazopyrimidine **407** was reacted with cinnamonitriles afforded the corresponding arylhydrazopyrimidine **408** is depicted in Scheme 146

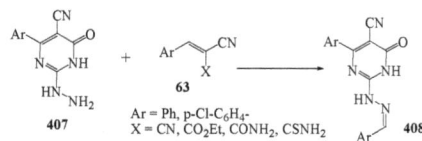

Scheme 146

Geies [311] has been reported that 6-aminouracil and 6-aminothiouracil **409** were reacted with benzylidenemalononitrile in ethanol in the presence of piperidine to afford **410a,b**, respectively. The reaction was assumed to proceed via Michael addition of the pyrimidine nucleus to the α,β-unsaturated nitriles and subsequent cyclization through nucleophilic addition of the amino group to one of the two cyano groups [312], as shown in Scheme 147.

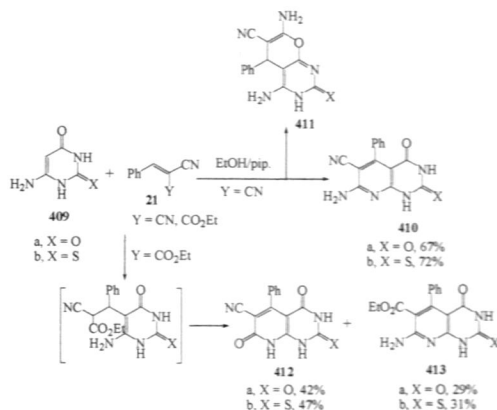

Scheme 147

The structure of compound **410a,b** was established as pyridopyrimidine rather than pyranopyrimidine **411** on the basis of 1H-NMR and IR spectra. On the other hand, the reaction of **409a,b** with benzylideneethylcyanoacetate under the same conditions results in a mixture of compound **412a,b** and/or **413a,b**, respectively. Aminopyrazole and aminoisoxazole derivatives have also been reported to react with acrylonitrile to yield either fused pyrimidines or ring N-cyanoethylated products, which readily cyclized to fused pyrimidines **414-416** [46,111,309,313-324], as outlined in Scheme 148.

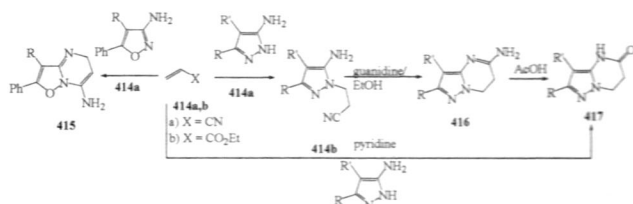

Scheme 148

A recent interesting pyrimidine synthesis has been reported and is summarized in Scheme 149. The utility of the resulting cyanopyrimidines for building up fused heterocycles has also been reported [106].

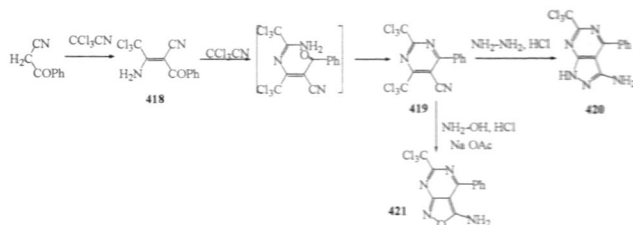

Scheme 149

The enaminonitrile **31** was utilized for synthesis of several new fused pyrimidine derivatives **422-427**, as described in Scheme 150.

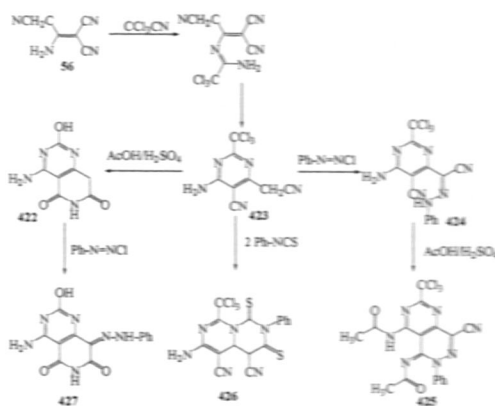

Scheme 150

Synthesis of several new ring system derived from pyrazolo[1,5-a]pyrimidines **429** and 1,2,4-triazolo[3,4-a]pyrimidines **430** has been recently reported via the reaction of enaminonitrile **428** with cyclic amidines. The mechanism of the reaction involved was discussed, as deicted in Scheme 151 [320].

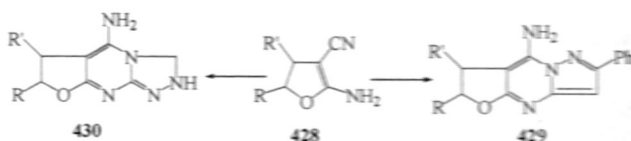

Scheme 151

Other syntheses of fused pyrimidines **431–433** from enaminonitriles are shown below [320-324], as outlined in Scheme 152.

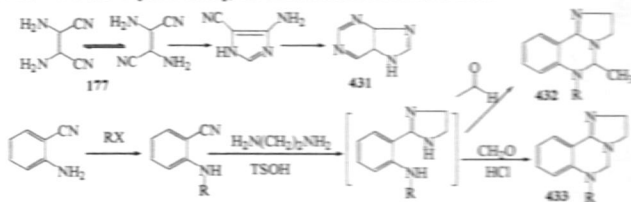

Scheme 152

**Synthesis of pyrazine derivatives:** Only few examples for synthesis of pyrazines **434, 435** utilizing nitriles as starting materials have been reported. A demonstrated example for this synthesis approach is shown below [325,326], as outlined in Scheme 153.

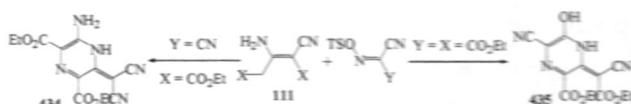

Scheme 153

**Synthesis of thiazine derivatives:** Several new pyrolidino[1,2-a]-3,1-thiazine-5,6-dione derivative **436** was synthesized via the reaction of 4-cyano-2,3-dioxo-5-thienopyrolidine **437** with a variety of activated nitriles, as depicted in Scheme 154 [327].

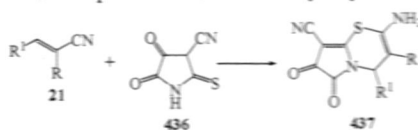

Scheme 154

Treatment of coumarinylmalononitrile **438** with bidentates, namely, guanidine hydrochloride, thiourea, thiosemicarbazide, thioacetamide and phenyl isothiocyanate in the presence of acetylacetone under phase transfer conditions gave the corresponding spiro coumarinyl-1,3-thiazine **439** and **440** [178], as outlined in Scheme 155.

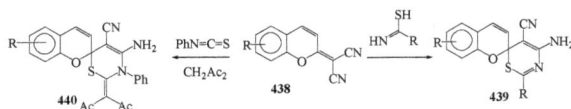

Scheme 155

**Six-membered rings with three heteroatoms:** Several triazine syntheses starting from α,β-unsaturated nitriles have appeared in some literature [41,287,288]. An interesting example of these syntheses is shown in Scheme 156 [328,329].

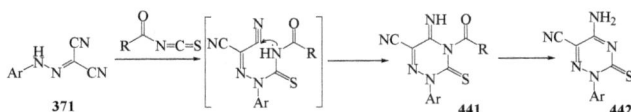

Scheme 156

## Synthesis of Other Heterocyclic Compounds

The recent wide applications of 2-propenoylamides, esters and 2-propenoyl chlorides in the synthesis of biologically and pharmacologically active compounds [330-340] and beside their uses in the synthesis of industriasl products make them worthy to be synthesized to obtain new structures of anticipated enhanced potency. Madkour et al. reported the reaction and uses of 3-(4'-methoxyphenyl) and 3-(2'-thienyl)-2-cyano-2-propenoyl chloride in heterocyclic synthesis, as described in Scheme 157.

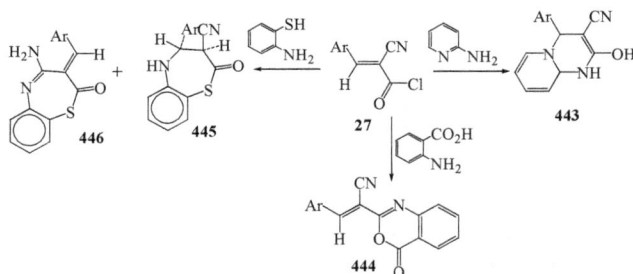

Scheme 157

## Conclusion

Alkenyl nitriles have proved to be a rich source of various heterocyclic compounds, and the discovery of potential biologically active heterocyclic compounds has become increasingly probable. Starting from alkenyl nitriles, our current work is focussed on synthesising novel heterocycles with or without sulphur that have biological activities against different diseases. The search for cheaper and simpler methods to synthesis such new compounds are continuing.

This review has summarised some of the achievements in the field of heterocyclic compounds derived from alkenyl nitriles. Our knowledge of the chemistry and reactions of alkenyl nitriles remains shallow, however, and this field needs to be explored in more detail. Further studies and investigations by us or other workers should continue to provide a strong background in the chemistry and reactions of alkenyl nitriles.

## References

1. MOWRY DT (1948) The preparation of nitriles. Chem Rev 42: 189-283.

2. Pelouze J (1834) Notiz über einen neuen Cyanäther. Annalen der Pharmacie 10: 249.

3. Fehling H (1844) Ueber die Zersetzung des benzoësauren Ammoniaks durch die Wärme. Justus Liebigs Annalen der Chemie 49: 91–97.

4. Fraser S Flemming (1999) Nitrile-containing natural products. Nat Prod Rep 16: 597-606.

5. Fleming FF, Yao L, Ravikumar PC, Funk L, Shook BC (2010) Nitrile-containing pharmaceuticals: efficacious roles of the nitrile pharmacophore. J Med Chem 53: 7902-7917.

6. Al-Azmi A, Elassar AZA, Booth BL (2003)The chemistry of diaminomaleonitrile and its utility in heterocyclic synthesis. Tetrahedron 59: 2749-2763.

7. Alves MJ, Booth BL, Proença MF (1990) Synthesis of 5-amino-4-(cyanoformimidoyl)-1H-imidazole: a reactive intermediate for the synthesis of 6-carbamoyl-1,2-dihydropurines and 6-carbamoylpurines. J Chem Soc Perkin Trans 2: 1705-1712.

8. Booth BL, Dias AM, Proença MF, Zaki ME (2001) The reactions of diaminomaleonitrile with isocyanates and either aldehydes or ketones revisited. J Org Chem 66: 8436-8441.

9. Ohtsuka Y (1978) Chemistry of diaminomaleonitrile. 3. Reaction with isocyanate: a novel pyrimidine synthesis. J Org Chem 43: 3231-3234.

10. Ohtsuka Y (1979) Chemistry of diaminomaleonitrile. 4. Nitrile hydration of the Schiff bases. J Org Chem 44: 827-830.

11. Shirai K, Matsuoka M, Fukunishi K (2000) New syntheses and solid state fluorescence of azomethine dyes derived from diaminomaleonitrile and 2,5-diamino-3,6-dicyanopyrazine. Dyes Pigm 47: 107-115.

12. Begland RW, Hartter DR, Jones FN, Saw DJ, Sheppard WA, et al. (1974) Hydrogen cyanide chemistry. VIII. New chemistry of diaminomaleonitrile. Heterocyclic synthesis. J Org Chem 39: 2341-2350.

13. Barton JW, Goodland MC, Gould KJ, McOmie JFW, Mound WR (1979) Biphenylenes-xxxi: Condensation of benzocyclobutene-1,2-dione with aliphatic and heterocyclic 1,2-diamines and the synthesis of cis-2-cyano-3-(2'-cyanovinyl)-1,4-diaz. Tetrahedron 35: 241-247.

14. Beall LS, Mani NS, White AJ, Williams DJ, Barrett AG, et al. (1998) Porphyrazines and Norphthalocyanines Bearing Nitrogen Donor Pockets: Metal Sensor Properties. J Org Chem 63: 5806-5817.

15. Woehrle D, Buttner P (1985) Polymeric schiff's base chelates and their precursors 8a, some cobalt chelates as catalysts for the isomerization of quadrycyclane to norbornadien. Polym Bull 13: 57-64.

16. Zwanenburg B, Peter ten H (2001) Thesynthetic potential of three-membered ring aza-heterocycles.Topics in Current Chemistry 216: 93-124.

17. Madkour HMF, Elgazwy ASH (2007) Utilities of some carbon nucleophiles in heterocyclic synthesis. Current Organic Chemistry 11: 853-908.

18. Ayman Wahba (1993) The chemistry of .beta.-enaminonitriles as versatile reagents in heterocyclic synthesis. Erian Chem Rev 93: 1991-2005.

19. Zil'berman EN (1962) Reactions of nitriles with hydrogen halides and nucleophilic reagents. Russ Chem Rev 31: 615.

20. Zil'berman EN (1960) Some reactions of nitriles with the formation of a new nitrogen–carbon bond.Russ Chem Rev 29: 331.

21. Keller CL, Dalessandro JD, Hotz RP, Pinhas AR (2008) Reactions in water: alkyl nitrile coupling reactions using Fenton's reagent. J Org Chem 73: 3616-3618.

22. Peter Pollak, Gérard Romeder, Ferdinand Hagedorn, Heinz-Peter Gelbke "Nitriles".

23. Kuo CW, Zhu JL, Wu JD, Chu CM, Yao CF, et al. (2007) A convenient new procedure for converting primary amides into nitriles. Chem Commun (Camb): 301-303.

24. Sharwan K, Dewan, Ravinder Singh, Anil Kumar (2006) One pot synthesis of nitriles from aldehydes and hydroxylamine hydrochloride using sodium sulfate (anhyd) and sodium bicarbonate in dry media under microwave irradiation. Arkivoc 2: 41-44.

25. Nagata W, Yoshioka M (1988) Diethylaluminum cyanide. Org Synth Coll 6: 436.

26. Nagata W, Yoshioka M, Murakami M (1988) Preparation of cyano compounds using alkylaluminum intermediates: 1-cyano-6-methoxy-3,4-dihydronaphthalene. Org Synth Coll 6: 307.

27. Reynold C Fuson, Oscar R Kreimeier, Gilbert L Nimmo (1930) Ring closures in the cyclobutane series. ii. cyclization of ?,?'-dibromo-adipic esters. J Am Chem Soc 52: 4074-4076.

28. Houben J, Walter Fischer (1930) Über eine neue Methode zur Darstellung cyclischer Nitrile durch katalytischen Abbau (I. Mitteil.) (p 2464-2472) Berichte der deutschen chemischen Gesellschaft (A and B Series) 63: 2464 – 2472.

29. Kukushkin V Yu, Pombeiro AJL (2005) Metal-mediated and metal-catalyzed hydrolysis of nitriles. Inorg Chim Acta 358: 1–21.

30. Smith Andri L, Tan Paula (2006) Creatine synthesis: an undergraduate organic chemistry laboratory experiment. J Chem Educ 83: 1654.

31. Jean-Marc M, Caroline M, Hassan H, Arkivoc MC (2006) The reductive decyanation reaction: chemical methods and synthetic applications.Arkivoc (iv): 90-118.

32. Nakao Y, Yada A, Ebata S, Hiyama T (2007) A dramatic effect of Lewis-acid catalysts on nickel-catalyzed carbocyanation of alkynes. J Am Chem Soc 129: 2428-2429.

33. Bruson HA (1949) Organic reactions. J Wiley and Sons 5: 101.

34. AI Meyers, JC Sircar (1990) Addition to the Cyano group to form Heterocycles. J Wiley and Sons New York, 341.

35. Freeman F (1981) Reactions of Malononitrile Derivatives. Synthesis 925.

36. El-Nagdy MH, El-Fahham HA, El-Gemeie GEH (1983) Heterocycles 20: 519.

37. El-Nagdy MH, El-Moghayar MRH, El-Gemeie GEH (1984) Synthesis 1.

38. Abdel-Galil FM, Sherief SM, El-Nagdy MH (1986) Heterocycles 24: 2023.

39. Freeman F (1969) The chemistry of malononitrile. Chem Rev 69: 591-624.

40. Freeman F (1980) Properties and reactions of ylidenemalononitriles. Chem Rev 80: 329-350.

41. El-Nagdy MH, Sherief SM, Mohareb RM (1987) Heterocycles 20: 497.

42. Mahmoud MR, El-Bordainy EAA, Hassan NF, Abu El-Azm FSM (2007) Utility of nitriles in synthesis of pyrido[2,3-d]pyrimidines, thiazolo[3,2-a] pyridines, pyrano[2,3-b]benzopyrrole and pyrido[2,3-d]benzopyrroles. J. P, S, Si and Related elements 182: 2507-2521.

43. Mahmoud MR, Madkour HMF, El-Bordainy EAA, Soliman EA (2009) Activated nitriles with ammonium benzyldithiocarbamate, synthesis of thietane derivatives. J.P, S, Si and Related elements 184: 1-8.

44. Gewald K, Bottcher H, Schlinke E (1966) Chem Ber 99: 94.

45. Goudic AG (1976) U.S. Patent 3: 750-963.

46. Arya VP (1972) Indian J Chem 10: 812.

47. Mawas MS, Sugiuta M, Chaula HPS (1978) Heterocyclic compounds. X. A new synthesis of thiadiazasteroids. J Heterocyclic Chem 15: 949-953.

48. Robba M, Lecomte JM, De Sevricowt MC (1974) Bull Soc Chim France 12: 2864.

49. Ishikawa T, Nagai K, Kudoh T, Saito S (1995) Synlett 1171.

50. El-Nagdy MH, Abed NH, El-Moghayar MRH, Fleita DN (1976) Indian J Chem 14B: 422.

51. El-Nagdy MH, El-Moghayar MRH, Hammam AG, Khallaf SA (1979) The reaction of malononitrile with thioglycolic acid. A novel procedure for the synthesis of thiazolone derivatives. J Heterocyclic Chem 16: 1541-1543.

52. Fiesselmann H, Thoma F (1956) Chem Ber 89: 1907.

53. Cairns TL, Carboni RA, Coffman DD, Engelhardt VA, Heckent RE, et al. (1958) J Amer Chem Soc 80: 2775.

54. WJ Middleton, VA Engelhardt, BS Fisher (1958) J Amer Chem Soc 80: 2822.

55. Augustin M, Dehne H, Rndolf WD, Krey P (1977) Ger Offen 124: 303.

56. El-Nagdy MH, Zayed EM, Khalifa MAE, Ghozlan SA (1981) Monatsh Chem 112: 245.

57. Konno T, Yamazaki T, Kitazume T (1996) Synthesis and application of α-trifluoromethylated aldehydes. Tetrahedron 52: 199-208.

58. Ahmed I Hashem, Alexander Senning, Abdel-Sattar S Hamad Elgazwy (1998) Photochemical Transformations of 2(5H)furanones. Org Prep Proceed Int 30: 403-427.

59. Schauman E, Mrolzek H (1979) Tetrahedron 35: 16.

60. Aran VJ, Soto JL (1983) Ann Chem 79: 340.

61. Hassel GH, Morgan BA (1970) J Chem Soc 1: 1345.

62. Bardsher CK, Hunt DA (1981) J Org Chem 46: 327.

63. Sherif SM (1985) Ph.D. Thesis, Cairo University.

64. Matsumoto K, Uchida T (1981) Cycloaddition reactions of cycloimmonium ylides with triphenylcyclopropene. J Chem Soc Perkin Trans 1: 73.

65. Matsumoto K, Uchida T, Paquette LA (1975) Synthesis 746.

66. Stachel HD, Hergel KK, Poscherieder H, Burghard H (1980) J Heterocyclic Chem 17: 1195.

67. Babler JH, Spina KP (1984) Tetrahedron Letters 25: 1659.

68. Gewald K, Martin A (1981) J Prakt Chem 323: 843.

69. Susse M, Johne S (1981) J Prakt Chem 323: 647.

70. Bayomi SM, Haddad DY, Sowell St JW (1984) One step synthesis of 6,7-dimethyl-2-oxo-1,2,3,4-tetrahydropyrrolo[1,2-a][1,3]pyrimidines. J Heterocyclic Chem 21: 1367-1368.

71. El-Moghayar MRH, Ibrahim MK, Ramiz MM, El-Nagdy MH, Naturforsch Z (1983) 38b: 724.

72. Akiyama Y, Abe J, Takano T, Kawasaki T, Sakamoto M (1984) Chem Pharm Bull 32: 2821.

73. Leete H, Leete SAS (1978) J Org Chem 43: 2122.

74. Gossauer G, Hinze (1978) RP J Org Chem 43: 283.

75. Verhe R, Dekimpe N, De Buyck L, Tilley M, Schamp N (1980) Tetrahedron 36: 131.

76. Lim M, Klein S (1981) Synthesis of "9-deazaadenosine"; a new cytotoxic c-nucleoside isostere of adenosine. Tetrahedron Letters 22: 25-28.

77. Smith LI, Dale WJ (1950) J Org Chem 15: 832.

78. Voloven MY, Svishchuk AA (1979) Khim Geterotsikl Soedin 1: 129.

79. Isomura K, Uto K, Taniguchi H (1977) J Chem Soc Perkin Trans 1: 664.

80. Noyori R, Umeda I, Kawauchi H, Takaya H (1975) Nickel(0)-catalyzed reaction of quadricyclane with electron-deficient olefins. J Amer Chem Soc 97: 812.

81. Newkome GR, Paudler WW (1982) Contemporary Heterocyclic Chemistry. J Wiley and Sons.

82. Garanti L, Zecchi G (1980) J Org Chem 45: 4767.

83. Davies JS, Davies VH, Hassall CH (1969) The biosynthesis of phenols. XX. Synthesis of anthraquinones through carbanions of ortho-substituted benzophenones. J Chem Soc Perkin 14: 1873-1879.

84. Khodairy, El-Sayed AM (2001) Synth Commun 31: 475-486.

85. Sobenina LN, Mikhaleva AL, Sergeeva MP, Petrova OV, Aksamentova TS, et al. (1995) Tetrahedron 51: 4223.

86. Sobenina LN, Mikhaleva AL, Sergeeva MP, Toryashinova DSD, OB Kozyreva, et al. (1996) Khim Geterotsikl Soedin 919.

87. Sobenina LN, Mikhaleva AL, Toryashinova DSD, Kozyreva OB, Trofimov BA (1997) Sulfur Lett 20: 205.

88. Sobenina LN, Demenev AP, Mikhaleva AI, Petrova OV, Larina LI, et al. (2000) Sulfur Lett 24: 1.

89. Sobenina LN, Demenev AP, Mikhaleva AI, Ushakov IA, Afonin AV, et al. (2002) Chem Heterocycl Compd 38: 86.

90. Dickstein JI, Miller SI, Patai S (1978) The chemistry of the carbon-carbon triple bond, part 2, Wiley, New York, 917.

91. J Burns (1983) J Prakt Chem 12: 2347.

92. Skotsh C, Kohlmeyer I (1979) Bveit E Synthesis 449.

93. Bauer L, Nambury CNV (1961) J Org Chem 26: 4917.

94. El-Nagdy MH (1974) Tetrahedron 30: 2791.

95. El-Nagdy MH, El-Moghayar MRH, Hafez EAA, Alnima HH (1975) J Org Chem 40: 2604.

96. Sadek KU (1977) M.Sc. Thesis, Cairo University.

97. Maeda K, Hosokawa T, Murahashi S, Moritani I (1973) Tetrahedron Lett 51: 5075.

98. Cusmano S, Sprio V, Trapani F (1952) Gazz Chim Ital 82: 98.

99. Abdou S, Fahmy SM, Sadek KU, El-Nagdy MH (1981) Activated nitriles in heterocyclic synthesis: a novel synthesis of pyrano[2,3-c]pyrazoles. Heterocycles 16: 2177-2180.

100. El-Moghayar MRH, Ibrahim MKA, El-Ghandour AHH, El-Nagdy MH (1981) Synthesis 635.

101. Daboun HAF, Abdou SE, Hussein MM, El-Nagdi MH (1982) Activated nitriles in heterocyclic synthesis: novel syntheses of pyrrolo[1,2-c]imidazole and pyrano[2,3-d]imidazole derivatives. Synthesis 1982: 502-504.

102. Aziz SI, Riad BY, Elfahham HA, El-Nagdi MH (1982) Activated nitriles in heterocyclic synthesis: a novel synthesis of pyrazolo[5,6:3',4']pyrano[5,4-b] isoxazoles. Heterocycles 19: 2251-2254.

103. Abdel-Sattar S, Elgazwy H (2003) The chemistry of isothiazoles. Tetrahedron 59: 7445-7463.

104. Khodairy, El-Sayed AM, Salah H, Abdel-Ghany H (2007) Synthetic Communications 37: 639-648.

105. Mayer R, Gewald K (1967) The action of carbon disulfide and sulfur on enamines, ketimines, and CH acids. Angewandte Chemie International Edition in English 6: 294-306.

106. El-Nagdi MH, Fahmy SM, Hafez EAA, El-Moghayar MRH, Amer SAR (1979) Pyrimidine derivatives and related compounds. A novel synthesis of pyrimidines, pyrazolo[4,3-d]pyrimidines and isoxazolo[4,3-d] pyrimidine. Journal of Heterocyclic Chemistry 16: 1109-1111.

107. El-Nagdi MH, Khalifa MAE, Ibrahim MKA, El-Moghayar MRH (1981) The reaction of nitriles with mercaptoacetic acid. A new synthesis of thiazole derivatives. Journal of Heterocyclic Chemistry 18: 877-879.

108. Cusmano S, Sprio V (1952) Gazz Chim Ital 82: 191.

109. Cusmano S, Sprio V (1952) Gazz Chim Ital 82: 420.

110. Kambe S, Sakurai A, Midorikawa H (1980) A Simplified Procedure for the Preparation of 2-Alkoxycarbonyl-5-aryl-4-cyano-3-hydroxy-3-phenyltetrahydrothiophenes. Synthesis 1980: 839-840.

111. El-Nagdi MH, Fleita DH, El-Moghayar MRH (1975) Reactions with β-cyanoethylhydrazine—II: Synthesis of some 4,5,6,7-tetrahydropyrazolo[1,5-a] pyrimidine derivatives. Tetrahedron 31: 63-67.

112. El-Nagdi MH, Hafez EAA, El-Fahham AH, Kandeel EM (1980) Reactions with heterocyclic amidines VIII. Synthesis of some new imidazo[1,2-b]pyrazole derivatives. Journal of Heterocyclic Chemistry 17: 73-76.

113. El-Nagdi MH, El-Moghayar MRH, El-Fahham HA, Sallam MM, Alnima HH (1980) Reactions with heterocyclic amidines. VI. Synthesis and chemistry of pyrazol-5-yl, and 1,2,4-triazol-5-ylhydrazonyl chlorides. Journal of Heterocyclic Chemistry 17: 209-212.

114. Takeuchi Y, Kawahara S, Suzuki T, Koizumi T, Shinoda H (1996) First electronic investigation of the structure of alkenes compatible to carbon radicals bearing a single fluorine atom. J Org Chem 61: 301-303.

115. Cusmano S, Sprio V (1952) Gazz Chim Ital 82: 373.

116. Coenen M, Faust J, Ringel C, Mayer R (1965) Synthesen mit Trichloracetonitril. Journal für Praktische Chemie 27: 239-250.

117. Gavrilenko BB, Momet VV, Bodnarchuck ND (1974) Zh Org Khim 10: 601 [C.A. 80, 133330 (1974)].

118. Bodnarchuck ND, Gavrilenko BB, Perkach GI (1968) Zh Org Khim 4: 1710 [C.A. 70, 19532 (1969)].

119. Josey AD (1964) The preparation of 1-amino-1-fluoroalkylethylenes by the addition of active methylene compounds to fluoroalkyl cyanides. J Org Chem 29: 707-710.

120. El-Nagdi MH, Fahmy SM, Zayed EM, Ilias MAN (1976) Z Nafurforsch 31b: 795.

121. Takagi K, Nagahara K, Ueda T (1970) Chem Pharm Bull 18: 2353.

122. Midorikawa H, Saito K, Hori I, Igarashi M (1974) The reaction of ethyl ethoxymethylenecyanoacetate with its hydrazino derivatives. Bulletin of the Chemical Society of Japan 47: 476-480.

123. Midorikawa H, Hori I, Saito K (1970) The formation of pyrazolo[1,5-a]pyrimidine derivatives. Bulletin of the Chemical Society of Japan 43: 849-855.

124. Lang SA, Lovell FM, Cohen E (1977) Synthesis of 4-(4-phenyl-3-pyrazolyl)-4H-1,2,4-triazoles. Journal of Heterocyclic Chemistry 14: 65-69.

125. Tam SYK, Klein RS, Wempen I, Fox JJ (1979) Nucleosides. 112. Synthesis of some new pyrazolo[1,5-a]-1,3,5-triazines and their C-nucleosides. J Org Chem 44: 4547-4553.

126. Rudorf WD, Augustin M (1978) Acylketen-S,S- und Acylketen-S,N-acetale als Bausteine für Heterocyclen: Pyrazole und Isoxazole. Journal für Praktische Chemie 320: 585-599.

127. Furukawa M, Yuki T, Hayashi S (1973) A convenient synthesis of pyrazolines from β-carbonylethylthiosulfates. Chemical & pharmaceutical bulletin 21: 1845-1846.

128. Ege G, Arnold P (1976) Aminopyrazole; II1. C-Unsubstituierte 1-Alkyl-3-aminopyrazole aus 2-Chloroacrylnitril bzw. 2,3-Dichloropropannitril und Alkylhydrazinen. Synthesis 1976: 52-53.

129. Burger K, Schickaneder H, Elguero J (1975) Eim einfacher zugang zum 1-NH-3-pyrazolin-system. Tetrahedron Letters 33: 2911-2914.

130. Peseke K (1975) [Preparation of substitution 5-amino-3-mercaptopyrazoles. 8. Syntheses with 1,3-dithietanes]. Pharmazie 30: 802.

131. El-Nagdi MH, El-Moghayar MRH, Fleita DH (1974) Reactions of α-Arylhydrazono-β-oxo-β-phenyl-propionitriles. Journal für Praktische Chemie 316: 975-980.

132. Taylor EC, Purdum WR (1977) Preparation of 3,4,5-trisubstituted pyrazoles from 2,2-dioxoketene-S,S-acetals Heterocycles 6: 1865-1869.

133. French J, Peseke K, Kristen H, Bräuniger H (1976) [Preparation of 1-(2-nitrovinyl)-pyrazole derivatives]. Pharmazie 31: 851-855.

134. Nagahara K, Takagi K, UedaT (1976) Reaction of ethoxymethylenemalononitrile with hydrazine hydrate. Chemical & pharmaceutical bulletin 24: 2880-2882.

135. Alcalde E, Mendoza JD, Elguero J, Marino1 J, Garcia-Marquina JM, et al. (1974) Elude de la réaction du β-aminocrotonitrile et du α-formyl phénylacétonitrile avec l'hydrazine: Synthèse d'amino-7 pyrazolo[1,5-a]pyrimidines. Journal of Heterocyclic Chemistry 11: 423-429.

136. Marsico JW, Joseph JP, Goldman L (1972) Patent US 3: 760

137. Senga K, Robins RK, O'Brien DE (1975) Synthesis of pyrazolo[1',5':1,2]-1,3,5-triazino[5,6-a]benzimidazoles. Journal of Heterocyclic Chemistry 12: 899-901.

138. Breuer H, Treuner UD (1974) Ger Offen 2: 408

139. Marsico JW, Joseph JP, Goldman L (1973) US Patent 3: 760.

140. Kreutzberger A, Burgwitz K (1980) Antibakterielle wirkstoffe. IV. Die nitrosubstitution im system der 5-amino-4-cyanpyrazole. Journal of Heterocyclic Chemistry 17: 265-266.

141. Kilpper G (1973) Ger Offen 2: 141.

142. Peseke K (1973) Ger (East) Patent 102: 382.

143. Schmidt R, Klemm K (1975) Ger Offen 2: 426.

144. Dickinson CL, Williams JK, Mckusick BC (1964) Aminocyanopyrazoles. J Org Chem 29: 1915-1919.

145. Hecht SM, Werner D, Traficante DD, Sundaralingam M, Prusiner P, et al. (1975) Structure determination of the N-methyl isomers of 5-amino-3,4-dicyanopyrazole and certain related pyrazolo[3,4-d]pyrimidines. J Org Chem 40: 1815-1822.

146. Earl RA, Pugmire RJ, Revankar GR, Townsend LB (1975) Chemical and

carbon-13 nuclear magnetic resonance reinvestigation of the N-methyl isomers obtained by direct methylation of 5-amino-3,4-dicyanopyrazole and the synthesis of certain pyrazolo[3,4-d]pyrimidines. J Org Chem 40: 1822-1828.

147. Ege G, Arnold P (1974) A simple synthesis of 3(5)-aminopyrazole. Angewandte Chemie International Edition in English 13: 206-207.

148. Shuman RF, Shearin WE, Tull RJ (1979) Chemistry of hydrocyanic acid. 1. Formation and reactions of N-(aminomethylene)diaminomaleonitrile, a hydrocyanic acid pentamer and precursor to adenine. J Org Chem 44: 4532-4536.

149. Aziz SI, Abd-Allah SO, Ibrahim NS (1984) Reactions with nitriles: a novel synthesis of furo[2,3-c]pyrazoles and pyrano[2,3-c]pyrazoles. Heterocycles 22: 2523-2527.

150. Girgis NS, El-Gemei GEH, Nawar GAM, El-Nagdi MH (1983) α, β-Unsaturated Nitriles in Heterocyclic Synthesis: The Reaction of β-(2-Furanyl)- and β-(2-Thienyl)acrylonitrile with Active Methylene Reagents. Liebigs Annalen der Chemie 1983: 1468-1475.

151. Abdel-Razek FM, Kandeel ZES, Hilmy KMH, El-Nagdi MH (1985) Substituted acrylonitriles in heterocyclic synthesis. The reaction of α-substituted β-(2-furyl)-acrylonitriles with some active-methylene heterocycles. Synthesis 1985: 432-434.

152. El-Gemei GEH, El-Faham HA, Ibrahim YR, El-Nagdi MH (1989) α, β-Unsaturated nitriles in heterocyclic synthesis: Synthesis of several arylpyridine and arylpyridazine derivatives. Archiv der Pharmazie 322: 535-539.

153. Abdel-Galil FM, Abdel-Motaleb RM, El-Nagdi MH (1988) Ann Chim 84: 19.

154. Otto HH (1974) Darstellung einiger 4H-pyrano[2.3-c]pyrazolderivate. Archiv der Pharmazie 307: 444-447.

155. Otto HH, Schmelz H (1980) Zur Darstellung von 6-Oxo-2H-pyrano[2,3—c] pyrazolen. Monatshefte für Chemie 111: 53-61.

156. Martin-Leon N, Queuteiro M, Seoane C, Soto JL (1990) J Chem Res(s) 156.

157. El-Nagdi MH, Taha NH, Elall FAMA, Abdel-Motaleb FM, Mahmoud FF (1989) Collect-Czech. Chem. Commun., 54, 1082.

158. El-Fahham HA, Abdel-Galil FM, Ibrahim YR, El-Nagdi MH (1983) Activated nitriles in heterocyclic synthesis. A novel synthesis of pyrazolo[1,5-a] pyrimidines and pyrano[2,3-c]pyrazoles. Journal of Heterocyclic Chemistry 20: 667-670.

159. El-Gemei GEH, Riad BY, Nawwar GA, El-Gamal S (1987) Nitriles in heterocyclic synthesis: Synthesis of new pyrazolo [1,5-a]pyrimidines, pyrano[2,3-c] pyrazoles and pyrano[3,4-c] pyrazoles. Archiv der Pharmazie 320: 223-228.

160. Kandeel ZE, Hilmy KHH, Abdel-Razek FM, El-Nagdi MH (1984) Chem. Ind. (London) 33.

161. Abdel-Hamid AO, Riad BY (1987) Arch Pharm 320: 1010.

162. El-Torgoman AM, El-Kousy SM, Kandeel ZE, Naturforsh Z (1987) 426: 107.

163. Mahmoud MR, Madkour HMF, Nassar MH, Habashy MM (2000) J Chinese Chem Soc 47: 937-941.

164. El-Nagdi MH, Ohta M (1973) Studies on 3,5-pyrazolidinediones. IV. Addition of 4-Arylazo-3,5-pyrazolidinediones to Ethyl Acrylate. Bulletin of the Chemical Society of Japan 46: 1830.

165. Maquestiau, Van Den Eynde JJ (1984) [C.A. 102, 112645k (1985)] Bull Soc Chim Belg 93: 451.

166. Mityurina KV, Kharachenko VG, Cherkesova LV (1981) Khim Geterotsikl Soedin [C.A. 95, 24909r (1981)] 403.

167. Metwally SA, Younes MI, Nour AM (1986) Synthesis of pyrano, thiopyrano, and pyrido[3,2-d]pyrazoles Heterocycles 24: 1631-1636.

168. El-Moghayar MRH, Khalifa MAE, Ibraheim MKA, El-Nagdi MH (1982) Reactions with heterocyclic β-enaminoesters: A novel synthesis of 2-amino-3-ethoxycarbonyl-(4H)-pyrans. Monatshefte für Chemie 113: 53-57.

169. Abdel-Sattar S Hamad, Hamed DAY (2001) A New Synthesis of 4-Cyano-1,3-dihydro-2-oxo-2H-imidazole-5-(N1-tosyl) carboxamide: A Reactive Precursor for ThioPurine Analogs Journal of Heterocyclic Chemistry 38: 939-944.

170. El-Nagdi MH, Erian AW (1990) Synthesis of substituted azaindenes: synthesis

of new pyrazolo[1,5-a]pyrimidine derivatives. Bulletin of the Chemical Society of Japan 63: 1854-1856.

171. Madkour HMF, Shiba SA, Sayed HM, Hamed AA (2000) Sulfur Letters 24: 151-179.

172. Shiba SA (1988) Arch Pharm 331: 91.

173. Shiba SA (1996) J P,S,Si, and Related elements 114: 29.

174. Shiba SA, Fahmy AFM, Abdel-Hamid HA, Massoud MS (1993) Egypt J Chem 36: 409.

175. Jaafar, Francis G, Danion-Bougot R, Danion D (1994) Synthesis 1: 56.

176. Mustafa A, Sammour A, Kira M, Hilmy MK, Anwar M, et al. (1965) Contributions to the chemistry of 3,5-pyrazolidinedione. Arch Pharm Ber Dtsch Pharm Ges 298: 516-532.

177. Kumar, HaH, Junjappa H (1976) Synthesis 324.

178. Khodairy (2000) J P,S,Si and Related elements 160: 159-180.

179. Ogdanowicz-Szweed KB, Rasodanska NK, Lipowska N, Rys B, Skonecka A (1993) Monatsheft fur Chemie 124: 721.

180. Al-Mousawi SM, El-Kholy YM, Mohammed MA, El-Nagdy MH (1999) Org Prep Proced Int 31: 305.

181. El-Nagdy MH, Fahmy SM, El-Moghayar MRH, Negm AN (1977) Naturforsh Z 32b: 1478.

182. El-Nagdy MH, Kandeel EM, Zayed EM, Kandeel ZK (1977) Pyrimidine derivatives and related compounds VI. A novel synthesis of 3,5-diacetamidopyrazole and of 2-aminopyrazolo [1,5-α] pyrimidines. J Heterocyclic Chem 14: 155-158.

183. Ibrahim NS, Sadek KU, Abdel-Al FA (1987) Studies on 3,5-Diaminopyrazoles: Synthesis of New Polyfunctionally Substituted Pyrazoloazines and Pyrazoloazoles. Arch Pharm 320: 240-246.

184. Chubb FL, Edward JT, Wong SC (1980) J Org Chem 45: 2315.

185. Cook AH, Downer JD, Heilbron SI (1948) J Chem Soc 2028.

186. Seng F, Ley K (1972) Synthesis 606.

187. Shawali AS, Sami M, Sherif SM, Parkanyi C (1980) Synthesis of some derivatives of imidazo[1,2-a]pyridine, pyrazolo[1,5-b]imidazole, and 4-(3H) quinazolinone from α-ketohydrazidoyl bromides. J Heterocyclic Chem 17: 877-880.

188. Riad SM, Abde-Hamid IA, Ibraheim HM, Al-Matar HM, El-Nagdy MH (2007) Heterocycles.

189. Ghozlan SAS, Abdel-Hamid SA, Ibraheim HM, El-Nagdy MH (2006) Studies with 2-arylhydrazononitriles: a new convenient synthesis of 2, 4-disubstituted-1,2,3-triazole-5-amines. Arkivoc 2006: 53-60.

190. El-Mousawi SM, Moustafa MS, El-Nagdy MH (2007) J Chem Res 9: 515-518.

191. Bruim P, Bikel AF, Koayman EC (1952) Rec Traw Chem 71: 1152.

192. Heckendorn R, Winkler T (1980) Helv Chem Acta 63: 1.

193. Huffman KR, Schaefer FC (1963) J Org Chem 28: 1816.

194. Gompper R, Nappel HE, Schaefer H (1963) Angew Chem 75: 918.

195. Paranik GS, Sushitzky H (1967) Syntheses of heterocyclic compounds. Part XVIII. Aminolysis of 3-aryl-4-bromosydnones, and acid hydrolysis of 3-arylsydnoneimines. J Chem Soc (C): 1006-1008.

196. Huffman KR, Schaefer FC (1963) J Org Chem 28: 1812.

197. L.S. Wittenbrook, J. Heterocyclic Chem., 12, 37 (1975).

198. Clapp LB (1976) 1,2,4-Oxadiazoles, in advances in Heterocyclic Chem Ed AR Katritzky, AJ Bouton, Academic Press, New York, 20, 72.

199. Schneider JP (1967) CR Acad Sci Paris Ser C 265: 638.

200. Person H (1967) CR Acad Sci Paris Ser C 265: 1007.

201. Ziegler E, Wimmer Th (1965) Synthesen von Heterocyclen, 72. Mitt.: 4-Phenyl-3,4-dihydro-carbostyrile aus α-Cyanzimtsäureamiden. Monatshefte für Chemie und verwandte Teile anderer Wissenschaften 96: 1252-1260.

202. Coutts RT (1969) Catalysed sodium borohydride reductions of ortho-nitrocinnamates. J Chem Soc C 713-716.

203. Wittmann H, Wohlkönig A, Sterk H, Ziegler E (1970) Synthesen von heterocyclen, 143. Mitt: Eine Synthese von Oxazolo[3,2-a]chinolinen. Monatshefte für Chemie 101: 383-386.

204. Matusch R, Hartke K (1972) Zur Kondensation von Acetonylaceton mit Cyanessigester. Chemische Berichte 105: 2594-2603.

205. Monti SA (1972) J Org Chem 37: 3834.

206. Jahine H (1973) Indian J Chem 11: 1122.

207. Rastogi RR, Ila H, Junjappa H (1975) A novel method for the synthesis of substituted and fused 3-cyano-4-methylmercapto-2(1H)-pyridones using α-oxoketen S,S-diacetals. J Chem Soc Chem Commun 645.

208. Hatada T, Sone M, Tominaga Y, Natsuki R, Matsuda Y (1975) [Reaction of beta-diketones with ketenethioacetals (author's transl)]. Yakugaku Zasshi 95: 623-628.

209. Zacharias G, Wolfbeis OS, Junek H (1974) Über Anilinomethylenverbindungen der Cyclohexandione. Monatsh Chem 105: 1283-1291.

210. Abdulla RF, Unger PL (1974) N-aryl-3-cyanoazetidin-2,4-diones. A correction. Tetrahedron Letters 15: 1781-1784.

211. Saito K, Kambe S, Sakurai A, Midorikawa H (1981) A Convenient Method for the Preparation of Some Condensed 2-Oxopyridine Derivatives from Ethyl 2-Cyanoacrylates, Cycloalkanones, and Ammonium Acetate. Synthesis 1981:211-213.

212. Duffy JL, Kurth JA, Kurth MJ (1993) Lithium, potassium and sodium alkoxides: Donors in the Michael addition reaction of α-nitroolefins. Tetrahedron Letters 34: 1259-1260.

213. Kambe S, Saito K, Sakurai A, Midorikawa H (1980) A Simple Method for the Preparation of 2-Amino-4-aryl-3-cyanopyridines by the Condensation of Malononitrile with Aromatic Aldehydes and Alkyl Ketones in the Presence of Ammonium Acetate. Synthesis 1980: 366-368.

214. El-Sayed AM, Khodairy A (1998) Synthesis of Some New Heterocycles Derived from Arylmethylenemalononitriles. Synthetic Communications: An International Journal for Rapid Communication of Synthetic Organic Chemistry 28: 3331-3343.

215. Soto JL, Seoane C, Zamorano P, Cuadrado JF (1981) A Convenient Synthesis of N-Amino-2-pyridones Synthesis 1981: 529-530.

216. Elmoghayar MRH, El-Agamey AGA, Nasr MYAS, Sallam M (1984) Activated nitriles in heterocyclic synthesis. Part III. Synthesis of N-amino-2-pyridone, pyranopyrazole and thiazolopyridine derivatives. Journal of Heterocyclic Chemistry 21: 1885-1887.

217. El-Nagdy MH (1981) 8th International Congress in Heterocyclic Chemistry Graz, Austria, 22.

218. Soto JL, Seoane C, Martin N, Perez H (1981) 8th International Congress in Heterocyclic Chemistry Graz, Austria, 12.

219. Joshi NN, Martin H, Hoffmann R (1986) Ultrasonics in the metal promoted cycloaddition of α,α'-dibromo ketones to 1,3-dienes. Tetrahedron Letters 27: 687-690.

220. Mohareb RM, Fahmy SM, Naturforsch Z (1986) 41b: 105.

221. Mohareb RM, Fahmy SM, Naturforsch Z (1985) 40b: 1537.

222. Fahmy SM, Mohareb RM (1985) Activated Nitriles in Heterocyclic Synthesis: A Novel Synthesis of Pyridine and Pyridazine Derivatives. Synthesis 1985: 1135-1137.

223. Daboun HA, Abdou SE, Khader MM (1982) Reactions with activated nitriles: some new approaches to the synthesis of pyridine derivatives. Heterocycles 19: 1925-1929.

224. El-Nagdi MH, Abdel-Razek FM, Ibrahim NS, Kandeel ZES (1984) Activated nitriles in heterocyclic synthesis: a new approach for the synthesis of pyridine and pyridinopyrimidine derivatives. Synthesis 1984: 970-972.

225. Daboun HAF, Riad BY (1984) Indian J Chem 23B: 675.

226. Soto JL, Encinas MJR, Seoane C (1984) Liebigs Ann Chem 17: 213.

227. Kambe S, Saito K, Kishi H, Sakurai A, Midorikawa H (1979) A one-step synthesis of 4-oxo-2-thioxopyrimidine derivatives by the ternary condensation of ethyl cyanoacetate, aldehydes, and thiourea. Synthesis 1987: 287-289.

228. Kandeel ZE, Hilmy KHH, Ismail NS, El-Nagdi MH (1984) J Prakt Chem 326: 248.

229. Corson BB, Stoughton RW (1982) J Amer Chem Soc 50: 2825.

230. Khalifa MAE, Tammam GH, Motaleb RMA, El-Nagdy MH (1983) Synthesis of azoloyl ketone and azoloylacetic acid derivatives: reactions of 4-arylazo-2-oxazolin-5-ones with active methylene compounds. Heterocycles 20: 45-49.

231. El-Gemei GEH, Elecs SA, El-Sakka I, El-Nagdy MH, Naturforsch Z (1983) 38b: 639.

232. Abdel-Galil FM, Riad BY, Sherif SM, El-Nagdy MH (1982) Chem Lett 1123.

233. Elaal FAEA, Hussein MM, El-Nagdy MH, El-Gemei GEH (1983) Monatsh Chem 115: 573.

234. Riad BY, Khalifa FA, Abdel-Galil FM, El-Nagdi MH (1982) Activated Nitriles in Heterocyclic Synthesis: A Novel Synthesis of 4-Hydroxy- and 4-Aminopyrano[2,3-c]pyrazoles. Heterocycles 19: 1637-1640.

235. Juergen, Horst H (1974) Ger Offen 106: 831.

236. Werner J, Jachim OH, Horst S (1980) Angew Chem 92: 390.

237. Gewald K, Hain H, Gruner M (1985) Chem Ber 2198.

238. Nadia SI, Mona MH, El-Nagdy MH (1987) Synthesis of New 3-(Pyridin-6-yl) pyrazolo[1,5-a]pyrimidines. Arch Pharm 320: 487-491.

239. El-Nagdy MH, Mona MH, Nadia SS (1987) Heterocycles 26: 4.

240. Geies AA, Kamal El-DeanAM, Bull Polish (1997) Academy Science Chem 45: 381-390.

241. Ukawa K, Ishiguro T, Wada Y, Nohara A (1986) Heterocycles 24: 1931.

242. Kikelj D, Neidlein R (1993) A convenient synthesis of 2-(1H-Tetrazol-5-yl)-2-cyanoacetate betaines. Synthesis 1993: 873-875.

243. Geies AA (1997) Synthesis of o-Aminonitriles of Indeno[2,1-b]pyrrole and Indeno[2,1-c]pyridine. Polish J Chem 71: 774-778.

244. Pawlak JW, Konopa J (1979) In vitro binding of metabolically activated [14C]-ledakrin, or 1-nitro-9-14C-(3'-dimethylamino-N-propylamino) acridine, a new antitumor and DNA cross-linking agent, to macromolecules of subcellular fractions isolated from rat liver and HeLa cells. Biochem Pharmacol 28: 3391-3402.

245. Pawlak JW, Pawlak K, Konopa J (1983) The mode of action of cytotoxic and antitumor 1-nitroacridines. II. In vivo enzyme-mediated covalent binding of a 1-nitroacridine derivative, ledakrin or nitracrine, with DNA and other macromolecules of mammalian or bacterial cells. Chemico-Biological Interactions 43: 151-173.

246. El-Kashef HS, Geies AA, El-Dean AMK, Abdel-Hafez AA (1993) Synthesis of thieno[2,3-c]pyridines and related heterocyclic systems. Journal of Chemical Technology and Biotechnology 57: 15-19.

247. Martin M, Quinterio JL, Segura C, Seoane JL, Soto M, et al. (1991) Ann Chem 827.

248. Suarez Y, Verdecia E, Ochoa N, Martin R, Martinez M, et al. (2000) Synthesis and structural study of novel 1,4,5,6,7,8-hexahydroquinolines. J Heterocyclic Chem 37: 735-742.

249. Midorikawa H, Kambe S, Saito K, Sakurai A (1981) Synthetic studies using α,β-unsaturated nitriles: facile synthesis of pyridine derivatives. Synthesis 1981: 531-533.

250. Madkour HMF, Mahmoud MR, Nassar MH, Habashy MM (1998) Sulfur Lett 21: 253-261.

251. Pretsch E, Clerc T, Seibl J, Simon W (1989) Tables of spectral data for Structure determination of organic compounds (2ndedn), Springer-Verlag, New York, USA.

252. Hafez EA, Abdel-Galil FM, Sherif SM, El-Nagdy MH (1986) J Heterocyclic Chem 23: 1375.

253. Junek H (1963) Monatsh Chem 94: 192.

254. Ghosh PB, Ternai B (1972) Reaction of 4- and 5-acetyloxazoles with malononitrile. J Org Chem 37: 1047- 1049.

255. Petrow VA (1946) New syntheses of heterocyclic compounds; 9-amino-6:8-dimethyl-7:10-diazaphenathrenes. J Chem Soc: 884-888.

256. Fuentes L, Soto JL, Vaquero JJ (1981) 8th International Congress in Heterocyclic Chemistry, Graz, Austria.

257. Carlos Seoane, Soto JL, Zamorano P, Quinteiro M (1981) Ring transformations of 4H-pyrans. Pyridines from 2-amino-4H-pyrans. J Heterocyclic Chem 18: 309-314.

258. Cocco MT, Congiu C, Maccioni A (1990) Synthesis of pyridines by heterocyclization of new dienamino esters. J. Heterocyclic Chem 27: 1143-1151.

259. Kametani T (1973) Japanese Patent 7307112.

260. Sowellim SZA, El-Taweel FMA, El-Agamey AA (1996) Bull Soc Chem. Fr 133: 229.

261. Khallh ZH, Abdel-Hafez AA, Geies AA, Kamal Eldeen AM (1991) Bull Chem. Soc Jpn 64: 668-670

262. Mahmoud MR, Madkour HMF, Sakr AM, Habashy MM (2001) Sci Pharm 69: 33-52.

263. Matsuda Y, Gotou H, Katou H, Matsumoto Y (1989) Chem Pharm Bull 37: 1188.

264. Benevsky P, Stille JR (1997) Aza-annulation as a versatile approach to the synthesis of non-benzodiazepene compounds for the treatment of sleep disorders. Tetrahedron Lett 38: 8475-8478.

265. Abdallah TA, Abdel-Hadi HM, Hassaneen HM (2002) Michael Reactions of Arylidenesulfonylacetonitriles. A New Route to Polyfunctional Benzo[a] quinolizines. Molecules 7: 540-548.

266. Czerney, Hartmann H, (1982) J Prakt Chem 324: 21.

267. Otto HH, Herbert (1979) Reactionen von Arylidentetralonen mit Cyanessigsäurederivaten. Monatsh Chem 110: 249-256.

268. Springer RH, Scholten MB, O'Brien DE, Novinson T, Miller JP, et al. (1982) Synthesis and enzymic activity of 6-carbethoxy- and 6-ethoxy-3,7-disubstituted-pyrazolo[1,5-a]pyrimidines and related derivatives as adenosine cyclic 3',5'-phosphate phosphodiesterase inhibitors. J Med Chem 25: 235-242.

269. Joachim OH, Horst S Ger Offen 2: 719

270. Bomika Z, Pelcers J, Arens A (1973) Akad Vestis Kim. Ser 2: 244.

271. Motaleb RMA (1986) Ph.D. Thesis, Cairo Unversity, Cairo.

272. Martín N, Martínez-Grau A, Seoane C, Marco JL, Albert A, et al. (2006) Michael Addition of Malononitrile to α-Acetylcinnamamides. Liebigs Ann Chem 801-804.

273. Marco JL, Martin N, Martinez-Grau A, Seoane C, Alberto A, et al. (1994) Development of methods for the synthesis of chiral, highly functionalized 2-amino-4-aryl-4H-pyrans. Tetrahedron 50: 3509-3528.

274. Jiménez B, Martín N, Martínez-Grau A, Seoane C, Marco JL (1995) Michael addition of benzoylacetonitrile to chiral α-cyanoacrylates derived from enantiomerically pure α-hydroxyaldehydes: Synthesis of 3-alkoxycarbonyl-4-alkyl-2-amino-5-cyano-6-phenyl-4H-pyrans. J Heterocycle Chem 32: 1381-1384.

275. Martin N, Martinez-Grau A, Seoane C, Marco J (1995) Tetrahedron: Asymmetry 6: 255.

276. El-Agamey AGA, El-Taweel FMA (1990) Ind J Chem 29: 885-886.

277. Al-Mousawi MS, El-Kholy MY, Mohamed AM, El-Nagdy MH (1999) Synthesis of New Condensed 2-Amino-4h-Pyran-3-Carbonitriles And Of 2-Aminoquinoline-3-Carbonitriles. Org Prep Proced Int 31: 305-313.

278. Yasuda H, Midorikawa H (1966) The Knoevenagel Reaction between Hydroxybenzaldehydes and Ethyl Cyanoacetate. Bull Chem Soc Jpn 39: 1754-1759.

279. Gupta D, Arunk, Paul MS, J Ind Chem Soc 47: 1017.

280. Gupta KAD, Amitava KRDG (1970) Ind J Chem 10: 32.

281. Kobayaski, Goro, Matsuda Yoshiro, Natsuki, Reiko Tominago, Yoshinori, et al. (1973) 93(7), 836 (1973), [C.A. 78, 91910].

282. Yallchenko, Koval RE, Krokhtyak VI, Yagupol, Skii LM, (1981) Zh Org Khim 17: 2630.

283. Wiener C, Schroeder CH, West BD, Link KP, (1962) Studies on the 4-Hydroxy-coumarins. XVIII.¹ª 3-[α-(Acetamidomethyl)benzyl]-4-hydroxycoumarin and Related Products¹ᵇ. J Org Chem 27: 3086-3088.

284. Kudo S, Masabuchi M Japan Patent Application 8324.

285. Rashed HMM, PhD Thesis (2002) Ain-Shams Unversity.

286. Afzal J, Vairdmani M, Hazara B, Das K (1980) A Novel Thermal Knoevenacel Condensation via A Thermal Michael Reaction. Synth Commun 10: 843-850.

287. Fahmy SM, Abed NM, R.M. Mohareb RM, El-Nagdy MH (1982) Synthesis 490.

288. Hafez EA, Khalifa MAE, Guda SKA, El-Nagdy MH (1980) Z Naturforsch 35b: 485.

289. Gewald K, Osmer C, Schafer H, Hain U (1984) Synthese von 4-Aminocinnolinen aus (Arylhydrazono)(cyan)-essigsäurederivaten. Liebigs Ann Chem 1390-1394.

290. Taylor EC (1978) Principals of Heterocyclic Chemistry (AC.S. Course Films).

291. Taylor EC, Mckipplo A (1970) The Chemistry of Cyclic Enaminonitriles and o-aminonitriles. J Wiley and Sons, New York, USA.

292. Aleksandrowicz P, Bukowska M, Maciejewski M, Prejzner J (1979) Cyanoethylation of the salts of cyansgnanidine in aprotic solvents. Can J Chem 57: 2593-2598.

293. Satoshi K, Saito K, Hiroshi K, Kishi H, Sakurai A, et Al. (1979) A One-Step Synthesis of 4-Oxo-2-thioxopyrimidine Derivatives by the Ternary Condensation of Ethyl Cyanoacetate, Aldehydes, and Thiourea. Synthesis 287-289.

294. Whilehead CW (1952) The Synthesis of 5-Carbethoxyuracils. J Amer Chem Soc 74: 4267-4271.

295. Whilehead CW (1955) J Amer Chem Soc 77: 5807 (1955).

296. Jones RG, Whilehead CW (1955) vic-dicarboxylic acid derivatives of pyrazole, isoxazole, and pyrimidine. J Org Chem 20: 1342-1347.

297. Whilehead CW, Traverso JJ (1956) Synthesis of 2-Thiocytosines and 2-Thiouracils. J Amer Chem Soc 78: 5294-5299.

298. Koyama T, Hirota T, Ito I, Toda M, Yamato M (1969) [Reactions of benzylcarbonyl compounds with formamide. II. A novel synthesis of isoquinolines]. Yakugaku Zasshi 89: 1492-1495.

299. Koyama T, Toda M, Hirota T, Hashimoto M, Yamato M (1969) [Reactions of benzylcarbonyl compounds with formamide. IV]. Yakugaku Zasshi 89: 1688-1690.

300. Koyama T, Toda M, Hirota T, Yamato M (1970) [Reactions of benzylcarbonyl compounds with formamide. V. On alpha-acyl-3,4-dimethoxyphenylacetonitriles]. Yakugaku Zasshi 90: 8-10.

301. Brown DJ (1962) The Pyrimidines. J Wiley and Sons, New York, USA.

302. Cupps TL, Wise DS, Townsend LB (1982) Use of allyltrimethylsilane in the formation of potential C-nucleoside precursors. J Org Chem 47: 5115-5120.

303. Abdel-Latif FF (1991) Indian J Chem, 30B: 263-265.

304. Rappoport Z (1970) The Chemistry of Cyano Group. J Wiley and Sons, New York, USA.

305. Fatiadi AJ (1983) Chemistry of C=X Compounds. J Wiley and Sons, New York, USA.

306. Freeman F (1982) The Chemistry of Malononitriles. California University Press, Los Angeles, USA.

307. Fatiadi AJ (1978) Synthesis 241.

308. Taylor EC, Borror AL (1961) J Org Chem 26: 2967.

309. El-Nagdy MH, Wamhoff H (1981) J Heterocyclic Chem 18: 1289.

310. Fawzy AA, Sanaa ME, Hanafi Eman ZA (1997) Reactions of pyrimidinonethione derivatives: Synthesis of 2-hydrazinopyrimidin-4-one, pyrimido[1,2-a]-1,2,4-triazine, triazolo-[1,2-a]pyrimidine, 2-(1-pyrazolo)pyrimidine and 2-arylhydrazonopyrimidine derivatives. Arch Pharmacol Res 20: 620-628.

311. Geies AA (1999) J Chinese Chem Soc 46: 69-75.

312. Youssif S, El-Bahaie S, Nabih E (1999) A facile one-pot synthesis of

pyrido[2,3-d]pyrimidines and pyrido[2,3-d:6,5-d']dipyrimidines. J Chem Res 112-113.

313. El-Nagdy MH, Sallam MMM, Ilias MAM (1975) Helv Chem Acta 58: 1944.

314. El-Nagdy MH, Abd-Allah SO (1973) Reactions with the arylhydrazones of some α-cyanoketones. J Prakt Chem 315: 1009-1016.

315. El-Nagdy MH, Kassab NAL, Fahmy SM, Elall FA (1974) Reactions with 3,5-pyrazolidinediones. III. Cyanoethylation of some 4-arylazo derivates of 3,5-pyrazolidinediones and 3-amino-2-pyrazolin-5-ones. J Prakt Chem 316: 177-184.

316. El-Nagdy MH, Ohta M (1973) Bull Chem Soc Jpn 46: 3813.

317. El-Nagdy MH, Kandeel EM, El-Moghayar MRH, Naturforsch Z (1977) 32b: 307.

318. Fahmy SM, Kandeel EM, El-Sayed ER, El-Nagdy MH (1978) Reactions with heterocyclic amidines 1. Cyanoethylation of cyclic amidines. J Heterocyclic Chem 15: 1291-1293.

319. El-Nagdy MH, Kandeel EM, Zayed EM, Kandil ZE (1978) Pyrimidine derivatives and related compounds. VIII [1]. Routes for the synthesis of 3,5-diaminopyrazoles, 2-aminopyrazolo[1,5-a]pyrimidines and 5-aminopyrazolo[1,5-a]pyrimidines. J Prakt Chem 320: 533-538.

320. Kagan J, Melnick B (1979) The synthesis and photochemistry of 4-amino-3-cyanopyrazole. J Heterocyclic Chem 16: 1113-1115.

321. Antonini, Cristalli G, Franchetti P, Grifantini M, Martelli S (1980) Synthesis of 2,3,5,6-tetrahydroimidazo[1,2-c]quinazoline derivatives as potential narcotic antagonists. J Heterocyclic Chem 17: 155-157.

322. Papadopouls EP (1981) Reactions of o-aminonitriles with isocyanates. 2. A facile synthesis of imidazo[1,2-c]quinazoline-2,5-(3H,6H)dione. J Heterocyclic Chem 18: 515-518.

323. Etson SR, Mottson RJ, Sowell JW (1979) Synthesis of substituted pyrrolo[2,3-D]pyrimidine-2,4-diones. J Heterocyclic Chem 16: 929-933.

324. Dave KG, Shishoo CJ, Devani MB, Kalyanaraman R, Ananthan S, et al. (1980) Reactions of nitriles under acidic conditions. Part I. a general method of synthesis of condensed pyrimidines. J Heterocyclic Chem 17: 1497-1500.

325. Fleury JP (1980) Heterocycles 14: 1581.

326. Perchais J, Fleury JP (1972) Tetrahedron 28: 2267.

327. Riad, Abde-Aziz M (1989) Sulfur Lett 9: 75-85.

328. Kandeel EM, Sadek KU, El-Nagdy MH, Naturforsch Z (1980) 35b: 91.

329. Wenternitz P (1978) Helv Chem Acta 61: 1175.

330. Kraemer W, Fischer R, Holmwood G, Hagemann H, Wachendroff-Neumann U, et al. Ger Offen DE 4: 401.

331. Bridges AJ, Denny WA, Dobrusin EM, Doherty AM, Fry DW, et al. (1997) PCT Int App1 Wo 97 38: 983 (C1. C07 D239/94), 23 Oct. (1997), [C.A. 128, 3695c (1998)].

332. Bombrun (1997) PCT Int App1 Wo 97 43: 287 (C1. C07 D471/04), 20 Nov., [C.A. 128, 34753z (1998)].

333. Purohit DM, Shah VH (1997) Heterocycl Commun 3: 437.

334. Mueller K, Wiegrebe W, Gurster D, Peters S (1995) PCT Int App1 Wo 95 03: 266.

335. Isobe Y, Katagiri T, Umerzawa J, Goto Y, Sasaki M, et al. (1996) Can Pat App1 CA 2 154: 293.

336. Courant J, Leblois D, Le Bout G, Venco C, Panconi E (1993) Eur J Med Chem 28: 821.

337. Haley GJ (1993) US 5, 270, 466.

338. Oozeki M, Kotado S, Yasuda K, Kudo K, Maeda K (1993) Jpn Kokai Tokkyo Koho JP 05, 194, 236.

339. Nishikawa Y, Shindo T, Ishii K, Nakamura H, Kon T, et al. (1989) Chem Pharm Bull 37: 100.

340. Chen Y, Wehrmann R, Koehler B (1996) Ger Offen DE 19: 505, 940 (C1. C07D311/16), 22 Aug. (1996), [C.A. 125, 221585w (1996)].

# 22

# Phytochemistry and Therapeutic Potentials of the Seed Essential Oil of *Eucalyptus maculata* Hook from Nigeria

**Ololade ZS[1]\*, Olawore NO[1] and Oladosu IA[2]**

[1]Department of Pure and Applied Chemistry, Ladoke Akintola University of Technology, P.M.B. 4000, Ogbomoso, Nigeria
[2]Department of Chemistry, University of Ibadan, Ibadan, Nigeria

## Abstract

The aim of this research was to study the phytochemicals, therapeutic potentials of the seed essential oil of *E. maculata* Hook grown in Nigeria. Composition of the seed essential oil was investigated by GC, GC-MS, MS and FT-IR. Analyses of the volatile oil resulted in the identification of fifty-eight phytocompounds representing 98.95% of the oil. Cyclofenchene (7.0%), $\alpha$-pinene (8.0%), 1R-$\alpha$-pinene (6.0%), 1S-$\alpha$-pinene (7.0%), DL-pinene (6.0%) and $\beta$-*trans*-Ocimene (8.0%) were detected as the principal components accounted for 42% of the seed essential oil. None of these main compounds has ever been detected in the leaves extracts of *E. maculata* that had been investigated before except $\alpha$-Pinene. The total phenolics content of the seed oil of *E. maculata* was estimated as 195.84 ± 0.002 µg/mg GAE. The results of free radical scavenging capacity using DPPH and FRAP methods showed that the oil possessed strong antioxidant activity with $IC_{50}$ 8.0 and 10.0 µg.ml$^{-1}$ respectively. The acute toxicity test showed that the seed oil extract is safe to be used *in vivo*. The seed oil at 1000 µg.kg$^{-1}$ (*p.o.*) gave 87.50% significant inhibition of paw edema. In the antinociceptive assay the oil inhibited the licking time by 86.46% in first phase (neurogenic pain) and 60.10% in the second phase (inflammatory pain). These results showed that the seed oil of *E. maculata* could be an active source of substances with antioxidant, anti-inflammatory and analgesic activities. The pharmacological potentials of the investigated seed oil could be explained by their antioxidant properties, due to their high phenolic and terpenoids contents.

**Keywords:** *Eucalyptus maculata*; Seed essential oil; Antioxidant; Analgesic; Edema

## Introduction

*Eucalyptus maculata* Hook commonly called spotted gum; a member of *Myrtaceae* family is an attractive with significantly larger vascular tissue was the better adaptation to a variety of environment types, fast growing and tall plant, usually about 35-45 metres in height and average width of 1.5 metres at breast height over bark. The immature leaves are glossy green and elliptic to ovate, while the adult leaves are lanceolate. The flowers are small, white and clustered which developed to ovoid or slightly urceolate fruits, which are disc depressed; valves enclosed, nectar from the flowers, even the seed was sometimes eaten. Trunks are relatively long and usually clean and straight without branches for more than half their height. Bark is smooth to ground level and greenish cream when fresh. It is shed in small irregular patches, leaving dimples that age to cream, grey, pink or coppery brown, giving trunks their characteristic spotted appearance. The wood is slightly greasy and gum veins are common. The plant is the adaptations to water limited environments like drought and salinity [1].

*E. maculata* is an odouriferous medicinal plant locally used for the treatment of asthma and chronic bronchitis [2]. The leaf extracts of this plant have antimicrobial properties [3]. The leaf essential oil of the plant is usually incorporated into soaps and disinfectants due to its antimicrobial and insecticidal potential of the oil. The oil is a natural source for the production of hydroxycitronellal, citronellylnitrile and menthol due to the high percentage of citronellal in the oil. The plant is considered a good source of natural antioxidants [4]. The resinous exudate from the trunk of the plant is also taken internally to cure bladder infections [5].

To best our knowledge, there is no literature on the phytochemicals, phenolic content, antioxidant, toxicity, anti-inflammatory and antinociceptive potentials of the seed essential oil of *E. maculata* so far.

Therefore, the present research is the first report on the phytochemical and therapeutic potentials of the seed essential oil of this plant.

## Materials and Methods

### Plant material

Seed of *E. maculata* were collected from Kaduna, Nigeria. The plant was authenticated at the Forest Research Institute of Nigeria (FRIN), Kaduna, Nigeria. The seeds were air-dried in a well ventilation place until the moisture content reduced to a minimum suitable for grinding.

### Isolation of essential oil

The essential oil was extracted by subjecting the air-dried seed of *E. maculata* to hydrodistillation using a Clevenger-type apparatus. Fresh seed (100 g) were chopped and mixed with distilled water in a 5 litre round bottom flask. The hydrodistillation lasted for 3 hours and the oil collected was dried with sodium sulphate and stored at 4°C in a refrigerator for further use [6].

### Instrumentation and analytical conditions

**Gas Chromatographic (GC) analysis:** The neat essential oil obtained from the seed of *E. maculata* was analysed using a GC-FTD

\*Corresponding author: Ololade ZS, Department of Pure and Applied Chemistry, Ladoke Akintola University of Technology, P.M.B. 4000, Ogbomoso, Nigeria
E-mail: suntolgroup@yahoo.com

with capillary column (30 m×0.25 mm, 0.25 μm film thickness). Oven temperature was kept at 60°C for 3 minutes initially, and then raised at the rate of 3°C/min to 250°C. Injector and detector temperatures were set at 250 and 290°C respectively. Ultra-high purity helium (flow rate: 1 ml/min) was used as carrier. Diluted samples 0.1 μl was auto-injected in the splitless mode. Peak area percent of each compound relative to the area percent of the entire spectrum (100%) were used for obtaining its quantitative data.

**Gas Chromatography-Mass Spectrometry (GC-MS) analysis:** GC-MS analysis of the neat seed essential oil was carried out on a gas chromatograph with a capillary column (30 m×0.25 mm, 0.25 μm film thickness) equipped with a mass selective detector in the electron impact mode (Ionization energy: 70 eV). Helium was the carrier gas at a flow rate 0.99 ml/min. The GC was interfaced with mass detector operating in the $EI^+$ mode. Elutes were automatically passed into a mass spectrometer with a dictator voltage set at 1.5 kv and sampling rate of 0.2 sec. The mass spectrum was also equipped with a computer fed mass spectra data bank. The GC-MS was operated under the same conditions as described for GC above. The mass spectra were generally recorded over 40-700 amu that revealed the total ion current (TIC) chromatograms. Temperature program was used as the same as described above for GC analysis. The temperatures of the injector and ion source were maintained at 250 and 200°C, respectively.

**Mass spectrophotometry analysis:** The compounds were separated on 30 m×0.25 mm×0.25 μm column. Injection volume, 1 μl; transfer temperature, 280°C MS parameters were as follows: EI mode, with ionization voltage 70 eV, ion source temperature, 200°C. The MS fragmentation pattern was compared with those of pure compound, by matching the MS fragmentation patterns with NIST mass spectra libraries and with those given in literature.

**Qualitative and quantitative analysis:** Identification of the individual component in the seed oil was made by matching their recorded mass spectra with the NIST library provided by the instrument software, and by comparing their retention indices with literature. Relative area percentages of the individual components were obtained from GC analysis [7,8].

**Fourier-Transform Infra-Red (FT-IR) analysis:** The FT-IR analysis of the seed oil was conducted using FT-IR spectrophotometer, absorbance and functional groups determined with the help of correlation charts. The samples were examined neat by placing them in between potassium bromide cells. 0.25 μl of the samples was deposited in the middle of a KBr pellet and the IR spectrum was recorded at different times until total evaporation.

## Total phenolic content and antioxidant capacity

Total phenolic content of the seed essential oil of *E. maculata* was measured by the Folin-Ciocalteu method [9]. A solution of the seed oil (0.2 ml) containing 1000 μg/ml of the oil in methanol was pipetted into a 50 ml volumetric flask, 46 ml distilled water and 1 ml Folin-Ciocalteu's phenol reagent were added, and the opaque flask was thoroughly shaken. 3 ml of (2% w/v) $Na_2CO_3$ (aq.) was added after 3 minutes and allowed to stand for 2 hours for incubation in dark with intermittent shaking. Absorbance values of the clear supernatants were measured at 760 nm against a blank (0.5 ml Folin-Ciocalteu's reagent + 1 ml $Na_2CO_3$) on a Spectrophotometer. The same procedure was repeated for the standard gallic acid solutions (0-1000 μg/0.1 ml); calculation of percentage total phenols content was based on Gallic Acid Equivalents (GAE). A standard curve obtained with the following equation:

$$Absorbance = 0.0008 \times gallic\ acid\ (\mu g) + 0.0068$$

### *In vitro* antioxidant assays

**In vitro DPPH free radical scavenging assay:** The 2,2′-diphenyl-1-picrylhydrazyl (DPPH) radical assay usually involves hydrogen atom transfer reactions and electron transfer mechanism, based on kinetic data. The free radical scavenging assay of seed oil against DPPH was determined spectrophotometrically. 1.0 ml of the seed *E. maculata* oils (10, 100 and 1000 μg.ml$^{-1}$) in methanol was added to 1.0 ml of a 0.004% w/v methanol solution of DPPH. The mixture was homogenized and the absorbance was monitored at 517 nm after 30 minutes of incubation, when the reaction reached a steady state. Ascorbic acid was used as reference compound. Assays were carried out in triplicate. The concentration which was responsible of half scavenging activity $IC_{50}$ (concentration causing 50% inhibition) value of each extract was determined graphically and expressed as in μg/ml by using this formula [10].

$$I\% = [(A_{blank} - A_{sample})/A_{blank}] \times 100$$

Where: $A_{blank}$ was constituted by a same amount of methanol and DPPH solution without the oil and $A_{sample}$ is the absorbance values of the test compounds.

**In vitro Ferric Reducing Antioxidant Power (FRAP):** The antioxidant activity has been reported to be concomitant with the development of reducing power. The FRAP assay of the seed oil extract was carried out according by the method of [11]. 1.0 ml of different concentrations of the oil extract (10, 100 and 1000 μg.ml$^{-1}$) were mixed with 2.5 ml of phosphate buffer (0.2 M, pH 6.6, 2.5 ml) and 2.5 ml of 1% aqueous potassium hexacyanoferrate $[K_3Fe(CN)_6]$ solution. After 20 minutes incubation at 50°C, 2.5 ml of trichloroacetic acid (10%) was added and the mixture was centrifuged at 1000 rpm for 10 minutes. Then, the supernatant (2.5 ml) was mixed with distilled water (2.5 ml) and a freshly prepared $FeCl_3$ solution (0.5 ml, 0.1%). Absorbances were read at 700 nm using an appropriate blank (containing all reagents except the test compound). Assays were carried out in triplicate. Ascorbic acid was used as a reference. The average values were plotted to obtain the half maximum effective concentration ($EC_{50}$) of $Fe^{3+}$ reduction by linear regression.

## Acute toxicity test

The acute toxicity of seed essential oil was determined according to [12]. Groups of rats received oral doses of the seed oil >1000 μg / kg. The groups were observed for 48 hours and mortality at end of this period was recorded for each group. The determination of $LD_{50}$ served to define the doses used in experiments of pharmacological assays.

### *In vivo* antiinflammatory assay

*In vivo* antiinflammatory potential was of the seed oil of *E. maculata* was tested in rat paw edema according to [13]. Healthy albino rats (200 ± 30 g) acclimatized to laboratory hygienic conditions were housed in clean cages under standard conditions of temperature (25 ± 2°C) and RH was % 55-60, 12 hours light/dark cycle were maintained in the quarantine and were fed with standard pellet diet and water *ad libitum*. The handling and uses of animals were in accordance to the institutional guidelines.

1% carrageenan (0.1 ml) was injected into the plantar surface of the rat hind paw 30 minutes after oral administration of the test compounds or vehicle. Indomethacin (25 mg.kg$^{-1}$) was used as reference drug. Paw volume was determined immediately after the injection of the phlogistic agent and again 2 and 4 hours later by means of a digital vernier calliper. The antiinflammatory activity of the seed oil was expressed as the percentage of inhibition calculated from the difference between the responses of the treated and the control groups. The inhibition percentage of the inflammatory reaction which was calculated by the formula given in equation (1) below was determined for each rat by the comparison of each group with controls.

$$\% I = 1 - (dt/dc) \times 100 \tag{1}$$

Where: I% = Percentage inhibition, 'dt' is the difference in paw volume in the drug-treated group and 'dc' is the difference in paw volume in control group.

### In vivo antinociceptive assay

In vivo antinociceptive activity of the seed oil of E. maculata was studied in rat according to [14]. The rats were divided into four groups of three animals each and fasted for 12 hours and deprived of water only during the experiment to ensure uniform hydration and minimize variability in edematous response. The rats were treated with 1000 $\mu g.kg^{-1}$ (p.o.) of essential oil and indomethacin. Thirty minutes later, the pain was induced by injecting 0.05 ml of 2.5% v/v formalin in distilled water into the sub-plantar right leg paw of the rats, the amount of time spent in licking the injected paw was monitored and was considered as an indicative of pain and frequency of the injected paw were recorded for 30 minutes. The number of lickings from 0-5 minutes (first phase) and 15-30 minutes (second phase) were counted after injection of formalin. The percentage inhibition (I) was calculated accordingly.

## Results and Discussion

### Identification and quantification of the seed essential oil

Hydrodistillation of the seed of E. maculata gave a pale-yellow coloured essential oil with about 1.60% v/w yield per 100 g of air-dried seed sample and with a pleasant odour characteristic of the oil. GC, GC-MS, MS and FT-IR analyses of the seed essential oil enabled us to identify fifty-nine components (Table 1) amounted 98.95% of the total content. In the seed oil extract different groups of compounds were present. The main compounds in E. maculata seed essential oil were: Cyclofenchene (7.0%), $\alpha$-pinene (8.0%), 1R-$\alpha$-pinene (6.0%), 1S-$\alpha$-pinene (7.0%), DL-pinene (6.0%) and $\beta$-trans-Ocimene (8.0). The major class of substances in the essential oil of E. maculata was monoterpene (48.50%), followed by oxygenated monoterpenoids (9.65%), sesquiterpenes (8.75%) and oxygenated sesquiterpenoids (4.60%).

The report from the literature showed that no part of this plant has been investigated before in Nigeria, some new phytocompounds were detected in the seed essential oil of this plant which were not present in the leaf essential oil of the same plant analysed from other part of the globe, these include compound: 1, 13, 14, 15, 16, 18, 28, 29, 31, 34, 35, 36, 37, 38, 39, 41, 45, 48, 49, 51, 52, 53, 54, 55 and 58.

According to [15], leaves essential oil of this plant grown in the western region of Cuba contained $\alpha$-pinene (49.7%), $\alpha$-eudesmol (18.0%) and $\beta$-eudesmol (11.3%) as major constituents. Moreover, [16] also reported their findings on the seasonal changes of leaf oil composition of the same plant from Khuzestan provinces in Iran where the major compositions were eucalyptol (in winter), citronellal (in spring) and citronellol (in summer). Elaissi et al. reported the constituent of E. maculata from Korbous arboreta Northeast Tunisia, whose leaf essential oil was characterized by two major components: eucalyptol and limonene [17].

From the mass spectrometry analysis of the principal compounds were discussed as following. Compound 1 is Cyclofenchene, a tricyclic monoterpene, with molecular formula of $C_{10}H_{16}$ (m/z 136), weak bond is broken to give a fragment at m/z 93 as the base peak, m/z 93 (M-43) is due to loss of $(H_3C)_2CH$ from the molecular ion m/z 136, m/z 121 (M-15) is due to loss of $CH_3$ group from m/z 136, other prominent peaks observed in Cyclofenchene occurred at 27, 39, 79, 105. Compound 5, 6 and 7 are pinene derivatives, they are bicyclic monoterpenes with molecular ion peak 136, the relatively low abundance of the molecular ion peak is consistent with the view that the molecular structure of the compound is rather strained or crowded; the base peak m/z 93 corresponds to the loss of 43 mass units and relatively abundance of the ion m/z 41 is about one quarter of the base peak. A point of distinction between the isomers arises from the abundance of the ion m/z 29 and 39 in 1S-$\alpha$-pinene which is not feature in $\alpha$-pinene and 1R-$\alpha$-pinene. The failure to detect the isopropyl ion strengthens that the loss of 43 mass units is not an entity. Therefore, the groups elided may be obtained by the breaking of two tertiary bonds with the removal of or concomitant hydrogen migration. The occurrence of gem dimethyl group as a part of ring system is common feature of many monoterpenes. Compound 8 is a monocyclic monoterpene, with molecular formula of $C_{10}H_{16}$ (m/z 136) is $\beta$-trans-Ocimene, m/z 43 $(C_3H_7^+)$ which is due to the detachment of isopropyl $[(H_3C)_2C-]$ group attached to the quaternary carbon of the compound and weak bond is broken to give a fragment at m/z 93 as the base peak.

The FT-IR analysis of the seed essential oil E. maculata gave the following absorption signals, especially in the regions around 3500 $cm^{-1}$, 2945 $cm^{-1}$ and 1650 $cm^{-1}$. Moreover, at frequency region of 1447-452 $cm^{-1}$, E. maculata showed more peaks. The band at 3600-3400 $cm^{-1}$ was due to OH stretching vibration, 3600 (sharp) was due to free OH, while 3400 $cm^{-1}$ (broad) was due to associated OH; both bands frequency present alkanol spectra; bands at 3400-3200 are due to N-H stretching vibrations, 3400 (sharp) was due to free N-H, while 3200 $cm^{-1}$ (broad) was due to hydrogen bonded N-H; 2945 $cm^{-1}$ was due OH stretching vibration, strongly hydrogen bonded, a very broad band in this region superimposed on the C-H stretching frequencies is characteristic of fatty acids. Peak at 1651 $cm^{-1}$ was attributed to C=O stretching vibration of amides or lactams. This band is lower in frequency by about 20 $cm^{-1}$ by conjugation. The frequency of the band is raised about 35-70 $cm^{-1}$ in lactam, this can be used as an indicative for the presence of esters in the E. maculata seed essential oil. 1447-1370 $cm^{-1}$ were due to the presence of -$CH_3$ stretching vibration.

### Phenolic content and antioxidant property

The result of absorbance value of the seed oils solution reacting with Folin-Ciocalteu phenol reagent and compared with the absorbance values of standard solutions of Gallic acid, total phenolics content of the seed oil of E. maculata was estimated as 195.84 ± 0.002 $\mu g/mg$ of GAE. This might be due to the presence of low molecular mass phenolic compound such as 4-Isopropenyl-1-cyclohexen-1-ylmethyl benzoate in the seed essential oil. The Folin-Ciocalteu assay is based on the transfer of electrons in alkaline medium from phenolic compounds to phosphomolybdic or phosphotungstic acid complexes to form blue complexes. Phenoloids constitute one of the major groups of compounds acting as primary antioxidants or free radical terminators [18]. Phenoloids have inhibitory effects on mutagenesis and carcinogenesis in humans and may contribute directly to antioxidative action, when included in daily meal mostly from plant source [19].

| SN | Compounds | Percentage composition | Retention index |
|---|---|---|---|
| 1 | Cyclofenchene | 7.0 | 729 |
| 2 | 3,4-Dimethyl-(Z,Z)-2,4-Hexadiene | 2.0 | 750 |
| 3 | 3,5-Dimethylcyclohexene | 4.0 | 838 |
| 4 | β,β-dimethyl-1H-Imidazole-4-ethanamine | 3.0 | 850 |
| 5 | D-Sabinene | 0.4 | 897 |
| 6 | 1R-α-pinene | 6.0 | 937 |
| 7 | 1S-α-pinene | 7.0 | 941 |
| 8 | DL-pinene | 6.0 | 943 |
| 9 | α-pinene | 8.0 | 948 |
| 10 | β-pinene | 1.0 | 970 |
| 11 | β-trans-Ocimene | 8.0 | 976 |
| 12 | L-β-pinene | 0.4 | 978 |
| 13 | 2,3,4-trimethylthiophene | 0.5 | 993 |
| 14 | Phenyltrimethylmethane | 0.4 | 1007 |
| 15 | 2-(1,1-Dimethyl-2-pentenyl)-1,1-dimethylcyclopropane | 2.0 | 1040 |
| 16 | 3,7,7-Trimethyl-1,3,5-cycloheptatriene | 0.5 | 1010 |
| 17 | α-Limonene | 0.6 | 1018 |
| 18 | 1,5-Dimethyl-1-vinyl-4-hexenylbutyrate | 0.1 | 1022 |
| 19 | o-Cymene | 0.5 | 1029 |
| 20 | L-Limonene | 0.5 | 1031 |
| 21 | D-(E)-Limonene oxide | 0.6 | 1030 |
| 22 | 2-Isopropylimidazole | 3.0 | 1038 |
| 23 | m-Cymene | 0.5 | 1042 |
| 24 | p-Cymene | 0.6 | 1045 |
| 25 | Limonene epoxide | 0.8 | 1048 |
| 26 | Eucalyptol | 1.45 | 1059 |
| 27 | L-Linalool | 0.2 | 1083 |
| 28 | 1,5-Dimethyl-1,5-cyclooctadiene | 0.5 | 1103 |
| 29 | 2-Ethyl-p-Xylene | 0.6 | 1119 |
| 30 | β-Terpineol | 0.7 | 1158 |
| 31 | 3,4,4-trimethyl-5-oxo-(Z)-2-hexenoic acid | 2.0 | 1200 |
| 32 | α-Copaene | 3.0 | 1221 |
| 33 | cis-Geraniol | 0.6 | 1228 |
| 34 | Ethenyldimethylester Phosphoric acid | 2.0 | 1233 |
| 35 | 2,3-Dimethylcyclohexanol | 2.0 | 1240 |
| 36 | 4,4-Dimethyl-2-cyclopenten-1-one | 3.0 | 1250 |
| 37 | 1,1,3-Trimethyl-1-silacyclo-3-pentene | 0.5 | 1260 |
| 38 | Phellandral | 3.0 | 1280 |
| 39 | 2-Cyclopentylcyclopentanol | 0.6 | 1289 |
| 40 | Nopol | 0.4 | 1290 |
| 41 | 2-Hydroxymethyl-5-(1-hydroxy-1-isopropyl)-2-cyclohexen-1-one | 0.5 | 1300 |
| 42 | α-Cubebene | 1.0 | 1344 |
| 43 | Aromadendrene | 0.2 | 1386 |
| 44 | L-allo-Aromadendrene | 0.8 | 1452 |
| 45 | (2E,6E)-2,6Dimethyl-2,6-octadiene-1,8-diol | 0.7 | 1471 |
| 46 | α-Selinene | 0.7 | 1474 |
| 47 | α-Bulnesene | 1.0 | 1490 |
| 48 | 1R,3Z,9s-4,11,11-Trimethyl-8-methylenebicyclo[7.2.0]undec-3-ene | 0.2 | 1494 |
| 49 | 2-isopropyl-5-methyl-9-methylenebicyclo[4.4.0]dec-1-ene | 0.6 | 1503 |
| 50 | trans-Z-α-Bisabolene epoxide | 1.0 | 1531 |
| 51 | 3-acetoxy-4-(1-hydroxy-1-methylethyl)-1-methylcyclohexene | 3.0 | 1533 |
| 52 | 1-(1-methylene-2-propenyl)cyclopentanol | 1.0 | 1560 |
| 53 | 2-isopropyl-5-methyl-9-methylenebicyclo[4.4.0]dec-1-ene | 0.5 | 1563 |
| 54 | N-Acetyl-3-propoxyamphetamine | 0.8 | 1580 |
| 55 | 5-Chloromethyl-2-oxazolidinone | 1.0 | 1600 |
| 56 | α-Eudesmol | 0.6 | 1650 |
| 57 | β-Eudesmol | 1.0 | 1652 |
| 58 | 4-Isopropenyl-1-cyclohexen-1-ylmethylbenzoate | 0.4 | 1958 |
| **Percentage Total** | | **98.95** | |

**Table 1:** Composition of the seed essential oil of *Eucalyptus* maculata Hook.

## *In vitro* antioxidant potentials by dpph and frap methods

In this research, two most widely used assays, namely, DPPH and FRAP methods, were applied to evaluate the antioxidant potentials of the seed essential oil. DPPH measures the oil's free radical scavenging, whereas FRAP measures the ability of the essential oil to reduce metal ions. DPPH involves single electron transfer (SET) and hydrogen atom transfer reactions (HAT), while in FRAP there is a single electron transfer (SET) [20].

The seed essential oil was able to inhibit the formation of DPPH radicals in a concentration dependent manner. The percentage inhibitions of the essential oil at various concentrations (10, 100 and 1000 $\mu g.ml^{-1}$) were $59.46 \pm 0.000$, $62.82 \pm 0.0006$ and $82.84 \pm 0.003\%$ respectively; while the $IC_{50}$ values was found to be 8.0 $\mu g.ml^{-1}$ in comparison to ascorbic acid which gave $54.37 \pm 0.00$, $84.51 \pm 0.001$ and $95.50 \pm 0.00$ as the percentage inhibitions with $IC_{50}$ value of 9.0 $\mu g.ml^{-1}$. The DPPH radical scavenging potential of the seed essential oil was at same range as observed for ascorbic acid (synthetic antioxidant) as shown in table 2.

The solution of radical was decolourized after reduction with the antioxidant (AH) or the radical (R·) in accordance with the following scheme [21]:

DPPH·+AH→DPPH·-H+A·

DPPH·+R·→DPPH·-R

The result of reducing power (FRAP) of the seed oil of *E. maculata* in comparison with ascorbic acid as a standard antioxidant is also shown in table 2. The reducing power of ascorbic acid used as standard in this study was ascorbic acid was $EC_{50}$: 20.00. The seed oil exhibited excellent reducing potential value at concentrations of 10, 100 and 1000 $\mu g.ml^{-1}$ with effective dose value at ($EC_{50}$: 10.00 $\mu g.ml^{-1}$). Reducing power of *E. maculata* seed oil increases from $0.31 \pm 0.005$ at 10 $\mu g/ml$ to $0.43 \pm 0.002$ at 100 $\mu g/ml$ and finally to $0.60 \pm 0.005$ at 1000 $\mu g.ml^{-1}$ in a concentration dependent manner. At tested concentrations the oil possessed the ability to reduce $Fe^{3+}$. It was observed that the seed oil of *E. maculata* showed high $Fe^{3+}$ reducing power comparable to ascorbic acid activity. The reducing power of the seed oil increased with concentrations in a strongly linear manner. The reducing power assay measures the electron donating ability of antioxidants using the potassium ferricyanide reduction method. Antioxidants cause the reduction of the $Fe^{3+}$/ferricyanide complex to the ferrous form and activity is measured as the increase in the absorbance at 700 nm.

The antioxidant power increased as the essential oil concentration increased, indicating some compounds in *E. maculata* are electron donors and could react with free radicals to convert them into more stable products and to terminate radical chain reactions.

Compared to reported extracts antioxidant of different *Eucalyptus* species, it was shown that the seed oil of this plant is more active, *E. maculata* gave lower $IC_{50}$ values than other species studied before; *E. oleosa* with $IC_{50}$: >1000 [22], *E. Globules* (leaf) $IC_{50}$: 57.00 $\mu g.ml^{-1}$ (DPPH); $IC_{50}$: 48.00 $\mu g.ml^{-1}$ (FRAP) [23].

The free radical scavenging potentials could be attributed to the presence of some components that have antioxidant activities, possible antagonistic and synergistic effects of compounds and functional groups in the seed oil could also be taken into consideration [24]. These results showed that the seed oil of *E. maculata* potentially exert its radical scavenging effects at a much lower concentration. This observed effect is certainly associated with high phenolic content, terpenes, terpenoids and carbonyl compounds in the seed oil. The body has several mechanisms to counteract oxidative stress by producing antioxidants, either naturally generated in situ (endogenous antioxidants), or externally supplied through natural products (exogenous antioxidants). The roles of antioxidants are to neutralize the excess of free radicals, to protect the cells against their toxic effects and to contribute to disease prevention. The results clearly showed that the seed oil of *E. maculata* possesses strong antioxidant activity and can be considered as good sources of natural antioxidants for therapeutic purposes such as reactive oxygen species ailments including chronic inflammatory joint disease such as rheumatoid arthritis.

## Acute toxicity test

The acute toxicity test showed that the seed oil of *E. maculata* was not toxic to rat at the doses administered per oral route (*p.o.*). During this experiment, no apparent behavioural side effects were observed in the animals; they were very active. This shows that the seed oil was relatively non-toxic and safe.

## *In vivo* antiinflammatory activity

The carrageenan assay was used because of its sensitivity in defecting orally active anti-inflammatory agents mainly in the acute phase of inflammation. The anti-inflammatory effects of the seed oil of *E. maculata* and standard antiinflammatory drug (indomethacin) on carrageenan induced edema in rats hind paws were presented in table 3. The antiinflammatory activity of oil was found to have effect in dose-time manner. There was a significant decrease in edema paw volume of rats in the test group. However, there was no reduction in inflammation found in case of control group. The results showed that the seed oil of *E. maculata* causes significant reduction in inflammation i.e. 87.50% (1000 $\mu g.kg.ml^{-1}$ *p.o*), while the standard anti-inflammatory drug indomethacin gave 93.75% (25 $mg.kg^{-1}$). The lower the paw volume the better the activity, the inhibitory activity of the oil is very close to indomethacin. There was no reduction in inflammation found in case of rats treated with 10% DMSO.

| Seed oil and reference compound | DPPH $IC_{50}$ $\mu gml^{-1}$ | FRAP $EC_{50}$ $\mu gml^{-1}$ |
|---|---|---|
| E. maculata | 8.00 | 10.00 |
| Ascorbic acid | 9.00 | 20.00 |

Data are presented as triplicate of the mean ± S.E.M

**Table 2:** $IC_{50}$ of the DPPH and FRAP assays of the seed essential oil of *E. maculate*.

| Seed Oil | 2 Hour | % I (2 Hr) | 4 Hour | % I (4 Hr) | Mean Paw (mm) | Mean % I |
|---|---|---|---|---|---|---|
| E. maculata | $4.50 \pm 0.67$ | 87.50 | $4.10 \pm 0.69$ | 62.50 | $4.30 \pm 0.53$ | 87.50 |
| Indomethacin | $4.70 \pm 0.21$ | 87.50 | $4.60 \pm 0.35$ | 99.65 | $4.65 \pm 0.29$ | 93.70 |

Data are presented as triplicate of the mean with standard deviation

**Key note:** I%=% Inhibition; 70-100: very strong activity; 40-69: moderate activity; ≤ 39: weak activity

**Table 3:** *In vivo* anti-inflammatory activity the seed essential oil of *E. maculate*.

| Seed oil and reference compound | Time of licking and biting percentage inhibition | | | |
|---|---|---|---|---|
| | Early phase (0-5) min. | Percentage inhibition | Late phase (5-30) min. | Percentage inhibition |
| E. maculata | 13.00 ± 9.19 | 86.46 | 46.70 ± 5.92 | 60.10 |
| Indomethacin | 34.33 ± 2.12 | 64.23 | 53.00 ± 2.12 | 54.70 |
| 10% DMSO | 96.00 | - | 117 | - |

Data are presented as triplicate of the mean ± S.E.M

**Table 4:** *In vivo* analgesic activity the seeds essential oil of *E. maculate*.

Inflammation is a complex physiopathological response to different stimuli. Formation of edema induced by carrageenan is commonly correlated with the early exudative stage of inflammation, one of the important processes of inflammatory pathology. In the early phase (one hour) of carrageenan injection, there is sudden elevation of paw volume as consequence of histamine, serotonin and related substances liberation from mastocyte cells. In the later phase (over one hour) the inflammation increases gradually and is elevated. This second phase is mediated by prostaglandins, cyclooxygenase, proteases and lysosome products. Continuity between the two phases is provided by kinins [25,26]. The seed oil extract promptly controlled both the phases of inflammation.

### *In vivo* antinociceptive activity

Chemically induced visceral pain and paw nociception are very useful models for the study of nociception and the assessment of analgesic drugs. The antinociceptive activity of the seed oil of *E. maculata* measured on abino rat by using injection of formalin solution as shown in table 4. Interestingly, the seed oil at the concentration of 1000 µg.kg⁻¹ exhibited high inhibitory effect 86.46 and 60.10% in early (neurogenic) and late (inflammatory) phases respectively, while the standard anti-inflammatory drug (indomethacin) gave 64.23 and 54.70% in first and second phase respectively. The results showed that the seed oil is more active than the synthetic drug (indomethacin) commonly used as analgesic and anti-inflammatory drug. Drugs that act primarily on the central nervous system inhibit both phases equally while peripherally acting drugs inhibit the late phase. The early phase is a direct result of stimulation of nociceptors in the paw which reflects centrally mediated pain while the late phase is due to inflammation with a release of serotonin, histamine, bradykinin and prostaglandins. The seed oil exhibited significant dose related reduction of paw licking caused by formalin and was able to block both phases of the formalin response. The formalin test represents a more valid model for clinical pain. The formalin test is a very useful method, not only for assessing antinociceptive drugs, but also helping in the elucidation of the action mechanism [27,28].

## Conclusion

The present study demonstrated for the first time the phytochemicals and therapeutic potentials of the seed essential oil of *E. maculata* grown in Nigeria. On the basis of this result the seed essential oil *E. maculata*, may be utilized as a source for the isolation of natural pinenes, cyclofenchene and other terpenoids. Pharmacological activities of the oil may be due to the synergetic effect of these chemical constituents. Terpenoids and esters are antiinflammatory, analgesic, antiseptic, expectorant and stimulating compounds. Some are antiviral and some help break down gallstones [29]. *β*-pinene, eucalyptol, *p*-menth-1-en-ol and ocimene, have strong antimicrobial activities and eucalyptol also used as therapeutic agent in cardiovascular effects [30]. The excellent antiinflammatory and remarkable antinociceptive potentials of the seed essential oil of *E. maculata* may provide supports in the treatment of pathologies such as painful and inflammatory disorder in which free

radical oxidation plays a fundamental role. Moreover, the seed oil gives no sign of toxicity, which indicates that the oil is therapeutic safety for the pharmacologically active doses. The information on essential oil profile can be used for the possible exploitation of this species for various research and pharmaceutical purposes. This shows that the seed essential oil of *E. maculata* would exert several beneficial effects by virtue of its phytochemicals and pharmacological potentials and could be harnessed in drug formulation.

### References

1. Ali I, Abbas SQ, Hameed M, Naz N, Zafar S, et al. (2009) Leaf Anatomical Adaptations in some Exotic Species of *Eucalyptus* L'hér (Myrtaceae). Pak J Bot 41: 2717-2727.

2. Rashwan OA (2002) New Phenylpropanoid Glucosides from *Eucalyptus maculate*. Molecules 7: 75-80.

3. Takahashi T, Kokubo R, Sakaino M (2004) Antimicrobial activities of eucalyptus leaf extracts and flavonoids from *Eucalyptus maculata*. Lett Appl Microbiol 39: 60-64.

4. Reddy LJ, Gopu S, Jose B, Jalli RD (2012) Evaluation of Antibacterial and DPPH Radical Scavenging Activities of The Leaf Essential Oils of *Pongamia pinnatta* and *Eucalyptus maculate*. Asian Journal of Biochemical and Pharmaceutical Research 3: 25-32.

5. Lassak EV, McCarthy T (2006) Australian Medicinal Plants, Reed publishers, Australia.

6. European Pharmacopoeia Commission (2004) Euro Pharmac. (5thedn).

7. Sibery D (2005) Starting your own consultancy. Into the great wide open. Leaving C-suite to start consultancy requires huge change in perspective. Mod Healthc 35: S8-9.

8. Ogunwande IA, Flamini G, Cioni PL, Omikorede O, Azeez RA, et al. (2010) Aromatic Plants growing in Nigeria: Essential Oil Constituents of *Cassia alata* (Linn.) Roxb. and *Helianthus annuus* L. Rec Nat Prod 4: 211-217.

9. Pourmorad F, Hosseinimehr SJ, Shahabimajd N (2006) Antioxidant activity, phenol and flavonoid contents of some selected Iranian medicinal plants. African Journal of Biotechnology 5: 1142-1145.

10. Ololade ZS, Olawore NO, Kolawole AS, Onipede OJ, Alao FO (2012) Phytochemicals, Free Radical Scavenging and Anti-inflammatory Activity of the Leaf Essential Oil of *Callitris columellaris* F. Muell from Plateau State, Nigeria. International Journal of Applied Research and Technology 1: 38-45.

11. Segev A, Badani H, Kapulnik Y, Shomer I, Oren-Shamir M, et al. (2010) Determination of 'polyphenols, flavonoids, and antioxidant capacity in colored chickpea (Cicer arietinum L.). J Food Sci 75: S115-119.

12. Santin JR, Silveira A, Muller E, Claudino VD, Cruz AB, et al. (2011) Evaluation of the acute toxicity, genotoxicity and mutagenicity of ethanol extract of *Piper aduncum*. Journal of Medicinal Plants Research 5: 4475-4480.

13. Igbe I, Ching FP, Eromon A (2010) Anti-inflammatory activity of aqueous fruit pulp extract of Hunteria umbellata K. Schum in acute and chronic inflammation. Acta Pol Pharm 67: 81-85.

14. Zeashana H, Amresha G, Raoa CV, Singh S (2009) Antinociceptive activity of *Amaranthus spinosus* in experimental animals. J Ethnopharmacol 122: 492–496.

15. Pino JA, Marbot R, Quert R, Garcia H (2002) Study of essential oils of *Eucalyptus resinifera* Smith, *E. tereticornis* Smith and *Corymbia maculata* (Hook.) KD Hill and L.A.S. Johnson, grown in Cuba. Flavour and Fragrance Journal 17: 1-14.

16. Assareh MH, Sedaghati M, Kiarostami K, Zare AG (2010) Seasonal changes of essential oil composition of Eucalyptus maculata Hook. Iranian Journal of Medicinal and Aromatic Plants 25: 580-588.

17. Elaissi A, Rouis Z, Mabrouk S, Salah KB, Aouni M, et al. (2012) Correlation between chemical composition and antibacterial activity of essential oils from fifteen Eucalyptus species growing in the Korbous and Jbel Abderrahman arboreta (North East Tunisia). Molecules 17: 3044-3057.

18. Cakir A, Mavi A, Yildirim A, Duru ME, Harmandar M, et al. (2003) Isolation and characterization of antioxidant phenolic compounds from the aerial parts of Hypericum hyssopifolium L. by activity-guided fractionation. J Ethnopharmacol 87: 73-83.

19. Gupta S, Kumar MNS, Duraiswamy B Chhajed, M Chhajed A (2012) In-vitro Antioxidant and Free Radical Scavenging Activities of Ocimum Sanctum. World Journal of Pharmaceutical Research 1: 78-94.

20. Prior RL, Wu X, Schaich K (2005) Standardized methods for the determination of antioxidant capacity and phenolics in foods and dietary supplements. J Agric Food Chem 53: 4290-4302.

21. Parejo I, Viladomat F, Bastida J, Rosas-Romero A, Flerlage N, et al. (2002) Comparison between the radical scavenging activity and antioxidant activity of six distilled and nondistilled mediterranean herbs and aromatic plants. J Agric Food Chem 50: 6882-6890.

22. Ben Marzoug HN, Romdhane M, Lebrihi A, Mathieu F, Couderc F, et al. (2011) Eucalyptus oleosa essential oils: chemical composition and antimicrobial and antioxidant activities of the oils from different plant parts (stems, leaves, flowers and fruits). Molecules 16: 1695-1709.

23. Noumi E, Snoussi M, Hajlaoui H, Trabelsi N, Ksouri R, et al. (2011) Chemical composition, antioxidant and antifungal potential of Melaleuca alternifolia (tea tree) and Eucalyptus globulus essential oils against oral Candida species. Journal of Medicinal Plants Research 5: 4147-4156.

24. Obame LC, Koudou J, Chalchat JC, Bassole I, Edou P, et al. (2007) Volatile components, antioxidant and antibacterial activities of Dacryodes buettneri H. J. Lam. essential oil from Gabon. Scientific Research and Essay 2: 491-495.

25. Perianayagam JB, Sharma SK, Pillai KK (2006) Anti-inflammatory activity of Trichodesma indicium root extract in experimental animals. J Ethnopharmacol 104: 410-414.

26. Fernandes JC, Spindola H, de Sousa V, Santos-Silva A, Pintado ME, et al. (2010) Anti-inflammatory activity of chitooligosaccharides in vivo. Mar Drugs 8: 1763-1768.

27. Le-Bars D, Gozariu M, Cadden SW (2001) Animal Models of Nociception. Pharmacol Rev 53: 597-652.

28. Garcia MD, Fernandez MA, Alvarez A, Saenz MT (2004) Antinociceptive and anti-inflammatory effect of the aqueous extract from leaves of Pimenta racemosa var. ozua (Myrtaceae). J Ethnopharmacol 91: 69-73.

29. Indraya AK, Garg SN, Rathiu AK, Sharma V (2007) Chemical composition and antimicrobial activity of the essential oil of alpinia officinarum rhizome. Indian Journal of Chemistry 46B: 2060-2063.

30. Nan P, Hu Y, Zhao J, Feng Y, Zhong Y, et al. (2004) Chemical Composition of the Essential Oils of Two Alpinia Species from Hainan Island, China. Z Naturforsch C 59: 157-160.

# 23

# Theoretical Descriptor for the Correlation of the Fish Embryo Toxicity test (FET) By QSAR Model of $LC_{50}$

**Jining L, Deling F, Lei W, Linjun Z and Lili S***

*Nanjing Institute of Environmental Sciences, Ministry of Environmental Protection, Nanjing, 210042, China*

**Abstract**

Quantitative structure–toxicity relationships were developed for the prediction of the fish embryo toxicity test using the CODESSA treatment, based on the linear heuristic method (HM) and support vector machines (SVM). Each kind of compound was represented by several calculated structural descriptors, derived for a diverse set of 97 compounds using DFT- B3LYP/6-31+G (d) level of the theory. A six-parameter correlation was found for the 97 compounds. In the HM method, the value of square of the correlation coefficient $r^2$ is 0.8142, s2 is 0.0380, in the SVM method, the value of the $r^2$ is 0.7105 and s2 is 0.0604. The HM model may be used for prediction of toxicity, safety and risk assessment of chemicals to achieve better ecotoxicological management.

**Keywords:** Quantitative Structure Activity Relationship (QSAR); Support Vector Machine (SVM); Heuristic method; Fish Embryo Toxicity Test (FET)

## Introduction

The fish acute toxicity is a mandatory component in the base set of data requirements for eco-toxicity testing but fish suffer distress and perhaps pain [1]. Animal alternative considerations have also been incorporated into new REACH regulations through strong advocacy for the reduction of testing with live animals. REACH, as a regulatory tool, was envisaged to revise the existing chemical policy and to harmonize chemical legislations in Europe and the major aim of REACH is to systematically evaluate the risk to human health and the environment of approximately 30,000 chemical substances [2]. The fish embryo toxicity (FET) testan alternative approach to classical acute fish toxicity testing with live fish, has been a mandatory component in routine whole effluent testing in Germany since 2005 and has already been standardized at the international level. A modified version has been submitted by the German Federal Environment Agency (UBA) as a draft guideline for an alternative to chemical testing with intact fish [1]. The fish embryo test is a surrogate for the OECD 203 acute fish toxicity test (or other guideline equivalent acute fish assays). Recently, the regulations of the People's Republic of China on Control over Safety of Hazard Chemicals revised and passed, therefore, FET is useful for the acute fish toxicity testing. In order to analyze the applicability of the FET also in chemical testing, a comparative re-evaluation was carried out for a total of 97 substances, and QSAR was developed to evaluate fish embryo toxicity data. QSAR to evaluate fish embryo toxicity data is described to be useful for future validation exercises.

The quantitative structure-property/activity relationship (QSPR/QSAR) method is based on the assumption that the variation of the behavior of the compounds, as expressed by many measured physicochemical properties, can be correlated with changes in molecular features of the compounds termed descriptors [3]. This method can be used for the prediction of the properties of new compounds. It can also be applied to identify and describe important structural features of the molecules that are relevant to variations in molecular properties. Computational models are useful because they rationalize a large number of experimental observations.

The QSAR model is widely used for the prediction of physicochemical properties and biological activities in chemical, environmental, and pharmaceutical areas [4,5], a potential or useful technique to estimate the toxicity especially for the compounds whose toxicity are not easy to test. The advantage of this approach over the experimental method lies in the fact that the descriptors used can be calculated from the structures alone and are not dependent on any experimental properties. Once the structure of a compound is known, a limited set of molecular properties, said descriptors, which can be obtained observing the molecular structure or semi-empirical quantum-mechanical calculations [6]. Therefore, once a reliable model is established, the model can be applied to help in identifying and describe important structural features of the molecules that are relevant to variations in molecular properties.

In this work we proposed a QSAR model to evaluate fish embryo toxicity data. 97 structures are calculated by the CODESSA software package, developed by Katritzky et al. [7,8] based on descriptors constitutional, topological, geometrical, geometrical, electrostatic and quantum chemical.

## Materials and Method

In order to develop the QSAR, a database of $LC_{50}$, reported in terms of micromolar ($\mu$M) concentration for correlation purposes, reported $LC_{50}$ values were converted to their molar units and subsequently to free energy related negative logarithmic state, i.e., log (1/ $LC_{50}$). These $LC_{50}$ values have been obtained from experiments and literature [9]. The selected series consists of total 97 numbers of compounds, divided in to two data sets, one was training set having 83 compounds and other was test set with the remaining 14 compounds based on random selection criteria. The biological activities were expressed in terms of median lethal concentration ($EC_{50}$).

***Corresponding author:** Lili S, Nanjing Institute of Environmental Sciences, Ministry of Environmental Protection, Nanjing, 210042, China
E-mail: sll@nies.org

## Calculation of the descriptors

The chemicals in the data set (Table 1) have been optimized and their electronic, geometric, and energetic parameters have been calculated with Gaussian 03 package [11]. The structures were optimized and

| NO. | Compounds | status | MW | Experimental (µg/L) | HM ( µg/L) | SVM (µg/L) |
|-----|-----------|--------|-----|---------------------|------------|------------|
| 1 | 2-Aminoethanol | Training | 61.08 | 6.5662 | 4.5114 | 4.8806 |
| 2 | 4-Aminophenol | Training | 109.13 | 2.6720 | 1.8662 | 2.2334 |
| 3 | Benzoicacid | Training | 122.12 | 4.3031 | 2.1385 | 2.3162 |
| 4 | 4-Bromoindole | Training | 195.95 | 3.6720 | 1.678 | 1.4798 |
| 5 | 5-Bromoindole | Training | 196.05 | 3.7250 | 1.7252 | 1.5358 |
| 6 | 6-Bromoindole | Training | 196.05 | 3.96426 | 2.3408 | 2.2578 |
| 7 | 2-Bromophenol | Training | 173.01 | 4.6543 | 2.0823 | 2.2993 |
| 8 | 3-Bromophenol | Training | 173.01 | 4.7829 | 3.2246 | 2.7912 |
| 9 | 4-Bromophenol | Training | 173.01 | 4.6657 | 3.3539 | 2.9999 |
| 10 | n-Butylamine | Training | 73.14 | 4.5552 | 3.743 | 2.408 |
| 11 | sec-Butylamine | Training | 73.14 | 4.9784 | 2.0591 | 2.2457 |
| 12 | Butyldiglycol | Training | 162.23 | 6.1077 | 1.8154 | 2.3225 |
| 13 | 2-Chloroaniline | Training | 127.6 | 4.4517 | 3.8754 | 2.8491 |
| 14 | 3-Chloroaniline | Training | 127.6 | 4.3222 | 3.0148 | 2.7056 |
| 15 | 4-Chloroaniline | Training | 127.6 | 4.3280 | 1.7793 | 1.5822 |
| 16 | 4-Chlorophenol | Training | 128.56 | 4.5800 | 1.6495 | 2.0511 |
| 17 | Cyclohexanol | Training | 100.16 | 6.1405 | 1.9123 | 2.32 |
| 18 | Cyclohexylamine | Training | 99.18 | 4.8019 | 2.7298 | 2.5952 |
| 19 | n-Decylamine | Training | 157.3 | 3.4977 | 1.4561 | 2.0633 |
| 20 | 2,4-Dibromophenol | Training | 251.9 | 3.9003 | 1.4761 | 2.0776 |
| 21 | 2,6-Dibromophenol | Training | 251.9 | 4.6208 | 3.8754 | 2.8491 |
| 22 | Dibutylamine | Training | 129.25 | 4.6069 | 3.3737 | 2.9902 |
| 23 | 2,4-Dichloroaniline | Training | 162.02 | 4.3324 | 5.1731 | 2.8854 |
| 24 | 3,4-Dichloroaniline | Training | 162.02 | 3.3031 | 3.3448 | 3.0048 |
| 25 | Dicyclohexylamine | Training | 181.32 | 4.4939 | 2.2432 | 2.8373 |
| 26 | Diethylamine | Training | 73.14 | 4.9696 | 2.0318 | 2.6624 |
| 27 | Diethyleneglycol | Training | 106.14 | 7.7059 | 3.6017 | 1.9352 |
| 28 | N,N-Diethylmethylamine | Training | 87.16 | 4.8450 | 3.3448 | 3.0048 |
| 29 | N,N-Diisopropylethylamine | Training | 129.25 | 5.0193 | 2.9621 | 2.8025 |
| 30 | Diisobutylamine | Training | 129.3 | 4.6738 | 2.3816 | 2.6403 |
| 31 | Diisopropylamine | Training | 101.2 | 4.9613 | 3.7035 | 3.1539 |
| 32 | N,N-Dimethylamine | Training | 45.09 | 5.5985 | 5.4871 | 4.7149 |
| 33 | N,N-Dimethylanilin | Training | 121.2 | 4.7342 | 1.1533 | 1.7563 |
| 34 | N,N-Dimethylbutylamine | Training | 101.19 | 4.7075 | 0.7109 | 1.3753 |
| 35 | N,N-Dimethylcyclohexylamine | Training | 127.23 | 4.7247 | 1.5988 | 1.7464 |
| 36 | N,N-Dimethylethylamine | Training | 73.14 | 4.9183 | 5.1469 | 5.2986 |
| 37 | N,N-Dimethylformamide | Training | 73.09 | 6.9746 | 4.8383 | 4.789 |
| 38 | 4,6-Dinitro-o-cresol | Training | 198.14 | 2.5910 | 4.5685 | 3.9195 |
| 39 | 2,4-Dinitrophenol | Training | 184.11 | 2.9542 | 2.4241 | 2.8188 |
| 40 | Dipentylamine | Training | 157.3 | 4.6313 | 2.7028 | 2.7623 |
| 41 | Dipropylamine | Training | 101.19 | 4.4936 | 2.4946 | 2.2544 |
| 42 | Ethanol | Training | 46.1 | 7.0625 | 3.2158 | 4.2721 |
| 43 | Ethylacetate | Training | 88.106 | 6.2692 | 2.9198 | 2.9311 |
| 44 | Ethylenediamine | Training | 60.1 | 5.5984 | 3.5851 | 4.5713 |
| 45 | 1-Ethylpiperidine | Training | 113.2 | 4.8531 | 6.14 | 5.7386 |
| 46 | 2-Ethylpiperidine | Training | 113.2 | 4.9729 | 3.7661 | 2.8756 |
| 47 | n-Heptylamine | Training | 115.22 | 4.4541 | 4.5401 | 4.1437 |
| 48 | n-Hexylamine | Training | 101.19 | 4.6261 | 5.746 | 5.3312 |
| 49 | Hydroquinone | Training | 110.11 | 3.8976 | 4.0228 | 4.1965 |
| 50 | Hydroxyurea | Training | 76.05 | 6.2530 | 5.1664 | 5.3695 |
| 51 | Isobutylamine | Training | 73.1 | 4.966694 | 4.201 | 4.4596 |
| 52 | Isopentylamine | Training | 87.16 | 4.7715 | 2.315 | 2.1711 |
| 53 | Isopropylamine | Training | 59.11 | 6.4425 | 2.6561 | 2.616 |
| 54 | Methanol | Training | 32.04 | 7.3443 | 2.668 | 2.6086 |
| 55 | Methoxyaceticacid | Training | 90.08 | 4.7298 | 4.6482 | 3.7387 |
| 56 | 2-Methoxyethanol | Training | 76.09 | 7.3125 | 1.4622 | 1.8345 |

| 57 | 1-Methoxy-2-propanol | Training | 90.12 | 7.1434 | 1.9533 | 2.2345 |
|---|---|---|---|---|---|---|
| 58 | 3-Methyl-1-butanol | Training | 88.17 | 6.0097 | 1.07 | 2.513 |
| 59 | N-Methylamine | Training | 31.1 | 5.8525 | 2.0118 | 1.8028 |
| 60 | N-Methylanilin | Training | 107.2 | 4.5339 | 3.1082 | 2.2898 |
| 61 | N-Methylformamide | Training | 59.07 | 7.0408 | 1.2202 | 2.3192 |
| 62 | 1-Methylpiperidine | Training | 99.18 | 4.8346 | 2.956 | 2.9168 |
| 63 | 2-Methylpiperidine | Training | 99.18 | 5.0100 | 1.5135 | 1.2284 |
| 64 | 4-Methylpiperidine | Training | 99.18 | 4.9681 | 3.0818 | 2.47 |
| 65 | Morpholine | Training | 87.12 | 5.7790 | 4.7142 | 5.0932 |
| 66 | 2-Nitroaniline | Training | 138.1 | 4.3324 | 2.3356 | 2.3419 |
| 67 | 4-Nitrobenzoicacid | Training | 167.12 | 4.4756 | 2.0638 | 2.3089 |
| 68 | 4-Nitrophenol | Training | 139.11 | 4.7573 | 1.2928 | 1.818 |
| 69 | n-Nonylamine | Training | 143.27 | 4.0592 | 4.6492 | 2.7816 |
| 70 | 1-Octanol | Training | 130.23 | 4.18752 | 4.3523 | 5.4814 |
| 71 | n-Octylamine | Training | 129.25 | 4.4058 | 0.5177 | 0.9196 |
| 72 | Pentachlorophenol | Training | 266.34 | 2.7403 | 2.8802 | 2.3573 |
| 73 | n-Pentylamine | Training | 87.16 | 4.4893 | 1.7798 | 1.2701 |
| 74 | 4-tert-Pentylphenol | Training | 164.24 | 3.5440 | 2.5337 | 1.7304 |
| 75 | Phenol | Training | 94.11 | 5.3222 | 2.7298 | 2.5952 |
| 76 | Piperidine | Training | 85.2 | 5.0433 | 3.9809 | 2.4011 |
| 77 | 2-Propanol | Training | 60.1 | 6.9721 | 5.3213 | 5.0889 |
| 78 | n-Propylamine | Training | 59.11 | 4.8984 | 3.5419 | 2.5303 |
| 79 | Salicylicacid | Training | 138.12 | 4.3577 | 4.8308 | 5.2073 |
| 80 | Tetrachloroethylene | Training | 165.83 | 4.4281 | 2.8316 | 2.9567 |
| 81 | 2,4,6-Tribromophenol | Training | 330.8 | 3.6454 | 2.0737 | 2.352 |
| 82 | Triethyleneglycol | Training | 150.2 | 7.7315 | 0.9536 | 2.5847 |
| 83 | Urea | Training | 60.06 | 7.3598 | 2.354 | 2.2464 |
| 84 | 2,5-Hexanedion | Test | 114.14 | 6.6691 | 6.0191 | 5.6191 |
| 85 | Acetone | Test | 58.08 | 4.0211 | 3.9211 | 3.0111 |
| 86 | Acrolein | Test | 56 | 2.5682 | 1.9682 | 1.5682 |
| 87 | Chloroacetaldehyde | Test | 78 | 3.5251 | 3.9551 | 2.9351 |
| 88 | Diazinon | Test | 304.34 | 4.4961 | 4.1161 | 4.0161 |
| 89 | Diethyleneglycoldimethylether | Test | 134.18 | 7.0436 | 6.5236 | 7.9436 |
| 90 | Dimethylsulfoxide | Test | 78.13 | 7.4638 | 6.4638 | 6.1238 |
| 91 | D-Mannitol | Test | 182.17 | 4.8908 | 5.0108 | 3.3508 |
| 92 | Formamide | Test | 45.041 | 6.9610 | 7.0610 | 2.9210 |
| 93 | p-tert-Butylphenol | Test | 150.22 | 3.2380 | 2.2380 | 1.2380 |
| 94 | Quinone | Test | 108.09 | 2.6739 | 1.9739 | 0.6739 |
| 95 | Saccharinsodiumsalthydrate | Test | 241.2 | 7.3211 | 6.2111 | 8.11211 |
| 96 | Tridecylmono-octylether | Test | 244.4 | 3.1568 | 2.1268 | 1.1128 |
| 97 | Valproicacid | Test | 144.21 | 4.3051 | 3.9151 | 2.3751 |

**Table 1:** The experimental values of all compounds and the predicted $LC_{50}$ by HM and SVM.

their vibrational frequencies calculated at the DFT level relying on the B3LYP functional. A comparison of all the methods, namely, AM1, MP3, HF, and DFT/B3LYP, indicates that the DFT/B3LYP method is more reliable than others and has a high predictive power [12]. After optimization, the CODESSA program was used to calculate five types of molecular descriptors constitutional (number of various types of atoms and bonds, number of rings, molecular weight, etc.), topological (Wiener index, Randic indices, Kier -Hall shape indices, etc.), geometrical (moments of inertia, molecular volume, molecular surface area, etc.), electrostatic (minimum and maximum partial charges, polarity parameter, charged partial surface area descriptors, etc.), and quantum chemical (reactivity indices, dipole moment, HOMO and LUMO energies, etc.). In the framework of the CODESSA program, the statistical structure-property correlation techniques can be used for the analysis of user-submitted experimental data in combination with the calculated molecular descriptors.

### Development of linear model by HM [13]

The heuristic procedure was used to obtain quantitative structure property relationship. Once the molecular descriptors are generated, the HM in CODESSA is used to pre-select the descriptors and build the near model. The advantages of the HM are the high speed, usually produces correlations 2–5 times faster than other methods with comparable quality and no software restrictions on the size of the data set [14]. The HM can either quickly give a good estimation about what quality of correlation to expect from the data, or derive several best regression models. HM of the descriptors selection proceeds with a pre-selection of the descriptors to eliminate: (1) those descriptors that are not available for each structure;(2) descriptors having a small variation in magnitude for all structures;(3)descriptors that provide a F-test value below 1.0 on the one-parameter correlation and (4) and the descriptors whose t-values are less than the user-specified value, etc. New descriptors were added one by one until the suitable number descriptors without losing any important information in the model is achieved [15]. The final result is a list of the 10 best models according to the values of the F-test and correlation coefficient. The statistical quality of the generated models was gauged by the F-test (F), squared correlation co-efficient ($R^2$), square of cross-validate coefficient regression ($R_{CV}^2$), t-test and the standard deviation ($s^2$). Fischer's value (F), which represents F-ratio between the variance of calculated and observed activity, and chance statistics assuring that the results are not merely based on chance correlations and t-test reflects significance of the parameter within the model. From the above processes, seven descriptors are selected from descriptors pool and the linear model is produced by the HM [16]. The experimental values of all compounds are summarized in Table 1.

## Development of linear model by SVM

The support vector machine algorithm was developed by Vapnik [17]. Support vector machine (SVM), as a novel type of learning machine is gaining rapid popularity due to its remarkable generalization performance [18], has gained much interest in pattern recognition and function approximation applications recently. In bioinformatics, SVMs have been successfully used to solve classification and correlation problems, with the basic idea of SVM is to map the original data into a higher dimensional feature space by a kernel function and then to do classification in this space by constructing an optimal separating hyperplane. Compared with traditional regression and neural networks methods, SVMs have some advantages, including the absence of local minima, good generalization ability, simple implementation, few free parameters, dimensional independence and its ability to condense information contained in the training set [19]. The flexibility in classification and ability to approximate continuous function make SVMs very suitable for QSAR and QSPR studies. Because it is difficult to predict in advance which descriptors are most relevant to the problem at hand, however, SVM is well known to tolerate irrelevant features [20-22]. In some cases, the best average performance was achieved when all the features were given to SVM [23].

## Results and Discussion

### Results of HM

The heuristic method is not made a prior choice of descriptors, but it performs an analysis on several intrinsic molecular properties available from the Gaussian output. We performed correlations for a growing number of descriptors in the range from 1 to 10.

Figure 1 shows the plots of $R^2$, cross-validated coefficient ($R_{CV}^2$) and the standard deviation ($s^2$) for the training set as a function of the number of descriptors models. $R^2$ and $R_{CV}^2$ are increased with the increasing number of descriptors. However, the values of $s^2$ decreased with the increasing number of descriptors [24,25]. When adding another descriptor did not improve significantly the statistics of a model, it was determined that the optimum subset size had been achieved. From Figure 1, it can be seen that six descriptors appear to be sufficient for a successful regression model. Then the corresponding descriptors are applied as inputs for the non-linear model. Best models were selected on the basis of their statistical significance. High $r^2$ and high F and low $s^2$ values in each model reflect their good predictive potential. Pair wise correlations between the descriptors used in the QSAR models obtained for series is summarized in Table 2. The molecular descriptors used in the selected QSAR models are defined in Table 3. The predict results are shown in Figure 2.

## Results of SVM

SVM is a machine-learning approach that allows one to learn from experimental data within a given set, and build a computer model to make predictions on new data sets. The associated parameters are c, p and g, –t kernel function is the RBF kernel function: $\exp(-gamma^*|uv|^{\wedge}2)$, the calculated optimal correlation parameters are c:3.9192, g:168.9664, p: 0.9892. The predict results are shown in Figure 3.

## Compare with the HM and SVM

In the HM method, the $R^2$ and $s^2$ are 0.8142 and 0.0380 in training set, the $R^2$ and $s^2$ are 0.9238 and 1.8503 in test set. At the same time, in the SVM method, the $R^2$ and $s^2$ are 0.7105 and 0.0604 in training set,

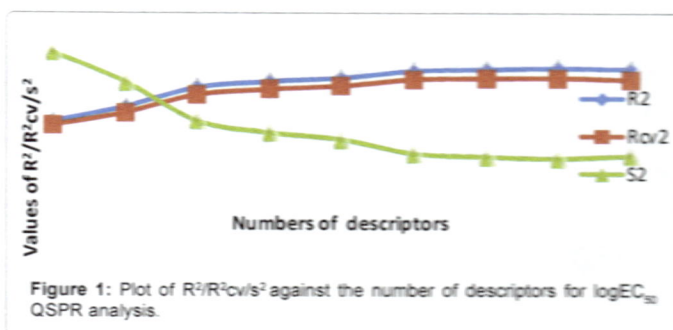

**Figure 1:** Plot of $R^2/R^2cv/s^2$ against the number of descriptors for logEC$_{50}$ QSPR analysis.

|   | A | B | C | D | E | F |
|---|---|---|---|---|---|---|
| A | 1.0000 | -0.0725 | 0.2135 | 0.3806 | 0.4239 | 0.2161 |
| B | -0.0725 | 1.0000 | -0.3540 | -0.1499 | -0.2106 | 0.1555 |
| C | 0.2135 | -0.3540 | 1.0000 | 0.2619 | 0.1297 | -0.2949 |
| D | 0.3806 | -0.1499 | 0.2619 | 1.0000 | 0.2398 | 0.2922 |
| E | 0.4239 | -0.2106 | 0.1297 | 0.2398 | 1.0000 | -0.2729 |
| F | 0.2161 | 0.1555 | -0.2949 | 0.2922 | -0.2729 | 1.0000 |

**Table 2:** Correlation matrix of the six descriptors used in this work.

| $R^2$ | $R^2_{cv}$ | F | X+DX | t-Test | Descriptors |
|---|---|---|---|---|---|
| 0.8142 | 0.7733 | 55.50 | 6.0388e+01 6.3153e+00 | 9.5622 | Intercept(A) |
|  |  |  | -1.3840e+01 1.5885e+00 | -8.7129 | Avg valency of a C atom(B) |
|  |  |  | -1.2775e+01 1.7259e+00 | -7.4017 | Relative number of double bonds(C) |
|  |  |  | -1.9631e-02 2.5005e-03 | -7.8508 | Molecular volume(D) |
|  |  |  | 4.9342e-01 1.0389e-01 | 4.7492 | HOMO-1 energy(E) |
|  |  |  | -5.9297e-01 1.2477e-01 | -4.7492 | HOMO energy(F) |
|  |  |  | -2.2411e+00 6.0338e-01 | -3.7143 | Average structural information content(order 1)(G) |

1. Avg valency of a C atom;2, Relative number of double bonds;3, PPSA-3 Molecular volume;4, HOMO-1 energy;5, HOMO energy;6, Average Structural Information content (order 1);
The fitted regression model is: logEC$_{50}$ = (6.0388e+01 6.3153e+00)*A+(-1.3840e+01 1.5885e+00)*B+(-1.2775e+01 1.7259e+00)*C+(-1.9631e-02 2.5005e-03)*D+(4.9342e-01 1.0389e-01)*E+(-5.9295e-01 1.2477e-01)*F+(-2.2411e+00 6.0338e-01)*G

**Table 3:** Correlations of toxicity by the heuristic method.

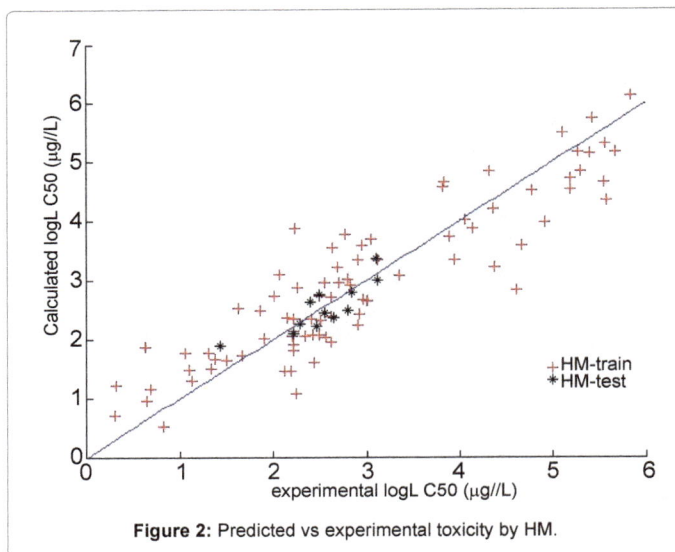

Figure 2: Predicted vs experimental toxicity by HM.

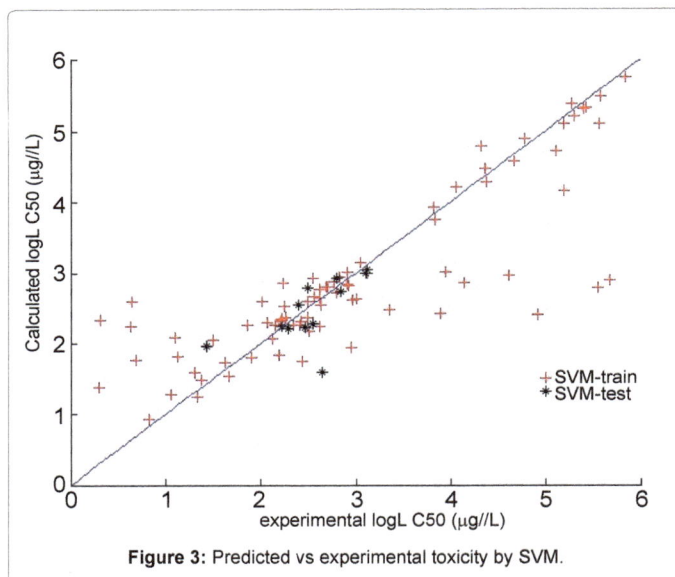

Figure 3: Predicted vs experimental toxicity by SVM.

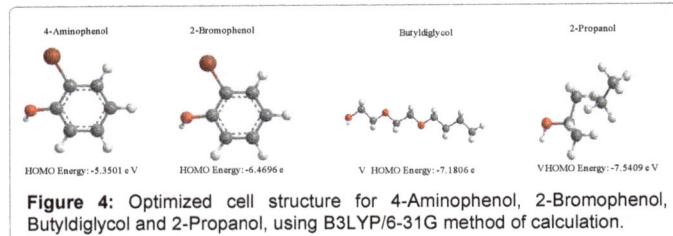

**Figure 4:** Optimized cell structure for 4-Aminophenol, 2-Bromophenol, Butyldiglycol and 2-Propanol, using B3LYP/6-31G method of calculation.

| | Training | | Test | |
|---|---|---|---|---|
| | **R²** | **s²** | **R²** | **s²** |
| HM | 0.8142 | 0.0380 | 0.9238 | 1.8503 |
| SVM | 0.7105 | 0.0604 | 0.7527 | 2.4351 |

**Table 4:** Performance comparison between HM and SVM.

in high-resolution structural determination of macromolecules. The HOMO-1, which is lower in energy than the HOMO, features a large contribution for toxicity.

The HOMO of 4-Aminophenol, 2-Bromophenol, Butyldiglycol and 2-Propanol, were calculated by B3LYP/6-31G method. 2-Propanol was the highest toxicity with the highest HOMO, as well as 4-Aminophenol was the lowest toxicity with the lowest HOMO (Figure 4). The structure of 4-Aminophenol was similar to 2-Bromophenol. The toxicity 2-Bromophenol, with higher HOMO, was higher than 4-Aminophenol. The Highest Occupied Molecular Orbital (HOMO) describes the effect of HOMO energies on the toxicity.

## Conclusions

A QSAR study for fish embryo toxicity test (FET) was performed using HM and SVM, based on electrostatic, CPSA, constitutional, and quantum-chemical descriptors and satisfactory results were obtained. Through analyzing the obtained results, the present study gives rise to QSAR with good statistical significance and predictive capacity. The 6-parameter model can be considered as the best linear modelthe correlations are satisfying. Additionally, HM produced good model with vigorous predictive ability. Furthermore, the proposed approach can also be extended in other QSAR investigations. Fish embryo testing of chemicals has matured to the point that international standardization, method validation, and broadening of chemical coverage is rapidly occurring. Fish embryo tests offer a reasonable alternative to increased use of fish in the future.

### Acknowledgment

Financial support from 2013 Commonweal and Environmental Protection Project of Ministry of Environmental Protection of the People's Republic of China(No.: 2013467028) and the National High Technology Research and Development Program of China(863 Program, No.: 2013AA06A308).

### References

1. EC (European Commission), 2007. Regulation ( EC) No 1907/2006 of the European Parliament and of the Council of 18 December 2006 concerning the Registration, Evaluation, Authorization and Restriction of Che- micals (REACH), establishing a European Chemicals Agency, amending Directive 1999/45/EC and repealing Council Regulation (EEC) No 793/93 and Commission Regulation (EC) No 1488/94 as well as Council Directive 76/769/EEC and Commission Directives 91/155/EEC, 93/67/EEC, 93/105/EC and 2000/21/EC. Off. J. Eur. Union L 136, 3–280.

2. Braunbeck T, Boettcher M, Hollert H, Kosmehl T, Lammer E, et al. (2005) Towards an alternative for the acute fish LC(50) test in chemical assessment: the fish embryo toxicity test goes multi-species -- an update. ALTEX 22: 87- 102.

the R² and s²are 0.7527 and 2.4351 in test set. The six descriptors are the most important physical chemistry properties for the construction of QSAR model for FET and the prediction of $LC_{50}$ with satisfied results. The predicted results of either training set or test set of the HM are better than those of SVM, both in the R² and s². Therefore, the HM has a good generalized performance (Table 4) as a non-linear method.

Unsaturated double bond, based on positive and negative regression coefficients, can be judged according to their contribution to the toxic compounds, The coefficients of number of double bonds is positive, the $LC_{50}$ increase with the double bond value increases. As for the electron donation of the elide, it is noted that the HOMO energy can be used a new method for showing the basicity of elides. The descriptors that matter most is on the Highest Occupied Molecular Orbital (HOMO) calculated from quantum chemical calculation, which describes the effect of HOMO energies on the toxicity. The results also show that the contribution from the HOMO-1 becomes less important than HOMO. Molecular volume is important to toxicity, the information contained

3. Yao XJ, Liu MC, Zhang XY (2002) Radial basis function network-based quantitative structure–property relationship for the prediction of Henry's law constant. Anal aChim Acta 462: 101–117.

4. Katritzky AR, Fara DC, Petrukhin RO, Tatham DB, Maran U, et al. (2002) The present utility and future potential for medicinal chemistry of QSAR/QSPR with whole molecule descriptors. Curr Top Med Chem 2: 1333-1356.

5. Katritzky AR, Maran U, Lobanov VS, Karelson M (2000) Structurally Diverse Quantitative Structure-Property Relationship Correlations of Technologically Relevant Physical Properties. J Chem Inf Comput Sci 4:1-18.

6. Arning J, Stolte S, Baschen A, Stock F, Pitner WB, et al. (2008) Qualitative and quantitative structure activity relationships for the inhibitory effects of cationic head groups, functionalised side chains and anions of ionic liquids on acetylcholinesterase[J]. Green Chem10: 47–58.

7. Katritzky AR, Lobanov VS, Karelson M (1995) CODESSA: Training Manual,University of Florida, Gainesville, FL.

8. Katritzky AR, Lobanov VS, Karelson M (1994) CODESSA: Reference Manual,University of Florida, Gainesville, FL.

9. Lammer E, Carr GJ, Wendler K, Rawlings JM, Belanger SE, et al. (2009) Is the fish embryo toxicity test (FET) with the zebrafish (Daniorerio) a potential alternative for the fish acute toxicity test? CompBiochemPhysiol C Toxicol Pharmacol 149: 196-209.

10. Kammann U, Vobach M, Wosniok W (2006) Toxic effects of brominated indoles and phenols on zebrafish embryos. Arch Environ ContamToxicol 51: 97-102.

11. Frisch MJ, Trucks GW, Schlegel HB (2003) Gaussian 03,revision A.1 [K]. Gaussian, Inc: Pittsburgh, PA.

12. Eroglu E, Palaz S, Oltulu O, Turkmen H, Ozaydın C (2007) Comparative QSTR Study Using Semi-Empirical and First Principle Methods Based Descriptors for Acute Toxicity of Diverse Organic Compounds to the Fathead Minnow[J]. Int J Mol Sci 8:1265-1283.

13. Katritzky AR, Lobanov VS, Karelson M (1994) CODESSA Reference Manual (version 2.0), Reference Manual, Version 2.0.

14. Katritzky AR, Lobanov VS, KarelsonM　CODESSA: Reference Manual, University of Florida, Gainesville, FL.

15. Devillers J, Chambon P (1986) Acute toxicity and QSAR of chlorophenols on Daphnia magna. Bull Environ Contam Toxicol 37: 599-605.

16. Katritzky AR, Petrukhin R, Jain R, Karelson M (2001) QSPR analysis of flash points. J Chem Inf Comput Sci 41: 1521-1530.

17. Vapnik VN (1998) Statistical Learning Theory. New York: John Wiley.

18. Liu HX, Zhang RS, Yao XJ, Liu MC, Hu ZD, et al. (2004) Prediction of the isoelectric point of an amino acid based on GA-PLS and SVMs. J ChemInf Comput Sci 44: 161-167.

19. Chen C, Zhou X, Tian Y, Zou X, Cai P (2006) Predicting protein structural class with pseudo-amino acid composition and support vector machine fusion network. Anal Biochem 357: 116-121.

20. Yang Y, Pederson JO (1995) A comparative study on feature selection in text categorization. In Fisher DH (ed): Proceedings of the ICML-95: 14th International Conference on Machine Leaning. San Francisco: Morgan Kaufmann: 412-420.

21. Rogati M, Yang (2002) High-performing feature selection for text classification. In Proceedings of the 11th International Conference on Information and Knowledge Management. New York: ACM Press: 659-661.

22. Brank J, Grobelnik M, Milic-Frayling N, Mladenic D (2002) Interaction of feature selection methods and linear classification models. Proceedings of the ICML-02 Workshop on Text Learning, Sydney, Australia.

23. Taira H, Haruno M (1999) Feature selection in SVM text categorization. In Proceedings of the Sixteenth National Conference on Artificial Intelligence and the Eleventh Innovative Applications of Artificial Intelligence Conference. Menlo Park, CA: American Association for Artificial Intelligence: 480-486.

24. Katritzky AR, Tatham DB, Maran U (2001) Theoretical descriptors for the correlation of aquatic toxicity of environmental pollutants by quantitative structure-toxicity relationships. J Chem Inf Comput Sci 41: 1162-1176.

25. Bruzzone S, Chiappe C, Focardi SE (2011) Theoretical descriptor for the correlation of aquatic toxicity of ionic liquids by quantitative structure–toxicity relationships, Chem Eng J 175:17–23.

# 24

# Tetra-N-Butyl Ammonium Hydroxide as Highly Efficient for the Acylation of Alcohols, Phenols and Thiols

**Mosstafa Kazemi\* and Mohammad Soleiman-Beigi**

*Department of Chemistry, Ilam University, Ilam, Iran*

### Abstract

Aqueous tetra-n-butyl ammonium hydroxide solution (TBAOH) is an efficient catalyst for the acylation of alcohols, phenols and thiols. This procedure is convenient, simple and suitable for the synthesis of esters and thioesters in high yields.

**Keywords:** Tetra-n-butyl ammonium hydroxide; Thiols; Esters; Thioesters; Acylation

## Introduction

One of the most important priorities within organic chemistry research is to find methods and processes more compatible and more economical compared to preceding methods. To accomplish this, we use aqueous solutions and ionic liquids as solvent in large scale. It should be noted reduction of reaction's time and the number of processes stage is to be taken into account.

Acylation of alcohols, phenols and thiols is one of very valuable and widely used transformations in organic synthesis because of their important role in the fields of biological, industry, synthetic and medicine chemistry [1-4].

In the past decades, many methods were described (reported) for the acylation of alcohols, phenols and thiols. Generally, acylation of alcohols and phenols is performed by means of acid anhydrides or acid chlorides in the presence of tertiary amines such as triethylamine, pyridine [1]. In addition to, other catalysts such as montmorillonite [5], ionic liquids [6], triflates [7-9], tributylphosphine [10], distannoxane [11], magnesium bromide [12], indium trihalides [13] and CsF–Celite have also been utilized to achieve the acylation of alcohols, phenols, thiols [14,15]. Some of the above described methods require the use of harshreaction conditions, hazardous materials, excess acylating agent, long reaction time, high temperature and low yields.However, still it is of great importance to find new useful and environmentally friendly methods with use of base catalysts for the acylation of alcohols, phenols and thiols.

Tetra-n-butyl ammonium hydroxide is a strong organic base, which also acts as a phase transfer reagent and a surfactant. It has been used as a base or additive in Aldol [16], Ullmann [17], non-Sonogashira [18] types and Knoevenagel [19] reactions, elimination [20], addition [21] reactions, as well as hydrolysis of esters and amides [22], alkylation [23], titration [24] and synthesis of nanoparticles [25] and titanium silicate [26].

In continuation of our interest in exploring new application of tetra-n-butyl ammonium hydroxide in organic synthetic methodologies [27-34] and attempts to develop previous methods, we wish to report acylation of alcohols, phenols under neat conditions in the presence of an aqueous solution of TBAOH at mild conditions.

## Results and Discussion

In order to optimize of the reaction conditions in terms of the amount of TBAOH (20% in water), time and temperature, the reaction of benzyl alcohol (1.5 mmol) with acetic anhydride (1.0 mmol) and benzyl mercaptan (1.0 mmol) with acetic anhydride (1.2 mmol) were studied without the presence of solvent (Scheme 1) as models reactions. As can be seen from Table 1, the rate and efficiency of reactions depend on the amount of TBAOH and temperature. The best results were obtained in the presence of 2 mL of TBAOH (Table 1, entry 4) at 50°C under air (Scheme 1 and Table 1).

Therefore, a diverserange of dialkyl (symmetric and unsymmetric) and aryl alkyl esters and thioesters was synthesized in good to excellent yields (80-92%) in the presence of 2 mL of aqueous solution of TBAOH (Scheme 2) in optimal reaction condition. The method is very general, which aliphatic alcohols as well as phenols react easily with acyl halides andacid anhydrides was converted to corresponding esters and thioesters exclusively (Scheme 2, Table 2 and 3).

## Products data

- Phenyl acetate ($C_8H_8O_2$) (Table 2, entry 15). Yield: 83%; colorless liquid;

$^1$H NMR (CDCl$_3$, 400 MHz): $\delta$ 2.08 (3H, s, COCH$_3$), 7.24-7.54 (5H, m, ArH). $^{13}$C NMR (100 MHz, CDCl$_3$): $\delta$ = 20.1, 113.8, 129.3, 130.7, 159.2, 170.5 ppm.

- Benzyl acetate ($C_9H_{10}O_2$) (Table 2, entry 1). Yield: 92%; colorless liquid;

$^1$H NMR (CDCl$_3$, 400 MHz): $\delta$ 2.17 (3H, s, COCH$_3$), 5.2 (2H, s, PhCH$_2$), 7.25-7.35 (5H, m, ArH). $^{13}$C NMR (100 MHz, CDCl$_3$): $\delta$ = 22,

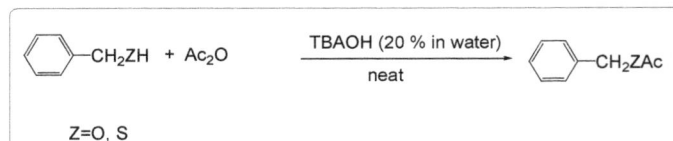

**Scheme 1:** The reaction of benzyl alcohol (1.5 mmol) with acetic anhydride (1.0 mmol) and benzyl mercaptan (1.0 mmol) with acetic anhydride (1.2 mmol).

**\*Corresponding author:** Mosstafa Kazemi, Department of Chemistry, Ilam University, PO Box 69315-516, Ilam, Iran, E-mail: mostaffa_kazemi@yahoo.com

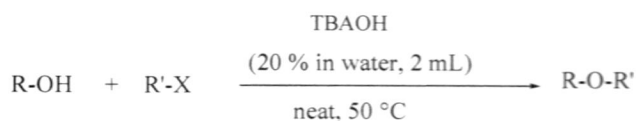

TBAOH

R-OH  +  R'-X  $\xrightarrow[\text{neat, 50 °C}]{\text{(20 \% in water, 2 mL)}}$  R-O-R'

R= alkyl, aryl

R'= alkyl

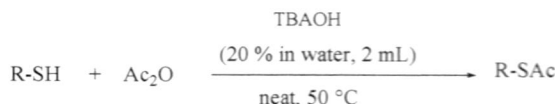

TBAOH

R-SH  +  Ac$_2$O  $\xrightarrow[\text{neat, 50 °C}]{\text{(20 \% in water, 2 mL)}}$  R-SAc

R= alkyl, aryl

Ac= (CH$_3$CO)$_2$O , PhCOCl

**Scheme 2:** The reaction of phenols with acyl halides and acid anhydrides forming thioesters.

65.4, 126.9, 128.5, 128.9, 138.7, 170.4 ppm.

- 4-Bromo Phenyl acetate (C$_8$H$_7$BrO$_2$) (Table 2, entry 18). Yield: 81%; colorless liquid;

$^1$HNMR (CDCl$_3$, 400 MHz): δ 2.45 (3H, s, COCH$_3$), 6.9 (2H, d, J=8.8, ArH), 7.29 (2H, d, J=8.8, ArH). $^{13}$C NMR (100 MHz, CDCl$_3$): δ = 21.1, 121.6, 125.9, 129.5, 156.8, 170.4 ppm.

- Isopropyl 4-methylbenzoate (C$_{11}$H$_{14}$O$_2$) (Table 2, entry 8). Yield: 87%; colorless liquid;

$^1$H NMR (CDCl$_3$, 400 MHz): δ 1.03 (6H, d, J=7.2, (CH$_3$)$_2$CH), 2.05 (3H, s, COCH$_3$), 4.2 (1H, m, J=7.2, ArH), 7.47 (2H, d, J=8.8, ArH).7.56 (2H, d, J=8.8, ArH). $^{13}$C NMR (100 MHz, CDCl$_3$): δ = 20.3, 59.1, 126.8, 128.8, 132.9, 135.5, 159.1 ppm.

- 2-Naphthyl acetate (C$_{12}$H$_{10}$O$_2$) (Table 2, entry 22). Yield: 89%; colorless liquid;

$^1$H NMR (CDCl$_3$, 400 MHz): δ 2.40(s, 3H, COCH$_3$), 7.27-7.30 (m, 1H, ArH), 7.49-7.5(m, 2H, ArH), 7.61 (d, 1H, ArH), 7.84-7.91 (m , 3H, ArH).$^{13}$C NMR (100 MHz, CDCl$_3$): δ = 21.2, 118.7, 121.3, 125.8, 126.7, 127.8, 127.9, 129.5, 131.6, 133.9, 148.5, 169.8 ppm.

IR (KBr) cm$^{-1}$: 1758(C=O)

- 4-nitrobenzyl acetate (C$_9$H$_9$NO$_4$) (Table 2, entry 3). Yield: 83%; colorless liquid;

$^1$H NMR (CDCl$_3$, 400 MHz): δ 2.17(s, 3H, COCH$_3$), 5.22(s, 2H, PhCH$_2$), 7.53 (2H, d, J=8.8Hz, ArH), 8.23(2H, d, J=8.8Hz, ArH) ppm.

IR (KBr) cm-1:1235(C-O), 1738(C=O)

- 3-Methoxybenzyl acetate (C$_{10}$H$_{12}$O$_3$) (Table 2, entry 11). Yield: 87%; colorless liquid;

$^1$H NMR (CDCl$_3$, 400 MHz): δ 2.13 (s, 3H, COCH$_3$), 3.83(s, 3H, OCH$_2$), 5.11(s, 2H, PhCH$_2$), 6.88-6.97(m, 3H, ArH), 7.30(t, 1H,ArH) ppm.

IR (KBr) cm$^{-1}$:1227(C-O), 1742(C=O)

- 2-phenyl propyl acetate (C$_{11}$H$_{14}$O$_2$) (Table 2, entry 6). Yield: 85%; colorless liquid;

$^1$H NMR (CDCl$_3$, 400 MHz): δ 1.34 (d, J=6.8Hz, 3H, CH$_3$), 2.05 (3H, J=6.8Hz, s, CH$_3$ COCH$_3$), 3.09-3.18(m, 1H, PhCH$_2$), 4.14-4.25(2H, m, J=6.8Hz, CH$_2$CO), 7.25-7.29 (m, 3H,), 7.33-7.37 (m, 2H). $^{13}$CNMR(10MHz, DMSO)=18.1, 20.9, 38.9, 69.4, 126.7, 127.3, 128.5, 143.2, 171.1 ppm.

IR (KBr) cm$^{-1}$:1233(C-O), 1741(C=O)

- phenylthioacetate (C$_8$H$_8$OS) (Table 3, entry 9). Yield: 88%; colorless liquid;

$^1$H NMR (CDCl$_3$, 400 MHz): δ 3.51 (3H, s, COCH$_3$), 7.29-7.44 (5H, m, ArH).$^{13}$C NMR (100 MHz, CDCl$_3$): δ = 29.8, 127.7, 127.8, 129.4, 137.2, 196.1 ppm.

- n-ButylThiobenzoate (C$_{11}$H$_{14}$OS) (Table 3, entry 12). Yield: 88%; colorless liquid;

$^1$H NMR (CDCl$_3$, 400 MHz): δ 0.93 (3H, t, J=7.2, CH$_3$), 1.42 (2H, sex, J=7.2, CH$_2$CH$_3$), 1.58 (2H, quint, J=7.2, CH$_2$CH$_2$), 2.45 (2H, t, J=7.2, CH$_2$S).7.26 (1H, quint, J=7.2, ArH) 7.34 (2H, t, J=7.2, ArH).7.53 (2H, d, J=7.2, ArH). $^{13}$C NMR (100 MHz, CDCl$_3$): δ = 13.7, 22.1, 31.1, 31.3, 127.7, 127.8, 128.5, 132.2, 137.4, 196.7 ppm.

- benzylthioacetate (C$_9$H$_{10}$OS) (Table 3, entry 1). Yield: 90%; colorless liquid;

$^1$H NMR (CDCl$_3$, 400 MHz): δ 2.17 (3H, s, COCH$_3$), 5.19 (2H, s, CH$_2$S), 7.3 (1H, t, J=7.2, ArH) 7.4 (2H, t, J=7.2, ArH). 7.53 (2H, quint, J=7.2, ArH). $^{13}$C NMR (100 MHz, CDCl$_3$): δ = 21.2, 43.8, 121.6, 127.6, 127.8, 128.5, 189.5 ppm.

- 4-methoxybenzyl thioacetate (C$_{10}$H$_{12}$O$_2$S) (Table 3, entry 2). Yield: 84%; colorless liquid;

$^1$H NMR (CDCl$_3$, 400 MHz): δ 2.02 (3H, s, COCH$_3$), 3.4 (2H, s, OCH$_3$), 4.71 (2H, s, CH$_2$S), 7.1 (2H, d, J=8, ArH), 7.4 (2H, d, J=8, ArH). $^{13}$C NMR (100 MHz, CDCl$_3$): δ = 15.3, 55.3, 65.4, 113.8, 129.3, 130.7, 159.1, 196, 196.7 ppm.

## Conclusions

In conclusion, we developed the utility of TBAOH as an efficient, versatile,commercially available, environmentally and economically friendlyorgano-basic catalyst for the preparation esters and thioesters. This method is applicable for the acylation of alcohols, phenols and thiols with acyl halides and acid anhydrides. In this way, we described a new, mild, simple and highly efficient method for the synthesis of esters and thioesters in excellent yields (80-92%) and short reaction times (70-110 min) under neat aqueous condition without using phase transfer reagent and organic solvent.

## Experimental

Chemicals were purchased from commercial suppliers and used without further purification. Yields refer to isolated products. Melting points were determined by an Electrothermal 9100 apparatus and are uncorrected. The IR spectra were obtained on a FT-IR Hartman- Bomen spectrophotometer as KBr disks, or neat. The $^1$H NMR (400 MHz) and $^{13}$C NMR (100 MHz) spectra were recorded on a BrukerAvance NMR spectrometer in CDCl$_3$ solution. The progress of the reaction was monitored by TLC using silica-gel SILG/UV 254 plates. All products are known and were characterized by comparing their physical and spectral data with those of the authentic samples.

### Typical procedure: benzyl acetate synthesis (Table 2, entry 1)

A mixture of TBAOH (2.0 mL, 20% in water) and benzyl alcohol

| Entry | Temp. (°C) | TBAOH (mL) | Thioether | | Ether | |
|---|---|---|---|---|---|---|
| | | | Time (min) | Yield (%)* | Time (min) | Yield (%)[b] |
| 1 | 50 | 0.5 | 400 | 70 | 380 | 72 |
| 2 | 50 | 1.0 | 240 | 77 | 260 | 75 |
| 3 | 50 | 1.5 | 110 | 85 | 100 | 81 |
| 4 | 50 | 2.0 | 70 | 90 | 80 | 92 |
| 5 | 50 | 2.5 | 70 | 90 | 80 | 91 |
| 6 | 25 | 2.0 | 400 | 30 | 400 | 25 |
| 7 | 40 | 2.0 | 400 | 74 | 400 | 65 |

[a]Model reaction conditions: molar ratio of benzyl alcohol (benzyl mercaptan)/ acetic anhydride(acetic anhydride) was 1.5 (1.0)/1.0 (1.5). The reactions run in the presence of TBAOH (20% in water) without any extra solvent under an air atmosphere conditions
[b]Yield refer to an isolated yield by preparative chromatography

**Table 1:** Optimization of TBAOH amount and temperature of reaction[a].

| Entry | Substrate | Reagent | Product | Time (min) | Yield[b] (%) |
|---|---|---|---|---|---|
| 1 | $CH_2OH$ (benzyl alcohol) | $(CH_3CO)_2O$ | $CH_2O_2CCH_3$ | 80 | 92 |
| 2 | cyclohexanol —OH | 4- $NO_2$PhCOCl | cyclohexyl 4-$NO_2$ benzoate | 95 | 85 |
| 3 | $CH_2OH$, $O_2N$— | $(CH_3CO)_2O$ | $CH_2O_2CCH_3$, $O_2N$— | 80 | 83 |
| 4 | $CH_2OH$, MeO— | $(CH_3CO)_2O$ | $CH_2O_2CCH_3$, MeO— | 90 | 86 |
| 5 | $(CH_3)_2CHOH$ | 4- $NO_2$PhCOCl | $CO_2CH(CH_3)_2$, $O_2N$— | 95 | 87 |
| 6 | $CH_2OH$ (1-phenylethyl) | $(CH_3CO)_2O$ | $CH_2O_2CCH_3$ | 85 | 85 |
| 7 | $CH_2OH$, Br— | $(CH_3CO)_2O$ | $CH_2O_2CCH_3$, Br— | 80 | 88 |
| 8 | $(CH_3)_2CHOH$ | 4- $CH_3$PhCOCl | $CO_2CH(CH_3)_2$, $H_3C$— | 90 | 87 |
| 9 | cyclopentanol —OH | PhCOCl | cyclopentyl benzoate | 100 | 84 |
| 10 | $(CH_3)_3CH_2OH$ | PhCOCl | $CO_2CH_2(CH_3)_3$ | 105 | 85 |
| 11 | $CH_2OH$, OMe | $(CH_3CO)_2O$ | $CH_2O_2CCH_3$, OMe | 95 | 87 |

| 12 | $CH_3CH_2OH$ | PhCOCl | $PhCO_2CH_2CH_3$ | 80 | 82 |
|----|----|----|----|----|----|
| 13 | $(CH_3)_4CH_2OH$ | PhCOCl | $CO_2CH_2(CH_3)_4$ | 95 | 80 |
| 14 | $PhCH_2OH$ | PhCOCl | $PhCH_2O_2CPh$ | 90 | 83 |
| 15 | OH | $(CH_3CO)_2O$ | $O_2CCH_3$ | 110 | 83 |
| 16 | OH $O_2N$ | $(CH_3CO)_2O$ | $O_2CCH_3$ $O_2N$ | 95 | 89 |
| 17 | OH Me | $(CH_3CO)_2O$ | $O_2CCH_3$ $H_3C$ | 105 | 84 |
| 18 | OH Br | $(CH_3CO)_2O$ | $O_2CCH_3$ Br | 100 | 81 |
| 19 | OH $NO_2$ | $(CH_3CO)_2O$ | $O_2CCH_3$ $NO_2$ | 90 | 83 |
| 20 | OH $NO_2$ | $(CH_3CO)_2O$ | $O_2CCH_3$ $NO_2$ | 95 | 80 |
| 21 | OH | PhCOCl | $CO_2Ph$ $O_2N$ | 80 | 83 |
| 22 | OH | $(CH_3CO)_2O$ | $O_2CCH_3$ | 100 | 89 |
| 23 | OH Me | PhCOCl | $O_2CPh$ Me | 110 | 86 |
| 24 | OH Cl | $(CH_3CO)_2O$ | $O_2CCH_3$ Cl | 95 | 81 |
| 25 | OH | PhCOCl | $O_2CPh$ | 105 | 82 |
| 26 | OH $O_2N$ | 4-OMePhCOCl | $O_2N$ $OCH_3$ | 110 | 83 |

| 27 | 2-nitrophenol (OH ortho to NO2) | 4-OMePhCOCl | aryl 4-methoxybenzoate ester (2-nitrophenyl ester) | 95 | 89 |

[a]All the products are known compounds and were characterized by comparison of their IR and NMR spectral data and physical properties with those reported in the literature [11,24-30]. [b]Isolated yield

**Table 2:** Acylation of alcohols and phenols in the presence of TBAOH (20% in water)[a].

| Entry | Substrate | Reagent | Product | Time (min) | Yield[b] (%) |
|---|---|---|---|---|---|
| 1 | Ph-CH2SH | (CH3CO)2O | Ph-CH2SOCCH3 | 70 | 90 |
| 2 | 4-MeO-C6H4-CH2SH | (CH3CO)2O | 4-MeO-C6H4-CH2SOCCH3 | 85 | 84 |
| 3 | 4-Cl-C6H4-CH2SH | (CH3CO)2O | 4-Cl-C6H4-CH2SOCCH3 | 80 | 86 |
| 4 | 4-MeO-C6H4-SH | 4-OMePhCOCl | H3CO-C6H4-S-CO-C6H4-OCH3 | 95 | 87 |
| 5 | 4-O2N-C6H4-CH2SH | PhCOCl | 4-O2N-C6H4-CH2-SOCPh | 80 | 89 |
| 6 | cyclohexyl-SH | (CH3CO)2O | cyclohexyl-S-CO-CH3 | 95 | 85 |
| 7 | 4-MeO-C6H4-SH | PhCOCl | 4-MeO-C6H4-SOCPh | 90 | 83 |
| 8 | Ph-SH | PhCOCl | Ph-SOCPh | 95 | 83 |
| 9 | Ph-SH | (CH3CO)2O | Ph-SOCCH3 | 90 | 88 |
| 10 | 4-MeO-C6H4-SH | (CH3CO)2O | 4-MeO-C6H4-SOCCH3 | 100 | 86 |
| 11 | 4-Cl-C6H4-SH | (CH3CO)2O | 4-Cl-C6H4-SOCCH3 | 110 | 84 |
| 12 | (CH3)3CH2SH | PhCOCl | Ph-COSCH2(CH3)3 | 105 | 88 |

[a]All the products are known compounds and were characterized by comparison of their NMR spectral data and physical properties with those reported in the literature [11,27,30]. [b]Isolated yield

**Table 3:** Acylation of thiols in the presence of TBAOH (20% in water)[a].

(1.5 mmol, 162 mg) was vigorously stirred at room temperature for 15 min. acetic anhydride (1.0 mmol, 102 mg) was then added to the mixture and stirring continued at 50°C for the appropriate times (Table 2) under air. The progress of reaction was monitored by TLC. After completion of the reaction, $CH_2Cl_2$ (15 mL) was added, and the mixture washed with $H_2O$ (3×10 mL). The organic layer was dried over anhydrous $Na_2SO_4$. The solvent was evaporated *in vacuo* to give benzyl acetate which was purified by preparative TLC (silica gel, eluent n-hexane: EtOAc = 4:1) to obtain 118 mg of the pure benzyl acetate (92%).

## References

1. Green TW, Wuts PG (1999) Protective Groups in Organic Synthesis. (3rdedn) Wiley, New York, p150.

2. Sano T, Ohashi K, Oriyama T (1999) Remarkably Fast Acylation of Alcohols with Benzoyl Chloride Promoted by TMEDA. Synthesis 7: 1141-1144.

3. Scriven EFV (1983) 4-dialkylaminopyridines: super acylation and alkylation catalysts. ChemSoc Rev 12: 129-161.

4. Hofle G, Steglich V, Vorbruggen H (1978) 4-Dialkylaminopyridines as Highly Active Acylation Catalysts. AngewChemInt Ed Engl 17: 569-583.

5. Li AI, Li TS, Ding TH (1997) Montmorillonite K-10 and KSF as remarkable acetylation catalysts.ChemCommun 1389-1390.

6. Lee SG, Park JH (2003) Metallic Lewis acids-catalyzed acetylation of alcohols with acetic anhydride and acetic acid in ionic liquids: study on reactivity and reusability of the catalysts. J MolCatal A: Chem 194: 49-52.

7. Procopiou PA, Baugh SPD, Flack SS, Inglis GGA (1998) An Extremely Powerful Acylation Reaction of Alcohols with Acid Anhydrides Catalyzed by TrimethylsilylTrifluoromethanesulfonate. J Org Chem 63: 2342-2347.

8. Ishihara K, Kubota M, Kurihara H, Yamamoto H (1996) Scandium trifluoromethanesulfonate as an extremely active lewis acid catalyst in acylation of alcohols with acid anhydrides and mixed anhydrides. J OrgChem 61:4560-4567.

9. Orita A, Tanahashi C, Kakuda A (2000) Highly Efficient and Versatile Acylation of Alcohols with Bi(OTf)(3) as Catalyst. J OteraAngewChemInt Ed 39: 2877-2879.

10. Vedejs E, Bennett NS, Conn LM, Diver ST, Gingres M, et al. (1993) Tributylphosphine-catalyzed acylations of alcohols: scope and related reactions. J OrgChem 58: 7286.

11. Orita A, Sakamoto K, Hamada Y, Mitsutome A (1999) Mild and practical acylation of alcohols with esters or acetic anhydride under distannoxane catalysis. J Otera Tetrahedron 55: 2899-2910.

12. Pansare SV, Malusara MG, Rai AN (2000) Magnesium Bromide Catalysed Acylation of Alcohols. Synth Commun 30: 2587-2592.

13. Ranu BC, Dutta P, Sarkar A (2000) Highly selective acylation of alcohols and amines by an indium triiodide-catalysedtransesterification process. J ChemSoc Perkin Trans 1: 2223-2225.

14. Shah STA, Khan KM, Hussain H, Anwar MU, Fecker M, et al. (2005) Cesium fluoride-Celite: a solid base for efficient syntheses of aromatic esters and ethers. Tetrahedron 61: 6652-6656.

15. Shah STA, Khan KM, Heinrich AM, Voelter W (2002) An alternative approach towards the syntheses of thioethers and thioesters using CsF–Celite in acetonitrile. Tetrahedron Lett 43: 8281-8283.

16. Varala R, Enugala R, Nuvula S, Adapa SR (2006) Catalytic aldol to type reaction of aldehydes with ethyl diazoacetate using quaternary ammonium hydroxide as the base. Tetrahedron Lett 47: 877-880.

17. Monopoli A, CalÃ² V, Ciminale F, Cotugno P, Angelici C, et al. (2010) Glucose as a clean and renewable reductant in the Pd-nanoparticle-catalyzed reductive homocoupling of bromo- and chloroarenes in water. J Org Chem 75: 3908-3911.

18. Mori A, Kawashima J, Shimada T, Suguro M, Hirabayashi K, et al. (2000) Non-Sonogashira-type palladium-catalyzed coupling reactions of terminal alkynes assisted by silver(I) oxide or tetrabutylammonium fluoride. Org Lett 2: 2935-2937.

19. Balalaie S, Bararjanian M (2006) Tetra-n-butylammonium Hydroxide (TBAH)–Catalyzed Knoevenagel Condensation: A Facile Synthesis of α-Cyanoacrylates, α-Cyanoacrylonitriles, and α-Cyanoacrylamides. Synth Commun 36: 533-539.

20. Li J, Huang P (2011) A rapid and efficient synthetic route to terminal arylacetylenes by tetrabutylammonium hydroxide- and methanol-catalyzed cleavage of 4-aryl-2-methyl-3-butyn-2-ols. Beilstein J Org Chem 7: 426-431.

21. Santana AS, Carvalho DB, Casemiro NS, Hurtado GR, Viana LH, et al. (2012) Improvement in the synthesis of (Z)-organylthioenynes via hydrothiolation of buta-1,3-diynes: a comparative study using NaOH or TBAOH as base. Tetrahedron Lett 53: 5733-5738.

22. Abdel-Magid AF, Cohen JH, Maryanoff CA, Shah RD, Villani FJ, et al. (1998) Hydrolysis of polypeptide esters with tetrabutylammonium hydroxide. Tetrahedron Lett 39: 3391-3394.

23. Meier MS, Bergosh RG, Gallagher ME, Spielmann HP, Wang Z (2002) Alkylation of dihydrofullerenes. J Org Chem 67: 5946-5952.

24. Buell BE (1967) Differential titration of acids and very weak acids in petroleum with tetrabutylammonium hydroxide and pyridine-benzene solvent. Anal. Chem. 39: 762-764.

25. Zou H, Li Z, Luan Y, Mu T, Wang Q, et al. (2010) Fast synthesis of nanostructured ZnO particles from an ionic liquid precursor tetrabutylammonium hydroxide. Current Opinion in Solid State and Materials Science14: 75-82.

26. Salehirad F, Aghabozorg HR, Manoochehri M, Aghabozorg H (2004) Synthesis of titanium silicalite-2 (TS-2) from methylamine-tetrabutylammonium hydroxide media.CatalCommun 5: 359-365.

27. Soleiman- Beigi M, Arzehgar Z, Movassagh B (2010) TBAH-Catalyzed One-Pot Synthesis of Symmetrical Trithiocarbonates from. Alkyl Halides and Carbon Disulfide under Neat Aqueous Conditions Synthesis 392-394.

28. Ratnam KJ, Reddy RS, Kantam ML, Figueras F (2007) Sulphated zirconia catalyzed acylation of phenols, alcohols and amines under solvent free conditions. J Molecular Catal A: Chem 276: 230-234.

29. Won JE, Kim HK, Kim JJ, Yim HS, Kim MJ, et al. (2007) Effective Esterification of Carboxylic Acids Using (6-Oxo-6H-pyridazin-1-yl)phosphoric Acid Diethyl Ester as Novel Coupling Agents. Tetrahedron 63: 12720-12730.

30. Wu XF (2012) Zinc-catalyzed oxidative esterification of aromatic aldehydes. Tetrahedron Lett 53: 3397-3399.

31. Shirini F,Zolfigol MA, Safari A (2005) A mild and efficient method for the acetylation of alcohols. IndianJChem44: 201-203.

32. Kadam ST, Lee H, Kim SS (2009) TMEDA: Efficient and Mild Catalyst for the Acylation of Alcohols, Phenols and Thiols under Solvent-free Condition. Bull Korean ChemSoc 30: 1071-1076.

33. Jin TS, Ma YR, Li TS, Zhang ZH, Duan GB (1999)An efficient and simple procedure for acetylation of alcohols and phenols with acetic anhydride catalysed by expansive graphite. IndianJChem 38: 109-110.

34. Kamal A, Khan MNA, Reddy KS, Srikanth YVV, Krishnaji T (2007) Al(OTf)3 as a highly efficient catalyst for the rapid acetylation of alcohols, phenols and thiophenols under solvent-free conditions. Tetrahedron Lett 48: 3813-3818.

# 25

# The Relationship between Fructose, Glucose and Maltose Content with Diastase Number and Anti-Pseudomonal Activity of Natural Honey Combined with Potato Starch

**Ahmed Moussa[1]\*, Djebli Noureddine[2], Aissat Saad[1] and Salima Douichene[2]**

[1]*Institute of Veterinary Sciences University Ibn-Khaldoun, Tiaret, Algeria*
[2]*Departments of Biology, Faculty of Sciences, Mostaganem University, Algeria*

## Abstract

Honey whose medicinal uses date from ancient times has been lately rediscovered as therapy for burns.

**Objective:** To evaluate the additive action of potato starch on the antipseudomonal activity of natural honey.

**Methods**: Physicochemical parameters of 6 samples of Algerian honeys were analysed; four parameters were measured, including Diastase, glucose, fructose and maltose. The antibacterial activity was tested using the well-agar diffusion assay.

**Results:** Six honey samples with initial diastase activity between 22.1 and 7.3 Schade units were tested. Glucose, fructose and maltose values range between 21, 45-30, 95%, 25, 20-37, 81% and 4, 72-78, 45% respectively. The zone inhibition diameter (ZID) for the six honey samples without starch against *P. aureogenosa* ranged between 26 and 31 mm. When starch was mixed with honey and then added to well, a zone inhibition increase diameter (ZIID) 27 and 32 mm. The percentage increase (PI %) was noticed with each variety and it ranged between 3, 57 and 18, 75%. Positive correlation has been established between the zone increase of inhibition and the Diastase number ($r$ value was 0.072 at $p<0.05$).

**Conclusion:** The use of potato starch allows honey benefit and would constitute an additive effect to the antibacterial activity of natural honey.

**Keywords:** Diastase number; Honey; Antibacterial activity; Potato starch

## Introduction

Many burn infections are treated with antibiotics that can be applied topically or administered orally or by injection. Unfortunately, due to the excessive use of antibiotics, some bacteria have evolved to become antibiotic resistant, and this has led to the present time being described as the "end of the antibiotic era" [1,2]. The dominant flora of burn wounds during hospitalization changes from Gram-positive bacteria such as *Staphylococcus* to Gram-negative bacteria like *Pseudomonas aeruginosa*. The majority of *P. aeruginosa*, an opportunistic human pathogen, isolates from burn patients were multidrug resistant (MDR) [3-5]. In wounds, *P. aeruginosa* has emerged as a multidrug-resistant organism that gives rise to persistent infections in burns patients [6,7] and chronic venous leg ulcers [8]. Novel antimicrobial interventions are needed.

The complexity of natural products, including honey, makes them very difficult to standardize and this can affect their acceptance in clinical medicine. However, this complexity also has benefits. Unlike conventional antibiotics it appears to be difficult for microorganisms to become resistant to the effects of honey, probably due to the action of the various active components in honey on multiple microbial targets [9]. Honey is the most famous rediscovered remedy that has been used to promote wound and burn healing and also to treat infected wounds [10]. The use of honey in modern clinical practice is based on its broad antimicrobial properties and its ability to stimulate rapid wound healing.

Several bioactive compounds have been identified in honey which contributed to its antibacterial action. The commonly accepted list of contributors includes osmolarity [11]. High osmolarity has been considered a valuable tool in the treatment of infections, because it

prevents the growth of bacteria and encourages healing [12]. Honey is a supersaturated sugar solution; and sugar content accounts for more than 95% of the dry matter. Honey is an extremely varying and complex mixture of sugars and other minor components. Fructose is the most dominant sugar followed by glucose in almost all types of honey [13]. Maltose content in natural honey is generally less than 30 mg/g [14,15]. Maltose in some honeys originating from certain plants can be up to 50 mg/g [16,17].

Honey contains small amounts of different enzymes, notably, diastase ($\alpha$-and $\beta$-amylase), invertase ($\alpha$-glucosidase), glucose-oxidase, catalase and acid phosphatase, which come from nectar sources, salivary fluids and the pharyngeal gland secretions of the honeybee [18]. A diastase is any one of a group of enzymes that catalyze the breakdown of starch into maltose [19]. Alpha amylase degrades starch to a mixture of the disaccharide maltose, the trisaccharide maltotriose and oligosaccharides known as dextrin's [20]. Diastase activity is expressed as the diastase number (DN) in Schade units and is defined as follows: one diastase unit corresponds to the enzyme activity of 1 g of honey,

**\*Corresponding author:** Ahmed Moussa, Institute of Veterinary Sciences University Ibn-Khaldoun, Tiaret, Algeria, E-mail: moussa7014@yahoo.fr

which can hydrolyse 0.01 g of starch in 1 h at 40°C. The range permitted for diastase number varies from 3 to 8 on Gothe's scale, depending on the climate prevailing in the place where the honey originates [21]. In previous studies, we have shown that there is an Additive action between honey and ginger starch in terms of antibacterial [22] and antifungal activity [23,24]. We suggested that α amylases present in honey originating from bees and pollen are responsible in the hydrolysis of starch chains to randomly produce dextrin and maltose that increase the osmotic effect of honey and consequently increase the antibacterial activity. But no starch is found in honey. The aim of this study is to evaluate the potential antibacterial activity of honey and potato starch when used jointly to manage superficial burn.

## Materials and Methods

### Preparation of honey sample

Six unifloral and multifloral honey samples were collected from beekeepers in different regions of the western Algeria during different seasons of the 2011 year depending on floral sources: Jujube, Citrus, Eucalyptus and Multifloral. All samples were collected in their original packages and were transferred to the laboratory and kept at 4-5°C until analysis. Honey was used within a few hours of preparation to avoid self-decomposition and decrease in diastase activity.

### Preparation of the stock starch solution

The stock starch solution was prepared by dissolving 0.5 g of dried soluble starch in deionised water in a volumetric flask. After heating and stirring the solution for approximately ten minutes, starch was completely dissolved, and the volumetric flask was filled with deionised water to the mark.

### Physicochemical analyses

All physicochemical tests were performed in triplicate.

### Determination of maltose, glucose and fructose contents

Sugar spectra (fructose, glucose, and maltose) were identified and determined by Bogdanov [25] for di- and oligosaccharides using high-performance liquid chromatography (HPLC).

### Diastase activity (Diastase number)

Diastase activity was measured with Phadebas, according to the Harmonized Methods of the European Commission of Honey [25]. An insoluble blue dyed cross-linked type of starch is used as the substrate. This is hydrolysed by the enzyme, yielding blue water-soluble fragments, determined photometrically at 620 nm. The absorbance of the solution is directly proportional to the diastatic activity of the sample. The diastase activity, expressed as DN or diastase number, was calculated from the absorbance measurements using Eqs. (1) and (2)

for high (8–40 diastase units) and low (up to 8 diastase units) activity values, respectively:

$$DN = 28.2 \times \Delta A620 - 2.64 \qquad (1)$$

$$DN = 35.2 \times \Delta A620 - 0.46 \qquad (2)$$

### Bacterial culture and inoculum preparation

Pure culture of *P. aeruginosa* ATCC 27853 was obtained from the Department of Biology, Faculty of Sciences, Mostaganem University, Algeria. The bacteria was grown on Nutrient Agar (NA; Merck Germany) slant, incubated at 37°C for 24 h, and kept at 4°C until further use. Bacterial suspension was prepared by inoculating one loopful of the 24-h-old bacterial colonies into 10.0 ml of sterilized distilled water. The inoculums size was adjusted to match the turbidity of McFarland 0.5 scale ($1\times10^8$ cells/ml) and diluted with sterilized distilled water to the inoculums size of $1\times10^7$ cells/ml.

### Measurement of zone of inhibition (Well diffusion assay)

A screening assay using well diffusion [26] was carried out with some minor modifications. Nutrient agar plates (Merck, Germany) were inoculated by rubbing sterile cotton swabs that were dipped into bacterial suspensions (over night) cultures grown at 37°C on nutrient agar and adjusted to 0.5 McFarland in sterile saline) over the entire surface of the plate. After inoculation 8.2 mm diameter wells were cut into the surface of the agar using a sterile cork borer. 50 µl of test honey was added to each well. Plates were incubated at 30°C for 24 h. A diffusion control of starch was used. Second step a mixture of starch-honey was prepared and incubated for one hour at 40°C. After inoculation 8.2 mm diameter wells were cut into the surface of the agar using a sterile cork borer. 50 µl of mixture (honey and starch) were added to each well. Zones of inhibition were measured using a Vernier caliper. The diameter of zones, including the diameter of the well, was recorded. Bioassay was performed in duplicate and repeated twice. The results were expressed in terms of the diameter of the inhibition zones:<5.5 mm, inactive; 5.5-9 mm, very low activity; 9-12 mm, low activity; 12-15 mm, average activity; and >15 mm, high activity.

### Statistical analysis

Each honey was analyzed in triplicate. Results are shown as mean values and standard deviation. Correlations were established using Pearson's correlation coefficient ($r$) in bivariate linear correlations ($p<0.01$). All statistical analyses were performed with the Statistica 7.0 software for Windows.

## Results and Discussion

### Physicochemical parameters

Table 1 reports the physico-chemical parameters of the honey

| Honey samples | glucose (g/100 g honey) | | fructose (g/100 g honey) | | maltose (g/100 g honey) | | Sugar total content (%) | | Diastase activity (Schade Number[b]) | |
|---|---|---|---|---|---|---|---|---|---|---|
| | Mean | DS | Mean | DS | Mean | DS | Mean | DS | Mean | DS |
| H1 | 21.45 | 2.14 | 25.20 | 3.32 | 0.00 | 1.32 | 46.65 | 2.26 | 15.1 | 2.8 |
| H2 | 26.18 | 2.14 | 37.81 | 3.32 | 7.13 | 1.32 | 71.12 | 2.26 | 23.5 | 2.8 |
| H3 | 25.78 | 2.14 | 33.92 | 3.32 | 8.45 | 1.32 | 68.45 | 2.26 | 11 | 2.8 |
| H4 | 28.84 | 2.14 | 36.84 | 3.32 | 7.01 | 1.32 | 72.69 | 2.26 | 26 | 2.8 |
| H5 | 27.24 | 2.14 | 35.41 | 3.32 | 4.72 | 1.32 | 67.37 | 2.26 | 7.3 | 2.8 |
| H6 | 30.95 | 2.14 | 35.18 | 3.32 | 7.10 | 1.32 | 73.23 | 2.26 | 16.4 | 2.8 |

[b]Schade number. corresponds with Gothe number, or 0.01 g starch hydrolysed 1h at 40°C per 1 g honey.

**Table 1:** The concentration of glucose, fructose and maltose in the honey samples (g/ 100 g) and diastase activity results represent the average of four measurements ± SD (n=3).

samples. The mean, standard deviation (SD) and the variable ranges are reported for comparison with international standards. Fructose, glucose and maltose values range between 25, 20-37, 81%, 21, 45-30, 95% and 4, 72-78, 45%, respectively.

## Analysis of amylase activity

The diastatic activity in honey is considered a quality factor. It decreases during storage, heat treatment and feeding of honeybees during honey flow; thus, it is an indicator of honey ageing, adulteration and overheating. The honey samples analyzed in the present work show a range of values, between 22.1 and 7.3 Schade units. One sample (H5) show values below 8 Schade units (Table 1). The explanation for the low content of diastatic activity found in this one sample could be accounted for an inadequate processing or storage conditions.

## Antibacterial activity

The six honey samples were studied in terms of antibacterial activity were performed in duplicate. (Table 2) and (Figures 1 and 2) summarize the zones of inhibition of the honey samples against the tested organism. The differences in inhibition were observed for six types of honey sample (H5) has the largest inhibition with an average diameter of 31 mm, followed by the sample (H6) in (30 mm), H4 (29 mm), H2 (28 mm), H3 (27 mm), and finally the sample H1 (26 mm). No zone of inhibition was determined with starch alone.

The differences in inhibition were observed for six types of honey

| Honeysample | Honey only | Starch and honey %(v/v) | Zone increase of inhibition Percentage increase (PI% ) |
|---|---|---|---|
| H1 | 26 | 32 | 18.75 |
| H2 | 28 | 28 | 3.57 |
| H3 | 27 | 27 | 3.57 |
| H4 | 29 | 30 | 3.57 |
| H5 | 31 | 31 | 00 |
| H6 | 30 | 30 | 00 |

<5.5 mm. inactive; 5.5–9 mm. very low activity; 9–12 mm. low activity; 12–15 mm. average activity; and >15 mm, high activity

**Table 2:** Mean Zones of Inhibition (diameter mm including well (8.2 mm).

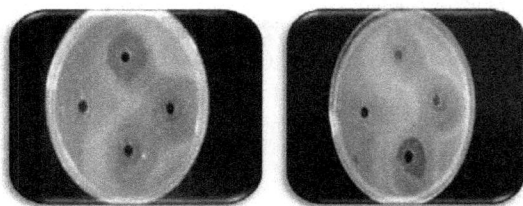

**Figure 1:** Inhibition Zone Diameters of natural honey only against *P.aeruginosa*.

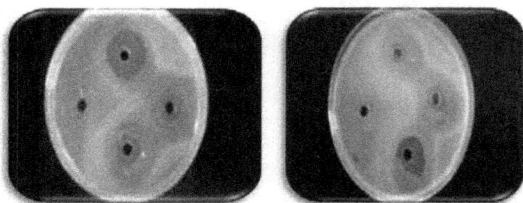

**Figure 2:** Inhibition Zone Diameters of natural honey with potato starch against *P. aeruginosa*.

with starch potato: the sample (H5) has the largest inhibition with an average diameter of 31 mm, followed by the sample (H1) in (31 mm), H5 (31 mm), H4 and H6) (30 mm), H2 (28 mm) and finally the sample H3 (27 mm). The percentage increase (PI%) was noticed with each variety and it ranged between 3, 57 and 18, 75%.

*Pseudomonas aeruginosa* is the predominant cause of fatal burn wound sepsis, and isolation of multidrug-resistant strains is a common problem in hospitals. With increasing interest in the use of alternative therapies and as the development of antibiotic resistant bacteria spreads. Many works was interested, during this last decade, with the products of the hive and in particular honey, efficient product against the germs secreted by the bees as a possible source of new pharmaceutical and medical agent. In Algeria, there are few types of alternative medicine such as honey. Recent experimental finding indicated that the amylase present in honey increases the osmotic effect in the media by increasing the amount of sugars and consequently increasing the antibacterial activity [22]. High osmolarity has been considered a valuable tool in the treatment of infections, because it prevents the growth of bacteria and encourages healing [27]. Use of sugar to enhance wound healing has been reported for several hundred patients [28].

Molan [29] has studied sugar syrups of the same water activity as honey and found them to be less effective than honey at inhibiting microbial growth *in vitro*. It was found that solutions of high osmolarity, such as honey, glucose, and sugar pastes, inhibit microbial growth because the sugar molecules tie up water molecules so that bacteria have insufficient water to grow [30]. Therefore, high osmolarity is valuable in the treatment of infections because it prevents the growth of bacteria and encourages healing. Sugar was used to enhance wound healing for several hundred patients [31]. It has been claimed that the sugar content of honey is responsible for its antibacterial activity, which is due entirely to the osmotic effect of its high sugar content [32].

In our study, the addition of starch potato honey showed a significant increase in the inhibition zone for honey studied against the strain tested except honey H5 and H6. Amylase present in honey was expected to split potato starch chains into randomly produced dextrin and maltose and probably increases the osmotic effect in the well by increasing the amount of sugar and consequently increases the antibacterial activity. Other results show that the addition ginger starch to honey could contribute to reducing the quantity of honey to be used without losing the expected effect [22-24]. In our study, the amount of the amylase present in honey is in positive correlation with the relative potency of starch and honey.

Neither honey nor starch has adverse effects on tissues, so they can be safely used in wounds, burns and inserted in cavities and sinuses to clear infection. A clinical trial would be carried out to validate these findings. The results will enable a systematic study of many varieties of honey on pathogens bacteria withincreased resistance opposite conventional antibiotics.

## Conflict of interest statement

We declare that we have no conflict of interest.

## Acknowledgement

Authors thank Staff of Tiaret University for providing material.

## References

1. Siegenthaler U (1975) Bestimmung der Amylase in Bienenhonig mit einem handelsublichen, farbmarkierten Substrat. Mitt Gebiete Lebensm Hyg 66: 393-399.

2. Bogdanov S (1984) Honigdiastase, Gegenüberstellungverschiedener Bestimmungsmethoden. Mitt Gebiete Lebensm Hyg 75: 214–220.

3.  Rastegar Lari A, Bahrami Honar H, Alaghehbandan R (1998) Pseudomonas infections in Tohid Burn Center Iran. Burns 24: 637-641.

4.  Sharma BR (2007) Infection in patients with severe burns: causes and prevention thereof. Infec Dis Clin North Am 21: 745-759.

5.  Lari AR, Alaghehbandan R (2000) Nosocomial infections in an Iranian burn care center. Burns 26: 737-740.

6.  Branski LK, Al-Mousawi A, Rivero H, Jeschke MG, Sanford AP, et al. (2009) Emerging infections in burns. Surg Infect 10: 389-397.

7.  Keen EF 3rd, Robinson BJ, Hospenthal DR, Aldous WK, Wolf SE, et al. (2010) Prevalence of multidrug resistant organisms recovered at a military burn center. Burns 36: 819-825.

8.  Jacobsen JN, Andersen AS, Sonnested MK, Laursen I, Jorgensen B, et al. (2011) Investigating the humoral immune response in chronic venous leg ulcer patients colonised with Pseudomonas aeruginosa. Int Wound J 8: 33-43.

9.  Blair SE, Cokcetin NN, Harry EJ, Carter DA (2009) The unusual antibacterial activity of medical -grade Leptospermum honey: antibacterial spectrum, resistance and transcriptome analysis. Eur J Clin Microbiol Infect Dis 28: 1199-1208.

10. Molan PC (2001) Potential of honey in the treatment of wounds and burns. Am J Clin Dermatol 2: 13-19.

11. Bose B (1982) Honey or Sugar in Treatment of Infected Wounds? Lancet 1: 963.

12. Archer HG, Barnett S, Irving S, Middleton KR, Seal DV (1990) A controlled model of moist wound healing: comparison between semi-permeable film, antiseptics and sugar paste. J Exp Pathol 71: 155-170.

13. Wang J, Li QX (2011) Chemical composition, characterization and differentiation of honey botanical and geographical origins. Adv Food Nutr Res 62: 89-137.

14. Cotte JF, Casabianca H, Chardon S, Lheritier J, Grenier-Loustalot MF (2003) Application of carbohydrate analysis to verify honey authenticity. J Chromatogr A 1021: 145-155.

15. Joshi SR, Pechhacker H, Willam W, von der Ohe W (2000) Physico-chemical characteristics of Apis dorsata, A. cerana and A. mellifera honey from Chitwan district, central Nepal. Apidologie 31: 367-375.

16. Costa LSM, Albuquerque MLS, Trugo LC, Quinteiro LMC, Barth OM, et al. (1999) Determination of non-volatile compounds of different botanical origin Brazilian honeys. Food Chem 65: 347-352.

17. Devillers J, Morlot M, Pham-Delegue MH, Dore JC (2004) Classification of monofloral honeys based on their quality control data. Food Chem 86: 305-312.

18. Huidobro JF, Santana FJ, Sanchez MP, Sancho MT, Muniategui S, et al. (1995) Diastase, invertase and a-glucosidase activities in fresh honey from north-west Spain. J Apic Res 34: 39-44.

19. Crane Honey E (1975) A Comprehensive Survey, International Bee Research Association, Heinemann, London, UK

20. Sakac N, Sak-Bosnar M (2012) A rapid method for the determination of honey diastase activity. Talanta 93: 135-138.

21. Tosi E, Martinet R, Ortega M, Lucero H, Re E (2008) Honey diastase activity modified by heating. Food Chem 106: 883-887.

22. Ahmed M, Aissat S, Djebli N, Boulkaboul A, Abdelmalek M, et al. (2011) The Influence of Starch of Ginger on the Antibacterial Activity of Honey of Different Types from Algeria against Escherichia coli and Staphylococcus aureus. I J M R 2: 258-262.

23. Ahmed M, Djebli N, Aissat S, Aggad H, Boucif Ahmed (2011) Antifungal Activity of a Combination of Algeria Honey and Starch of Ginger against Aspergillus niger. I J M R 2: 263-266.

24. Ahmed M, Djebli N, Hammoudi SM, Aissat S, Akila B, et al. (2012) Additive potential of ginger starch on antifungal potency of honey against Candida albicans. Asian Pac trop Biomed 2: 253-255.

25. Bodganov S, Martin P, Lüllmann C (1997) Harmonised methods of the European Honey Commission. Apidologie.

26. Al Somal N, Coley KE, Molan PC, Hancock BM (1994) Susceptibility of Helicobacter pylori to the antibacterial activity of manuka honey. J R Soc Med 87: 9-12.

27. Archer HG, Barnett S, Irving S, Middleton KR, Seal DV (1990) A controlled model of moist wound healing: comparison between semipermeable film, antiseptics and sugar paste. J Exp Pathol 71: 155-170.

28. Knutson RA, Merbitz LA, Creekmore MA, Snipes HG (1981) Use of sugar and povidone-iodine to enhance wound healing: five year's experience. South Med J 74: 1329-1335.

29. Molan PC (1992) The antibacterial activity of honey. 1. The nature of the antibacterial activity. Bee World 73: 5-28.

30. Chirife J, Scarmato G, Herszage L (1982) Scientific basis for use of granulated sugar in treatment of infected wounds. Lancet 1: 560-561.

31. Green A (1988) Wound healing properties of honey. Br J Surg 75: 1278.

32. Somerfield S (1991) Honey and healing. J R Soc Med 84: 179.

# 26

# Rational Design of Star-Shaped Molecules with Benzene Core and Naphthalimides Derivatives End Groups as Organic Light-emitting Materials

**Ruifa Jin***

*College of Chemistry and Chemical Engineering, Chifeng University, Chifeng 024000, China*

## Abstract

A series of star-shaped molecules with benzene core and naphthalimides derivatives end groups have been designed to explore their optical, electronic, and charge transport properties as charge transport and/or luminescent materials for organic light-emitting diodes (OLEDs). The frontier molecular orbitals (FMOs) analysis has turned out that the vertical electronic transitions of absorption and emission are characterized as intramolecular charge transfer (ICT). The calculated results show that the optical and electronic properties of star-shaped molecules are affected by the substituent groups in N-position of 1,8-naphthalimide ring. Our results suggest that star-shaped molecules with n-butyl (**1**), Benzene (**2**), Thiophene (**3**), thiophene $S',S'$-dioxide (**4**), benzo[c][1,2,5]thiadiazole (**5**), and 2,7a-dihydrobenzo[d] thiazole (**6**) fragments are expected to be promising candidates for luminescent and electron transport materials for OLEDs. This study should be helpful in further theoretical investigations on such kind of systems and also to the experimental study for charge transport and/or luminescent materials for OLEDs.

**Keywords:** 1,8-Naphthalimide derivatives; Optical and electronic properties; Charge transport property; Luminescent materials; Organic light-emitting diodes (OLEDs)

## Introduction

Organic light-emitting diodes (OLEDs) have received considerable interest due to their promising applications in the large-area flat-panel displays and solid-state lighting [1-6]. The devices using organic materials have shown several advantages over their inorganic counterparts, for example, light weight, potentially low cost, capability of thin-film, large-area, and flexible device fabrication, and wide selection of emission colors via molecular design of organic materials. However, the lower efficiency of OLEDs is a thorny obstacle to the application of efficient light-emitting devices. Since the first report on OLEDs in 1987, the light generation efficiencies of OLEDs have been steadily increased by using novel materials and the different device structures [7-9]. Unfortunately, most OLEDs emitters are still not satisfactory. Therefore, the design and synthesize for new emitting materials with high efficiency and thermal stability remain one of the most active areas of the studies. A number of studies demonstrate that the interplay between theory and experiment is capable of providing useful insights into the understanding of the the nature of molecules [10,11]. Among the various kinds of OLEDs materials, 1,8-naphthalimide (**NI**) derivatives usually exhibit strong fluorescence and good photostability [12-14]. They have been widely used as the most important materials for fabrication of OLEDs. Furthermore, **NI** derivatives have high electron affinity and excellent transport property due to the existence of an electron-deficient centre. Thus, **NI** derivatives have been extensively applied in many fields such as coloration and brightening of polymers [15], potential photosensitive biologically units [16], fluorescent markers in biology [17], light emitting diodes [18,19], fluorescence sensors and switchers [20], and electroluminescent materials [21]. A large variety of auxochromic groups in **NI** derivatives may be easily grafted to fine tune the absorption and emission wavelengths. Naphthalimides comprise a class of fluorophore whose electronic absorption and emission depend upon the properties of the surrounding medium. The emission spectrum can be tuned by introducing different electron-donating substituent groups, such as N-substituted groups [22], C-substituted groups [23], and O-substituted groups [24]. Furthermore, substitution of electron-donating groups usually increases the intensity of the fluorescence emission, particularly when a methoxy or amino group at C-4 position is used. Recently, some starburst amorphous molecules 1,3,5-Tris(1,8-naphthalimide-4-yl)benzenes have been reported [25]. It was found that the devices using these molecules performance are better than using the most prevalent tris(8-quinolinato)aluminum (Alq3) as a counterpart.

With the above considerations, in this work, we investigated a series of star-shaped molecules with benzene as core and **NI** derivatives as end groups for OLEDs applications (Scheme 1). An in-depth interpretation of the optical and electronic properties of these compounds has been presented. Several derivatives (**1–6**), as shown in Scheme 1, have been designed to provide a demonstration for the rational design of novel luminescent and charge transporting materials for OLEDs (Scheme 1).

## Computational Methods

All calculations have been performed using Gaussian 09 code [26]. Generally, the B3LYP method appeared notably adapted to **NI** derivatives [27-31]. Therefore, The geometry optimization of designed molecules in ground states (S$_0$) were carried out by the B3LYP method using the 6-31G(d,p) basis set. The corresponding geometry in the first excited singlet state (S$_1$) were optimized using the TD-B3LYP with 6-31G (d,p) basis set. The harmonic vibrational frequency calculations using the same methods as for the geometry optimizations were used to ascertain the presence of a local minimum. The absorption and fluorescent properties of **1–6** have been predicted using the TD-B3LYP/6-31G(d,p) method based on the S$_0$ and S$_1$ optimized geometries, respectively. To investigate the influence of solvents on the optical properties for the S$_0$ and S$_1$ states of the molecular systems

***Corresponding author:** Ruifa Jin, College of Chemistry and Chemical Engineering, Chifeng University, Chifeng 024000, China
E-mail: Ruifajin@163.com

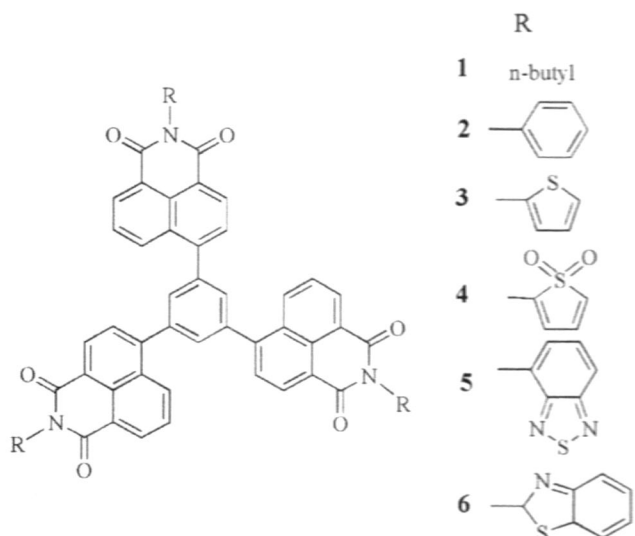

**Scheme 1:** Geometries of Designed Molecules 1–6.

correspond to a π–π* excited singlet state. For **1**, **2**, and **5**, the HOMOs are distributed on the 1,8-naphthalimide (**NI**) and benzene (**BZ**) moieties, with minor contributions from N-substituent groups (**SG**). The sum contributions of **NI** and **BZ** fragments of HOMOs are larger than 97.2%, while the corresponding contributions of **SG** fragments are within 2.8%, respectively. For **3** and **6**, the HOMOs are mainly localized on the **SG** fragments with only minor contributions from **NI** and **BZ** fragments. The contributions of **SG** fragments of HOMOs are larger than 95%, while the corresponding sum contributions of **NI** and **BZ** fragments are within 4.9%, respectively. For **4**, the HOMOs are distributed on the **NI** and **SG** fragments, with minor contributions from **BZ** fragment. However, the LUMOs of **1–6** are mainly composed

in chloroform (dielectric constant: 2.0906) solvent, we performed the polarized continuum model (PCM) [32] calculations at the TD-DFT level.

The stability is a useful criterion to evaluate the nature of devices for charge transport and luminescent materials. To predict the stability of **1–6** from a viewpoint of conceptual density functional theory, the absolute hardness, $\eta$, of **1–6** were calculated using operational definitions [33,34] given by:

$$\eta = \frac{1}{2}\left(\frac{\partial \mu}{\partial N}\right) = \frac{1}{2}\left(\frac{\partial^2 E}{\partial N^2}\right) = \frac{IP - EA}{2} \qquad (1)$$

Where, $\mu$ is the chemical potential and $N$ is the total electron number. In this work, the values for $IP$ (ionization potential) and $EA$ (electron affinity) were determined according to the equation $IP = E_{cr} - E_p$ and $EA = E_p - E_{ar}$, where p, cr, and ar indicate the parent molecule and the corresponding cation and anion radical generated after electron transfer.

## Results and Discussion

### Frontier molecular orbitals

To characterize the optical transitions and the abilities of electron and hole transport, it is useful to examine the frontier molecular orbitals (FMOs) of the compounds under investigation. The origin of the geometric difference introduced by excitation can be explained, at least in qualitative terms, by analyzing the change in the bonding character of the orbitals involved in the electronic transition for each pair of bonded atoms. An electronic excitation results in some electron density redistribution that affects the molecular geometry [35]. We calculated the distribution patterns of FMOs for **1–6** in $S_0$ (Figure 1). The total and partial densities of states (TDOS and PDOS) on each fragment of the investigated molecules around the HOMO – LUMO gaps were calculated based on the current level of theory. The FMOs energies $E_{HOMO}$ and $E_{LUMO}$, HOMO – LUMO gaps, and the contributions of individual fragments (in %) to the FMOs of **1–6** are given in Table 1. As shown in Figure 1, the $S_0 \rightarrow S_1$ excitation process can be mainly assigned to the HOMOs → LUMOs and HOMOs-1 → LUMOs transitions, which

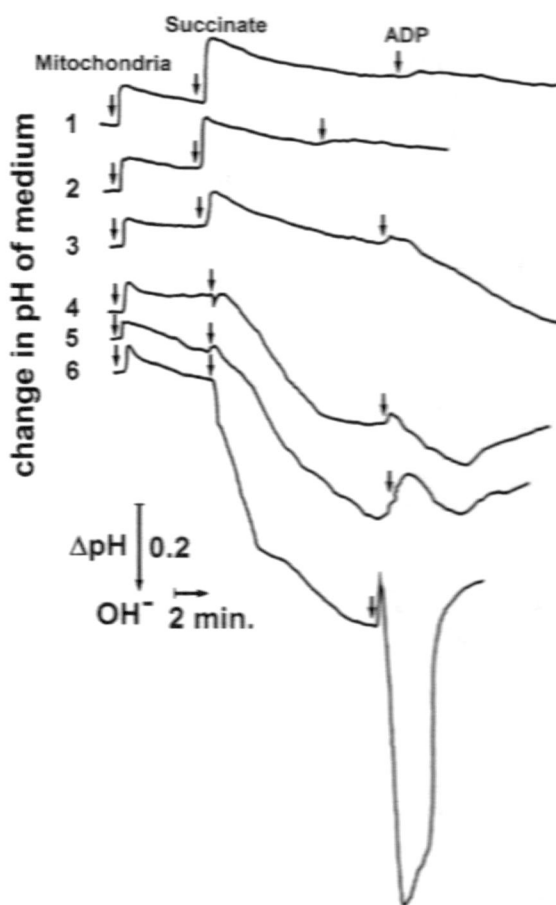

**Figure 1:** Electronic density contours of the frontier molecular orbitals for investigated derivatives.

| Species | $E_{HOMO}$ | HOMO | | | $E_{LUMO}$ | LUMO | | | $E_g$ |
| | | NI[a] | BZ[b] | SG[c] | | NI | BZ | SG | |
|---|---|---|---|---|---|---|---|---|---|
| 1 | -6.54 | 85.5 | 14.4 | 0.1 | -2.69 | 92.7 | 7.3 | 0.0 | 3.85 |
| 2 | -6.54 | 84.4 | 14.2 | 1.4 | -2.69 | 92.6 | 7.4 | 0.0 | 3.85 |
| 3 | -6.41 | 4.9 | 0.0 | 95.1 | -2.77 | 93.0 | 7.0 | 0.0 | 3.64 |
| 4 | -6.81 | 53.6 | 8.5 | 37.9 | -3.02 | 92.8 | 6.9 | 0.2 | 3.79 |
| 5 | -6.59 | 83.8 | 13.4 | 2.8 | -2.74 | 92.9 | 6.9 | 0.2 | 3.85 |
| 6 | -6.44 | 1.9 | 0.1 | 98.1 | -2.87 | 92.9 | 7.1 | 0.0 | 3.57 |

[a] NI: 1,8-naphthalimide moieties; [b] BZ: benzene moieties; [c] SG: substituent groups

**Table1:** The FMOs Energies $E_{HOMO}$ and $E_{LUMO}$, HOMO–LUMO gaps (eV), and HOMOs and LUMOs Contributions (%) of 1–6.

| Species | $\lambda_{ab}$ | $f$ | Assignment |
|---|---|---|---|
| 1 | 365 | 0.85 | HOMO-1 → LUMO (0.70)<br>HOMO-1 → LUMO+2 (0.12) |
| 2 | 365 | 0.62 | HOMO → LUMO (-0.31)<br>HOMO → LUMO+1 (-0.32)<br>HOMO-1 → LUMO (0.43) |
| 3 | 365 | 0.52 | HOMO → LUMO (0.60)<br>HOMO → LUMO+2 (0.14) |
| 4 | 370 | 0.60 | HOMO → LUMO (0.59)<br>HOMO → LUMO+2 (0.12) |
| 5 | 365 | 0.57 | HOMO → LUMO (0.58)<br>HOMO → LUMO+1 (0.17) |
| 6 | 368 | 0.59 | HOMO → LUMO (0.45)<br>HOMO → LUMO+1 (0.20)<br>HOMO → LUMO+2 (0.22) |
| Exp[a] | 360 | | |

[a] Experimental data for 1 in chloroform [25].
**Table 2:** The absorption wavelengths $\lambda_{abs}$ (in nm), the oscillator strength $f$, and main assignments (coefficient) of 1–6 in chloroform obtained at the TD-B3LYP/6-31G(d,p)//B3LYP/6-31G(d,p) level, along with available experimental data.

of contributions of **NI**, with minor contributions from **SG** and **BZ** fragments. The contributions of **NI** fragments of LUMOs are larger than 92.6%, while the corresponding sum contributions of **SG** and **BZ** fragments are within 7.4%, respectively.

The distribution patterns of the FMOs also provide a remarkable signature for the charge-transfer character of the vertical $S_0 \rightarrow S_1$ transition. Analysis of the FMOs indicates that the excitation of the electron from the HOMO to LUMO leads the electronic density to flow mainly from the **SG** and **BZ** fragments to **NI** fragments for **1, 2, 4**, and **5**. The percentages of charge transfer are the differences between the contributions of fragments for LUMOs and the corresponding contributions for HOMOs in the compounds under investigation. The percentages of charge transfer from **SG** and **BZ** fragments to **NI** fragments are 7.2, 8.2, 39.2, and 9.1%, respectively. On the contrary, for **3** and **6**, the excitation of the electron from the HOMO to LUMO leads the electronic density to flow mainly from **SG** fragments to **BZ** and **NI** fragments. The percentage of charge transfer of 3 and 6 are 95.1 and 98.1%, respectively.

Another way to understand the influence of the optical and electronic properties is to analyze the $E_{HOMO}$, $E_{LUMO}$, and $E_g$ values. From Table 1, one can find that the $E_{HOMO}$ values of 2, 4, and 5 decreases, while the corresponding value of 3 and 6 increase compared with that of 1. The HOMOs energies are in the order of 3 > 6 > 1 ≈ 2 > 5 4. However, the values of $E_{LUMO}$ and HOMO–LUMO gaps $E_g$ for 2–6 decrease compared with those of 1. The sequence of LUMOs energies is 1 ≈ 2 ≈ 5 > 3 6 > 4. The $E_g$ values are in the order of 1 ≈ 2 ≈ 5 > 4 3 > 6. It implies that the introduction of different donor groups to the 1 leads to the change of the $E_{HOMO}$, $E_{LUMO}$, and $E_g$ values for its derivatives.

The absorption and fluorescence spectra can be tuned by donor groups, providing a powerful strategy for prediction of the optical properties of novel electroluminophores.

## Absorption and Fluorescence Spectra

The absorption $\lambda_{abs}$ and fluorescence $\lambda_{fl}$ wavelengths, main assignments, and the oscillator strength $f$ for the most relevant singlet excited states in each molecule are listed in Tables 2 and 3, respectively. The $\lambda_{abs}$ and $\lambda_{fl}$ values of 1 are all in agreement with experimental results [25], the deviations are 5 and 25 nm, respectively. The Stokes shift of 1 is 36 nm, which is comparable to the experimental 66 nm. Thus, this result credits to the computational approach, so appropriate electronic transition energies can be predicted at these levels for this kind of system.

For the absorption spectra, the excitation to the $S_1$ state corresponds mainly to the HOMO-1 → LUMO for 1, while the corresponding excitations for 2–6 correspond mainly to the HOMOs → LUMOs and HOMOs → LUMOs+1 and/or HOMOs → LUMOs + 2. From Table 2, one can find that the $\lambda_{abs}$ values of 1–6 are almost equal to that of 1. It suggests that the substituent effects do not significantly affect the absorption spectra of 2–6 compared with those of 1. Moreover, 2–6 have nearly equal values of oscillator strengths, being smaller slightly than the value of 1. The oscillator strength for an electronic transition is proportional to the transition moment [36]. In general, larger oscillator strength corresponds to larger experimental absorption coefficient or stronger fluorescence intensity. This implies that these bipolar molecules shown large absorption intensity.

For the fluorescence spectra, the HOMO ← LUMO+1 and HOMO-1 ← LUMO excitations play a dominant role for 1. The fluorescence peaks of 2, 3, and 5 are mainly correspond to HOMOs-1 ← LUMOs excitations. The $\lambda_{fl}$ value of 2 is almost equal to that of 1, while the $\lambda_{fl}$ values of 3–6 show bathochromic shifts 5, 27, 8, and 53 nm compared with that of 1, respectively. The Stokes shifts of 3–6 are 41, 58, 44, and 86 nm, respectively. Furthermore, the $f$ values 2–6 are almost equal to that of **MEBN**, corresponding to strong fluorescence spectra. This implies that 2–6 have large fluorescent intensity and they are promising luminescent materials for OLEDs. As shown in Table 3, it clearly shows that the substituent groups can affect the fluorescence spectra of these molecules. The emissions color of molecules can be tuned by the N-substituent groups. Furthermore, all the substituted derivatives show stronger fluorescence intensity (Tables 2 and 3).

| Species | $\lambda_{flu}$ | $f$ | Assignment |
|---|---|---|---|
| 1 | 401 | 0.76 | HOMO ← LUMO+1 (0.55)<br>HOMO-1 ← LUMO (-0.43) |
| 2 | 399 | 0.83 | HOMO ← LUMO+1 (-0.49)<br>HOMO-1 ← LUMO (0.49) |
| 3 | 406 | 0.52 | HOMO-1 ← LUMO (0.67)<br>HOMO-2 ← LUMO (0.14) |
| 4 | 428 | 0.59 | HOMO ← LUMO+4 (0.47)<br>HOMO ← LUMO+3 (0.34) |
| 5 | 409 | 0.70 | HOMO ← LUMO+5 (0.56)<br>HOMO-1 ← LUMO (-0.40) |
| 6 | 454 | 0.50 | HOMO ← LUMO+1 (-0.41)<br>HOMO-1 ← LUMO (0.67) |
| Exp[a] | 426 | | |

[a] Experimental data for 1 in chloroform [25].
**Table 3:** The fluorescence wavelengths $\lambda_{flu}$ (in nm), the oscillator strength $f$, and main assignments (coefficient) of 1–6 in chloroform obtained at the TD-B3LYP/6-31G(d,p)//TD-B3LYP/6-31G(d,p) level, along with available experimental data.

| Species | $\lambda_h$ | $\lambda_e$ | |
|---------|-------------|-------------|--------|
| 1 | 0.142 | 0.150 | 2.827 |
| 2 | 0.328 | 0.180 | 2.837 |
| 3 | 0.292 | 0.156 | 2.790 |
| 4 | 0.315 | 0.216 | 2.670 |
| 5 | 0.304 | 0.110 | 2.893 |
| 6 | 0.294 | 0.122 | 2.866 |

**Table 4:** Calculated molecular $\lambda_e$, $\lambda_h$, and $\eta$ (all in eV) of 1–6 at the B3LYP/6-31G(d,p) level.

## Charge Transport Properties

The charge transfer rate can be described by Marcus theory [37,38] via the following equation:

$$K = \left(V^2 / \hbar\right)\left(\pi / \lambda k_B T\right)^{1/2} \exp\left(-\lambda / 4k_B T\right) \tag{2}$$

Where, T is the temperature, $k_B$ is the Boltzmann constant, $\lambda$ represents the reorganization energy due to geometric relaxation accompanying charge transfer, and V is the electronic coupling matrix element (transfer integral) between the two adjacent species dictated largely by orbital overlap. It is clear that two key parameters are the reorganization energy and electronic coupling matrix element, which have a dominant impact on the charge transfer rate, especially the former.

For the reorganization energy $\lambda$, they can be divided into two parts, external reorganization energy ($\lambda_{ext}$) and internal reorganization energy ($\lambda_{int}$). $\lambda_{ext}$ represents the effect of polarized medium on charge transfer, which is quite complicated to evaluate at this stage. $\lambda_{int}$ is a measure of structural change between ionic and neutral states [39,40]. Our designed molecules are used as charge transport materials for OLEDs in the solid film; the dielectric constant of the medium for the molecules is low. The computed values of the external reorganization energy in pure organic condensed phases are not only small but also are much smaller than their internal counterparts [41,42]. Moreover, there is a clear correlation between $\lambda_{int}$ and charge transfer rate in literature [43,44]. Therefore, we mainly study the $\lambda_{int}$ of the isolated active organic π-conjugated systems owing to ignoring the environmental changes and relaxation in this work. Hence, the $\lambda_e$ and $\lambda_h$ can be defined by equations (3) and (4): [45]

$$\lambda_e = \left(E_0^- - E_-^-\right) + \left(E_-^0 - E_0^0\right) \tag{3}$$

$$\lambda_h = \left(E_0^+ - E_+^+\right) + \left(E_+^0 - E_0^0\right) \tag{4}$$

Where, $E_0^+$ ($E_0^-$) is the energy of the cation (anion) calculated with the optimized structure of the neutral molecule. Similarly, $E_+^+$ ($E_-^-$) is the energy of the cation (anion) calculated with the optimized cation (anion) structure, $E_+^0$ ($E_-^0$) is the energy of the neutral molecule calculated at the cationic (anionic) state. Finally, $E_0^0$ is the energy of the neutral molecule in ground state. For comparing with the interested results reported previously [46,47], the reorganization energies for electron ($\lambda_e$) and hole ($\lambda_h$) of the molecules were calculated at the B3LYP/6-31G (d,p) level on the basis of the single point energy.

The calculated reorganization energies for hole and electron are listed in Table 4. It is well-known that, the lower the reorganization energy values, the higher the charge transfer rate [37,38]. The results displayed in Table 4 show that the calculated $\lambda_e$ values of 1–6 (0.110 – 0.180 eV) are larger than that of tris(8-hydroxyquinolinato) aluminum(III) (Alq3) ($\lambda_e$ = 0.276 eV), a typical electron transport material [46]. It indicates that their electron transfer rates might be higher than that of Alq3, suggesting that 1–6 could be good electron

transfer materials from the stand point of the $\lambda_e$ values. On the other hand, the calculated $\lambda_h$ values of 2–6 (0.292 – 0.328 eV) are larger than that of N,N'-diphenyl-N,N'-bis(3- methylphenyl)-(1,1'-biphenyl)-4,4'-diamine (TPD), which is a typical hole transport material ($\lambda_h$ = 0.290 eV) [47]. It indicates that their whole transfer rates might be lower than that of TPD. It indicates that 1–6 can be used as promising electron transport materials in OLEDs from the stand point of the smaller reorganization energy.

As the stability is a useful criterion to evaluate the nature of devices for charge transport and luminescent materials. The absolute hardness $\eta$ is the resistance of the chemical potential to change in the number of electrons. As expected, inspection of Table 4 reveals clearly that the $\eta$ values of 2–6 are almost equal to that of values of 1. These results reveal that the different π-conjugated bridges do not significantly affect the stability of these bipolar molecules.

## Conclusions

In this paper, a series of star-shaped molecules with benzene core and naphthalimides derivatives end groups have been systematically investigated. The FMOs analysis have turned out that the vertical electronic transitions of absorption and emission are characterized as intramolecular charge transfer (ICT). The calculated results show that their optical and electronic properties are affected by their substituent groups in N-position of 1,8-naphthalimide. The study of substituent effects suggest that the $\lambda_{abs}$ values of 2–6 are almost equal to that of the parent compound 1, while the $\lambda_g$ of 2–6 show bathochromic shifts compared with that of 1. Furthermore, 2–6 have large fluorescent intensity. The different substituent groups do not significantly affect the stability of these molecules. Our results suggest that 2–6 are expected to be promising candidates for luminescent materials and electron transport materials for OLEDs.

### Acknowledgement

Financial support from the Inner Mongolia Key Laboratory of Photoelectric Functional Materials and Natural Science Foundation of Inner Mongolia Autonomous Region (No. 2011ZD02) are gratefully acknowledged.

### References

1. Mullen K, Scherf U (2006) Organic light-emitting devices, synthesis, properties, and applications, Wiley-VCH , Weinheim C.

2. Grimsdale AC , Chan KL, Martin RE, Jokisz PG, Holmes AB (2009) Synthesis of light-emitting conjugated polymers for applications in electroluminescent devices. Chem Rev 109: 897-1091.

3. Minaev B, Baryshnikov G, Agren H (2014) Principles of phosphorescent organic light emitting devices. Phys Chem Chem Phys 16: 1719-1758.

4. Sasabe H , Kido J (2013) Development of high performance OLEDs for general lighting. J Mater Chem 1: 1699-1707.

5. Dimitrakopoulos CD, Malenfant PRL (2002) Organic thin film transistors for large area electronics. Adv Mater 14: 99-117.

6. Yao JH, Zhen C, Loh KP, Chen ZK (2008) Novel iridium complexes as high-efficiency yellow and red phosphorescent light emitters for organic light-emitting diodes. Tetrahedron 64: 10814-1082.

7. Tang CW, VanSlyke SA (1987) Organic electroluminescent diodes. Appl Phys Lett 51: 913-915.

8. Walzer K, Maennig B, Pfeiffer M, Leo K (2007) Highly efficient organic devices based on electrically doped transport layers. Chem Rev 107: 1233-1271.

9. Koh TW, Choi JM, Lee S, Yoo S (2010) Optical out coupling enhancement in organic light-emitting diodes: highly conductive polymer as a low-index layer on micro structured ITO electrodes. Adv Mater 22 : 1849-1853.

10. Liu YL, Feng JK, Ren AM (2008) Theoretical study on photophysical properties of bis-dipolar diphenylamino-endcapped oligoarylfluorenes as light-emitting materials. J Phys Chem A 112: 3157-3164.

11. Zou LY, Ren AM, Feng JK, Liu YL, Ran XQ, et al. (2008) Theoretical study on photophysical properties of multifunctional electroluminescent molecules with different pi-conjugated bridges. J Phys Chem A 112: 12172-12178.

12. Ramachandram B, Saroja G, Sankaran NB, Samanta A (2000) Unusually high fluorescence enhancement of some 1,8-naphthalimide derivatives induced by transition metal salts. J Phys Chem B 104: 11824-11832.

13. Ivanov IP, Dimitrova MB, Tasheva DN, Cheshmedzhieva DV, et al. (2013) Synthesis, structural analysis and application of a series of solid-state fluorochromes-aryl hydrazones of 4-hydrazino-N-hexyl-1,8-naphthalimide. Tetrahedron 69: 712-721.

14. Li Y, Xu Y, Qian X , Qu B (2004) Naphthalimide-thiazoles as novel photonucleases: molecular design, synthesis, and evaluation. Tetrahedron Lett 45: 1247-1251.

15. Hrdlovi P, Chmela S, Danko M, Sarakha M, Guyot G (2008) Spectral properties of probes containing benzothioxanthene chromophore linked with hindered amine in solution and in polymer matrices. J. Fluoresc 18: 393-402.

16. Tao ZF, Qian X (1999) Naphthalimide hydroperoxides as photonucleases: substituent effects and structural basis. Dyes Pigm 43: 139-145.

17. Martin E, Weigand R, Pardo A (1996) Solvent dependence of the inhibition of intramolecular charge-transfer in N-substituted 1,8-naphthalimide derivatives as dye lasers. J Lumin 68: 157-164.

18. Grabchev I, Chovelon JM, Qian X (2003) A copolymer of 4-N,N dimethylaminoethylene-N-allyl-1,8-naphthalimide with methylmethacrylate as a selective fluorescent chemosensor in homogeneous systems for metal cations. J Photochem Photobiol A158: 37-43.

19. Morgado J, Gruner J, Walcott SP, Yong TM, Cervini R, et al. (1998) 4-AcNI—a new polymer for light-emitting diodes. Synth Met 95: 113-117.

20. Tamanini E1, Katewa A, Sedger LM, Todd MH, Watkinson M (2009) A synthetically simple, click-generated cyclam-based zinc(II) sensor. See comment in PubMed Commons below Inorg Chem 48: 319-324.

21. Chatterjee S, Pramanik S, Hossain SU, Bhattacharya S, Bhattacharya SC (2007) Synthesis and photoinduced intramolecular charge transfer of N-substituted 1,8-naphthalimide derivatives in homogeneous solvents and in presence of reduced glutathione. J Photochem Photobiol A 187: 64-71.

22. Islam A, Cheng CC, Chi SH, Lee SJ, Hela GP, et al.( 2005) Amino naphthalic anhydrides as red-emitting materials:? Electro luminescence, crystal structure, and photo physical properties. J Phys Chem B 109: 5509-5517.

23. Yang JX, Wang XL, Wang XM, Xu LH (2005) The synthesis and spectral properties of novel 4-phenylacetylene-1,8-naphthalimide derivatives. Dyes Pigm 66 : 83-87.

24. Magalhaes JL, Pereira RV, Triboni ER, Berci Filho P, Gehlen MH, et al. (2006) Solvent effect on the photo physical properties of 4-phenoxy-N-methyl-1,8-naphthalimide. J Photochem Photobiol A 183: 165-170.

25. Liu Y, Niu F, Lian J, Zeng P, Niu H (2010) Synthesis and properties of starburst amorphous molecules: 1,3,5-Tris(1,8-naphthalimide-4-yl)benzenes. Synthetic Met 160 : 2055-2060.

26. Frisch MJT, Trucks GW, Schlegel HB, Scuseria GE, Robb MA, et al. (2009) Gaussian 09; Gaussian Inc. Wallingford CT, USA

27. Mancini G, Zazza C, Aschib M, Sannaa N (2011) Conformational analysis and UV/Vis spectroscopic properties of a rotaxane-based molecular machine in acetonitrile dilute solution: when simulations meet experiments. Phys Chem Chem Phys 13: 2342-2349.

28. Li H, Li N, Sun R, Gu H, Ge J, et al. (2011) Dynamic random access memory devices based on functionalized copolymers with pendant hydrazine naphthalimide group. J Phys Chem C 115 : 8288-8294.

29. Dhar S, Roy SS, Rana DK, Bhattacharya S, Bhattacharya S, et al. (2011) Tunable solvatochromic response of newly synthesized antioxidative naphthalimide derivatives: intramolecular charge transfer associated with hydrogen bonding effect. J Phys Chem A 115: 2216-2224.

30. Meng X, Zhu W, Zhang Q, Feng Y, Tan W, et al. (2008) Novel bisthienylethenes containing naphthalimide as the venter ethene bridge: photochromism and solvatochromism for combined NOR and INHIBIT logic gates. J Phys Chem B 112: 15636-15645.

31. Li Z, Yang Q, Chang R, Ma G, Chen M, et al. (2011) N-Heteroaryl-1,8-naphthalimide fluorescent sensor for water: Molecular design, synthesis and proper. Dyes Pigm 88: 307-314.

32. Gudeika D, Michaleviciute A, Grazulevicius JV, Lygaitis R, Grigalevicius S, et al. (2012) Structure properties relationship of donor-acceptor derivatives of triphenylamine and 1,8-naphthalimide. J Phys Chem C 116: 14811-14819.

33. Pearson RG (1985) Absolute electronegativity and absolute hardness of Lewis acids and bases. J Am Chem Soc107: 6801-6806.

34. Stark MS (1997) Epoxidation of alkenes by peroxyl radicals in the gas phase:? structure? activity relationships. J Phys Chem A 101: 8296-8301.

35. Fores M, Duran M, Sola M, Adamowicz L (1999) Excited-state intramolecular proton transfer and rotamerism of 2-(2'-hydroxyvinyl)benzimidazole and 2-(2'-hydroxyphenyl)imidazole. J Phys Chem A 103: 4413-4420.

36. Schleyer P, Von R, Allinger NL, Clark T, Gasteiger J, et al. (1998) Chichester UK.

37. Marcus RA (1993) Electron transfer reactions in chemistry theory and experiment. Rev Mod Phys 65: 599-610.

38. Marcus RA (1964) Chemical and electrochemical electron-transfer theory. Annu Rev Phys Chem 15: 155-196.

39. Lemaur V, Steel M, Beljonne D, Bredas JL, Cornil J (2005) Photoinduced charge generation and recombination dynamics in model donor /acceptor pairs for organic solar cell applications:? a full quantum-chemical treatment. J Am Chem Soc 127: 6077-6076.

40. Hutchison GR1, Ratner MA, Marks TJ (2005) Hopping transport in conductive heterocyclic oligomers: reorganization energies and substituent effects. See comment in PubMed Commons below J Am Chem Soc 127: 2339-2350.

41. Martinelli NG, Ide J, Sánchez-Carrera RS, Coropceanu V, Bredas JL, et al. (2010) Influence of structural dynamics on polarization energies in anthracene single crystals. J Phys Chem C 114: 20678-20685.

42. McMahon DP, Trois A (2010) Evaluation of the external reorganization energy of polyacenes. J Phys Chem Lett 1: 941-946.

43. Köse ME1, Long H, Kim K, Graf P, Ginley D (2010) Charge transport simulations in conjugated dendrimers. J Phys Chem A 114: 4388-4393.

44. Sakanoue K, Motoda M, Sugimoto M, Sakaki S (1999) A molecular orbital study on the hole transport property of organic amine compounds. J Phys Chem A 103: 5551-5556.

45. Köse ME, Mitchell WJ, Kopidakis N, Chang CH, Shaheen SE, et al. (2007) Theoretical studies on conjugated phenyl-cored thiophene dendrimers for photovoltaic applications. J Am Chem Soc 129: 14257-14270.

46. Lin B, Cheng CP, You ZQ, Hsu CP (2005) Charge transport properties of tris(8-hydroxyquinolinato)aluminum(III): why it is an electron transporter. J Am Chem Soc 127: 66-67.

47. Gruhn NE, da Silva Filho DA, Bill TG, Malagoli M, Coropceanu V, et al. (2002) The vibrational reorganization energy in pentacene: molecular influences on charge transport. J Am Chem Soc 124: 7918-7919.

# 27

# Palladium-Catalyzed Regioselective 2-Carbethoxyethylation of 1H – Indoles By C-H Activation: One-Step Synthesis of Ethyl 2-(1H-Indol-2-Yl) Acetates

**Yi Zhang[1,2], Jiang Wang[2], Yong Nian[1,2], Hongbin Sun[1] and Hong Liu[1,2]\***

[1]State Key Laboratory of Natural Medicines, China Pharmaceutical University, 24 Tongjiaxiang, Nanjing 210009, P.R. China
[2]CAS Key Laboratory of Receptor Research, Shanghai Institute of Materia Medica, Chinese Academy of Sciences, 555 Zuchongzhi Road, Shanghai 201203, P. R. China

**Abstract**

An efficient and convenient method was developed for the one-step synthesis of various substituted ethyl 2-(1H-indol-2-yl) acetates via a palladium-catalyzed regioselective cascade C-H activation reaction. Importantly, this practical approach can be carried out with readily accessible starting materials and exhibits excellent functional group compatibility.

**Keywords:** Ethyl 2-(1H-Indol-2-yl) Acetates; C-H Activation; Palladium; Norbornene

## Introduction

2-substituted indoles are important structural motifs present in diverse biologically active molecules [1] and are precursors for a wide variety of alkaloids, such as vindoline [2], vindorosine [2], ellipticine [3], etc. Among them, 2-(1H-indol-2-yl)acetic acids constitute a valuable class of building blocks for natural product and natural product analogue syntheses (Figure 1) [4-7], combinatorial [8], diversity-oriented syntheses [9,10], and medicinal chemistry [11-18]. In addition, they can serve as highly attractive precursors for various chemical transformations, such as diazomethylation to 1-diazo-3-(2-indolyl)-2-propanone [19] and reduction to 2-(2-hydroxyethyl) indole [20]. Therefore, the development of efficient synthetic methods for these compounds has received much attention.

Hydrolysis of ethyl 2-(1H-indol-2-yl) acetates is an important route to 2-(1H-indol-2-yl) acetic acids. Accordingly, the literature describes several preparations of ethyl 2-(1H-indol-2-yl) acetates. For example, Capuano et al. demonstrated that the intramolecular Wittig reaction of 2-[(ω-alkoxycarbonylacyl) amino] benzyltri-phenylphosphonium salts produced 2-(1H-indol-2-yl)-acetate with a 78% yield [21]. Furthermore, Moody et al. reported that reductive cyclisation of the ethyl 4-(2-nitropheny)-acetoacetate using titanium(III) chloride in aqueous acetone gave 2-(1H-indol-2-yl) acetate with a 75% yield [22]. Despite producing good yields, these methods suffer from indispensable multi-step pre-transformations of commercially available starting materials, and are therefore of limited synthetic scope [23]. Wilkens et al. described a one-pot reaction of N-phenylhydroxylamine, benzaldehyde and ethyl 2,3-butadienoate followed by hydrolysis-mediated production of 2-(1H-indol-2-yl)acetate [24]. In spite of a 49% yield, this approach also had its drawbacks, including the use of unstable and relatively rare reactant (i.e. allenes) that might limit their broad applications. Osornio YM et al. and Guerrero MA et al. demonstrated that the direct intermolecular oxidative radical alkylation of indole under xanthate-mediated radical conditions afforded 2-(1H-indol-2-yl) acetate in 60% yield [25,26]. Nevertheless, the radical mechanism of this approach might limit its broad synthetic applications.

If indoles can be directly functionalized at the C2-position, the preparation of the target compound class would be more facile. Owing to the development of the transition-metal catalyzed C-H activation

chemistry, methods for regioselective direct C2/C3-alkenylation [27-33], alkynylation [34-41], cyano [42,43] and arylation [44-52] of indole nucleus have been well developed to date. However, regioselective C2-alkylation of indole is more challenging.

The Catellani reaction, a palladium-catalyzed norbornene-mediated cascade reaction, has been modified to achieve the direct functionalization of indoles [53-55]. Recently, Bressy et al. reported an intramolecular direct arylation of indoles based on modified Catellani conditions [56]. More recently, Jiao et al. reported a direct 2-alkylation reaction of indoles relying on a Pd(II)-catalyzed norbornene-mediated direct alkylation method for indoles, which regioselectively installs an alkyl group to the C2-position of free N-H indoles [57,58]. However, the original publication did not report the use of 2-bromoacetate as an alkylating reagent. To the best of our knowledge, a straight-forward approach for direct C2- alkoxycarbonylalkylation of indoles has not yet been well established. Given its high C2-regioselectivity and excellent

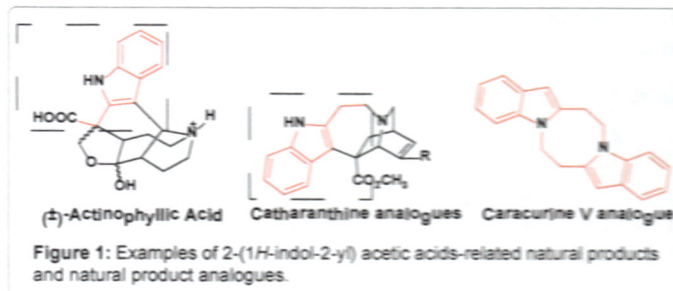

**Figure 1:** Examples of 2-(1H-indol-2-yl) acetic acids-related natural products and natural product analogues.

**\*Corresponding author:** Hong Liu, State Key Laboratory of Natural Medicines, China Pharmaceutical University, 24 Tongjiaxiang, Nanjing 210009, P.R. China, E-mail: hbsun2000@yahoo.com; hliu@mail.shcnc.ac.cn

functional group tolerance, we envisioned to apply Bach's method to the synthesis of 2-(1H-indol-2-yl) acetates by employing indoles and ethyl 2-bromoacetate as the starting materials. As part of our continuing effort to assemble indole-based drug scaffolds [59-61], we here present our findings on one-step synthesis of various substituted ethyl 2-(1H-indol-2-yl)acetates via a palladium-catalyzed regioselective cascade C-H activation reaction. Various substituents are tolerated in this system in moderate to good yields. Our protocol highlights a facile one-step transformation from easily available starting material and excellent functional group compatibility.

## Experimental

Unless otherwise noted, the reagents (chemicals) were purchased from commercial sources, and used without further purification. Water was deionized before used. Analytical thin layer chromatography (TLC) was HSGF 254 (0.15-0.2 mm thickness). Compound spots were visualized by UV light (254 nm). Column chromatography was performed on silica gel FCP 200-300. NMR spectra were run on 300 or 400 MHz instrument. Chemical shifts were reported in parts per million (ppm, $\delta$) downfield from tetramethylsilane. Proton coupling patterns are described as singlet (s), doublet (d), triplet (t), quartet (q), multiplet (m), and broad (br). Low- and high-resolution mass spectra (LRMS and HRMS) were measured on spectrometer.

### General procedure for synthesis of Ethyl 2-(1H-indol-2-yl) acetates

A vial equipped with a magnetic stir bar and a rubber stopper was charged with Pd(PhCN)$_2$Cl$_2$ (65 mg, 0.171 mmol), indole substrate (1.71 mmol), norbornene (321 mg, 3.41 mmol), NaHCO$_3$ (574 mg, 6.84 mmol) and capped with septa. The vial was evacuated and backfilled with argon and the process was repeated three times. A solution of water in DMF (0.5 M) was added via syringe as the solvent. Under argon, ethyl bromoacetate (0.41 mL, 3.41 mmol) was added via syringe, and then the resulting mixture was stirred at room temperature for 10 minutes. After that, the reaction mixture was then placed in a preheated oil bath at 70°C for appropriate time and vigorous stirring was applied. The reaction was monitored by TLC. Upon completion, the reaction mixture was cooled to room temperature, diluted with ethyl acetate, washed with water (twice) and brine (once), dried over Na$_2$SO$_4$, and concentrated. The residue was directly submitted to flash column chromatography (by dry loading) to afford the ethyl 2-(1H-indol-2-yl) acetates products as yellow oils.

## Results and Discussion

We initiated our study by reacting indole with ethyl bromoacetate in the presence of norbornene and Pd(MeCN)$_2$Cl$_2$. However, the direct application of the reported optimized reaction conditions yielded diethyl 2,2'-(1H-indole-2,3-diyl) diacetate, a 2,3-disubstitued-indole product, as the major product.

According to the postulated catalytic cycle of Bach's approach [57,58], and distincts from the previous Catellani reaction, the arylpalladium(II) species in the terminating step undergoes proto-depalladation (i.e. hydrogenolysis), instead of a further palladium catalyzed carbon functionalization, such as Heck, Suzuki and Sonagahira reactions. Consequently, we speculated that for more active alkyl halide substrates, such as ethyl bromoacetate, it might be possible to quench the further substitution reaction and improve the yields of mono-2-substitued-indole products by the use of different

combination of solvent and base. Furthermore, we postulated that the phosphine ligands might have an effect on the rate of the substitution process as well.

With the aforementioned considerations in mind, we screened different reaction conditions for optimal results (Table 1). We conducted the first runs (Table 1, entries 1-6) employing the reported optimized Pd source [Pd(OAc)$_2$], base (K$_2$CO$_3$), reaction temperature (70°C) and time (14 hours) in different solvents (Table 1, entries 1-3). Notably, the higher water content in the solvents was found to significantly increase the selectivity for the mono-2-substitued-indole product 1 (Table 1), but led to a compromised conversion rate. These results suggest that water might serve as the hydrogen source in the catalytic system, which accelerates the norbornene-mediated cascade reaction by hydrolysing the final arylpalladium(II) species. As DMF/5 M water produced a better yield than the other solvents, it was selected as the solvent in the following tests. The effect of the Pd source was then investigated (Table 1, entries 7-10). Palladium species other than Pd(MeCN)$_2$Cl$_2$ and Pd(PhCN)$_2$Cl$_2$ were substantially less effective. Further screening of catalysts did not lead to better yields and demonstrated that the reaction may not proceed well in the presence of phosphine ligands which may reduce the palladium(II) species to palladium(0) species and interferes the normal palladium(II-IV) catalytic cycle [62]. Pd(PhCN)$_2$Cl$_2$ was chosen as the preferred catalyst because it provided higher conversion rates than the other catalysts. Given the possibility that excessive water in the solvent might hydrolyse the arylpalladium(II) species formed in the first oxidative addition step and lower the solubility of the intermediates, resulting in a cessation of the catalytic cycle and undesired conversion rate, we switched the solvent back to DMF with 0.5M H$_2$O. Finally, a brief screen of bases showed an interesting correlation between the alkalinity of the base and the yield of the 2,3-disubstitued-indole byproduct 2 (Table 1, entries

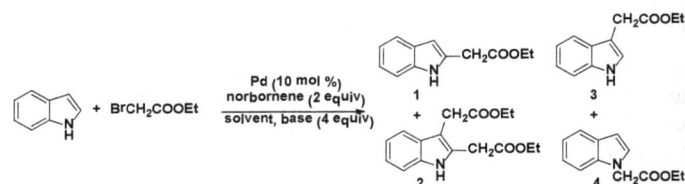

| Entry | Pd Source | Solvent | Base | Yield (%)[b] | |
|-------|-----------|---------|------|------|------|
| | | | | 1 | 2 |
| 1 | Pd(OAc)$_2$ | DMA / 0.5 M H$_2$O | K$_2$CO$_3$ | trace[c] | 56 |
| 2 | Pd(OAc)$_2$ | DMF / 0.5 M H$_2$O | K$_2$CO$_3$ | 7 | 62 |
| 3 | Pd(OAc)$_2$ | DMA / 5 M H$_2$O | K$_2$CO$_3$ | 19 | 26 |
| 4 | Pd(OAc)$_2$ | DMF / 5 M H$_2$O | K$_2$CO$_3$ | 34 | 0 |
| 5 | Pd(OAc)$_2$ | DMF / 2 M H$_2$O | K$_2$CO$_3$ | 24 | 10 |
| 6 | Pd(OAc)$_2$ | CH$_3$CN / 5 M H$_2$O | K$_2$CO$_3$ | 15 | 21 |
| 7 | Pd(Ph$_3$P)$_2$Cl$_2$ | DMF / 5 M H$_2$O | K$_2$CO$_3$ | trace[c] | 0 |
| 8 | Pd(dppf)$_2$Cl$_2$·CH$_2$Cl$_2$ | DMF / 5 M H$_2$O | K$_2$CO$_3$ | trace[c] | 0 |
| 9 | Pd(MeCN)$_2$Cl$_2$ | DMF / 5 M H$_2$O | K$_2$CO$_3$ | 46 | 0 |
| 10 | Pd(PhCN)$_2$Cl$_2$ | DMF / 5 M H$_2$O | K$_2$CO$_3$ | 52 | 0 |
| 11 | Pd(PhCN)$_2$Cl$_2$ | DMF / 0.5 M H$_2$O | K$_2$CO$_3$ | 8 | 79 |
| 12 | Pd(PhCN)$_2$Cl$_2$ | DMF / 0.5 M H$_2$O | KHCO$_3$ | trace[c] | 67 |
| 13 | Pd(PhCN)$_2$Cl$_2$ | DMF / 0.5 M H$_2$O | NaHCO$_3$ | 85 | 7 |
| 14 | Pd(PhCN)$_2$Cl$_2$ | DMF / 0.5 M H$_2$O | KOAc | 58 | 0 |
| 15 | Pd(PhCN)$_2$Cl$_2$ | DMF / 0.5 M H$_2$O | NaOAc | 42 | 0 |

[a]General reaction conditions: Pd (0.171 mmol), indole (1.71 mmol), norbornene (3.41 mmol), base (6.84 mmol), ethyl bromoacetate (3.41 mmol), solvent (8 mL, total), 70 °C, 14 h. [b]Isolated yield. [c]Monitored by LC-MS and TLC

**Table 1:** Optimization of the catalysis conditions[a].

11-15). Specifically, the yields of the 2,3-disubstitued-indole products increased significantly as the alkalinity of the base increased, with NaHCO$_3$ affording the target products in desirable yields. On the other hand, insufficient basicity led to undesired conversion rates (Table 1, entries 14 and 15). Thus, DMF with 0.5M H$_2$O as the solvent, NaHCO$_3$ as the base, a reaction temperature of 70˚C, and a reaction time of 14 hours were selected as the optimal conditions. It was noteworthy that in all tests, the other possible substituted byproducts, N- and 3-substituted products 3, 4 (Table 1) were not detected.

Encouraged by these initial results, we next proceeded to examine the general utility of the Pd(PhCN)$_2$Cl$_2$/norbornene catalytic system for synthesis of a wide range of ethyl 2-(1H-indol-2-yl) acetates (Table 2). Indoles with electron-donating and electron-drawing substituents at the various positions smoothly participated in the C2-ethoxycarbonylmethylation reaction. Four regioisomeric methylindoles effectively participated in the cascade reactions to afford equally good levels of yields (Table 2, entries 1-4). The reaction worked well with different halogens on various positions on the benzene ring

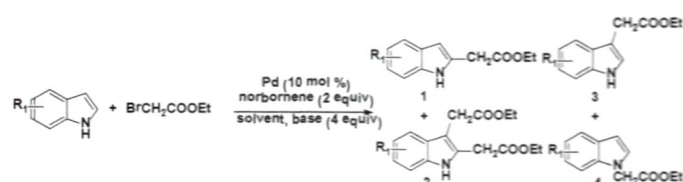

Scheme 1: Postulated Catalytic Cycle for a Direct 2-alkoxycarbonyl-alkylation of Indole by a Norbornene-Mediated Cascade C-H Activation.

(Table 2, entries 6-12), which has three main advantages. First, halogen substituents could be used to adjust the polarity, lipophilicity, and metabolic stability of dyes or pharmaceuticals. Second, halogens allow further modification using cross-coupling reactions for the elaboration of molecule libraries. Third, halogen substituents on various positions can serve as invaluable building blocks for enrichment of Structure-Activity Relationships in medicinal chemistry. It was noteworthy to mention that 5-iodoindole, a potential substrate for classic Catellani reaction, was readily tolerated in our catalytic system and gave the desired product 6l in good yield. Interestingly, electron-deficient indoles reacted better than their electron-rich counterparts to give good yields of 2-(1H-indol-2-yl) acetates under the optimal conditions (Table 2, entries 6-15). Moreover, the reaction could be extended to functionalized azaindoles to give the expected product 6o in modest yield. We also noted that a more bulky 3-position blocked the substrate, 3-methyl-indole, produced the desired product 3 in a 35% yield (Table 2, entry 16). Control experiment showed that this catalytic system did not work in the absence of norbornene. By contrast, N-methylindole failed to participate in the reaction and remained mostly unchanged even after extension of reaction time (Table 2, entry 17). These results are consistent with Bach and Jiao's latest work regarding the mechanism of regioselective Pd-catalyzed, norbornene-mediated alkylation of indoles. It has been proved that N1-norbornene type palladacycle rather than originally proposed C3-norbornene type palladacycle is formed as the key intermediate in this catalytic cycle [58].

On the basis of above observations, recent publications on the mechanism of the Catellani reaction [63] and extensive mechanistic work by Jiao et al. [58], a plausible mechanism for the observed reaction is proposed in Scheme 1. The norbornene-mediated cascade C-H activation process proceeds as follows: a. N1-position direct palladation; b. syn-aminopalladation of norbornene; c. irreversible palladacycle formation, leading to C2-position ortho-C–H palladation; d. oxidative addition with an alkyl halide to generate palladium IV species; e. reductive elimination of the palladium IV species; f. norbornene expulsion; and g. release of the 2-alkyl indole product and

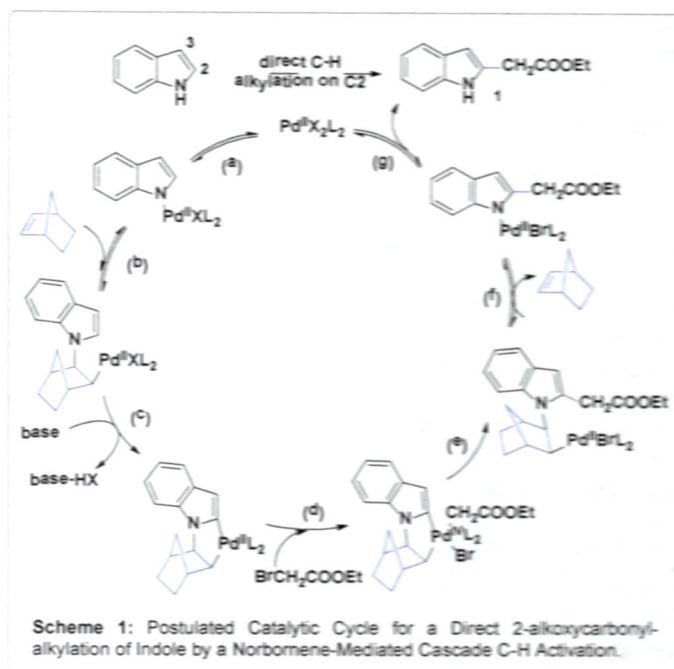

| Entry | R$_1$ | Product | Yield[b] |
|-------|-------|---------|----------|
| 1 | 4-Me (5a) | 6a | 67% |
| 2 | 5-Me (5b) | 6b | 71% |
| 3 | 6-Me (5c) | 6c | 69% |
| 4 | 7-Me (5d) | 6d | 65% |
| 5 | 5-OMe (5e) | 6e | 80% |
| 6 | 4-F (5f) | 6f | 75% |
| 7 | 5-F (5g) | 6g | 83% |
| 8 | 6-F (5h) | 6h | 82% |
| 9 | 7-F (5i) | 6i | 78% |
| 10 | 5-Cl (5j) | 6j | 85% |
| 11 | 5-Br (5k) | 6k | 84% |
| 12 | 5-I (5l) | 6l | 71% |
| 13 | 5-CN (5m) | 6m | 89% |
| 14 | 5-NO$_2$ (5n) | 6n | 91% |
| 15 | (5o) | 6o | 58% |
| 16 | (5p) | 6p | 35% |
| 17 | (5q) | 6q | trace[c] |

[a]Reaction conditions: Pd(PhCN)$_2$Cl$_2$ (0.171 mol), indole (1.71 mmol), norbornene (3.41 mmol), NaHCO$_3$ (6.84 mmol), ethyl bromoacetate (3.41 mmol), DMF with 0.5 M H$_2$O (8 mL, total), 70 °C, 14 h. [b]Isolated yield. [c]Monitored by LC-MS

Table 2: Palladium-catalyzed synthesis of various ethyl 2-(1H-indol-2-yl) acetates[a].

regeneration the Pd(II) species (Scheme 1). The 2-monosubstituted product of this catalytic cycle might undergo further electrophilic substitution process [56] to afford the 2,3-disubstitued-indole product.

## Conclusion

To conclude, a practical Pd(PhCN)$_2$Cl$_2$/norbornene catalytic system for the versatile and one-step synthesis of ethyl 2-(1$H$-indol-2-yl)acetates has been developed. This practical method features the ability to rapidly and efficiently synthesize various substituted ethyl 2-(1$H$-indol-2-yl)acetates as precursors to a class of valuable synthetic building blocks, 2-(1$H$-indol-2-yl)acetic acids. In addition, the protocol exhibits excellent functional group compatibility, leading to valuable derivatives that are not readily available by conventional methods. The present reaction also enables a new access to 2,3-dialkoxycarbonylalkylated indole derivatives which can be applied to indole fused heterocycles and indole alkaloid synthesis.

### Acknowledgements

We gratefully acknowledge financial support from the National Natural Science Foundation of China (Grants 21021063, 91229204, and 81025017), National S&T Major Projects (2012ZX09103101-072, 2012ZX09301001-005, and 2013ZX09507-001), Sponsored by Program of Shanghai Subject Chief Scientist (Grant 12XD1407100).

### References

1. Katritzky AR, Li J, Stevens CV (1995) Facile Synthesis of 2-Substituted Indoles and Indolo[3,2-b]carbazoles from 2-(Benzotriazol-1-ylmethyl)indole. J Org Chem 60: 3401-3404.

2. Kuehne ME, Podhorez DE, Mulamba T, Bornmann WG (1987) Biomimetic alkaloid syntheses. 15. Enantioselective syntheses with epichlorohydrin: total syntheses of (+)-, (-)- and (.+-.)- vindoline and a synthesis of (-)-vindorosine. J Org Chem 52: 347-353.

3. Modi SP, Zayed AH, Archer S (1989) Synthesis of 6-substituted 7H-pyrido[4,3-c]carbazoles. J Org Chem 54: 3084-3087.

4. Martin CL, Overman LE, Rohde JM (2008) Total synthesis of (+/-)-actinophyllic acid. J Am Chem Soc 130: 7568-7569.

5. Sundberg RJ, Hong J, Smith SQ, Sabat M, Tabakovic I (1998) Synthesis and Oxidative Fragmentation of Catharanthine Analogs. Comparison to the Fragmentation - Coupling of Catharanthine and Vindoline. Tetrahedron 54: 6259-6292.

6. Gul W, Hammond NL, Yousaf M, Peng J, Holley A, et al. (2007) Chemical transformation and biological studies of marine sesquiterpene (S)-(+)-curcuphenol and its analogs. Biochim Biophys Acta 1770: 1513-1519.

7. Sripha K, Zlotos DP, Holzgrabe U, Ruchirawat S (2011) A New Synthetic Approach to Pentacyclic Ring, 6,7,14,15-Tetrahydro[1,5]diazocino[1,2-a:6,5-a']-8-dihydrodiindole. HETEROCYCLES 83: 2627-2633.

8. Boger DL, Goldberg J, Satoh S, Ambroise Y, Cohen SB, et al. (2000) Non-Amide-Based Combinatorial Libraries Derived from N-Boc-Iminodiacetic Acid: Solution-Phase Synthesis of Piperazinone Libraries with Activity Against LEF-1/ß-Catenin-Mediated Transcription. Helv Chim Acta 83: 1825-1845.

9. Royer D, Wong YS, Plé S, Chiaroni A, Diker K, et al. (2008) Diastereodivergence and appendage diversity in the multicomponent synthesis of aryl-pyrrolo-tetrahydrocarbazoles. Tetrahedron 64: 9607-9618.

10. Waldmann H, Kühn M, Liu W, Kumar K (2008) Reagent-controlled domino synthesis of skeletally-diverse compound collections. Chem Commun (Camb) : 1211-1213.

11. Méndez-Andino J, Colson AO, Denton D, Mitchell MC, Cross-Doersen D, et al. (2007) MCH-R1 antagonists based on an arginine scaffold: SAR studies on the amino-terminus. Bioorg Med Chem Lett 17: 832-835.

12. Potter AJ, Ray S, Gueritz L, Nunns CL, Bryant CJ, et al. (2010) Structure-guided design of alpha-amino acid-derived Pin1 inhibitors. Bioorg Med Chem Lett 20: 586-590.

13. Weber D, Berger C, Eickelmann P, Antel J, Kessler H (2003) Design of selective peptidomimetic agonists for the human orphan receptor BRS-3. J Med Chem 46: 1918-1930.

14. Harikrishnan LS, Srivastava N, Kayser LE, Nirschl DS, Kumaragurubaran K, et al. (2012) Identification and optimization of small molecule antagonists of vasoactive intestinal peptide receptor-1 (VIPR1). Bioorg Med Chem Lett 22: 2287-2290.

15. Yang K, Wang Q, Su L, Fang H, Wang X, et al. (2009) Design and synthesis of novel chloramphenicol amine derivatives as potent aminopeptidase N (APN/CD13) inhibitors. Bioorg Med Chem 17: 3810-3817.

16. Rode HB, Sos ML, Grütter C, Heynck S, Simard JR, et al. (2011) Synthesis and biological evaluation of 7-substituted-1-(3-bromophenylamino)isoquinoline-4-carbonitriles as inhibitors of myosin light chain kinase and epidermal growth factor receptor. Bioorg Med Chem 19: 429-439.

17. McInnis CE, Blackwell HE (2011) Thiolactone modulators of quorum sensing revealed through library design and screening. Bioorg Med Chem 19: 4820-4828.

18. Meng Q, Zhao B, Xu Q, Xu X, Deng G, et al. (2012) Indole-propionic acid derivatives as potent, S1P3-sparing and EAE efficacious sphingosine-1-phosphate 1 (S1P1) receptor agonists. Bioorg Med Chem Lett 22: 2794-2797.

19. Salim M, Capretta A (2000) Intramolecular Carbenoid Insertions: the Reactions of a-Diazoketones Derived from Pyrrolyl and Indolyl Carboxylic Acids with Rhodium(II) Acetate. Tetrahedron 56: 8063-8069.

20. Sripha K, Zlotos DP, Buller S, Mohr K (2003) 6,7,14,15-Tetrahydro-15aH-azocino[1,2-a:6,5-b']diindole. Synthesis of a novel pentacyclic ring system. Tetrahedron Lett 44: 7183-7186.

21. Capuano L, Ahlhelm A, Hartmann H (1986) New Syntheses of 2-Acylbenzofurans, 2-Acylindoles, 2-Indolylcarboxylates, and 2-Quinolones by Intramolecular Wittig Reaction. Chem Ber 119: 2069-2074.

22. Moody CJ, Rahimtoola KF (1990) Diels–Alder reactivity of pyrano[4,3-b]indol-3-ones, indole 2,3-quinodimethane analogues. J Chem Soc, Perkin Trans 1 673-679.

23. Attia MI, Guclu D, Hertlein B, Julius JP, Witt-Enderby A, et al. (2007) Synthesis, NMR conformational analysis and pharmacological evaluation of 7,7a,13,14-tetrahydro-6H-cyclobuta[b]pyrimido[1,2-a : 3,4-a ']diindole analogues as melatonin receptor ligands. Organic & Biomolecular Chemistry 5: 2129-2137.

24. Wilkens J, Kühling A, Blechert S (1987) Hetero-cope rearrangements - vl1 short and stereoselective syntheses of 2-vinylindoles by a tandem-process. Tetrahedron: 43, 3237-3246.

25. Osornio YM, Cruz-Almanza R, Jiménez-Montaño V, Miranda LD (2003) Efficient, intermolecular, oxidative radical alkylation of heteroaromatic systems under "tin-free" conditions. Chem Commun (Camb) : 2316-2317.

26. Guerrero MA, Miranda LD (2006) Et3B-Mediated radical alkylation of pyrroles and indoles. Tetrahedron Letters 47: 2517-2520.

27. Grimster NP, Gauntlett C, Godfrey CR, Gaunt MJ (2005) Palladium-catalyzed intermolecular alkenylation of indoles by solvent-controlled regioselective C-H functionalization. Angew Chem Int Ed Engl 44: 3125-3129.

28. García-Rubia A, Gómez Arrayás R, Carretero JC (2009) Palladium(II)-catalyzed regioselective direct C2 alkenylation of indoles and pyrroles assisted by the N-(2-pyridyl)sulfonyl protecting group. Angew Chem Int Ed Engl 48: 6511-6515.

29. Ding Z, Yoshikai N (2012) Mild and efficient C2-alkenylation of indoles with alkynes catalyzed by a cobalt complex. Angew Chem Int Ed Engl 51: 4698-4701.

30. Kandukuri SR, Schiffner JA, Oestreich M (2012) Aerobic palladium(II)-catalyzed 5-endo-trig cyclization: an entry into the diastereoselective C-2 alkenylation of indoles with tri- and tetrasubstituted double bonds. Angew Chem Int Ed Engl 51: 1265-1269.

31. Baran PS, Richter JM (2004) Direct coupling of indoles with carbonyl compounds: short, enantioselective, gram-scale synthetic entry into the hapalindole and fischerindole alkaloid families. J Am Chem Soc 126: 7450-7451.

32. Pan S, Ryu N, Shibata T (2012) Ir(I)-catalyzed C-H bond alkylation of C2-position of indole with alkenes: selective synthesis of linear or branched 2-alkylindoles. J Am Chem Soc 134: 17474-17477.

33. Byers JH, Campbell JE, Knapp FH, Thissell JG (1999) Radical aromatic substitution via atom-transfer addition. Tetrahedron Lett 40: 2677-2680.

34. Gu Y, Wang X (2009) Direct palladium-catalyzed C-3 alkynylation of indoles. Tetrahedron Lett 50: 763-766.

35. Brand JP, Charpentier J, Waser J (2009) Direct alkynylation of indole and pyrrole heterocycles. Angew Chem Int Ed Engl 48: 9346-9349.

36. Brand JP, Chevalley C, Waser J (2011) One-pot gold-catalyzed synthesis of 3-silylethynyl indoles from unprotected o-alkynylanilines. Beilstein J Org Chem 7: 565-569.

37. Brand JP, Chevalley C, Scopelliti R, Waser J (2012) Ethynyl benziodoxolones for the direct alkynylation of heterocycles: structural requirement, improved procedure for pyrroles, and insights into the mechanism. Chemistry 18: 5655-5666.

38. de Haro T, Nevado C (2010) Gold-catalyzed ethynylation of arenes. J Am Chem Soc 132: 1512-1513.

39. Yang L, Zhao L, Li CJ (2010) Palladium-catalyzed direct oxidative Heck-Cassar-Sonogashira type alkynylation of indoles with alkynes under oxygen. Chem Commun (Camb) 46: 4184-4186.

40. Ruano JL, Alemán J, Marzo L, Alvarado C, Tortosa M, et al. (2012) Arylsulfonylacetylenes as alkynylating reagents of Csp2-H bonds activated with lithium bases. Angew Chem Int Ed Engl 51: 2712-2716.

41. García Ruano JL, Alemán J, Marzo L, Alvarado C, Tortosa M, et al. (2012) Expanding the scope of arylsulfonylacetylenes as alkynylating reagents and mechanistic insights in the formation of Csp2-Csp and Csp3-Csp bonds from organolithiums. Chemistry 18: 8414-8422.

42. Yan G, Kuang C, Zhang Y, Wang J (2010) Palladium-catalyzed direct cyanation of indoles with K(4)[Fe(CN)(6)]. Org Lett 12: 1052-1055.

43. Ghosh K, Sarkar AR, Samadder A, Khuda-Bukhsh AR (2012) Pyridinium-based fluororeceptors as practical chemosensors for hydrogen pyrophosphate (HP2O7(3-)) in semiaqueous and aqueous environments. Org Lett 14: 4314-4317.

44. Lane BS, Sames D (2004) Direct C-H bond arylation: selective palladium-catalyzed C2-arylation of N-substituted indoles. Org Lett 6: 2897-2900.

45. Wang X, Lane BS, Sames D (2005) Direct C-arylation of free (NH)-indoles and pyrroles catalyzed by Ar-Rh(III) complexes assembled in situ. J Am Chem Soc 127: 4996-4997.

46. Lane BS, Brown MA, Sames D (2005) Direct palladium-catalyzed C-2 and C-3 arylation of indoles: a mechanistic rationale for regioselectivity. J Am Chem Soc 127: 8050-8057.

47. Deprez NR, Kalyani D, Krause A, Sanford MS (2006) Room temperature palladium-catalyzed 2-arylation of indoles. J Am Chem Soc 128: 4972-4973.

48. Stuart DR, Fagnou K (2007) The catalytic cross-coupling of unactivated arenes. Science 316: 1172-1175.

49. Lebrasseur N, Larrosa I (2008) Room temperature and phosphine free palladium catalyzed direct C-2 arylation of indoles. J Am Chem Soc 130: 2926-2927.

50. Phipps RJ, Grimster NP, Gaunt MJ (2008) Cu(II)-catalyzed direct and site-selective arylation of indoles under mild conditions. J Am Chem Soc 130: 8172-8174.

51. Cornella J, Lu P, Larrosa I (2009) Intermolecular decarboxylative direct C-3 arylation of indoles with benzoic acids. Org Lett 11: 5506-5509.

52. Potavathri S, Pereira KC, Gorelsky SI, Pike A, LeBris AP, et al. (2010) Regioselective oxidative arylation of indoles bearing N-alkyl protecting groups: dual C-H functionalization via a concerted metalation-deprotonation mechanism. J Am Chem Soc 132: 14676-14681.

53. Catellani M (2008) Novel methods of aromatic functionalization using palladium and norbornene as a unique catalytic system. Top Organomet Chem 14: 21-53.

54. Catellani M, Motti E, Della Ca' N (2008) Catalytic sequential reactions involving palladacycle-directed aryl coupling steps. Acc Chem Res 41: 1512-1522.

55. Martins A, Mariampillai B, Lautens M (2010) Synthesis in the key of Catellani: norbornene-mediated ortho C-H functionalization. Top Curr Chem 292: 1-33.

56. Bressy C, Alberico D, Lautens M (2005) A route to annulated indoles via a palladium-catalyzed tandem alkylation/direct arylation reaction. J Am Chem Soc 127: 13148-13149.

57. Jiao L, Bach T (2011) Palladium-catalyzed direct 2-alkylation of indoles by norbornene-mediated regioselective cascade C-H activation. J Am Chem Soc 133: 12990-12993.

58. Jiao L, Herdtweck E, Bach T (2012) Pd(II)-catalyzed regioselective 2-alkylation of indoles via a norbornene-mediated C-H activation: mechanism and applications. J Am Chem Soc 134: 14563-14572.

59. Ye D, Wang J, Zhang X, Zhou Y, Ding X, et al. (2009) Gold-catalyzed intramolecular hydroamination of terminal alkynes in aqueous media: efficient and regioselective synthesis of indole-1-carboxamides. Green Chem 11: 1201-1208.

60. Wang J, Zhou S, Lin D, Ding X, Jiang H, et al. (2011) Highly diastereo- and enantioselective synthesis of syn-ß-substituted tryptophans via asymmetric Michael addition of a chiral equivalent of nucleophilic glycine and sulfonylindoles. Chem Commun 47: 8355-8357.

61. Feng E, Zhou Y, Zhao F, Chen X, Zhang L, et al. (2012) Gold-catalyzed tandem reaction in water: an efficient and convenient synthesis of fused polycyclic indoles. Green Chem 14: 1888-1895.

62. Tolnai GL, Ganss S, Brand JP, Waser J (2013) C2-selective direct alkynylation of indoles. Org Lett 15: 112-115.

63. Maestri G, Motti E, Della Ca' N, Malacria M, Derat E, et al. (2011) Of the ortho effect in palladium/norbornene-catalyzed reactions: a theoretical investigation. J Am Chem Soc 133: 8574-8585.

# 28

# Thermal, Spectroscopic and Chemical Characterization of Biofield Energy Treated Anisole

Mahendra Kumar Trivedi[1], Alice Branton[1], Dahryn Trivedi[1], Gopal Nayak[1], Gunin Saikia[2] and Snehasis Jana[2*]

[1]Trivedi Global Inc., 10624 S Eastern Avenue Suite A-969, Henderson, NV 89052, USA
[2]Trivedi Science Research Laboratory Pvt. Ltd., Bhopal, Madhya Pradesh, India

## Abstract

The objective of the present study was to evaluate the impact of biofield energy treatment on the thermal, spectroscopic, and chemical properties of anisole by various analytical methods such as gas chromatography-mass spectrometry (GC-MS), high performance liquid chromatography (HPLC), differential scanning calorimetry (DSC), Fourier transform infrared (FT-IR) spectroscopy, and ultraviolet-visible (UV-Vis) spectroscopy. The anisole sample was divided into two parts, control and treated. The control part was remained same while the other part was treated with Mr. Trivedi's unique biofield energy treatment. Mass spectra showed the molecular ion peak with five fragmented peaks in control and all treated samples. The isotopic abundance ratio of $^2H/^1H$, and $^{13}C/^{12}C$ [(PM+1)/PM] in treated sample was increased by 154.47% (T1) as compared to the control [where, PM- primary molecule, (PM+1)-isotopic molecule either for $^{13}C$ or $^2H$]. The HPLC chromatogram showed retention time of treated anisole was slightly decreased as compared to the control. Moreover, the heat change in the sharp endothermic transition of treated anisole was increased by 389.07% in DSC thermogram as compared to the control. Further, C-C aromatic stretching frequency of treated sample was shifted by 2 cm$^{-1}$ to low energy region in FT-IR spectroscopy. The UV-Vis spectra of control sample showed characteristic absorption peaks at 325 nm, which was red shifted and appeared as shoulder in the treated sample. These results suggested that biofield treatment has significantly altered the physical and spectroscopic properties of anisole, which could make them stable solvent for organic synthesis and as a suitable reaction intermediate in industrial applications.

**Keywords:** Biofield energy treatment; Anisole; Gas chromatography-Mass spectrometry; High performance liquid chromatography

**Abbreviations:** GC-MS: Gas chromatography-Mass spectrometry; PM: Primary molecule; PM+1: Isotopic molecule either for $^{13}C/^{12}C$ or $^2H/^1H$

## Introduction

Anisole, phenyl ether (Ph-O-Me, Figure 1) usually used as a starting material for various pharmaceutical/flavonoid products and as solvent in organic synthesis and physical studies [1]. It is used for the synthesis of raw material for drugs such as cyclofenil, in treating Raynaud's phenomenon in people with scleroderma [2], and tramadol hydrochloride, that is used for the relief of moderate or severe pain [3]. 4-hydroxyanisole and t-butylhydroxyanisole are the most popular derivatives of anisole and used extensively as depigmenting agents and antioxidant respectively [4,5]. The anisole was used selectively in various steps as a solvent in the synthesis of cefoxitin [6] and latamoxef an antibiotic administered intravenously has a broad spectrum of activity against Gram-positive and Gram-negative bacteria [7]. Anisole is a weakly polar aprotic solvent with lower electric permittivity ($\varepsilon s=4.33$), which allows its usability in cyclic voltammetric studies as an alternative to tetrahydrofuran and dichloromethane [8]. The use of hazardous and toxic solvents in synthesis and characterization is regarded as a very important point for the safety of lab-workers and pollution. Nowadays, a relatively green solvent, anisole, have been successfully applied to process organic/polymer solar cells [9]. Hence, the stability of anisole is important to perform organic reactions at moderate temperature, and its applicability in pharmaceutical products. The chemical and physical property could be altered by Mr. Trivedi's unique biofield energy treatment which is well known to modify the physical, and structural characteristics of living and non-living substances [10,11]. The electrical current generates through internal physiological processes like blood flow, brain activity, and heart function *etc.* exists inside the human body in the form of vibratory energy particles like ions, protons, and electrons and they generate magnetic field in the human body [12]. The energy fields that purportedly surround the human body is called the biofield. Currently, researchers have been exploring the potential benefits of integrative energy medicine in a variety of situations to promote overall health and wellness of individuals. The energy medicine is regarded as one of the complementary and alternative medicine (CAM) and defined under the subcategory of energy therapies by National Center for Complementary and Alternative Medicine (NCCAM) [13]. The practitioner of energy medicine can harness the energy from the environment/universe and can transmit into any object around the globe. The object(s) receive the biofield energy and responded in a useful way. This process is called as biofield energy treatment. Mr. Trivedi's unique biofield treatment is also called as The Trivedi Effect®. The Trivedi Effect® has been well studied in various research fields like microbiology [10,14] biotechnology [15,16] and agricultural research [17]. Based on the outstanding results achieved by biofield treatment on microbiology and biotechnology, an attempt was made to evaluate the impact of biofield treatment on various properties of anisole.

---

***Corresponding author:** Snehasis Jana, Trivedi Science Research Laboratory Pvt. Ltd., Hall-A, Chinar Mega Mall, Chinar Fortune City, Hoshangabad Rd. Bhopal-462 026, Madhya Pradesh, India
E-mail: publication@trivedisrl.com

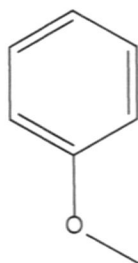

Figure 1: Structure of anisole.

## Experimental

### Materials and methods

Anisole was procured from Genuine Chemical, India. The samples were characterized using, gas chromatography-mass spectrometry (GC-MS), high performance liquid chromatography (HPLC), differential scanning calorimetry (DSC), Fourier transform infrared (FT-IR) spectroscopy, and ultraviolet-visible (UV-Vis) spectroscopy.

### Biofield treatment modalities

Anisole was taken in this experiment for biofield treatment. The compound was divided into two parts named as control and treated. No treatment was given to the control set. The second set of anisole was handed over to Mr. Trivedi for biofield energy treatment under laboratory conditions. Mr. Trivedi provided the biofield treatment through his energy transmission process to second sets of samples without touching. After treatment, the treated samples were stored at standard conditions for GC-MS analysis as per the standard protocol. The experimental results in treated groups were analyzed and compared with the untreated (control) set.

### GC-MS

The GC-MS analysis was done on Perkin Elmer/auto system XL built with Turbo mass, USA. The detection limit of the detector is upto 1 picogram. For GC-MS analysis, the treated sample was further divided into three parts as T1, T2, and T3. The GC-MS spectrum was plotted as the % abundance vs. mass to charge ratio ($m/z$). The isotopic abundance ratio of $^{13}C/^{12}C$ or $^{2}H/^{1}H$, (PM+1)/PM, and $^{18}O/^{16}O$, (PM+2)/PM was expressed by its deviation in the treated sample as compared to the control. The percentage changes in isotopic ratio (PM+1)/PM and (PM+2)/PM was calculated on a percentage scale from the following formula:

$$\text{Percentage changes in isotopic ratio} (PM+1/PM) = \frac{R_{Treated} - R_{Control}}{R_{Control}} \times 100$$

Where, $R_{Treated}$ and $R_{Control}$ are the ratios of intensity at (PM+1) to PM in mass spectra of treated and control samples respectively.

### HPLC

The HPLC analysis was performed on a Knauer High Performance Liquid Chromatograph (Berlin, Germany), equipped with Smartline Pump 1000 and a UV 2600 detector. Chromatographic separation was performed on a $C_{18}$ column (Eurospher 100) with a dimension of 250 × 4 mm and 5 μm particle size. The mobile phase used was methanol with a flow rate of 1 mL/min at 25°C. The solutions of standard and the sample were prepared in methanol for both GC-MS and HPLC studies.

The method development for this assay of anisole based on its chemical properties. Anisole is polar molecule and therefore, a polar solvent methanol was used as the diluent.

### DSC

The DSC was done with Perkin Elmer/Pyris-1, USA, a heating rate of 10°C/min and nitrogen flow of 5 mL/min was used. The change in latent heat (ΔH) of control and treated ethanol was recorded from their respective DSC curves. The percent change in boiling point and latent heat of vaporization was computed using following equations: Percent change was calculated using following equations:

$$\% \text{ change} = \frac{[T_{Treated} - T_{Control}]}{T_{Control}} \times 100$$

Where, $T_{Control}$ and $T_{Treated}$ are the peak point of degradation control and treated samples, respectively.

### FT-IR spectroscopy

For FT-IR spectra, Shimadzu's Fourier transform infrared spectrometer (Japan) was used in the frequency region of 500-4000 cm$^{-1}$. The samples were prepared by crushing anisole crystals with spectroscopic grade KBr into fine powder and then pressed into pellets.

### UV-Vis spectroscopy

UV-Vis spectra of control and treated samples of anisole were acquired from Shimadzu UV-2400 PC series spectrophotometer within the wavelength region of 200-400 nm. Quartz cell with 1 cm and a slit width of 2.0 nm were used for analysis.

## Results and Discussion

### GC-MS analysis

The GC-MS spectra of control and treated samples of anisole are presented in Figure 2. For GC-MS study of the treated sample was divided into three parts T1, T2, and T3. Mass spectra showed the PM peak at $m/z$=108 in control and all the treated anisole samples (T1, T2, and T3) same intensities. The intensity ratio of (PM+1)/PM is presented in Table 1 and percent change in isotopic abundance ratio was calculated and shown in Figure 3. Five major peaks at $m/z$=108, 93, 78, 65, 51, and 39 were observed in both control and treated samples of anisole due to the following ions respectively: $C_7H_8O^+$, $C_6H_5O^+$, $C_6H_6^+$, $C_5H_5^+$, $C_4H_3^+$, and $C_3H_3^+$ ions. Peaks at $m/z$=93, 78, 65, 51, and 39 were observed due to the fragmentation of anisole to phenol, benzene, cyclopentadiene, buten-3-yne and propyne ions. All peaks were same for both treated and control samples and well matched with reported literature [18]. The isotopic abundance ratio of (PM+1)/PM of anisole sample was increased in all the treated samples T1=154.47%, T2=21.71%, and T3=34.44%. However, the isotopic abundance ratio of (PM+2)/PM was not seen in the mass spectrum. The increased isotopic abundance ratio of (PM+1)/PM in the treated anisole may increase the effective mass (μ) and binding energy in this molecules with heavier isotopes. This may alter the property of the chemical bond and eventually property of the molecule. The transformation may be happened in nuclear level due to the biofield energy treatment. It is expected that some of $^{1}H$ and $^{12}C$ may be interconverted to $^{2}H$, and $^{13}C$ atoms inside the molecule, respectively. The bond strength also plays an important role in kinetic effects due to the greater strength of the isotopic bonds such as $^{2}H$-$^{12}C$ bond relative to a $^{1}H$-$^{12}C$ bond [19].

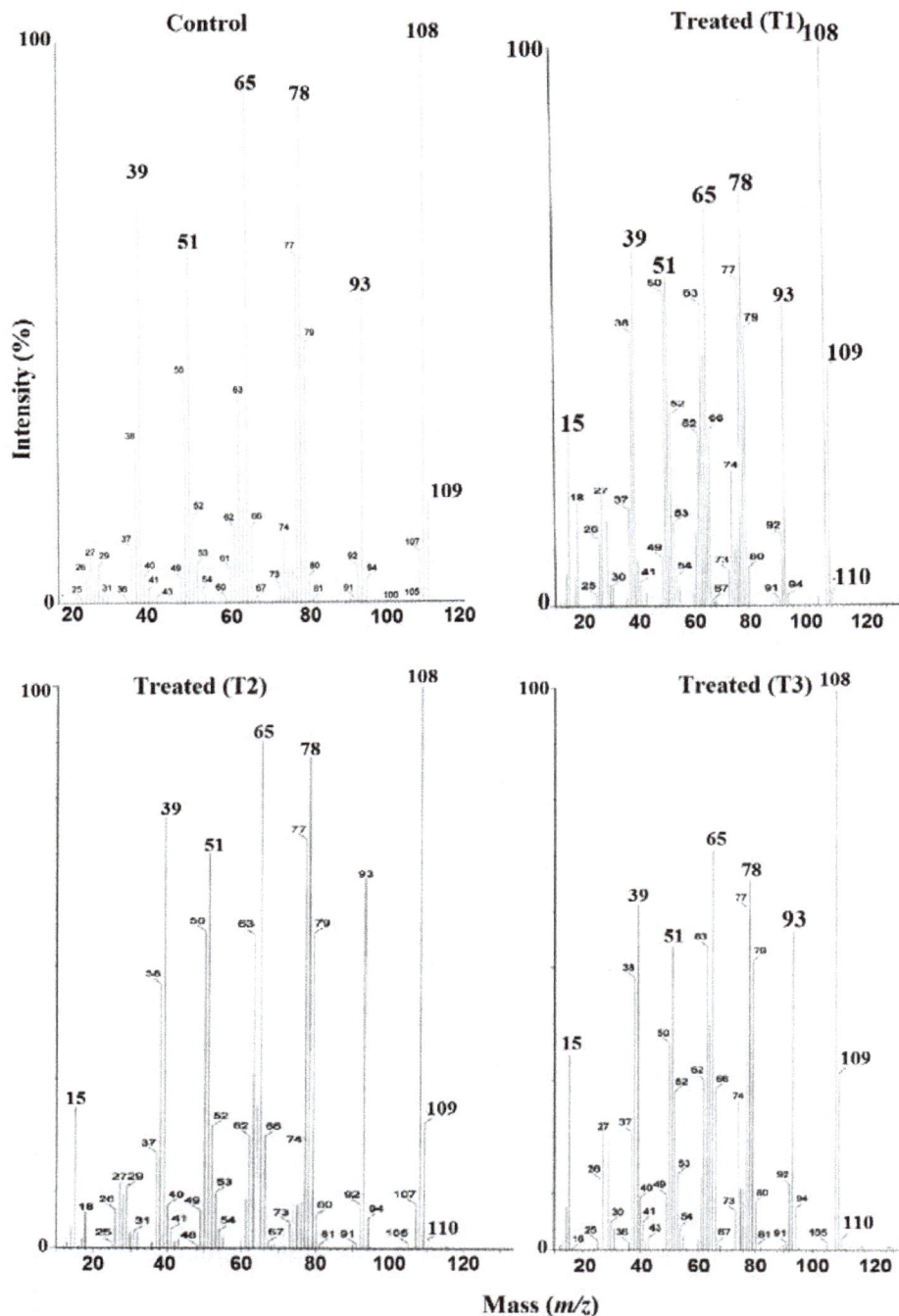

**Figure 2:** GC-MS spectra of control and treated anisole samples.

| Peak Intensity | Control | Treated | | |
|---|---|---|---|---|
| | | T1 | T2 | T3 |
| $m/z$=(PM) | 100 | 100.00 | 82.62 | 98.19 |
| $m/z$=(PM+1) | 23.11 | 23.76 | 13.00 | 20.25 |
| $m/z$=(PM+2) | 1.05 | 2.05 | 1.00 | 1.80 |

**Table 1:** GC-MS isotopic abundance analysis result of anisole.

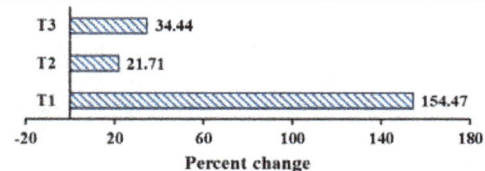

**Figure 3:** Percent change in isotopic abundance (PM+1)/PM and (PM+2)/PM of anisole under biofield treatment as compared to control.

## HPLC analysis

HPLC chromatogram is the detector response as a function of time is shown in Figure 4 (control and treated anisole). The compounds have been separated into two peaks in both control and treated samples. The retention time ($T_R$) of control sample was 2.58 min and 2.96 min with relative intensity 22.51% and 77.49%, respectively and the $T_R$ of treated anisole showed at 2.41 min and 2.71 min with relative intensity 29.85% and 70.10%, respectively. These decreased in $T_R$ may be attributed to the increased polarity of anisole after biofield energy treatment.

## DSC analysis

DSC was used for thermal analysis of control and treated anisole samples. The change in heat ($\Delta H$) of control and treated samples is shown in Figure 5. One strong endothermic transition was observed at 114.09°C in control anisole sample. However, it was observed at 112.27°C after biofield energy treatment. The heat absorbed in this process was found to be 550.35 J/g in treated anisole sample. The increase in $\Delta H$ in the treated sample was significantly increased by 389.08% as compared to the control, (Table 2). The increase in $\Delta H$ after biofield treatment could be due to alteration of intermolecular interaction in anisole that might increase the thermal stability of the treated samples.

## FT-IR spectroscopic analysis

The FT-IR spectrum of control and biofield energy treated anisole is shown in Figure 6. The vibrational peaks at 3003 cm$^{-1}$ and 2955 cm$^{-1}$ were assigned to aromatic C-H stretching of the phenyl ring for both control and treated samples. The vibrational peak at 2904 cm$^{-1}$ was observed due to the methyl (aliphatic) C-H stretching for both the control and treated samples. The C-C stretching frequency (in-ring, aromatic) was observed at 1603 cm$^{-1}$ and 1496 cm$^{-1}$ for control sample whereas it was shifted to 1601 cm$^{-1}$ and 1494 cm$^{-1}$ (the lower energy region) after biofield energy treatment. The characteristic C-O stretching for ethers was seen at 1302 cm$^{-1}$ and 1077 cm$^{-1}$ for aliphatic (R-O) and aromatic (Ar-O) parts respectively in both control and treated samples. The shifting of aromatic stretching frequency to the lower wavenumber may be attributed to the increased number of higher isotopes (PM+1) after biofield energy treatment. The vibration energy of a bond (in wavenumbers) in a molecule is represented by the following equation

$$\bar{\upsilon} = \frac{1}{2\pi C} \sqrt{\frac{K}{\mu}} \qquad (1)$$

Where, $K$ is a constant that varies from one bond to another, C is the speed of light, and $\mu = m_1 m_2/(m_1 + m_2)$, is the reduced mass or effective mass ($m_i$ is the mass of atom $i$). According to the equation (1), $\bar{\upsilon}$ is inversely proportional to the reduced mass [20]. It exhibited that reduced mass is higher in case of heavier isotope as compared to lighter one. This showed that anisole with heavier isotope after biofield energy treatment has low vibration energy as compared to lighter in control. Due to the increased isotopic abundance ratio of (PM+1) in the treated anisole, the effective mass ($\mu$) was increased and subsequently wavenumber was decreased. The FT-IR results after biofield energy treatment as compared to the control have shown changes in chemical properties of the molecule, which may lead to the changes in polarity as well as thermal property of anisole molecule.

## UV-Vis spectroscopic analysis

The UV-vis spectra of anisole in methanol are shown in Figure 7. The UV spectrum of control sample showed characteristic absorption

peaks at $\lambda_{max} = 325$ nm, however after biofield treatment the peak was red shifted and exhibited a shoulder. Both the control and treated samples showed characteristic strong absorption in the range of 200-275 nm, which was saturated at that particular concentration. Reported

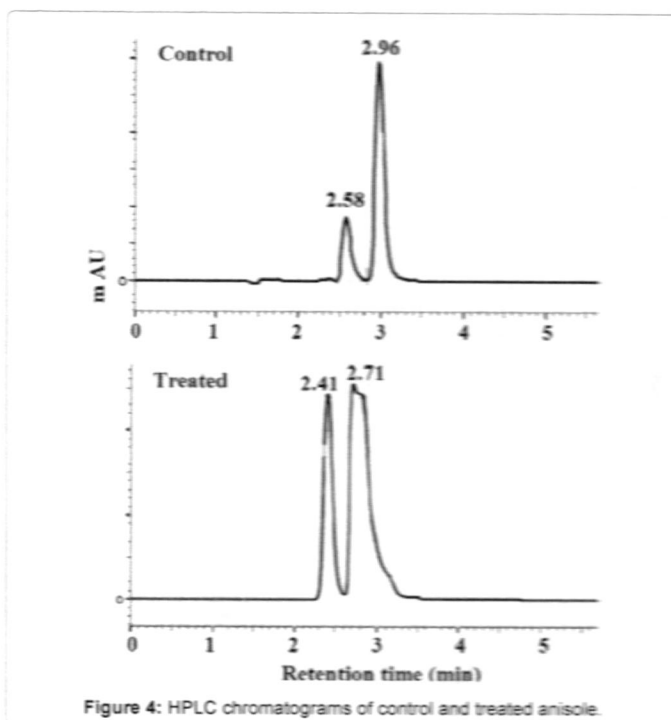

**Figure 4:** HPLC chromatograms of control and treated anisole.

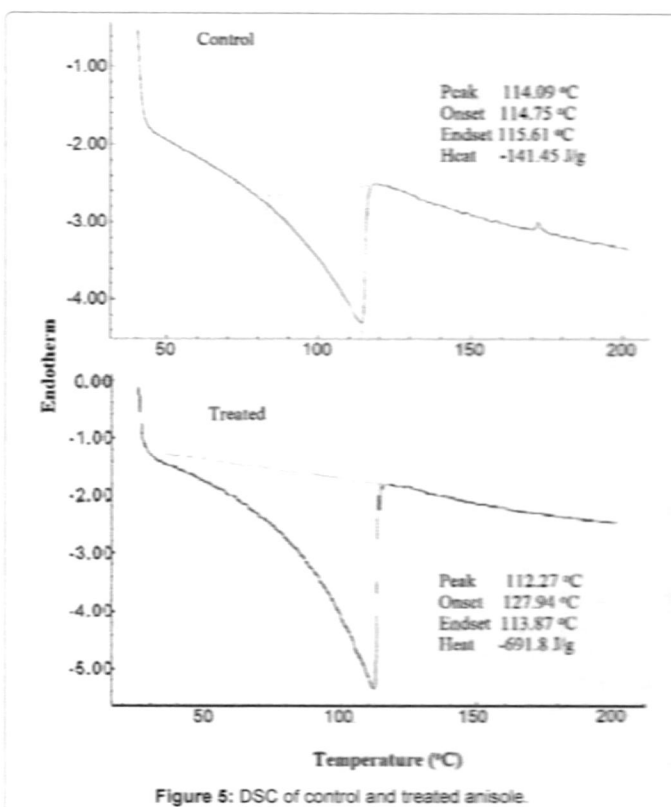

**Figure 5:** DSC of control and treated anisole.

|  | Peak position (°C) | Heat change J/g |
|---|---|---|
| Control | 114.09 | -141.45 |
| Treated | 112.27 | -691.80 |
| Percent Change | -1.59 | 389.07 |

**Table 2:** DSC analysis of anisole.

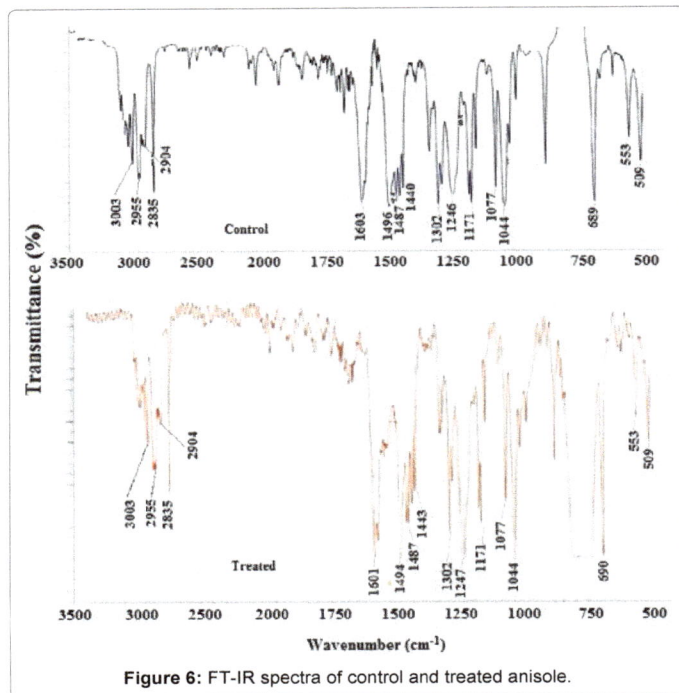

**Figure 6:** FT-IR spectra of control and treated anisole.

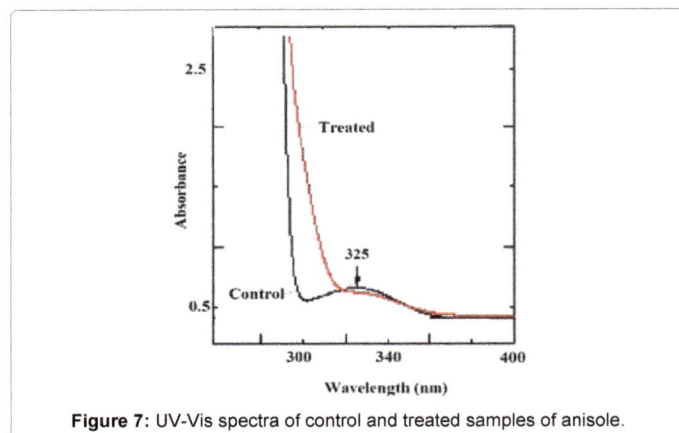

**Figure 7:** UV-Vis spectra of control and treated samples of anisole.

literature have suggested that the first strong band in the region of 300-350 nm that corresponds to $\pi$-$\pi^*$ transitions and the weak n-$\pi^*$ band in the region of 320-370 nm [21]. The n-$\pi^*$ bands show a distinct shift in the treated sample, it was observed with the varied polarity of the medium/solvent and reported in the literature [22]. Hence, it is hypothesized that biofield treatment might induce some polarity to the system that makes shift of the absorption positions. This observation is also supported by the lower $T_R$ of the treated sample in HPLC chromatogram.

## Conclusions

Anisole was studied to see the influence of biofield energy treatment and isotopic abundance ratio of treated anisole was compared with the control sample. After performing various analytical studies on biofield treated anisole, it was found that anisole has shown a strong response to biofield energy treatment. GC-MS data showed the isotopic abundance ratio of (PM+1)/PM was increased by 154.47% in the treated sample as compared to the control. Biofield energy may also alter the polarity of the molecule, which eventually alters the absorption position in UV-Vis and the retention time peak in HPLC chromatogram. Higher isotopic ratio (PM+1)/PM in treated samples correlated with the higher $\Delta H$ in DSC thermogram and shifting of aromatic C-C stretching frequency in FT-IR spectra to the lower wavenumber region as compared to the untreated sample. It is assumed that biofield energy treatment may enhance the stability of the organic small molecule of industrial importance, which ultimately affects the shelf-life and efficacy of the final product.

### Acknowledgment

The authors would like to acknowledge the whole team of Sophisticated Analytical Instrument Facility (SAIF), Nagpur for providing the instrumental facility. We are very grateful for the support of Trivedi Science, Trivedi Master Wellness and Trivedi Testimonials in this research work.

### References

1. Ullmann (2002) Encyclopedia of industrial chemistry, Wiley–VCH Verlag, New York.

2. Torres MA, Furst DE (1990) Treatment of generalized systemic sclerosis. Rheum Dis Clin North Am 16: 217-241.

3. Katz WA (1996) Pharmacology and clinical experience with tramadol in osteoarthritis. Drugs 52 Suppl 3: 39-47.

4. Fleischer AB Jr, Schwartzel EH, Colby SI, Altman DJ (2000) The combination of 2% 4-hydroxyanisole (Mequinol) and 0.01% tretinoin is effective in improving the appearance of solar lentigines and related hyperpigmented lesions in two double-blind multicenter clinical studies. J Am Acad Dermatol 42: 459-467.

5. Sasikumar JM, Mathew GM, Priya Darsini DT (2010) Comparative studies on antioxidant activity of methanol extract and flavonoid fraction of Nyctanthes arbortristis leaves. EJEAF Che 9: 227-233.

6. Zimmerman SB, Stapley EO (1976) Relative morphological effects induced by cefoxitin and other beta-lactam antibiotics in vitro. Antimicrob Agents Chemother 9: 318-326.

7. Carmine AA, Brogden RN, Heel RC, Romankiewicz JA, Speight TM, et al. (1983) Moxalactam (latamoxef). A review of its antibacterial activity, pharmacokinetic properties and therapeutic use. Drugs 26: 279-333.

8. Jan SJ, Cembor M, Orlik M (2005) Anisole as a solvent for organic electrochemistry. J Electroanal Chem 582: 165-170.

9. Venkatesan S, Chen Q, Ngo EC, Adhikari N, Nelson K, et al. (2014) Polymer solar cells processed using anisole as a relatively nontoxic solvent. Energy Technology 2: 269-274.

10. Trivedi MK, Patil S, Shettigar H, Bairwa K, Jana S (2015) Phenotypic and biotypic characterization of Klebsiella oxytoca: An impact of biofield treatment. J Microb Biochem Technol 7: 203-206.

11. Trivedi MK, Nayak G, Patil S, Tallapragada RM, Latiyal O, et al. (2015) An evaluation of biofield treatment on thermal, physical and structural properties of cadmium powder. J Thermodynamics Catal 6: 147.

12. Planck M (1903) Treatise on thermodynamics. (3rd edn), English translated by Alexander OGG, Longmans, Green, London, UK.

13. NIH (2008) National Center for Complementary and Alternative Medicine. CAM Basics. Publication 347.

14. Trivedi MK, Patil S, Shettigar H, Gangwar M, Jana S (2015) Antimicrobial sensitivity pattern of Pseudomonas fluorescens after biofield treatment. J Infect Dis Ther 3: 222.

15. Patil S, Nayak GB, Barve SS, Tembe RP, Khan RR (2012) Impact of biofield treatment on growth and anatomical characteristics of Pogostemon cablin (Benth.). Biotechnology 11: 154-162.

16. Nayak G, Altekar N (2015) Effect of biofield treatment on plant growth and adaptation. J Environ Health Sci 1: 1-9.

17. Shinde V, Sances F, Patil S, Spence A (2012) Impact of biofield treatment on growth and yield of lettuce and tomato. Aust J Basic & Appl Sci 6: 100-105.

18. http://www.pherobase.com/ms-popup.html?anisole

19. Rieley G (1994) Derivatization of organic-compounds prior to gas-chromatographic combustion-isotope ratio mass-spectrometric analysis: Identification of isotope fractionation processes. Analyst 119: 915-919.

20. Smith BC (2011) Fundamentals of Fourier transform infrared spectroscopy, CRC Press, Taylor and Francis Group, Boka Raton, New York.

21. Gerrard DL, Maddams WF, Tucker PJ (1978) Solvent effects in U.V. absorption spectra. IV. substituted phenols, anisole and phenetole. Spectrochim Acta A: Mol Spectros 34: 1225-1230.

22. Balfour WJ, Ram SR (1982) The near-ultraviolet vapour phase absorption spectrum of anisole. Chem Phys Lett 92: 279-282.

# 29

# Simple and Efficient Synthesis of Novel Fused Bicyclic Heterocycles Pyrimido-Thiazine and Their Derivatives

**Sirsat Shivraj B and Vartale Sambhaji P***

*P.G. Research centre, Department of Chemistry, Yeshwant Mahavidyalaya, Nanded 431602 (MS), India*

### Abstract

We report simple and efficient synthesis of novel fused bicyclic heterocyclic compounds 3 using bis (methylthio) methylene malononitrile 1 and thiourea 2 with potassium carbonate in DMF at reflux condition. The molar ratios of these substrates are 2:1 for the preparation of 2,6-dihydro-2,6-diimino-4,8-bis(methylthio) pyrimido[2,1-b][1,3]thiazine-3,7-dicarbonitrile. This newly synthesized pyrimido thiazine acts as bis-electrophilic species reacting with various nucleophiles yielding 2,6-dihydro-2,6-diimino-4,8-(disubstituted)-pyrimido[2,1-b][1,3]thiazine-3,7-dicarbonitrile in good yields.

**Keywords:** Thiourea; Bis (methylthio) methylene malononitrile; Bis-electrophilic species; Various nucleophiles

## Introduction

In recent years, the synthesis of fused bicyclic heterocyclic compounds possessing pyrimido-thiazine central core has been the focus of great interest. This type of compounds shows various biological properties such as antibacterial, antiallergic, anti-inflammatory, antitumor, phsphodiesterase inhibition and antiparkinsonism [1-6], many workers have synthesized different 1, 3-thiazines [7,8]. Thiazines are very useful units in the fields of medicinal and pharmaceutical chemistry and have been reported to exhibit a variety of biological activities [9,10]. Recently, substituted thiazine are prepared using α, β- unsaturated carbonyl system and that are very versatile substrates for the evolution of various reactions [11] and physiologically active compounds [12]. The reaction of thiourea with α,β- unsaturated system (Michael acceptor) results in 1,3 thiazine [13,14]. It has been well focused that the presence of pyrimido-thiazine with various chemically reactive moieties is an important structural feature and also substituted imino group present in thiazine ring, and the resulting molecule would exhibit promising biological activities in continuation of our work [15-21]. In the present study, we synthesize pyrimido-thiazine containing more reactive functional groups using thiourea and bis methylthio methylene malanonitrile which is used for further cyclisation and derivatization. The synthesized compounds act as bis-electrophilic species reacting with various nucleophiles such as substituted aromatic amines, aromatic phenol, various active methylene compound and alicyclic heterocyclic compound construct 2,6-dihydro-2,6-diimino-4,8-disubstitutedpyrimido[2,1-b][1,3]thiazine-3,7-dicarbonitrile in good yields.

## Experimental Section

Melting points were determined by open capillary tubes and were uncorrected. The silica gel $F_{254}$ plates were used for thin layer chromatography (TLC) in which the spots were examined under UV light and then developed by an iodine vapor. Column chromatography was performed with silica gel (BDH 100-200 mesh). Solvents were purified according to standard procedures. The spectra were recorded with the following instruments; IR: Perkin-Elmer RX1 FT-IR spectrophotometer; NMR: Varian Gemini 200 MHz ($^1$H) and 50 MHz ($^{13}$C) spectrometer; ESIMS: VG-Autospec micromass. Elemental analysis was performed on a Heraeus CHN-O rapid analyzer.

## General procedure for the synthesis 2,6-dihydro-2,6-diimino-4,8-bis(methylthio)pyrimido[2,1-b][1,3]thiazine-3,7-dicarbonitrile (3)

A mixture of (methylthio) methylene malononitrile (1) (2 mmol) and thiourea (2) (1 mmol) in DMF and anhydrous potassium carbonate (10 mg) was refluxed for 12 hours on oil bath. The reaction was monitored by TLC. After completion, the reaction mixture was cooled at room temperature then washed with water (3×10 mL) and subsequently extracted with ethyl acetate (3×10 mL). The extract was concentrated and the residue was subjected to column chromatography (silica gel, n-hexane- ethyl acetate 8:2) to obtain pure solid compound 3. The compound 3 confirmed by IR, $^1$H and $^{13}$C NMR and MS analytical data compound is given below.

Yellow solid (yield 76%). Mp: 145-147°C. IR (KBr): 3380 (=NH), 2250(-CN) cm$^{-1}$.

$^1$H NMR (CDCl3) δ 3.15 (s, 6H, SCH$_3$), 9.17 (s, 1H, =NH), 9.81 (s, 1H, =NH), $^{13}$C NMR (CDCl$_3$) δ 180 (2), 145.1 (2), 131.9, 136.6, 45(2) MS *m/z*: 343(M$^+$ Na 100%) 240,212, 198, 140 Anal. Calcd for C$_{11}$H$_8$N$_6$S$_3$: C-41.23, H-2.52, N-26.23, S-30.02. Found: C- 41.01, H-2.91, N-26.15, S-30.0.

## 2,6-dihydro-2,6-diimino-4,8-bis (substituted) pyrimido [2,1b][1,3]thiazine-3,7- dicarbonitrile (4a-e), (5a-e)

A mixture of 3 (1 mmol) and, independently, various substituted aromatic amines, and substituted aromatic phenol (0.002 mol) in N, N'- dimethyl formamide (10 mL) and anhydrous potassium carbonate (10 mg) was refluxed for 4 to 6 hours. The reaction mixture was cooled to room temperature and poured into ice cold water. The separated

*****Corresponding author:** Vartale Sambhaji P, P.G. Research Centre, Department of Chemistry, Yeshwant Mahavidyalaya, Nanded-431602 (MS), India
E-mail: spvartale@gmail.com

solid product was filtered, washed with water and recrystallized using ethyl alcohol.

## 2,6-dihydro-2,6-diimino-4,8-bis(phenylamino) pyrimido [2,1-b][1,3]thiazine-3,7-dicarbonitrile (4a)

Colourless solid (yield 65%). Mp: 155-156°C. IR (KBr): 3450 (=NH), 2220 (-CN) cm$^{-1}$. $^1$H NMR (CDCl$_3$) δ 5.15 (s, 2H,-NH), 9.27 (s, 1H,=NH), 9.61 (s, 1H,=NH), 7.26.-7.5 (s, 10H Ar-H). MS m/z: 411(M$^+$ 100%) 323, 240, 212, 198, 140 Anal. Calcd for C$_{21}$H$_{14}$N$_8$S: C-61.23, H-3.52, N-27.23, S-7.82. Found: C- 61.01, H-3.91, N-27.15, S-8.0.

## 4,8-bis(4-bromophenylamino)-2,6-dihydro-2,6-diiminopyrimido[2,1-b][1,3]thiazine-3,7-dicarbonitrile (4b)

Yellow solid (yield 70%). Mp: 185-186°C. IR (KBr): 3435 (=NH), 2255 (-CN) cm$^{-1}$. $^1$H NMR (CDCl$_3$) δ 5.0 (s, 2H,-NH), 9.37 (s, 1H, =NH), 9.71 (s, 1H, NH), 7.2-7.5 (dd, 8H J=7.5-8HzAr-H).MS m/z: 568(M$^+$ 2 80%) 411, 323, 240, 212, 198, 140 Anal. Calcd for C$_{21}$H$_{12}$N$_8$SBr$_2$: C-44.40, H-2.12, N-19.70, S-5.5, Br-28.12. Found: C- 44.12, H-2.15, N-20.00, S-5.9, Br-28.00.

## 4,8-bis(4-methoxyphenylamino)-2,6-dihydro-2,6-diiminopyrimido[2,1-b][1,3]thiazine-3,7-dicarbonitrile (4c)

Brown solid (yield 60%). Mp: 150-152°C. IR (KBr): 3350 (=NH), 2240 (-CN) cm$^{-1}$. $^1$H NMR (CDCl$_3$) δ 4.8 (s, 2H, -NH-Ar), 9.47 (s, 1H,=NH), 9.81 (s, 1H,=NH), 7.2-7.5 (dd, 8H J=7.5-8Hz Ar-H) , 3.5 (s 6H-OCH$_3$). MS m/z: 471(M$^+$ 100%) 411, 323, 212, 190, 140 Anal. Calcd for C$_{23}$H$_{18}$N$_8$SO$_2$: C-58.80, H-3.86, N-23.82, S-6.5. Found: C- 58.50, H-3.90, N-24.00, S-6.4.

## 4,8-bis(4-methylphenylamino)-2,6-dihydro-2,6-diiminopyrimido[2,1-b][1,3]thiazine-3,7-dicarbonitrile (4d)

White solid (yield 65%). Mp: 145-148°C. IR (KBr): 3380 (-NH), 3400 (=NH), 2255 (-CN) cm$^{-1}$. $^1$H NMR (CDCl3) δ 5.0 (s, 2H, -NH-Ar), 9.23 (s, 1H,=NH), 9.71 (s, 1H,=NH), 7.0-7.3 (dd, 8H J=7.5-8Hz, Ar-H), 1.8 (s 6H, CH$_3$). MS m/z: 439(M$^+$ 60, 323, 212, 190, 140 Anal. Calcd for C$_{23}$H$_{18}$N$_8$S: C-63.00, H-4.15, N-25.55 S-7.1 Found: C- 62.5, H-4.10, N-25.00, S-7.5.

## 4,8-bis(3-nitrophenylamino)-2,6-dihydro-2,6-diiminopyrimido[2,1-b][1,3]thiazine-3,7-dicarbonitrile (4e)

Yellow solid (yield 70%). Mp: 190-192°C. IR (KBr): 3370 (-NH), 3410 (=NH), 2220 (-CN) cm$^{-1}$. $^1$H NMR (CDCl$_3$) δ 4.5 (s, 2H -NH-Ar), 9.12 (s, 1H, =NH), 9.67 (s, 1H,=NH), 7.0-7.2 (s, 2H Ar-H) , 7.5-7.8 (m 6H Ar-H). MS m/z: 501(M$^+$ 1), 315, 212, 166, 140 Anal. Calcd for C$_{21}$H$_{12}$N$_{10}$SO$_4$: C-50.35, H-2.42, N-27.99, S-6.41 Found: C- 50.50, H-2.5, N-28.00, S-6.5.

## 2,6-dihydro-2,6-diimino-4,8-diphenoxypyrimido[2,1-b][1,3] thiazine-3,7-dicarbonitrile (5a)

White solid (yield 75%). Mp: 145-146°C. IR (KBr): 3410 (=NH), 2240 (-CN) cm$^{-1}$. $^1$H NMR (CDCl$_3$) δ 9.27 (s, 1H, =NH), 9.61 (s, 1H, =NH), 7.1-7.5 (s, 10H, Ar-H). MS m/z: 413(M$^+$1 60%) 306 , 220, 212, 198, Anal. Calcd for C$_{21}$H$_{12}$N$_6$SO$_2$: C-61.16, H- 2.94, N-20.23, S-7.82. Found: C-61.01, H-3.00, N-27.15, S-8.00.

## 4,8-bis(4-bromophenoxy)-2,6-dihydro-2,6-diiminopyrimido[2,1-b][1,3]thiazine-3,7-dicarbonitrile (5b)

Yellow solid (yield 70%). Mp: 190-193°C. IR (KBr): 3410 (=NH), 2240 (-CN) cm$^{-1}$. $^1$H NMR (CDCl$_3$) δ 9.14 (s, 1H, =NH), 9.32 (s, 1H, =NH),

7.4-7.7 (dd, 8H J=7-7.5Hz, Ar-H). MS m/z: 570(M+ 2 67%) 405, 305, 212, 125, 140 Anal. Calcd for C$_{21}$H$_{10}$N$_6$SO$_2$Br$_2$: C-44.30, H-1.77, N-14.70, S-5.5, Br-28.23. Found: C- 44.12, H-2.15, N-20.00, S-5.9, Br-28.00.

## 4,8-bis(4-methoxyphenoxy)-2,6-dihydro-2,6-diiminopyrimido[2,1-b][1,3]thiazine-3,7-dicarbonitrile (5c)

Brown (yield 60%). Mp: 145-152°C. IR (KBr): 3412 (=NH), 2222 (-CN) cm$^{-1}$. $^1$H NMR (CDCl$_3$) δ 9.33 (s, 1H,=NH), 9.67 (s, 1H,=NH), 7.4-7.8 (dd, 8H J=7.5-8Hz, Ar-H), 3.1 (s 6H,-OCH$_3$). MS m/z: 472(M$^+$ 70%) 411, 397, 381, 212, 190, Anal. Calcd for C$_{23}$H$_{16}$N$_6$SO$_4$: C-58.47, H-3.46, N-17.82, S-6.79. Found: C- 58.50, H-3.50, N-17.50, S-6.4.

## 4,8-bis(4-chlorophenoxy)-2,6-dihydro-2,6-diiminopyrimido[2,1-b][1,3]thiazine-3,7-dicarbonitrile (5d)

Solid (yield 65%). Mp: 145-148°C. IR (KBr): 3420 (=NH), 2240 (-CN) cm$^{-1}$. $^1$H NMR (CDCl$_3$) δ 9.24(s, 1H,=NH), 9.32 (s, 1H,=NH), 7.1-7.5 (dd, 8H J=7.5-8HzAr-H). MS m/z: 482(M$^+$ 2 67%) 405, 307, 212, 140 Anal. Calcd for C$_{21}$H$_{10}$N$_6$SO$_2$Cl$_2$: C-52.40, H-2.09, N-17.70, S-6.66, Cl-28.23. Found: C-52.30, H-2.00, N-17.90, S-6.76, Cl-28.00.

## 4,8-bis(3-nitrophenoxy)-2,6-dihydro-2,6-diiminopyrimido[2,1-b][1,3]thiazine-3,7-dicarbonitrile (5e)

Yellow (yield 70%). Mp: 180-183°C. IR (KBr): 3410 (=NH), 2245(-CN) cm$^{-1}$. $^1$H NMR (CDCl$_3$) δ 9.45 (s, 1H, =NH), 9.87 (s, 1H, =NH), 7.3-7.5 (s, 2H, Ar-H), 7.5-7.8 (m 6H, Ar-H). MS m/z: 477(M$^+$), 318, 212, 166, 140 Anal. Calcd for C$_{21}$H$_{10}$N$_8$SO$_6$: C-50.20, H-2.02, N-22.30, S-6.41 Found: C- 50.50, H-2.0, N-22.00, S-6.5.

## Result and Discussion

The fused heterocyclic compounds 2,6-dihydro-2,6-diimino-4,8-bis(methylthio) pyrimido[2,1-b][1,3]thiazine-3,7-dicarbonitrile (3) was prepared from bis (methylthio) methylene malononitrile 1 and thiourea 2 with catalytic amount of potassium bicarbonate (1 mmol) in DMF at reflux condition and the molar ratios of these substrates are 2:1 (Scheme 1).

Proposed pathway for formation of 2,6-dihydro-2,6-diimino-4,8-bis(methylthio) pyrimido [2,1-b][1,3]thiazine-3,7-dicarbonitrile (Scheme 2).

The compound 3 posses a replaceable active methylthio group (-SCH$_3$) at 4, 8- position which is activated by nitrogen atom and electron withdrawing cyano group. Compound 3 reacted with selected various nucleophilles like substituted aryl amines hetryl amines, substituted phenols and activated methylene compound in DMF and catalytic amount of anhydrous potassium carbonate, to afford 2,6-dihydro-2,6-diimino-4,8-bis(substituted) pyrimido[2,1-b] [1,3]thiazine-3,7-dicarbonitrile (4a-e) and (5a-d) Table 1 (Scheme 3) respectively.

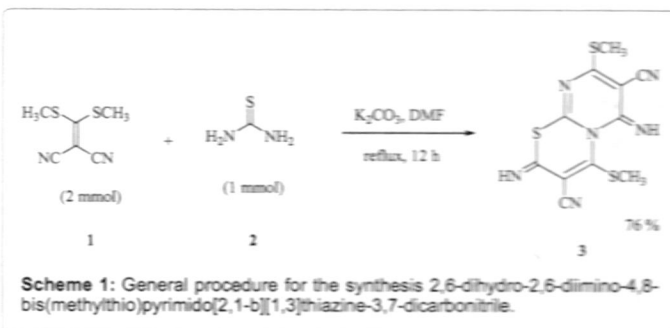

**Scheme 1:** General procedure for the synthesis 2,6-dihydro-2,6-diimino-4,8-bis(methylthio)pyrimido[2,1-b][1,3]thiazine-3,7-dicarbonitrile.

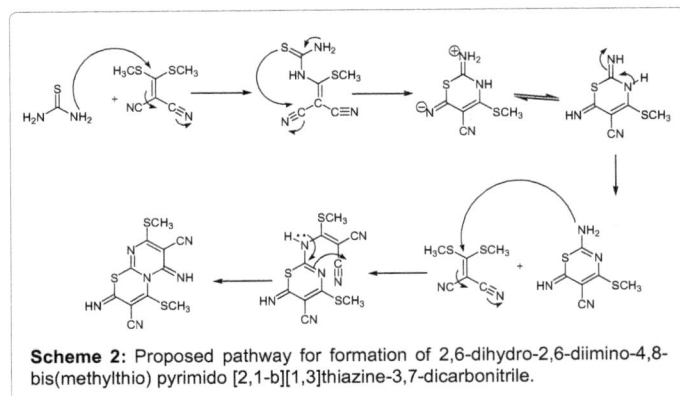

**Scheme 2:** Proposed pathway for formation of 2,6-dihydro-2,6-diimino-4,8-bis(methylthio) pyrimido [2,1-b][1,3]thiazine-3,7-dicarbonitrile.

| Compound | 4a | 4b | 4c | 4d | 4e |
|----------|-----|------|--------|--------|--------|
| -R | -H | P-Br | P-OCH₃ | P-CH₃ | P-NO₂ |
| Compound | 5a | 5b | 5c | 5d | 5e |
| -R¹ | -H | P-Br | P-OCH₃ | P-Cl | M-NO₂ |

**Table 1:** 2,6-dihydro-2,6-diimino-4,8-bis (substituted) pyrimido [2,1b][1,3] thiazine-3,7- dicarbonitrile.

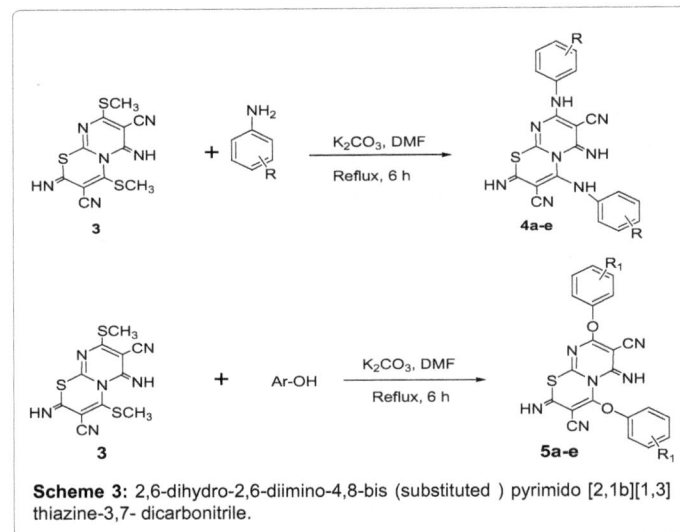

**Scheme 3:** 2,6-dihydro-2,6-diimino-4,8-bis (substituted ) pyrimido [2,1b][1,3] thiazine-3,7- dicarbonitrile.

## Conclusion

In conclusion, we have synthesised simple and efficent novel fused bicyclic heterocycles pyrimido-thiazine having bis-electrophilic species reacting with various nucleophiles.

### Acknowledgement

The authors are grateful to Dr. N. V. Kalyankar, Principal, Yeshwant Mahavidyalaya, Nanded for providing laboratory facilities, To UGC New Delhi for financial assistance under major research project (F.N 39-834/2010 (SR)) and Director, Indian Institute of Chemical Technology, Hyderabad for providing spectra.

### References

1.  Gupta S, Ajmera N, Gautam N, Sharma R, Gautam DC (2009) ChemInform

Abstract: Novel Synthesis and Biological Activity Study of Pyrimido[2,1-b] benzothiazoles. ChemInform 40: 42.

2.  Dash B, Patra M, Mahapatra PK (1980) Synthesis of thiazolo pyrimidine derivatives. J Inst Chem 52: 92.

3.  Glennon RA, Gaines JJ, Rogers ME (1981) Benz-fused mesoionic xanthine analogues as inhibitors of cyclic-AMP phosphodiesterase. J Med Chem 24: 766-769.

4.  Heter HL (1972) Patent US 3704303 Chem Abstr 78: 43513x.

5.  Covington RR, Temple DL, Yevich JP (1982) "Antial-lergics: 3-(1H-tetrazol-5-yl)-4H-pyrimido [2,1- b] ben-zothi- azol-4-ones". Journal of Medicinal Chemistry 25: 864-868.

6.  Wade JJ, Toso CB, Matson CJ, Stelzer VL (1983) Synthesis and antiallergic activity of some acidic derivatives of 4H-pyrimido[2,1-b]benzazol-4-ones. J Med Chem 26: 608-611.

7.  Ramekar MA, Chincholkar MM (1995) "Synthesis of Some New 4,6-Diaryl-5-aroyl-2-imino-6H-2,3- dihydro- 1,3- thiazines". ChemInform 26.

8.  Raut AW (2001) "Synthesis of Schiff bases of thiophene-2-carboxaldehyde and its antimicrobial activity". Orient J Chem 17:131–133.

9.  Rajesh V, Prakash C, Hariom S, Varma BL (2008) "Microwave-assisted synthesis of 6H-2-amino-4,6- diaryl-1,3-thiazines". Indian J Heterocyclic Chem 17: 237.

10. Jhala YS, Pradhuman, Ranawat S, Dulawat SS, Varma BL (2005) "Microwave assisted synthesis of chalcones using Claisen-Schmidt condensation in dry media". Indian J Heterocyclic Chem 14: 357.

11. Kelly DR, Caroff E, Flood RW, Heal W, Roberts SM (2004) The isomerisation of (Z)-3-[2H1]-phenylprop-2-enone as a measure of the rate of hydroperoxide addition in Weitz-Scheffer and Julia-Colonna epoxidations. Chem Commun (Camb) : 2016-2017.

12. Iwata S, Nishino T, Inoue H, Nagata N, Satomi Y, et al. (1997) Antitumorigenic activities of chalcones (II). Photo-isomerization of chalcones and the correlation with their biological activities. Biol Pharm Bull 20: 1266-1270.

13. Jain AC, Prasad AK (1995) Reaction of chalcone 2-4,-dimethoxychalcone, a-bromo and methoxy chalcone with thiourea. Indian J Chem 34B: 496.

14. Sambhaji PV, Vijay NB, Sandeep V, Khansole, Ramdas NK (2009) A convenient one pot synthesis of 3-cyno-9-methyl-2-methylthio-4-oxo-4H-pyrimidi-[2,1-b] [4,5-b]quinoline and its reaction with selected nucleophiles. Letters in organic chemistry 6: 544-548.

15. Galabov AS, Galabov BS, Neykova NA (1980) Structure-activity relationship of diphenylthiourea antivirals. J Med Chem 23: 1048-1051.

16. Baheti KG, Kapratwar SB, Kuberkar SV (2002) A convenient synthesis of 2,3 disubstituted derivatives of 4H-pyrimido [2,1-b]benzotjiazole-4-one Synth Comm 32: 2237

17. Baheti KG, Kuberkar SV (2003) Novel Synthesis of 3-Amino-4-oxo-(2H)-pyrazolo[3',4':4,5]pyrimido- [2,1- b]benzothiazole and its 2- and 3-Substituted Derivatives. J Het Chem 40: 547-551.

18. Baheti KG and Kuberkar SV (2003) "Synthesis and biological activity of 4H-pyrimido [2,1-b] benzothiazole-8-substituted-2-thiomethyl-3-cyano-4-ones". Ind J Het Chem 12: 343.

19. Baheti KG, Jadhav JS, Suryawanshi AT, Kuberkar SV (2005) Novel synthesis and antibacterial activity of 15-iminobenzothiazole[2,3b]pyrimido[5,6e] pyrimido[2,3]benzothiazole-14-H ones and its 3,10 disubstituted derivatives. Ind J Chem 44B: 834.

20. Pingle MS, Vartale SP, Bhosale VN, Kuberkar SV (2006) A convenient synthesis of 3-cyano-4-imino-2- methylthio-4Hpyrimido [2,1-b] [1,3] benzothiazole and its reactions with selected nucleophiles. Arkivoc X, 190.

21. Vartale SP, Jadhav JS, Kale MA, Kuberkar SV (2006) Synthesis and antimicrobial activity of 6/7/8- substituted-1-[aryl/6'-substituted-2'-benzothiazolyl]-pyrazolo [4,5-b] quinolines. Ind J Chem.

# Permissions

# List of Contributors

**Aydın Aktaş and Yetkin Gök**
Department of Chemistry, Faculty of Arts and Sciences, Inönü University, Malatya, Turkey

**Mehmet Akkurt**
Department of Physics, Faculty of Sciences, Erciyes University, Kayseri, Turkey

**Namık Özdemir**
Department of Physics, Faculty of Arts and Sciences, Ondokuz Mayis University, Samsun, Turkey

**Linda-Lucila Landeros-Martinez, Erasmo Orrantia-Borunda and Norma Flores-Holguin**
NANOCOSMOS Virtual Lab, Advanced Materials Research Center (CIMAV), Miguel de Cervantes 120, Complejo Industrial Chihuahua, Chihuahua 31190, México

**Balram Prasad Baranwal, Abhay Kumar Jain and Alok Kumar Singh**
Coordination Chemistry Research Laboratory, Department of Chemistry, D.D.U. Gorakhpur University, Gorakhpur 273 009, India

**Venkata Sai Prakash Chaturvedula, Rafael Ignacio San Miguel and Indra Prakash**
The Coca-Cola Company, Organic Chemistry Department, Global Research and Development, One Coca-Cola Plaza, Atlanta, GA 30313, USA

**Sanny Verma and Suman L. Jain**
Chemical Sciences Division, CSIR-Indian Institute of Petroleum, Mohkampur, Dehradun-248005, India

**N. M. El-Moein, E. A. Mahmoud and Emad A. Shalaby**
Biochemistry Department, Faculty of Agriculture, Cairo University, Giza, Egypt, 12613

**Mahendra Kumar Trivedi, Alice Branton, Dahryn Trivedi and Gopal Nayak**
Trivedi Global Inc., 10624 S Eastern Avenue Suite A-969, Henderson, NV 89052, USA

**Omprakash Latiyal and Snehasis Jana**
Trivedi Science Research Laboratory Pvt. Ltd., Hall-A, Chinar Mega Mall, Chinar Fortune City, Hoshangabad Rd., Bhopal, Madhya Pradesh, India

**Rajendra D Patil and Yoel Sassona**
Casali Institute of Applied Chemistry, The Institute of Chemistry, The Hebrew University of Jerusalem, Jerusalem, Israel

**L.I. Chernogor, N.N. Denikina and S.I. Belikov**
Limnological Institute of the Siberian Branch of the Russian Academy of Sciences, Ulan-Batorskaya 3, Irkutsk 664033, Russia

**A.V. Ereskovsky**
Department of Embryology, Faculty of Biology and Soils, Saint-Petersburg State University, Universitetskaja nab. 7/9, St. Petersburg 199034, Russia
Centre d'Océanologie de Marseille, Station marine d'Endoume - CNRS UMR 6540 DIMAR, rue de la Batterie des Lions, 13007 Marseille, France

**Mahendra Kumar Trivedi, Alice Branton, Dahryn Trivedi and Gopal Nayak**
Trivedi Global Inc., 10624 S Eastern Avenue Suite A-969, Henderson, NV 89052, USA

**Ragini Singh and Snehasis Jana**
Trivedi Science Research Laboratory Pvt. Ltd., Hall-A, Chinar Mega Mall, Chinar Fortune City, Hoshangabad Rd., Bhopal, Madhya Pradesh, India

**K. L. Ameta, Nitu S. Rathore and Biresh Kumar**
Department of Chemistry, FASC, Mody Institute of Technology & Science, Lakshmangarh-332311, Rajasthan, India

**Edith S. Maalaga M and Manuela Veraastegui**
Department of Microbiology, Faculty of Science and Philosophy, Universidad Peruana Cayetano Heredia, Lima, Peru

**Robert H. Gilman**
Department of International Health, Bloomberg School of Public Health, Johns Hopkins University, Baltimore, Maryland-21205, USA

**Mukerjea R, Sheets RL, Gray AN and Robyt JF**
Laboratory of Carbohydrate Chemistry and Enzymology, Department of Biochemistry, Biophysics and Molecular Biology, Iowa State University, Ames, IA 50011, USA

**Mahendra KT, Shrikant P and Rakesh KM**
Trivedi Global Inc., 10624 S Eastern Avenue Suite A-969, Henderson, NV 89052, USA

**Snehasis J**
Trivedi Science Research Laboratory Pvt. Ltd., Hall-A, Chinar Mega Mall, Chinar Fortune City,Hoshangabad Rd., Bhopal- 462026, Madhya Pradesh, India

**Rafat M. Mohareb**
Department of Chemistry, Faculty of Science, Cairo University, Giza, A. R. Egypt
Department of Organic Chemistry, Faculty of Pharmacy, October University for the Modern science and Arts (MSA), Elwahaat Road, October City, Egypt

**Fatma O. Al-farouk**
Department of Chemistry, Faculty of Science, American University in Cairo, 5th Settlement, A.R., Egypt

**Moussa Ahmed, Saad Aissat, Baghdad Khiati, Abdelmalek Meslem and Abdelkader Berrani**
Institute of Veterinary Sciences University Ibn-Khaldoun, Tiaret, Algeria

**Noureddine Djebli and Salima Douichene**
Departments of Biology, Faculty of Sciences, Mostaganem University, Algeria

**Pranita P. Kore, Snehal D. Kachare, Sandip S. Kshirsagar and Rajesh J. Oswal**
Department of Pharmaceutical Chemistry, JSPM's Charak College of Pharmacy and Research, Gat No. 720/1&2, Wagholi, Pune-Nagar Road, Pune-412207, Maharashtra, India

**Kabir OO, Abdulfatai TA and Akeem AJ**
Chemistry Unit, Department of Chemical, Geological and Physical Sciences, Kwara State University, Malete, Ilorin, Nigeria

**Al-Omran F and El-Khair A**
Department of Chemistry, Faculty of Science, Kuwait University, P.O. Box 12613, Safat 13060, Kuwait

**Jining Liu, Deling Fan, Lei Wang, Linjun Zhou and Lili Shi**
Nanjing Institute of Environmental Sciences, Ministry of Environmental Protection, Nanjing, China

**Chen Tang**
Nanjing Entry-Exit Inspection and Quarantine Bureau, Nanjing, China

**Natvar A Sojitra and Bharat C Dixit**
Chemistry Department, V. P. & R. P. T. P. Science College, Affiliated to Sardar Patel University, Gujarat, India

**Ritu B Dixit**
Ashok & Rita Patel Institute of Integrated Studies and Research in Biotechnology and Allied Sciences, Gujarat, India

**Rajesh K Patel**
Department of Life Sciences, Hemchandracharya North Gujarat University, Gujarat, India

**Abdel-Sattar S Hamad Elgazwy and Mahmoud R Mahmoud Refaee**
Department of Chemistry Faculty of Science, Ain Shams University, Abbassia 11566, Cairo, Egypt

**Ololade ZS and Olawore NO**
Department of Pure and Applied Chemistry, Ladoke Akintola University of Technology, P.M.B. 4000, Ogbomoso, Nigeria

**Oladosu IA**
Department of Chemistry, University of Ibadan, Ibadan, Nigeria

**Jining L, Deling F, Lei W, Linjun Z and Lili S**
Nanjing Institute of Environmental Sciences, Ministry of Environmental Protection, Nanjing, 210042, China

**Mosstafa Kazemi and Mohammad Soleiman-Beigi**
Department of Chemistry, Ilam University, Ilam, Iran

**Ahmed Moussa and Aissat Saad**
Institute of Veterinary Sciences University Ibn-Khaldoun, Tiaret, Algeria

**Djebli Noureddine and Salima Douichene**
Departments of Biology, Faculty of Sciences, Mostaganem University, Algeria

**Ruifa Jin**
College of Chemistry and Chemical Engineering, Chifeng University, Chifeng 024000, China

**Hongbin Sun**
State Key Laboratory of Natural Medicines, China Pharmaceutical University, 24 Tongjiaxiang, Nanjing 210009, P.R. China

**Yi Zhang, Yong Nian and Hong Liu**
State Key Laboratory of Natural Medicines, China Pharmaceutical University, 24 Tongjiaxiang, Nanjing 210009, P.R. China
CAS Key Laboratory of Receptor Research, Shanghai Institute of Materia Medica, Chinese Academy of Sciences, 555 Zuchongzhi Road, Shanghai 201203, P. R. China

**Jiang Wang**
CAS Key Laboratory of Receptor Research, Shanghai Institute of Materia Medica, Chinese Academy of Sciences, 555 Zuchongzhi Road, Shanghai 201203, P. R. China

**Gunin Saikia and Snehasis Jana**
Trivedi Science Research Laboratory Pvt. Ltd., Bhopal, Madhya Pradesh, India

**Sirsat Shivraj B and Vartale Sambhaji P**
P.G. Research centre, Department of Chemistry, Yeshwant Mahavidyalaya, Nanded 431602 (MS), India

# Index